FLUOROPOLYMERS

HIGH POLYMERS

A SERIES OF MONOGRAPHS ON THE CHEMISTRY, PHYSICS, AND

TECHNOLOGY OF HIGH POLYMERIC SUBSTANCES

VOLUME XXV

FLUOROPOLYMERS

EDITED BY

LEO A. WALL

Polymer Chemistry Section
National Bureau of Standards

WILEY-INTERSCIENCE

A Division of John Wiley & Sons, Inc.
New York · London · Sydney · Toronto

Library of Congress Catalog Card Number: 74-165023

ISBN 0-471-39350-9

Printed in the United States of America.

10 9 8 7 6 5 4 3 2 1

CONTRIBUTORS TO VOLUME XXV

JOSEPH M. ANTONUCCI, *Polymer Chemistry Section, National Bureau of Standards, Washington, D.C.*

TERENCE W. BATES, *Department of Chemistry, Dartmouth College, Hanover, New Hampshire*

ANTHONY J. BUR, *Polymer Dielectrics Section, National Bureau of Standards, Washington, D.C.*

JAMES E. FEARN, *Polymer Chemistry Section, National Bureau of Standards, Washington, D.C.*

LEWIS J. FETTERS, *Institute of Polymer Science, The University of Akron, Akron, Ohio*

ROLAND E. FLORIN, *Polymer Chemistry Section, National Bureau of Standards, Washington, D.C.*

JEROME HOLLANDER, *Narmco Research and Development Division, Whittaker Corporation, San Diego, California*

GEORGE P. KOO, *Allied Chemical Corporation, Plastics Division, Morristown, New Jersey*

NORMAN L. MADISON, *Dow Chemical Company, Midland, Michigan*

KAZIMINERA L. PACIOREK, *Dynamic Science, Irvine, California*

ALLEN G. PITTMAN, *Agricultural Research Service, Albany, California*

WALTER J. PUMMER, *Polymer Chemistry Section, National Bureau of Standards, Washington, D.C.*

WILLIAM H. SHARKEY, *E. I. du Pont de Nemours and Company, Experimental Station, Wilmington, Delaware*

LEO A. WALL, *Polymer Chemistry Section, National Bureau of Standards, Washington, D.C.*

JOHN A. YOUNG, *Denver Research Institute, University of Denver, Denver, Colorado*

PREFACE

In this book we have endeavored to review critically in depth, although not exhaustively, all available knowledge in the field of fluoropolymers. Both the chemical and the physical aspects of these materials have received attention. The first nine chapters discuss the synthesis of monomers and polymerization processes. Here our emphasis is on various new polymers and the status of the many approaches that have been reported by fluorine chemists in their search for new materials. The last seven chapters cover the cross-linking, radiation, and thermal chemistry of these polymers and important physical properties.

Starting from the pioneering work of Frederic Swarts (1890–1930) and into the 1940s the fluorosubstances synthesized were mainly saturated or olefinic compounds. In 1947 much fluorine research that had been carried out during World War II was made public. At this time the syntheses of hexafluorobenzene (McBee et al.) and other perfluorinated aromatics were first reported. Since then the earlier work by Desirant of Belgium on the synthesis of hexafluorobenzene came to light and fluorinated aromatics became a substantial area of fluorocarbon chemistry.

It is evident that in spite of their relatively higher cost, fluorine-containing compounds comprise a field which, although still undeveloped, is equivalent to that of hydrogen-containing compounds in numbers and in properties that will prove to be useful and beneficial for mankind. Although we do not anticipate that the production of any one fluoropolymer will approach in tonnage the current commercial polymers such as polystyrene, polyethylene, and polyvinylchloride, it can be expected that in future they will be the most important because of the superior or unique properties of some of these compounds.

Fluorination usually will not produce thermal stability or oxidative resistance; however, some fluoropolymers are extremely stable or inert, whereas some are extremely unstable. The principal general value of fluorine substituents is the great variability in properties that can be achieved. This is of the utmost importance for the materials scientist.

My thanks go to the authors of each chapter for their invaluable contributions and patient cooperation and to Dr. Roland E. Florin and Professor Mary Aldridge for their kindness in consenting to critically read the various chapters and for their many helpful suggestions for improving the clarity of the language and the accuracy of the technical content. Also with deep appreciation, the aid of Mrs. Judy D. Harne, who typed many of the chapters, drew many of the chemical formulas, mathematical equations, and figures, and was in general extremely helpful, is gratefully acknowledged.

LEO A. WALL

Washington, D.C.
April 1971

CONTENTS

1. POLYMERIZATION OF FLUOROOLEFINS

JAMES E. FEARN, *Polymer Chemistry Section, National Bureau of Standards, Washington, D.C.*

Contents

I. MONOOLEFINS

Fluoroolefin chemistry had its genesis near the end of the nineteenth century with the pioneering work of Swarts (1–3) whose developments included the reaction

$$C_nCl_{2n+2} \xrightarrow[\text{SbCl}_5]{\text{SbF}_3} C_nCl_{2n+2-a}F_a$$

where α is the number of chlorine atoms converted to fluorine atoms by the reaction. This reaction enabled early fluorine chemists to prepare many fluoroalkanes, which became precursors for fluoroalkenes (4–6). It enabled early investigators to prepare all of the chlorofluoroethanes from hexachloroethane. The literature contains a number of methods of obtaining hexachloroethane. The two methods following are illustrative of the general approach to this synthesis (7):

$$CCl_4 + AlCl_3 + Al \xrightarrow{60\text{–}75°C} CCl_3CCl_3$$

$$CCl_4 + C \longrightarrow CCl_3CCl_3$$

1

In the second method carbon tetrachloride was passed over electrically heated carbon at the decomposition temperature of the carbon tetrachloride (8). With an ample supply of hexachloroethane and armed with the Swarts reaction, resourceful investigators were able to produce all of the compounds which are now known commercially as Freons and Genetrons plus many others that may have importance in the future. These materials have subsequently become the precursors of most of the fluoroolefins. In 1934 Henne and his co-workers (9) published a comprehensive paper covering all of the chlorofluoro and fluoroethanes and ethenes, as listed in Table 1. The physical constants shown in the table differ only slightly from those reported in 1934.

TABLE 1

Fluoroalkenes and Some Precursors

Compound	B.P. (°C)	M.P. (°C)	n_D^{20}	d
CF_2CF_2	−76			
$CClF_2CClF_2$	3.8			
$CBrF_2CBrF_2$	46.4	−112		2.149
CF_2CFCl	−26.8			
$CClF_2CCl_2F$	46.5	−36.5	1.35572	1.56354
$CBrF_2CBrClF$	93.1	−72.9	1.4278	2.2478
CF_2CCl_2	18.9	−116		
$CClF_2CCl_3$	91.5	40.6		
$CFClCFCl$ (cis)	21.1	−130.5	1.3777	1.4950
$CFClCFCl$ (trans)	22	−110.3	1.3798	1.4936
$CBrClFCBrClF$	138.9	45.5		
CCl_2FCCl_2F	92.8	24.65	1.41297	1.64470
$CFClCCl_2$	71.0	−118.9	1.4379	1.5460
CCl_3CCl_2F	137.9	101.3		1.74
CF_2CHF	−51			
CF_2CH_2	·−82			
$CFHCFH$ (cis)	−26.0			
$CFHCFH$ (trans)	−53.1			
$CFHCH_2$	−72			
CF_2CFCF_3	−29.4			
$CF_2CFCF_2CF_3$	$\frac{1}{740 \text{ mm}}$		1.5443	
$CF_3CFCFCF_3$	$\frac{0}{740 \text{ mm}}$		1.5297	
$CF_2CF(CF_2)_4CF_3$	81		1.2782	
⌐¯¯¯¯¯¬ $CF_2CFCFCF_2$	6	−60		
$CF_3CF_2CF_2CHCH_2$	31			
⌐¯¯¯¯¯¬ CF_2CF_2O	−62	−121		

Chlorotrifluoroethylene (CTFE) was the first fluoroolefin that achieved industrial importance. Its early preparation, as Table 1 suggests, involved either the dechlorination of 1,1,2-trichloro-1,2,2-trifluoroethane or the debromination of 1,2-dibromo-1-chlorotrifluoroethane using zinc and ethanol. The dechlorination of 1,1,2-trichloro-1,2,2-trifluoroethane, now known commercially as Freon 113, remains the principal industrial preparation for this monomer. This dechlorination is achieved in a number of ways. It has been carried out using zinc and methanol, sodium and methanol, sodium or potassium amalgam and various solvents (10), magnesium alloy with butanol and methanol (11), wet methanol and a trace of hydrochloric acid (12), or by passing heated vapors of the ethane over nickel, iron, copper silicate, chromium oxide-calcium fluoride, etc. After dechlorination the monomer obtained may be purified by washing with a dilute terpene solution followed by a careful distillation (13). When CTFE is treated in this way, polymer samples prepared from it possess good thermal and oxidative stability plus many other desirable properties. This is a consequence of increased molecular weight. Removal of termination and chain transfer agents during purification is the principal cause of the increased molecular weight. Tetrafluoroethylene (TFE) is probably the most widely used fluorocarbon polymer at present. This monomer may be prepared from 1,2-dichlorotetrafluoroethane or 1,2-dibromotetrafluoroethane by simple dehalogenation as suggested by Table 1, but it is usually prepared commercially through pyrolysis of chlorodifluoromethane through use of one of the techniques described below. Both involve passing the starting material through hot tubes.

$$CF_2HCl \xrightarrow{650°C} CF_2CF_2 + HCl$$

$$\underset{200\ kg}{CF_2HCl} + \underset{15\ kg}{H_2O} \xrightarrow[150°C]{\underset{Al_2O_3}{steam}} \underset{71\ kg}{CF_2CF_2} + \underset{1.8\ kg}{CF_2CFCF_3} + \underset{1.2\ kg}{\overset{\displaystyle CF_2-CF_2}{\underset{\displaystyle CF_2-CF_2}{|\qquad|}}}$$

In the second procedure 30% of the starting material is recovered and recycled (15). TFE prepared in this way may be purified in the manner described earlier for CTFE. Great care must be exercised when dealing with pure TFE at room temperature and above and with even the slightest superpressure. It is not entirely safe under pressure even when it contains an inhibiter. In the latter of the two syntheses outlined above it may be noted that hexafluoropropylene (HFP) is one of the by-products. HFP may be prepared in a number of other ways; in addition to the first reaction above (14) is the pyrolysis of sodium perfluorobutyrate (16), thermal cracking of TFE polymer under special conditions (17), and the passing of TFE monomer under reduced pressure over a platinum wire at 1400°C (18). HFP may be purified by the scrubbing technique previously described or by vapor phase chromatography (VPC).

Fluoroolefins containing hydrogen are prepared by conventional organic processes. Vinyl fluoride (VF) is obtained in good yield in the three ways outlined below (19, 20, 17, respectively):

$$CH_2CH_2 + HF \xrightarrow[Hg^{++}]{} CH_3CH_2F \xrightarrow[400°C]{Al_2(SO_4)_3} CH_2CHF$$

$$CHCH + HF \xrightarrow[Hg^{++}]{500°C} CH_2CHF + CH_3CHF_2$$

$$CH_3CH_2CF_3 \xrightarrow{830°C} CH_2CHF + CH_2CF_2$$

In the third synthesis it may be observed that vinylidene fluoride (VF$_2$) is a by-product. VF$_2$ may also be prepared as follows (21, 22):

$$CH_3CF_2Cl \xrightarrow{870°C} CH_2CF_2 + CH_2CFCl$$

$$CF_3CH_2I \xrightarrow{Mg} CH_2CF_2$$

Trifluoroethylene has been prepared by dehalogenation of 1,2-dichloro-1,1, 3-trifluoroethane (23). Another synthesis involves electrolyzing an aqueous ethanol solution of 1,1,2-trichloro-1,2,2-trifluoromethane containing zinc chloride (24). This brought about either simultaneous or sequential dechlorination and reduction, giving the compound in good yield. Other investigators (25,26), including the author, have prepared trifluoroethylene using simple organic reactions such as the Grignard reaction outlined below:

$$CF_2CFI + Mg \xrightarrow[ether]{anhydrous} CF_2CFMgI \begin{cases} \xrightarrow{H_2O} CF_2CFH \\ \\ \xrightarrow{D_2O} CF_2CFD \end{cases}$$

The reactions discussed in this and the preceding paragraphs along with direct fluorination techniques (27) permit the preparation of all the fluoroolefins listed in Table 1, as well as many others whose synthesis may become desirable. Tetrafluoroethylene oxide is prepared by irradiating a mixture of TFE and oxygen with ultraviolet under conditions that prevent a buildup of peroxide concentration (28). It must be stored at liquid nitrogen temperature since it decomposes rapidly at room temperature.

The polymerization and copolymerization of TFE, CTFE, trifluoroethylene, VF$_2$, VF, and HFP are discussed in the first part of this chapter; the second part deals with the preparations of perfluorodienes, which were accomplished largely through telomerization reactions involving these monomers. The procedures used in telomerizing or homopolymerizing one fluoroolefin are usually applicable to many other systems with only slight alteration. In this work each monomer is discussed separately, avoiding repetition if possible.

A. Chlorotrifluoroethylene

Space does not permit the listing of all of the ways in which CTFE has been polymerized using free radical initiation. In some of the earlier systems a peroxide catalyst at 100–150°C at pressures approximating 1000 atm was employed (29). A copolymer of CTFE and ethylene prepared under these conditions showed a tensile strength of 2700 lb psi and was found to be useful in films, foils, fibers, and as a coating for electric wires because of its excellent tear resistance. This peroxide theme was thoroughly developed by many investigators (30,31). Variations were achieved by altering conditions of time, temperature, and pressure and by using a wide variety of organic, inorganic, and occasionally completely fluorinated peroxides (32). Other free radical sources, such as ultraviolet, have seen service (31), yielding polymer samples similar to those obtained using peroxide catalyst.

Primitive polymerization led inevitably to more sophisticated approaches, more involved techniques, and eventually rate studies (33). These studies showed that the polymerization rate of CTFE in pentachloroethane is three to five times as great as that in bulk at corresponding temperatures. A careful study of Fig. 1 will disclose this. Figure 2, in similar fashion, suggests the dependence of the polymerization rate at 80.1, 70.3, and 60.1°C upon the

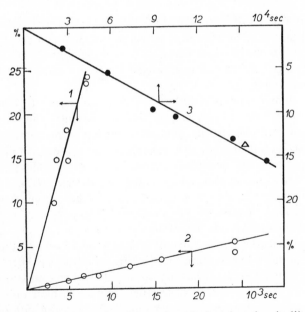

Fig. 1. Percentage chlorotrifluoroethylene polymerized against time in (1) pentachloroethane at 70.3°C, (2) bulk at 70.3°C, and (3) bulk at 60.1°C. Concentration of benzoyl peroxide was 0.0209 mole kg^{-1} (33).

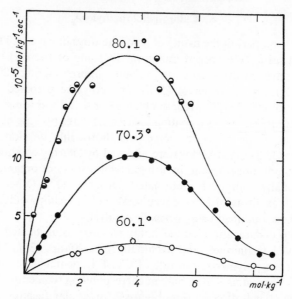

Fig. 2. Polymerization rate of chlorotrifluoroethylene in pentachloroethane solution against concentration at various temperatures (33).

concentration of CTFE in pentachloroethane solution. Dependence of this rate on the nature of the solvent is shown in Fig. 3. A comprehensive study of these three graphs provides a clear picture of the polymerization of CTFE and how this polymerization is affected by solvents and concentrations as well as temperature.

Addition of an oxidation-reduction system to an organic peroxide catalyst permitted other types of rate studies (34). By carefully adjusting peroxide concentration and pH, the pH rate of conversion of monomeric CTFE to polymer may be maximized. Gamma initiation has been employed in the bulk polymerization of many halogenated ethylenes, including CTFE. This work was carried out at temperatures ranging between -15 and $15°C$ (35). In other gamma-initiated systems CTFE and TFE have been copolymerized under controlled conditions in chlorinated hydrocarbon solvents (36). Polymerization studies utilizing gamma initiation tend to suggest that the rate of polymerization varies directly with the square root of the radiation intensity at a given temperature. Electrolytic initiation has been attempted with some limited success. It has been found that when a mixture of CTFE and hydrogen fluoride is electrolyzed in a certain fashion polymerization results and that the degree of polymerization (DP) is a function of temperature and current density (37). At low temperatures and current densities the DP is high, whereas at high temperatures and current densities low DP

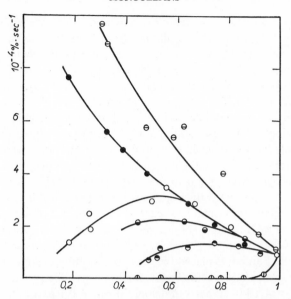

Fig. 3. Rate of chlorotrifluorethylene polymerization against mole fraction of monomer in ($\ominus$) pentachloroethane, ($\bullet$) carbon tetrachloride, ($\bigcirc$) 1,2-dichloroethane, ($\ominus$) 1,1,2-trichlorotrifluoroethane, ($\ominus$)benzene, and ($\oplus$) 1,3,5-trimethylbenzene. Temperature was 60.1°C; peroxide concentration 0.0209 mole kg^{-1} (33).

materials are produced. These low molecular weight materials find many uses as softeners and waxes.

Later investigations along the same line tend to confirm these findings and venture to postulate mechanisms (38). These investigations indicate that initiation at the anode results from electrons or hydrogen atoms produced by electrolysis, whereas at the cathode the centers of initiation are free radicals resulting from the discharge of ions. As in any electrolytic reaction, the net effect may be altered by changes in the material of the anode, by changes in pH, voltage, temperature, and agitation, or by the nature of impurities in the system. Ziegler-type catalyst systems, employing titanium tetrachloride and metal alkyls, have proved very effective. They produce high conversion both in homopolymerization and in copolymerizations and are useful in grafting fluorinated ethylenes such as CTFE and TFE on hydrocarbon-based polymers such as polyalphamethylstyrene (39,40).

The most widely used technique in the practical polymerization of CTFE as well as many other fluoroolefins is the colloidal dispersion method. In early efforts with this method (41) water-soluble initiators such as alkali persulfates were used in combination with alkali bisulfites in aqueous media. Later investigators (42) added silver salts to these mixtures as accelerators. In this way it was possible to increase the polymerization rate without lowering the melt viscosity of samples obtained from a given system. Many other additives have been used with varying effects. Experiments with a large number of organic acids have been carried out without greatly altering the polymer samples produced by persulfate systems (43). Addition of either dichlorobenzene or methyl acrylate to an aqueous persulfate system resulted in a latex with a particle size of 1800 Å rather than the coagulated masses usually obtained (44). Upon adding a C_{5-20} perfluorocarboxylic acid, which served largely as an emulsifier, samples were obtained which possessed superior shore hardness and high tensile strength (45). Other critical factors in persulfate systems were temperature and pH (46). When these conditions approached the optimum values conversions from 80 to 100% were obtained. By 1964 the persulfate system had reached a high degree of sophistication and had been used by Bolstad (47) for a great deal of polymerization and copolymerization work. A sample polymerization utilizing this procedure follows.

Into a metal bomb was placed 150 parts of carefully deionized water and 1 part of perfluorooctanoic acid. To this mixture was added 1 part of potassium persulfate dissolved in 50 parts of deionized water, 0.1 part of ferrous sulfate-heptahydrate and 0.4 part of sodium sulfate. The bomb was cooled to $-195°C$ and after carefully and rigorously exhausting the bomb of air the monomer was distilled into it. It was then rocked at 25°C for 21 hr. Conversions with this general system approximate 90% regardless of the fluorocarbon monomer used.

CTFE has been copolymerized with all of the fluorovinyl monomers and with a great many monomers that are basically hydrocarbon, e.g., ethylene, isobutylene, and methyl methacrylate, and with chlorinated ethylenes. Copolymers containing vinyl chloride and vinylidene chlorides have shown properties that stimulated limited commercial interest. Copolymers of CTFE with hydrocarbon dienes such as butadiene and diallyl maleate have been of similar interest. Polychlorotrifluoroethylene may be treated with a solution of sodium in liquid ammonia in a manner such that the resulting surface is amicable to adhesives and is suitable for the grafting thereon of other monomeric species (48). A copolymer of CTFE and an alkyl vinyl ether was treated in this way. Upon the treated surface vinyl chloride was grafted, producing a terpolymer which was useful in coating fabrics (49). A terpolymer has been prepared from a mixture of 75% CTFE, 20% isobutylene, and 5% vinyl acetate using the much discussed persulfate system (50). This material was dissolved in p-cymene and applied to copper plate. The plate was then heated at 130–140°C for 2 hr. The resulting coating was strongly adherent and completely transparent. Interpolymers, materials containing four or more monomers, have been prepared which contained CTFE, TFE, VF_2, etc. (51). Many of these are of industrial importance.

Low-molecular-weight poly-CTFE products are useful as plasticizers and as additives for various other purposes. They are also useful as oils and solvents when good thermal stability is a requisite. A series of acids having the general formula $CF_2Cl(CFClCF_2)_nCOOH$, where n may be 1,2, ..., may be prepared from a CTFE telomer. Other CTFE and TFE telomers have been used in the preparation of the fluorodienes. This will be discussed later in this chapter.

The uses of high-molecular-weight poly-CTFE are many and diverse. It is used as a protective coating in the fabrication of many articles in which thermal and chemical stability is desirable. It can be compounded with metals or other inorganic materials in the preparation of self-lubricating bearings. It can be sintered and used in the preparation of alundumlike cells, which are extremely useful in electrochemistry. It has been used in the actual construction of electrodes (52) and in components of storage batteries (53). The use of CTFE as a binder in the preparation of gas generating propellants (54) and pyrotechnic compositions (55) is firmly established. The chemical inertness of poly-CTFE has made practical its use as a food container (56) and in recent years it has been pressed into service as a nonporous membrane for separating helium from mixtures containing oxygen, nitrogen, carbon monoxide, etc., through differences in their rates of diffusion (57). In the construction of VPC columns, low-molecular-weight poly-CTFE has been used as a stationary phase material and high-molecular-weight polymer has been employed as a support medium in place of celite, firebrick, or diatomaceous earth. Suspension of powdered graphite or various powdered metals in low-molecular-weight $CTFE-VF_2$ copolymer may be used as "ink" in printing electrical circuits on high-molecular-weight fluorocarbon polymers. After "baking," these circuits are "set" and may be used in many forms of electrical apparatus. If organic dyes are substituted for the graphite or powdered metal inks which are useful in marking both fluorocarbon and hydrocarbon polymeric materials may be obtained (58,59).

B. Tetrafluoroethylene

The polymerization of TFE has developed in a manner parallel to that of CTFE in that most of the systems that produce excellent samples of CTFE will produce samples of comparable quality of TFE as well as copolymers of the two of various compositions, depending on the mole ratios of the two monomers. The polymerization of TFE may be inadvertently photo-initiated when the monomer is stored in glass containers, and it has been known to polymerize spontaneously in dark metal containers. The polymerization reaction is strongly exothermic and occasionally occurs with great force. At room temperature the monomer is extremely unsafe to handle unless it is inhibited with 1–2% by weight of limonene or other radical scavenger. Even then it is not considered safe for shipment. Companies that once supplied this material now recommend that it be prepared *in situ* from 1,2-dibromotetrafluoroethane. In the laboratory it is desirable that reactions involving the transfer of pure uninhibited TFE be carried out at temperatures as low as possible. If chain transfer agents are vigorously excluded, the refined persulfate systems (47) are excellent for the preparation of poly-TFE.

Homopolymers of TFE and copolymers of TFE and CTFE have been prepared in the solid state using actinic radiation as initiator (60). In the preparation of these copolymers very careful measurements were made in order to show the relationship of monomer composition to polymer composition. The plot of TFE in the initial mixture versus TFE in the copolymer

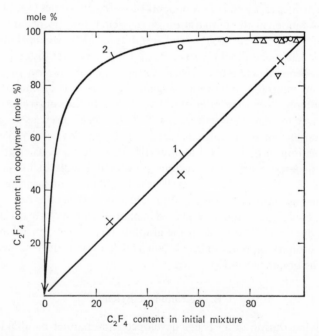

Fig. 4. Percentage of tetrafluoroethylene in copolymer against percentage of tetrafluoroethylene in initial tetrafluoroethylene-chlorotrifluoroethylene mixture (60).

Phase	Temp. (°C)	Dose rate (rad/sec)	Total dose (Mrad)	Conv. (%)
▽ gas	62	110	2	16
× liquid	0	200	2	5–85
△ solid solution	−170	130	3.3	1–4
○ solid solution	−145	130	3.3	1–4

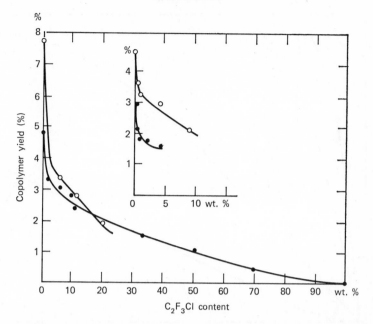

_Fɪɢ. 5.   Copolymer yield against percentage of chlorotrifluoroethylene in mixtures with tetrafluoroethylene at a radiation intensity of 130 rad sec^{-1}, 3.3-Mrad dose, 2.0-Mrad dose (60); (●) at a temperature of $-170°C$; (○) at a temperature of $-140°C$.

when the polymerization was carried out in the solid state at $-170°C$ and at $-145°C$ with a radiation dose of 130 rad/sec and a total dose of 33 Mrad was made and compared with the plots of initial concentration versus copolymer composition when the polymerization was carried out in the liquid phase at $62°C$. These plots are compared in Fig. 4. The composition of the initial mixture showed a pronounced effect not only on the copolymer composition but also on the copolymer yield, as Fig. 5 shows. A qualitative picture of what takes place during this copolymerization of TFE and CTFE may be obtained through a careful study of Figs. 4 and 5.

Poly-TFE has also been prepared through gamma irradiation of the gas, which produces samples that are finely divided powders, or of the liquid, which produces homogeneously solid material (61). Gamma radiation for this set of experiments was obtained from spent fuel elements from an atomic reactor. Kinetic studies of the homopolymerization of TFE using X-ray initiation (62) have been undertaken at various pressures of the monomer. Figure 6 shows the dependence of percentage conversion of TFE to poly-TFE on irradiation time at a constant dose rate of 4050 rad/min. In this experiment the pressure of monomeric TFE was kept constant at 1.0 mm of mecury. The data for curve 2 (Fig. 6) was obtained from polymerization carried out in the

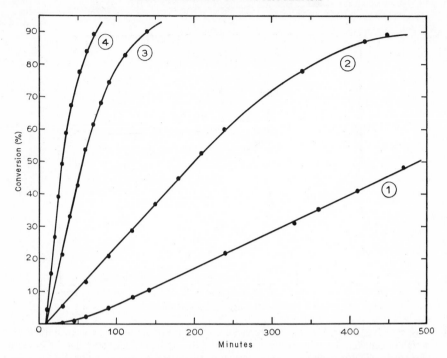

Fig. 6. Gas-phase polymerization of tetrafluoroethylene at 4050 rad/min dose rate. Conversion as a function of time, initial pressure of 1.0 mm of Hg. Data for curve 1 were taken from polymerization in a clean cell, curve 2 from a cell containing polymer prepared in experiment 1, curve 3 from the same cell containing polymer from both earlier experiments, curve 4 from the three previous experiments (62). (Courtesy of La Chimica et l'Industria, Milan.)

presence of polymer produced in obtaining the data for curve 1. Data for curves 3 and 4 were obtained in similar fashion, i.e., with polymer produced in each of the earlier studies present. When the pressure of monomeric TFE was increased, the curves of percentage conversion versus radiation time became steeper, as shown in Fig. 7. Dose rate was again rad/min. The result of these kinetic studies is the hypothesis that the polymerization of TFE takes place according to a free radical mechanism and that this polymerization may be expressed by the following scheme:

$$M \rightarrow R^* \qquad\qquad v_1 = K_1 I[M] \qquad\qquad (1)$$

$$R^* + M \rightarrow RM_1^* \qquad v_2 = K_2(R^*)[M] \qquad (2)$$

$$R_n^* + M \rightarrow RM_{n+1}^* \qquad v_3 = \frac{dM}{dt} = K_p RM^*[M] \qquad (3)$$

$$RM_n^* + RM^* \rightarrow polymer \qquad v_4 = 2K_4[RM^*]^2 \qquad (4)$$

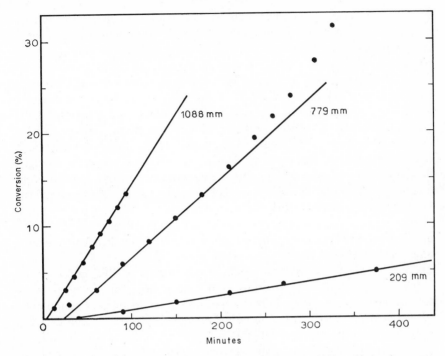

Fig. 7. Gas-phase polymerization of tetrafluoroethylene at 4050 rad/min dose rate, conversion as a function of time at various pressures (62). (Courtesy of La Chimica et l'Industria, Milan.)

$$RM_n + R^* \rightarrow \text{polymer} \qquad v_5 = K_5[RM^*][R^*] \qquad (5)$$

$$R^* + R^* \rightarrow Q \qquad v_6 = 2K_6[R^*]^2 \qquad (6)$$

From the data and the scheme above the following expressions for the initial velocity of polymerization were developed:

$$-\frac{dM}{dt} = K_p \left(\frac{K_1 I}{2K_4} \right)^{1/2} [M]^{3/2} \qquad (7)$$

$$-\frac{dM}{dt} = \frac{K_p K_2}{K_5} [M]^2 \qquad (8)$$

Both of the derived rate equations indicate the dependence of rate of monomer disappearance on monomer pressure as shown in Fig. 7.

In other preparations of poly-TFE a system was used which consisted of a fluorinated peroxygenated acid from anhydride with 30 % aqueous hydrogen peroxide, the excess anhydride reacting with the water formed (63). By using these media, polymerizations have been made at temperatures between zero and 100°C at pressures which varied from monomer to monomer. In addition, CTFE, HFP, and VF were employed, as were various combinations of these. A copolymer of 20% HFP and 80% TFE prepared in this way has many properties that make it superior to poly-TFE for specific applications.

Copolymerizations of TFE and CTFE have been attempted with the Ziegler catalyst in an effort to produce polymer samples with a high degree of stereo regularity (64). Good protective coatings and adhesives for bonding halogenated polymers have been prepared by copolymerizing TFE and diallyl maleate using the persulfate system (65). Copolymers with VF_2 were found to make excellent films, fibers, and molded articles, with a high degree of strength, flexibility, and transparency (66). Other comonomers investigated in the same comprehensive study include VF, vinyl chloride, and 3,3,3-trifluoropropylene. Other copolymerization work has been done with TFE and trifluoromethyl vinyl ether (67), 2-chloroethylvinyl ether, and various other vinyl ethers (68), and has produced rubbery polymer samples with good chemical resistance and electrical properties.

Considerable study has been done on copolymers of TFE and ordinary ethylene (69). Graft copolymers of TFE and various monomers have interested many investigators. In the preparation of semipermeable membranes, acrylic acid has been grafted on poly-TFE (70). Styrene has been carefully polymerized then subjected to graft polymerization by preirradiated TFE (71), yielding a poly-TFE coated styrene. Approaching from the opposite direction, styrene has been carefully diffused into poly-TFE in discrete amounts. This poly-TFE-styrene system was then irradiated in such fashion that the liquid was grafted to the polymer homogeneously (72). This enabled the scientists doing this work to change the properties of poly-TFE in a predictable manner.

The uses of poly-TFE are many and varied. It has seen widespread use in lining cooking utensils (73) because of its thermal and chemical stability. Many patents have been issued for operations in which it is compounded with other plastics or with metals in the preparation of self-lubricating bearings (74, 75) and for its compounding with aluminum silicate, glass, asbestos, or other nonmetallic fibrous materials in the construction of energy transmitting devices such as clutch plates (76). Low-molecular-weight TFE is used in the preparation of high-temperature lubricating greases (77) and adhesives (78). By mixing finely divided TFE polymer with graphite, granulating the mixture, and coating the resultant material first with molybdenum sulfide then with silver, gold, or nickel, a conducting plastic material is produced (79). In a very unusual process, powdered poly-TFE was added to monomeric lactam. Upon polymerization of this suspension, using sodium hydride catalyst and a diisocyanate initiator, a material was produced which possessed greater flexibility than poly-TFE and more chemical stability than nylon (80).

TFE homopolymer is considered the finest insulation for electric wires used in space environments (81) because of its thermal stability, electrical inertness, and nonflammability. Through recently developed electrochemical techniques poly-TFE may be copper plated (82). Properly treated poly-TFE may be used as a support material in gas-liquid chromatography (83), in the manufacture of sintered plastic articles, in the preparation of gaseous fuel cells (84), and as a substitute for leather (85). Like CTFE, it can be used in the preparation of membranes used in separating helium from gaseous mixtures (57). A dropping mercury electrode composed of poly-TFE has identical polarographic properties with one made of glass (86). Its use in the fabrication of laboratory apparatus, tubing, pipes, gaskets, etc., is well known and firmly established.

C. Vinyl Fluoride

VF may be polymerized using any of the methods already discussed (e.g., 35, 44, 64). The homopolymers of VF as well as its copolymers with TFE, vinyl chloride, HFP, and 3,3,3-trifluoropropylene have been prepared in a comprehensive study (66). These materials have considerable strength and may be cold-drawn very easily. Poly-VF has been fabricated with other materials into siding sheets which see extensive use in the construction industry.

In the same laboratory in a later investigation (87), kinetic studies were made of the polymerization and copolymerization of the monomers mentioned above, employing Ziegler catalyst and using various solvents at temperatures ranging between 0 and 50°C. When triisobutyl boron was employed as initiator in ethyl acetate and in cyclohexane at 25°C, the homopolymerization was found to be first order as to monomer concentration, as is shown in Fig. 8. In these investigations it was found that the presence of oxygen or an oxidizing agent accentuated polymer conversion and that the degree of accentuation depended on the ratio of oxygen to boron alkyl. The highest polymerization conversion occurred at an O_2/BR_3 ratio of 0.5, as shown in Fig. 9. This series of plots also shows the considerable effect on the polymerization rate with this catalyst system when the nature of the solvent is changed. Copolymer studies involving VF and various other monomers were carried out so that the percentage composition of VF in the monomer composition could be plotted against the percentage in the copolymer (Fig. 10). A cursory examination of Fig. 10 presents a qualitative picture of the reactivity ratios of VF and the other monomers covered in this work. Table 2 gives the copolymerization parameters of VF with the other monomers shown in Fig. 10. The consensus of this work was that the Ziegler-Natta catalyst did not greatly improve the stereo regularity of poly-VF and that the polymerizations in these kinetic studies were free radical in nature.

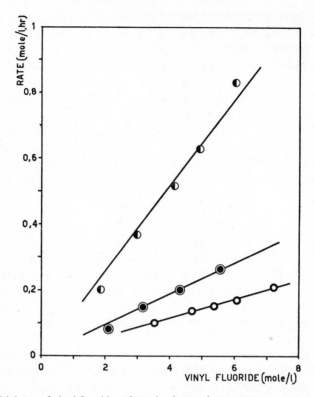

FIG. 8.   Initial rate of vinyl fluoride polymerization against monomer concentration under the following conditions (87):

- ◑ 0.25×10^{-3} mole triisobutylborane with a ratio of oxygen to borane of 0.5 at 30°C in ethyl acetate. Overall volume 38 ml.
- ○   Same as for ◑, but with cyclohexane solvent.
- ◉ 0.5×10^{-3} mole triethylborane-ammonia with an oxygen to borane ratio of zero at 25°C in ethyl acetate. Overall volume 9.2 ml.

D. Trifluoroethylene

Trifluoroethylene polymerizes so readily that cylinders containing this monomer must be stored at low temperatures. Any of the methods used for polymerizing the fluoroolefins mentioned earlier may be used for this monomer. Preparation of short-chain telomers or oligomers, which proved quite feasible with CTFE and TFE, proved extremely difficult with trifluoroethylene because of its tendency to form high polymer. The polymer is a tough, white solid with generally good properties, but it is neither as chemically inert nor as thermally stable as poly-TFE or poly-CTFE. The homopolymer and many of the copolymers of this compound have widespread academic interest but very few commercial applications.

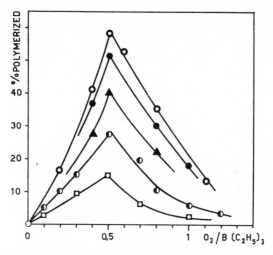

Fig. 9. Polymer conversion against oxygen to triethylboxane mole ratio ($\bigcirc$) in bulk, ($\bullet$) in dimethyl sulfoxide, ($\blacktriangle$), in ethyl acetate, ($\pmb{\Complex}$) in methylene chloride, and ($\square$) in cyclohexane. Each curve represents a 5-hr reaction at 30°C with 0.25×10^{-3} mole of triethylborane and 11.5 g of monomer (87).

E. Vinylidene Fluoride and Hexafluoropropylene

VF_2 and HFP have both been homopolymerized by previously discussed methods, but the homopolymers have very little commercial significance and only limited academic interest. Many eloquent claims appear in the patent literature for copolymers of each of these monomers with TFE, CTFE, VF, etc. (88), but the most important polymer involving either of these monomers is probably the copolymer composed of the two of them. One material of this type, known commercially as Viton rubber, is a tough, flexible substitute for natural rubber. Copolymers of these two monomers of various percentage may be prepared by the peroxide and persulfate systems discussed above. In addition to general copolymerization studies, many investigations have been undertaken in which the copolymerization was directed toward specific objectives. In some the objective was to determine the effect of pressure and the presence of silica on the polymerization scheme (89), whereas in others the goal was to correlate the mode of preparation of copolymer samples with their anticipated uses (90). Efforts to develop a superior caulking compound have resulted in the devotion of much time to this type of copolymerization (91).

Rubbery samples of VF_2-HFP copolymer may be cross-linked (vulcanized) by a number of agents and in a number of ways, a few of which are briefly discussed here. Dicumyl peroxide produces a material which is insoluble in methyl ethyl ketone (92). Carbon black and diallyl amine produce essentially the same effect (93), as does calcium or magnesium oxide when used in

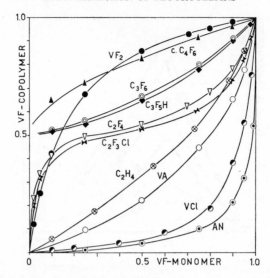

FIG. 10. Mole fraction of vinyl fluoride in copolymer versus mole fraction in initial mixture for various comonomers at 30°C using an oxygen-triisobutylborane system as iniator and ethyl acetate as solvent. The oxygen triisobutylborane mole ratio was 0.5 (87).

TABLE 2

Copolymerization Parameters for Vinyl Fluoride (M_1) (87)

Monomer 2	r_1	r_2	$r_1 r_2$	e_2	Q_2	e_1	Q_1
$CH_2{=}CH_2$	0.3	1.7	0.51	-0.21	0.010	$(+0.61)$ -1.03	(0.005) 0.0065
$CH_2{=}CHCl$	0.05	11	0.55	$+0.20$	0.044	$(+0.97)$ -0.57	(0.0046) 0.0035
$CF_2{=}CF_2$	0.27	0.05	0.013	$+1.22$	0.049	$(+3.29)$ -0.85	(12.6) 0.08
$CF_2{=}CFCl$	0.18	0.06	0.011	$+1.48$	0.020	$(+3.60)$ -0.64	(7.7) 0.015
$CH_2{=}CH{-}CN$	$ca.\ 1 \times 10^{-3}$	24	$ca.\ 0.024$	$+1.20$	0.60	$(+3.13)$ -0.73	(0.25) 0.0025
$CH_2{=}CHOCOCH_3$	0.16	2.9	0.46	-0.22	0.026	$(+0.66)$ -1.10	(0.007) 0.010

conjunction with an organic peroxide and an alkylene diamine carbamate (94, 95). In the calcium oxide vulcanization, diallyl terephthalate or tetraallyl phthalamide may be substituted for the alkylene diamine carbonate if the compounded stock is heated to 100–200°C (96). Vulcanization for Viton polymers has been summarized (97) and a study of model compounds related to VF$_2$-HFP copolymers appears in the scientific literature (98). Differential thermal analysis (DTA) tends to confirm the concept that the cross-linking process occurs betweeen 100 and 200°C (99). This is discussed in another chapter.

Specific uses of VF$_2$-HFP copolymer are numerous. In general it is used as a sealant and caulking material and in place of ordinary rubber when chemical and thermal stability are at a premium. It is an excellent partitioning " fluid " for use as a stationary phase in gas-liquid chromatography, and it has proved satisfactory in the preparation of pyrotechnic mixtures (55). Many materials, especially porous materials, may be coated with these copolymers, thanks to extensive studies of effective solvent mixtures (100). Such materials include firebrick, celite, fabrics, paper, felt, wood, and concrete.

F. Other Monoolefins

The polymerization and copolymerization of a great many more fluorinated or partially fluorinated monoolefins have been extensively studied. Partially fluorinated propylenes have been polymerized at high pressure (101,102), showing considerable promise. Of even greater interest are the copolymers of these materials with TFE. Some of these copolymers are soluble, so that solutions of them may be used in coating fabrics or solid surfaces. Upon heating to certain temperatures the hydrocarbon area of the polymer coating may be burned away, leaving virtually a poly-TFE-impregnated fabric or a poly-TFE-coated solid that would be difficult to prepare with insoluble poly-TFE. Interesting copolymers with TFE of 3,3,4,4,5,5,5-heptafluoro-1-pentene have been prepared and studied. Homopolymers of this material and perfluoro-cyclobutene have also been prepared and characterized. Work on these materials is continuing as of this writing.

In recent years much study has been given to the preparation and polymerization of perfluoroepoxides (28). These materials are prepared by reaction of oxygen or ozone with the parent perfluoroolefin. Perfluoropropylene oxide is reasonably stable, but perfluoroethylene oxide must be stored at liquid nitrogen temperature. Its half-life at room temperature approximates 48 hr. The ethylene oxide may be converted to high polymer by gamma irradiation at −196°C. In this polymerization molecular weight varies inversely with temperature. The high-molecular-weight polymer is extremely stable thermally, more so even than poly-TFE. Copolymers of perfluoroethylene oxide and TFE have also been prepared and studied (103).

II. DIOLEFINS

A. Tetrafluoroallene

Tetrafluoroallene has been prepared from dibromodifluoromethane and vinylidene fluoride by the following reactions (104):

$$CF_2Br_2 + CF_2CH_2 \xrightarrow{200°C} CF_2BrCH_2CF_2Br \xrightarrow{KOH} CF_2{=}C{=}CF_2$$

Polymerization was effected by sealing the monomer in thick-walled glass tubes which were rocked at room temperature for three days. The polymer was completely insoluble in all hydrocarbon or other conventional organic solvents and in such fluorinated solvents as perfluoroheptane. Pyrolyses even at 700°C under reduced pressure did not regenerate monomer; it yielded instead a mixture of polymeric products. When tetrafluoroallene containing a small amount of terpene B was heated at 40°C for several days, a liquid dimer was obtained which was assigned the structure

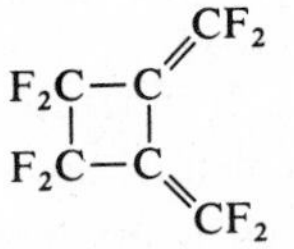

$$
\begin{array}{c}
F_2C{-}C\diagup^{CF_2} \\
\;\;|\;\;\;\;| \\
F_2C{-}C\diagdown_{CF_2}
\end{array}
$$

This dimer was readily oxidized to perfluorosuccinic acid with potassium permanganate, but with ozone a waxy, white material with the formula C_3F_4O was obtained. These and other studies led to the conclusion that the structure of the polymer is

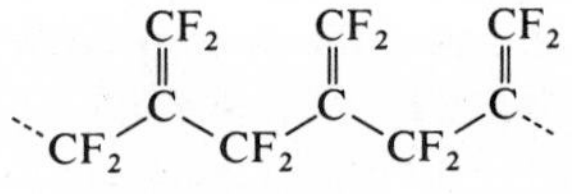

$$
\cdots CF_2\diagup\overset{\displaystyle CF_2}{\underset{\displaystyle C}{\|}}\diagdown CF_2\diagup\overset{\displaystyle CF_2}{\underset{\displaystyle C}{\|}}\diagdown CF_2\diagup\overset{\displaystyle CF_2}{\underset{\displaystyle C}{\|}}\diagdown\cdots
$$

B. Hexafluoro-1,3-Butadiene

Homopolymers of hexafluoro-1,3-butadiene and copolymers of this monomer with vinyl chloride, chloroprene, and many other monomers have been prepared (105). In many cases precise measurements were made so that activity ratios could be determined. In another study (106) it was found that hexafluoro-1,3-butadiene undergoes the following transformation:

$$CF_2{=}CF{-}CF{=}CF_2 \rightleftharpoons CF_2{=}C{=}CF{-}CF_3 \rightleftharpoons CF_3C{\equiv}CCF_3$$

It will also form both saturated and unsaturated dimers:

$$\text{ICF}_2\text{CFI} \xrightarrow{250°\text{C}} \text{ICF}_2\text{CF}_2^* + \text{I}^*$$

$$2\text{ICF}_2\text{CF}_2^* \longrightarrow \text{I}(\text{CF}_2\text{CF}_2\text{CF}_2\text{CF}_2)\text{I}$$

$$\text{I}(\text{CF}_2\text{CF}_2)_n\text{CF}_2\text{CF}_2\text{I} \xrightarrow{250°\text{C}} \text{I}(\text{CF}_2\text{CF}_2)_n\text{CF}_2\text{CF}_2^* + \text{I}^*$$

$$\text{I}(\text{CF}_n\text{CF}_2)_n\text{CF}_2\text{CF}_2^* + \text{ICF}_2\text{CF}_2^* \longrightarrow \text{I}(\text{CF}_2\text{CF}_2)_{n+2}\text{I}$$

$$2\text{I}^* \longrightarrow \text{I}_2$$

Knunyants found that the greater the value of n, the more stable the molecule, which is consistent with the findings of the author. For the investigations in our laboratories, telomers with $n > 4$ were useless. The compounds listed in Table 3 were starting materials for the synthesis of the dienes discussed below.

D. Perfluoro-1,5-Hexadiene and Perfluoro-1,7-Octadiene

By cross-coupling the ethane and butane in Table 3, using zinc and methylene chloride (121), we may obtain 1,2,3,4-tetrachlorohexafluorobutane, 1,2,5,6-tetrachlorodecafluorohexane, and 1,2,7,8-tetrachloroperfluorooctane from which simple dechlorination, utilizing zinc and a high-boiling ether (106), produces perfluoro-1,3-butadiene, perfluoro-1,5-hexadiene, and perfluoro-1,7-octadiene, respectively. It is at once evident from a study of Table 3 that selective homocoupling or cross-coupling will produce all of the perfluorinated terminal dienes with even numbers of carbon atoms from $C = 4$ to $C = 20$, but those with more than eight carbon atoms are not expected to undergo the inter-intramolecular polymerization mechanisms which was the basis for these studies.

Perfluoro-1,5-hexadiene may also be prepared by treating 1,2-dichloro-6-iododecafluorohexane with methyllithium in ether at $-80°\text{C}$ and permitting this system to slowly warm to room temperature. This brings about deiodofluorination, yielding 5,6-dichloroperfluoro-1-hexane, which on dechlorination produces perfluoro-1,5-hexadiene (122). This monomer has also been prepared through the coupling of perfluoroallyl chloride with sodium or perfluoroallyl iodide with zinc in a nonhydroxyl solvent (123).

E. Perfluoro-1,6-Heptadiene

The synthesis of perfluoro-1,6-heptadiene, a hitherto unreported compound, has proceeded along several lines in these laboratories. The first effort used as starting material a commercially available telomer alcohol which itself was prepared by a telomerization reaction (124):

$$\text{CH}_3\text{OH} + \text{CF}_2\text{CF}_2 \longrightarrow \text{H}(\text{CF}_2\text{CF}_2)_n\text{CH}_2\text{OH}$$

The telomer alcohol in which $n = 3$ was subjected to the following sequence of reactions:

$$H(CF_2CF_2)_3CH_2OH \xrightarrow[H_2SO_4 + H_2O]{Na_2Cr_2O_7} \overset{O}{\overset{\|}{H(CF_2CF_2)_3C}}-H \cdot H_2O \qquad (17)$$
$$(70\%)$$

$$\overset{O}{\overset{\|}{H(CF_2CF_2)C}}-H \cdot H_2O \xrightarrow{SF_4} H(CF_2CF_2)_3CF_2H \qquad (18)$$
$$(4.3\%)$$

$$H(CF_2)_7H \longrightarrow CF_2{=}CF(CF_2)_3CF{=}CF_2 \qquad (19)$$
$$(0.0\%)$$

The double dehydrofluorination could not be effected in any of the ways usually utilized for dehydrohalogenation. Refluxing with alcoholic potassium hydroxide or refluxing the compound itself with powdered potassium hydroxide had no effect. Refluxing with tri-*n*-butylamine proved equally ineffective, and passing the material through tubes packed with sodium fluoride, potassium hydroxide, steel wool, nickel turnings, etc., at successively higher temperature resulted finally in degrading the carbon chain. No unsaturated material, as evidenced by infrared monitoring, was ever obtained. This is explained by the inductive effect of all of the highly electronegative fluorines in this aliphatic chain. The effect is to shorten the C—H bond and make infinitely more difficult any effort to dislodge the hydrogen.

The second effort at synthesis of perfluoro-1,6-heptadiene was carried out in the following manner using 1,2-dichlorotrifluoroiodoethane and 5,6-dichloro-1-iodoperfluorohexane as starting materials:

$$CF_2ClCFCl(CF_2CF_2)I \xrightarrow[150°C]{HSO_3Cl} CF_2ClCFClCF_2CF_2CF_2COOH$$

$$NH_4OH \Big\downarrow AgNO_3$$

$$CF_2ClCFClCF_2CF_2CF_2I \overset{I}{\underset{\Delta}{\longleftarrow}} CF_2ClCFClCF_2CF_2CF_2COOAg$$

$$CF_2ClCFClCF_2CF_2CF_2I + CF_2ClCFClI \xrightarrow[CH_2Cl_2]{Zn} CF_2ClCFCl(CF_2)_3CFClCF_2Cl$$

$$Zn \Big\downarrow \text{high-boiling ether}$$

$$CF_2{=}CF(CF_2)_3CF{=}CF_2$$

The yield from the chlorosulfonic acid reaction (125) averaged 73% while conversion to the silver salt was virtually quantitative. Pyrolysis produced 44% of the desired iodo compound and the cross-coupling reaction, a weak link in any synthesis chain, yielded 30% of the immediate precursor. Dechlorination produced about 75% of the final product from the tetrachloroheptane. The overall yield was thus about 7.2%. The high-boiling ether referred to in all of these syntheses is bis[2(2-methoxy ethoxy)ethyl] ether.

The third effort used 3,4-dichloroheptafluoro-1-iodobutane and proceeded accordingly:

$$CF_2ClCFClCF_2CF_2I \xrightarrow[\Delta]{SO_3} CF_2ClCFClCF_2COOH$$

$$\text{NH}_4\text{OH} \Big| \text{AgNO}_3$$

$$CF_2ClCF{=}CF_2 + CF_2ClCFClCF\ I \xleftarrow[\Delta]{I} CF_2ClCFClCF_2COOAg$$
$$(75\%) \qquad\qquad (15\%)$$

$$CF_2ClCFClCF_2I + CF_2ClCFClCF_2CF_2I \xrightarrow[\text{CH}_2\text{Cl}_2]{\text{Zn}} CF_2ClCFCl(CF_2)_3CFClCF_2Cl$$

$$\text{Zn} \Big| \begin{array}{l}\text{high-boiling}\\ \text{ether}\end{array}$$

$$CF_2{=}CF(CF_2)_3CF{=}CF_2$$

The sulfur trioxide reaction that initiated this scheme (126) gave 50% yield of the butyric acid based on the iodobutane. This reaction passes through a butyryl fluoride intermediate which is not shown. Conversion to the silver salt was again quantitative. Yield in the cross-coupling reaction was 25% and the dechlorination was again 75%. The overall yield was 1.1% or considerably lower than that of the preceding method.

The fourth and most efficient method employed up to this writing starts with 7,8-dichloroperfluoro-1-iodooctane. It is outlined below:

$$CF_2ClCFCl(CF_2)_6I \xrightarrow[150^\circ\text{C}]{\text{HSO}_3\text{Cl}} CF_2ClCFCl(CF_2)_5COOH$$

$$\text{NaOH} \Big|$$

$$CF_2ClCFCl(CF_2)_3CF{=}CF_2 \xleftarrow[350^\circ\text{C}]{} CF_2ClCFCl(CF_2)_5COONa$$

$$\text{Zn} \Big| \begin{array}{l}\text{high-boiling}\\ \text{ether}\end{array}$$

$$CF_2{=}CF(CF_2)_3CF{=}CF_2$$

The sulfonic acid reaction averaged 50% with yields ranging from 44 to 56% from the iodo compound to the sodium caprylate. Pyrolysis yielded 85% and dechlorination 78%. The overall yield was therefore 33%, making it by far the best synthesis presently known for this monomer.

A fifth effort to prepare perfluoro-1,6-heptadiene was mounted with some success, but it involves a number of complications. This synthesis starts with the free radical telomerization of HFP with 3,4-dichloroheptafluorobutane. The outline of this preparation follows:

$$CF_2ClCFClCF_2CF_2I + CF_2{=}CF \longrightarrow CF_2ClCFClCF_2CF_2CF_2CF_2I$$
$$\underset{CF_3 \; \rightarrow}{|} \quad \underset{\begin{array}{c}\text{ethyl}\\ \text{ether}\end{array}}{} \Big| CH_3Li \quad \underset{CF_3}{|}$$

$$CF_2{=}CF(CF_2)_3CF{=}CF_2 \xleftarrow[\begin{array}{c}\text{high-boiling}\\ \text{ether}\end{array}]{\text{Zn}} CF_2ClCFClCF_2CF_2CF_2CF{=}CF_2$$

The telomerization yielded over 50% of the desired monoadduct, but the methyllithium reaction produced a mixture from which separation of the desired heptene could be effected only through use of VPC. When the methyllithium reaction was carried out on 5,6-dichloroperfluoro-1-iodo-hexane, conversion to 5,6-dichloroperfluoro-1-hexene was obtained to the extent of 65%; however, with the 1-iodohexane, deiodofluorination could take place only in one direction, whereas with the 2-iodoheptane deiodo-fluorination can take place toward the end of the molecule, producing the desired 1-heptene, or toward the center of the molecule, producing a 2-heptene which upon dechlorination would produce a 1,5-heptadiene, which in all probability would not undergo the cyclic mechanism. Through VPC techniques the desired heptene and the undesired heptene have been separated and identified through infrared spectroscopy, but the undesired isomer plus many other compounds that have not been identified comprise the major yield. It is conjectured that these compounds result from the isomerization and cyclization of 5,6-dichloroperfluoro-2-heptene. As of this writing positive proof of this hypothesis is lacking.

The five terminal fluorodienes whose syntheses have been discussed in the latter part of this chapter are listed at the bottom of Table 4. They are all

TABLE 4

Fluorodienes

Compound	B.P., °C
CF_2CCF_2	-38
$CF_2CFCHCH_2$	8.5
$CF_2CHCFCH_2$	16.9
$CF_2CFCFCH_2$	-12 / 300 mm
$CF_2CFCFCF_2$	7.5
$CF_2CFCF_2CFCF_2$	36
$CF_2CF(CF_2)_2CFCF_2$	59
$CF_2CF(CF_2)_3CFCF_2$	84
$CF_2CFCF_2CFClCF_2CFCF_2$	112
$CF_2CF(CF_2)_4CFCF_2$	106

clear, colorless liquids with pleasant odors when pure, but upon standing in air they rapidly become cloudy and acrid due to oxygen attack on the double bonds, producing acid fluorides which hydrolyze to produce fluorinated acids. It is necessary to subject samples slated for polymerization to preparative VPC. Even then purity is a problem since some of these monomers undergo structure changes under polymerization conditions. Polymer samples of these pure, unconjugated perfluorodienes are tough, white solids with thermal stability comparable with that of poly-CTFE. Irradiation of terminal perfluorodiolefins at low temperatures at autogenous pressure is known to produce only dimers and trimers (127), but at temperatures between 100 and 200°C and under pressures up to 20,000 atm high polymer is obtained. Molecular weights on the order of 10^5 are not unusual. The polymerization techniques and the nature of polymer samples derived from this work are discussed in Chapter 4.

References

1. F. Swarts, *Bull. Acad. Roy. Bel.*, **24**(3): 474 (1892).
2. F. Swarts, *Bull. Acad. Roy. Bel.*, **29**(3): 874 (1895).
3. F. Swarts, *Centralblat*, **1**: 3 (1903).
4. T. Midgley, Jr., and A. L. Henne, *Ind. Eng. Chem.*, **22**: 542 (1930).
5. H. S. Booth, W. L. Mong, and P. E. Burchfield, *Ind. Eng. Chem.*, **24**: 328 (1932).
6. H. S. Booth, P. E. Burchfield, E. M. Bixby, and J. B. McKelvey, *J. Am. Chem. Soc.* **55**: 2231 (1933).
7. G. M. Bartlett, U.S. Patent 1,800,371 (Apr. 14, 1931).
8. C. Strosacker and C. C. Schwegler, U.S. Patent 1,930,350 (Oct. 10, 1934).
9. E. G. Locke, W. R. Brode, and A. L. Henne, *J. Am. Chem. Soc.*, **56**: 1726 (1934).
10. Farbenfabriken Bayer, British Patent 681,067 (Oct. 15, 1952).
11. H. Madai, East German Patent 12,182 (Oct. 6, 1956).
12. L. B. Smith and C. B. Miller, U.S. Patent 2,635,121 (Apr. 14, 1953).
13. B. F. Landrum and R. F. Herbst, U.S. Patent 2,753,379 (July 3, 1956).
14. F. B. Downing, A. F. Benning, and R. C. McHarness, U.S. Patent 2,384,821 (Sept. 18, 1945).
15. Farbwerke Hoechst, Netherlands Patent 302,391 (CIC-07c) (June 25, 1964), German Patent Application (Dec. 22, 1962).
16. J. D. LaZerte, L. J. Hals, et. al., *J. Am. Chem. Soc.*, **75**: 4525 (1953).
17. F. B. Downing, A. F. Benning, and R. C. McHarness, U.S. Patent 2,480,560 (Aug. 30, 1949).
18. A. F. Benning, F. B. Downing, and J. D. Park, U.S. Patent 2,394,582 (Feb. 12, 1946).
19. B. F. Skiles, U.S. Patent 2,674,631 (Apr. 6, 1954).
20. J. C. Hillyer and J. F. Wilson, U.S. Patent 2,471,525 (May 31, 1949).
21. A. L. Henne and C. J. Fox, *J. Am. Chem. Soc.*, **76**: 479 (1954).
22. H. Gilman and R. G. Jones, *J. Am. Chem. Soc.*, **65**: 2037 (1943).
23. F. Swarts, *Bull. Acad. Roy. Bel.*, **1899**: 357.
24. R. Ehrenfeld, U.S. Patent Application 762,873; *Official Gaz.*, **646**: 91377 (1951).
25. R. N. Haszeldine and B. R. Steele, *J. Chem. Soc.*, **1954**: 3747.
26. J. W. Clark, British Patent 698,386 (Oct. 14, 1953).
27. A. M. Lovelace, D. A. Rousch, and W. Postelnik, "Aliphatic Fluorine Compounds," Rheinhold, New York, 1958, pp. 1–30.
28. J. L. Warnell, U.S. Patent 3,125,599 (Mar. 17, 1965).
29. W. E. Hanford, U.S. Patent 2,392,378 (Jan. 8, 1946).
30. F. G. Pearson, British Patent 578,168 (Jan. 18, 1946).
31. F. G. Pearson, British Patent 584,742 (Jan. 22, 1947).
32. W. J. Miller, A. L. Dittman, and S. K. Reed, U.S. Patent 2,586,550 (Feb. 19, 1952).
33. M. Lazar, *J. Polymer Sci.*, **29**: 573 (1958).
34. G. J. Raedel, U.S. Patent 2,613,202 (Oct. 7, 1952).

35. J. W. Borland, C. B. Miller, and J. H. Pearson, U.S. Patent 2,865,824 (Dec. 23, 1958).
36. R. U. Zimakov, E. V. Volkova, A. V. Fokin, A. D. Sorakin, and V. M. Balikov, *Trudy Usesayuz Soveschaniya Riga*, **1**: 219 (1960).
37. D. Goerrig, H. Jonas, and W. Moschel, German Patent 935,867 (Dec. 1, 1955).
38. Chin-Yung Wu and Chang-Yu Hu, *Hsueh Tung Pro* **6**: 329 (1964).
39. G. H. Crawford, U.S. Patent 3,084,144 (Apr. 2, 1963).
40. O. W. Burke, Jr., British Patent 821,971 (Oct. 14, 1959).
41. K. L. Berry, U.S Patent 2,559,751 (July 10, 1951),
42. J. Hamilton, U.S. Patent 2,569,524 (Oct. 2, 1951).
43. I. Dennstedt and W. Becker, German Patent 959,060 (Feb. 28, 1957).
44. D. A. Rousch and H. G. Hahn, U.S. Patent 3,072,589 (Jan. 8, 1963).
45. G. Bier, H. H. Frey, R. Schaef, and H. Fritz, German Patent 1,139,977 (Nov. 22, (1962).
46. S. Bandes, U.S. Patent 3,018,276 (Jan. 23, 1962).
47. A. N. Bolstad, U.S. Patent 3,163,628 (Dec. 29, 1964).
48. J. A. Sateras, Spanish Patent 282,731 (Nov. 27, 1962).
49. R. W. Perry, U.S. Patent 2,779,025 (Jan. 29, 1957).
50. T. Yamamato, Japan Patent 10,499 (Dec. 14, 1957).
51. F. J. Honn and J. M. Hoyt, U.S. Patent 3,053,818 (Sept. 1, 1962).
52. O. Metzler and W. Kaus, Belgian Patent 616,596 (Aug. 15, 1962).
53. J. Euler and P. Scholz, German Patent 1,174,380 (July 23, 1964).
54. W. A. Gey, U.S. Patent 3,067,074 (Dec. 4, 1962).
55. M. H. Kaufman and J. S. Davidson, U.S. Patent 3,122,462 (Feb. 25, 1964).
56. Anon, *Federal Register*, **29**: 3394 (Mar. 14, 1964).
57. S. A. Stern, P. H. Mohr, and P. J. Gareis, German Patent 1,139,474 (Nov. 15, 1962).
58. F. W. West and F. N. Roberts, U.S. Patent 2,866,764 (Dec. 30, 1958).
59. F. W. Troester, U.S. Patent 2,875,105 (Feb. 24, 1959).
60. M. A. Bruk, A. D. Abkin, P. M. Khomikovskii, G. A. Gol'der, and Hsiang-Ling Chu, *Dokl. Akad. Nauk SSSR*, **157** (6): 1399 (1964).
61. P. J. Manno, *Am. Chem. Soc. Polymer Preprints*, **4**(1): 78 (1963).
62. D. Cordischi, A. Delle Site, M. Lenzi, and A. Mele, *La Chimica e l'Industria* (*Milan*), **44**:1101 (1962).
63. M. I. Bro, R. J. Connery, and C. Schreyer, British Patent 840,080 (July 6, 1960).
64. J. C. MacKenzie, French Patent 1,375,985 (Oct. 23, 1964).
65. B. F. Landrum and R. L. Herbst, Jr., U.S. Patent 2,951,783 (Sept. 6, 1960).
66. D. Sianesi, G. Caprisco, and E. Strepparola, Belgian Patent 635,081 (Nov. 18, 1963).
67. P. E. Aldrich, U.S. Patent 3,162,622 (Dec. 22, 1964).
68. D. B. Pattison, French Patent 1,365,581 (July 3, 1964).
69. K. Hirose, M. Umehara, S. Heyachi, and T. Kawomata, Japan Patent 22,586 (Sept. 12, 1964).
70. A. Chapiro and P. Seider, French Patent 1,371,843 (Sept. 11, 1964).
71. J. Dobo, *Radiation Chem. Proc., Tehony Hung.*, 195 (1962). (pub. 1964 in Eng.).
72. M. Kondo and H. Matsuo, *Rev. Elec. Commun. Lab.* (*Tokyo*) **12**(7–8): 410 (1964).
73. A. Fieevez, French Patent 1,399,713 (May 21, 1965).
74. W. Barthel, K. Long, and W. Uhlman, East German Patent 38,065 (Apr. 15, 1965).
75. R. E. Geller and H. J. Couch, U.S. Patent 3,194,702 (July 13, 1965).
76. H. W. Christenson, R. H. Schaefer, and F. C. Flowers, British Patent 1,003,246 (Sept. 2, 1965).
77. H. A. Ambrose and P. R. McCarthy, U.S. Patent 3,159,577 (Dec. 1, 1964).
78. F. Krebs, R. Lohman, F. Solditt, German Patent 1,193,629 (May 26, 1965).

79. A. H. Lopez, Spanish Patent 294,783 (Jan. 25. 1964).
80. Polymer Corporation, British Patent 999,262 (July 21, 1965).
81. R. A. Suess and G. R. Neff, *Sci. Tech. Aerospace Report*, **2**(12): 1504 (1964).
82. W. Goldie, *Metal Finishing*, **62**(12): 50 (1964).
83. A. B. Boerman and B. O. Ayers, U.S. Patent 3,167,946 (Feb. 2, 1965).
84. L. W. Niedrich and H. R. Alford, Belgian Patent 644,187 (Mar. 3, 1964).
85. J. Hochberg, U.S. Patent 3,202,540 (Aug. 24, 1965).
86. H. P. Rouen, *Anal. Chem.*, **36**(13): 2420 (1964).
87. D. Sianesi and G. Caporiccio, *J. Polymer Sci. A*1 **6**: 335 (1968).
88. C. L. Sandberg and J. M. Mullins, U.S. Patent 3,080,347 (Mar. 5, 1963).
89. E. Shin Lo, U.S. Patent 3,023,187 (Feb. 27, 1962).
90. W. Koehler, *Werkstoff Korrosion*, **13**: 331 (1962).
91. E. I. DuPont de Nemours, British Patent 888,766 (Feb. 7, 1962).
92. E. L. Yuon, U.S. Patent 3,025,183 (Mar. 13, 1962).
93. W. R. Griffin, U.S. Patent 3,041,316 (June 26, 1966).
94. J. F. Smith, U.S. Patent 3,090,775 (May 21, 1963).
95. J. F. Smith, U.S. Patent 3,039,992 (June 19, 1962).
96. J. F. Smith, U.S. Patent 3,011,995 (April 12, 1960).
97. W. Koblizek, *Kautschuke u Gummi*, **14**: 308 (1961).
98. R. D. Chambers, J. Hutchinson, and W. K. R. Muskgrave, *Tetrahedron Letters*, **1963**: 619.
99. K. L. Paciorek, W. G. Liginess, and C. J. Lank, *J. Polymer Sci.*, **60**: 141 (1962).
100. I. Roche, U.S. Patent 3,109,750 (1963).
101. D. W. Brown and L. A. Wall, *Am. Chem. Soc. Polymer Preprints*, **7**(2): 116 (1966).
102. E. M. Sullivan, E. W. Wise, and F. P. Reding, U.S. Patent 3,110,705 (1963).
103. E. I. DuPont de Nemours, British Patent 953,152 (1964).
104. T. L. Jacobs and R. S. Bauer, *J. Am. Chem. Soc.*, **81**: 606 (1959).
105. A. L. Klehanski and O. A. Timofeev, *J. Polymer Sci.*, **52**: 23 (1961).
106. W. T. Miller, Jr., W. Frass, and P. R. Resnick, *J. Am. Chem. Soc.* **83**: 1767 (1961).
107. H. Iserson, M. Hauptschein, and F. E. Lawlor, *J. Am. Chem. Soc.*, **81**: 2676 (1959).
108. G. B. Butler and R. J. Agnello, *J. Am. Chem. Soc.*, **79**: 3128 (1957).
109. J. D. Park and J. R. Lacher, *W.A.D.C. Tech. Rept.* 56–590, Part 1, pp. 21–22 (1957).
110. J. E. Fearn and L. A. Wall, *SPE Trans.*, **3**: 231 (1963).
111. J. E. Fearn, D. W. Brown, and L. A. Wall, *J, Polymer Sci. A*1, **4**: 131 (1966).
112. D. W. Brown, J. E. Fearn, and R. E. Lowry, *J. Polymer Sci. Part A*1, 3:1640 (1965).
113. W. T. Miller, W. Frass, and P. R. Resnick, *J. Am. Chem. Soc.*, **83**:1767 (1961).
114. W. T. Miller, private communication.
115. R. N. Haszeldine, *J. Chem. Soc.*, **1955**: 4291.
116. M. Hauptschein, M. Braid, and A. M. Fainburg, *J. Am. Chem. Soc.*, **83**: 2495 (1961).
117. M. Hauptschein, M. Braid, and F. E. Lawlor, *J. Am. Chem. Soc.*, **79**: 2549 (1957).
118. M. Hauptschein, M. Braid, and F. E. Lawlor, *J. Am. Chem. Soc.*, **80**: 846 (1958).
119. I. L. Knunyants, Li Chai-Yuan, and V. V. Shokina, *Doklady. Akad. Nauk SSSR*, **136**: 610 (1961).
120. I. L. Knunyants, S. P. Khrlakyan, and Yu V. Zeifman, *Doklady. Akad. Nauk SSSR*, **2**: 384 (1964).
121. I. L. Knunyants, Li Chai-Yuan, and V. V. Shokina, *Izvestiya, Akademie Nauk SSSR Otdelenie Khimicheskikh Nauk*, 10: 1462 (1961).
122. J. E. Fearn, *J. Res. NBS*, **75A**: 41 (1971).
123. W. T. Miller, U.S. Patent 2,668,182 (Feb. 2, 1954).
124. R. M. Joyce, U.S. Patent 2,559,628 (July 10, 1951).

125. M. Hauptschein and M. Braid, *J. Am. Chem. Soc.*, **83**: 2500 (1961).
126. I. L. Knunyants, Li Chai-Yuan, and V. V. Shokina, *Isvestiya Akademie Nauk SSSR Otdelenie Khimicheskikh Nauk*, **10**: 1910 (1961).
127. V. A. Kramchenkov and A. V. Zimin, *Proc. 2nd All Union Conf. on Radiation Chem.*, Moscow (1962).

2. THE SYNTHESIS AND POLYMERIZATION OF FLUOROSTYRENES AND FLUORINATED VINYL PHENYL ETHERS

Joseph M. Antonucci, *Polymer Chemistry Section, National Bureau of Standards, Washington, D.C.*

Contents

I. INTRODUCTION

In science a fortuitous event frequently serves as the genesis of an entirely new area of research. Thus the accidental discovery that tetrafluoroethylene readily undergoes polymerization can be envisioned as marking the beginning of fluorocarbon polymer chemistry (1,2).

The discovery of polytetrafluoroethylene (Teflon), which is still the most chemically and thermally stable addition polymer known, led to an ever-widening search for other fluorinated polymers possessing similar outstanding chemical and physical properties. Not unexpectedly, a great deal of this research has been directed to the synthesis and polymerization of other ethylenic monomers of varying fluorine content, e.g., vinyl fluoride, vinylidene fluoride, trifluoroethylene, and chlorotrifluoroethylene. Much work also has been directed to copolymerization studies involving tetrafluoroethylene and other fluoroolefins.

Until recently, however, comparatively little research has been devoted to the synthesis and polymerization of fluorinated styrenes, particularly those of high fluorine content. The reasons for this seeming lack of interest in fluorinated styrenes compared to the other fluorinated vinyl monomers probably can be ascribed to two factors: (1) the lack of, or the difficulty in obtaining the necessary starting materials such as bromopentafluorobenzene, and (2) the state of the art of fluorocarbon synthesis and polymerization in the early days of fluorocarbon chemistry. This situation has been remedied by the discovery since 1960 of new methods of synthesis and polymerization. It is now possible to prepare and polymerize almost any fluorinated styrene or related monomer.

II. STYRENES WITH FLUORINATED VINYL GROUPS

A. β-Fluorostyrene

In 1919 the first fluorostyrene, β-fluorostyrene, was synthesized (3). In a study exploring the behavior of the Grignard reagent of bromobenzene with 1,1-dihalo-2,2-difluoroethanes, β-fluorostyrene was found among the products of these reactions. The mechanism of the reaction was envisioned as follows:

$$C_6H_5MgBr + CX_2HCF_2H \xrightarrow{(C_2H_5)_2O} [C_6H_5CXHCF_2H] \xrightarrow{-XF} C_6H_5CH=CFH$$
$$\text{unisolated}$$
$$\text{intermediate}$$

where X = Cl or Br. Moreover, when 1,1-dichloro-2,2-difluoroethane was the coreagent, the Grignard reaction also gave rise to another fluorostyrene, α-fluoro-β-chlorostyrene. The mechanisms of these reactions remain obscure.

A more clear-cut synthesis of β-fluorostyrene involves the decarboxylation of α-fluorocinnamic acid (4,5). The α-fluorocinnamic acid is made conveniently by the Perkins reaction of benzaldehyde with methyl monofluoroacetate as shown:

$$C_6H_5CHO + CFH_2CO_2CH_3 \xrightarrow[\text{(2) hydrolysis}]{\text{(1) Na } (C_2H_5)_2O} C_6H_5CH=CFCO_2H$$

$$\xrightarrow[\substack{\Delta \\ -CO_2}]{} C_6H_5CH=CFH$$

Another good synthetic route to β-fluorostyrene and similar halostyrenes is illustrated below (yields in parenthesis) (6,7):

$$C_6H_5COCHFX \xrightarrow[\substack{\text{or} \\ \text{NaBH}_4}]{\text{LiAlH}_4} C_6H_5CHOHCHFX \xrightarrow[\text{base}]{\text{SOCl}_2}$$
$$(72\text{–}92\%)$$

$$C_6H_5CHClCHFX \xrightarrow[\substack{\text{solvent} \\ \text{(eg., acetamide)}}]{\text{Zn}} C_6H_5CH=CHF$$
$$(71\text{–}80\%) \qquad\qquad\qquad (60\text{–}65\%)$$

where X = F, Cl, or Br.

The starting acetophenones may be prepared by methods such as the following:

$$C_6H_6 \xrightarrow[\substack{\text{Friedel-Crafts} \\ \text{catalysis}}]{\text{CHF}_2\text{COX}} C_6H_5COCHF_2 \xleftarrow{C_6H_5Li} CHF_2CO_2Li$$

Other halostyrenes, such as β-chloro-β-fluorostyrene, may be prepared by similar methods (8):

$$CFCl_2CO_2Na \xrightarrow{C_6H_5MgBr} C_6H_5COCFCl_2 \xrightarrow{\text{LiAlH}_4} C_6H_5CHOHCF_2Cl$$

$$\xrightarrow[\text{pyridine}]{\text{SOCl}_2} C_6H_5CHClCFCl_2 \xrightarrow[\text{C}_2\text{H}_5\text{OH}]{\text{Zn}} C_6H_5CH=CFCl$$

B. α-Fluorostyrene

The last unknown isomer of monofluorostyrene has only recently been synthesized. In 1962 the synthesis of α-fluorostyrene was achieved by the addition of anhydrous hydrogen fluoride to either phenylacetylene or α-chlorostyrene followed by the subsequent thermal dehydrofluorination of the intermediate α,α-difluoroethylbenzene (9):

$$\left.\begin{array}{c} C_6H_5C\equiv CH \\ \text{or} \\ C_6H_5CCl=CH_2 \end{array}\right\} + 2HF \xrightarrow[\substack{13\text{--}14\% \text{ HgO} \\ \text{on act. carbon} \\ \text{at } 150°C}]{\substack{\text{ether} \\ 0°C \\ \text{or}}} \underset{(11\text{--}45\%)}{C_6H_5CF_2CH_3}$$

$$C_6H_5CF_2CH_3 \xrightarrow[-HF]{350\text{--}400°C} \underset{(53\%)}{C_6H_5CF=CH_2}$$

C. α,β,β-Trifluorostyrene

The earliest synthesis of α,β,β-trifluorostyrene involved the following multistep procedures (10,11):

$$C_6H_6 + CF_2XCOCl \xrightarrow[CS_2]{AlCl_3} C_6H_5COCF_2X + HCl \tag{1}$$

$$C_6H_5COCF_2X + PCl_5 \longrightarrow \underset{I}{C_6H_5CCl_2CF_2X} + POCl_3 \tag{2}$$

where X = F, Cl, H. Dehalogenation of the intermediate I gives α-chloro-β,β-difluorostyrene:

$$\underset{I}{C_6H_5CCl_2CF_2X} \xrightarrow[\text{abs. ethanol}]{Zn} \underset{(70\%)}{C_6H_5CCl=CF_2} \tag{3}$$

where X = Cl.

The synthesis of α,β,β-trifluorostyrene requires an additional step, the monofluorination of intermediate I:

$$\underset{I}{C_6H_5CCl_2CF_2X} + \tfrac{1}{3}SbF_3 \xrightarrow[140\text{--}170°C]{Br_2} C_6H_5CClFCF_2X + \tfrac{1}{3}SbCl_3 \tag{4}$$

$$C_6H_5CClFCF_2X \xrightarrow[\text{abs. ethanol}]{Zn} \underset{(48\%)}{C_6H_5CF=CF_2} \tag{5}$$

where X = Cl. The same monomer may be obtained by the dehydrochlorination of the intermediate α-chloro-α,β,β-trifluoroethylbenzene (X = H), although this results in poorer yield (12):

$$C_6H_5CClFCF_2H \xrightarrow[230\text{--}250°C]{\text{NaOH}} \underset{(17\%)}{C_6H_5CF=CF_2} \tag{6}$$

β-Chloro-α,β-difluorostyrene can be prepared by a similar sequence of reactions (11):

$$C_6H_5COCH_3 \xrightarrow{Cl_2} C_6H_5COCCl_3 \xrightarrow[HF]{AgF} C_6H_5COCCl_2F \qquad (7)$$

$$C_6H_5COCCl_2F \xrightarrow{PCl_5} C_6H_5CCl_2CCl_2F \xrightarrow{SbF_3} C_6H_5CClFCCl_2F \qquad (8)$$

$$C_6H_5CClFCCl_2F \xrightarrow[\text{abs. ethanol}]{Zn} \underset{(48\%)}{C_6H_5CF=CFCl} \qquad (9)$$

D. β,β-Difluorostyrene

The synthesis of β,β-difluorostyrene can be accomplished in the low overall yield by the following set of reactions (12):

$$C_6H_5COCHF_2 \xrightarrow{Al(i\text{-prop})_3} C_6H_5CHOHCHF_2 \qquad (10)$$

$$C_6H_5CHOHCHF_2 \xrightarrow{C_6H_5COCl} C_6H_5CHOCOC_6H_5CHF_2 \qquad (11)$$

$$C_6H_5CHOCOC_6H_5CHF_2 \xrightarrow{625\text{-}636°C} C_6H_5CH=CF_2 \qquad (12)$$

The difficult benzoate ester pyrolysis proceeded in only 13% yield.

An improved synthesis of β,β-difluorostyrene is shown below (7):

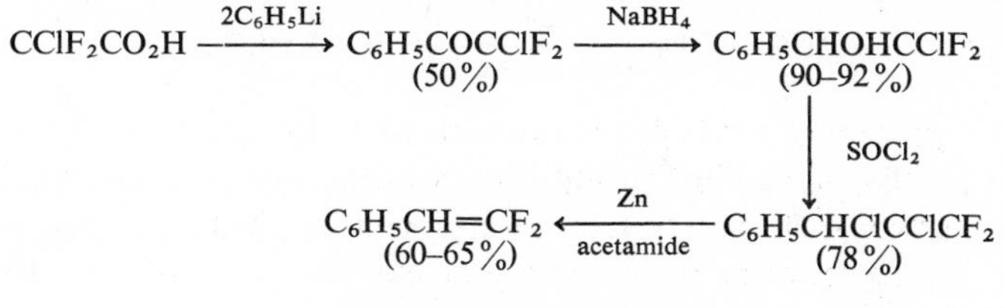

$$CClF_2CO_2H \xrightarrow{2C_6H_5Li} \underset{(50\%)}{C_6H_5COCClF_2} \xrightarrow{NaBH_4} \underset{(90\text{-}92\%)}{C_6H_5CHOHCClF_2}$$

$$\downarrow SOCl_2$$

$$\underset{(60\text{-}65\%)}{C_6H_5CH=CF_2} \xleftarrow[\text{acetamide}]{Zn} \underset{(78\%)}{C_6H_5CHClCClF_2}$$

E. α,β-Difluorostyrene

The preparation of the isomeric α,β-difluorostyrene involves dechlorofluorination of the intermediate α-chloro-α,β,β-trifluoroethylbenzene:

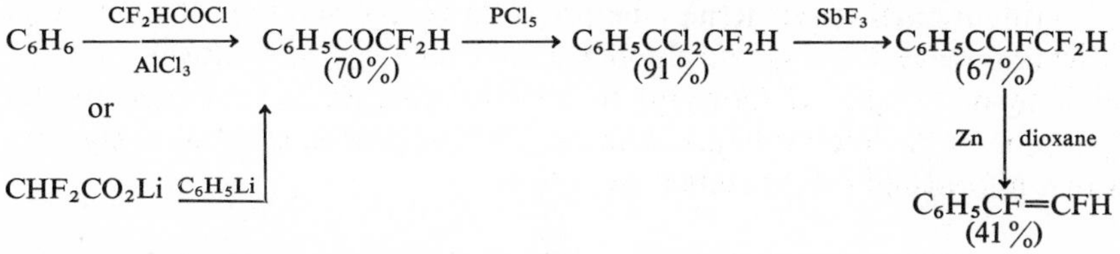

$$C_6H_6 \xrightarrow[AlCl_3]{CF_2HCOCl} \underset{(70\%)}{C_6H_5COCF_2H} \xrightarrow{PCl_5} \underset{(91\%)}{C_6H_5CCl_2CF_2H} \xrightarrow{SbF_3} \underset{(67\%)}{C_6H_5CClFCF_2H}$$

or

$$CHF_2CO_2Li \xrightarrow{C_6H_5Li}$$

$$Zn \mid dioxane$$

$$\underset{(41\%)}{C_6H_5CF=CFH}$$

The last step in this synthesis is an example of the unexpectedly facile elimination of chlorine and fluorine from certain neighboring carbon atoms. In the field of aliphatic fluorocarbon chemistry, it is usually considered difficult to effect such a dehalogenation; however, some precedent exists for such eliminations as witnessed by the following reaction involving the dechlorofluorination of 1,2,3,4-tetrachlorooctafluorocyclohexane (13):

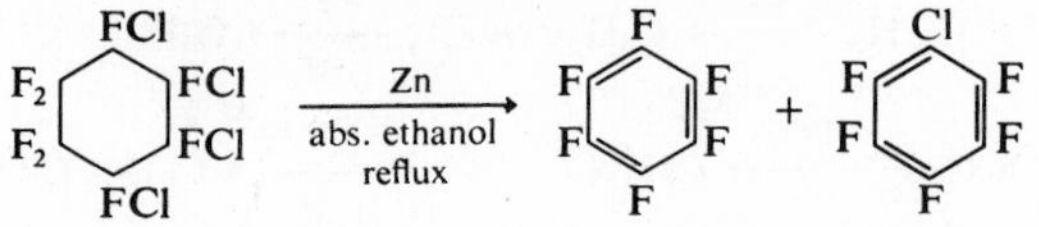

$$\underset{\substack{FCl \\ F_2\text{---}FCl \\ F_2\text{---}FCl \\ FCl}}{} \xrightarrow[\substack{\text{abs. ethanol} \\ \text{reflux}}]{Zn} \underset{\substack{F \\ F\text{---}F \\ F\text{---}F \\ F}}{} + \underset{\substack{Cl \\ F\text{---}F \\ F\text{---}F \\ F}}{}$$

It was further observed that 1-phenoxy-1,1-difluoro-2,2-dichloroethane could be converted in good yield to the corresponding vinyl phenyl ether by dechlorofluorination (14):

$$C_6H_5OCF_2CCl_2H \xrightarrow[\text{abs. ethanol}]{Zn} \underset{(70\%)}{C_6H_5OCF=CClH}$$

Some more recent examples involve the preparation of α-chloro-β-fluorostyrene (12), α-fluoro-β-chlorostyrene (7), and α,β,β-trifluorostyrene (15), respectively:

$$C_6H_5CCl_2CF_2H \xrightarrow[\text{dioxane}]{Zn} \underset{(52\%)}{C_6H_5CCl=CFH} \tag{13}$$

$$C_6H_5CF_2CCl_2H \xrightarrow[\text{acetamide}]{Zn} \underset{(80\%)}{C_6H_5CF=CClH} \tag{14}$$

$$C_6H_5CClFCF_3 \xrightarrow[\text{ethylene glycol}]{Zn} C_6H_5CF=CF_2 \tag{15}$$

From the preceding examples it appears that dechlorofluorination may occur with facility if the vinyl group that is being formed is in conjugation with some electron-rich moiety such as a phenyl or phenoxy group. It may also occur in other systems that offer the proper steric orientation of vicinal halogens (16).

F. α-Trifluoromethylstyrene

α-Trifluoromethylstyrene may be prepared by the addition of the methyl Grignard reagent to α,α,α-trifluoroacetophenone or, alternatively, by the addition of the phenyl Grignard to 1,1,1-trifluoroacetone followed by dehydration of the intermediate carbinols (17). The general scheme for styrenes with α-fluorinated substituents is as follows:

$$C_6H_5MgBr + R_fCOCH_3 \longrightarrow \underset{\overset{|}{R_f}}{C_6H_5COHCH_3} \xrightarrow{P_2O_5} \underset{\overset{|}{R_f}}{C_6H_5C=CH_2}$$

where $R_f = CF_3$, CF_2H, etc.

Another modification is the synthesis of α-trifluoromethylstyrenes with trifluoromethyl substituents on the phenyl ring as well (18):

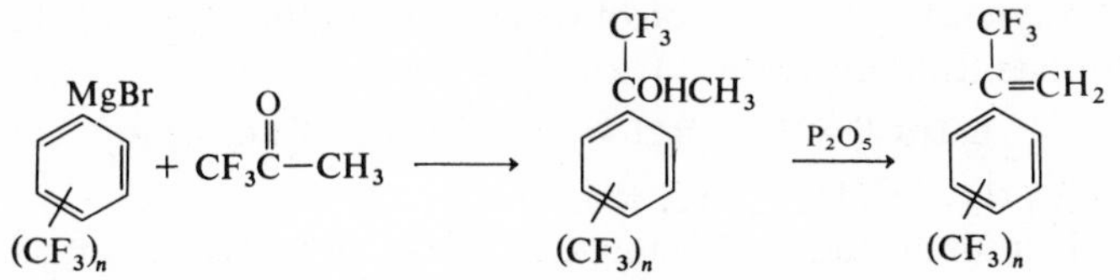

$$\underset{(CF_3)_n}{\overset{MgBr}{\bigcirc}} + CF_3\overset{O}{\overset{\|}{C}}-CH_3 \longrightarrow \underset{(CF_3)_n}{\overset{\overset{CF_3}{\underset{|}{COHCH_3}}}{\bigcirc}} \xrightarrow{P_2O_5} \underset{(CF_3)_n}{\overset{\overset{CF_3}{\underset{|}{C=CH_2}}}{\bigcirc}}$$

A somewhat different approach to trifluoromethylated styrenes uses the following reaction scheme (19):

$$R_{Ar}COR_f + CH_3Li \longrightarrow R_{Ar}\overset{\overset{R_f}{\underset{|}{}}}{COHCH_3} \xrightarrow{-H_2O} R_{Ar}\overset{\overset{R_f}{\underset{|}{}}}{C}=CH_2$$

where $R_{Ar} = C_6H_5$, $CH_3C_6H_4$, $CF_3C_6H_4$, etc., and $R_f = CF_3$, CF_2H, etc.

Some of the preceding multistep syntheses involving the use of fluorinated acetophenones as intermediates could possibly be improved by replacing the two-step classical phosphorus pentachloride-antimony trifluoride sequence by the one-step conversion of $-C=O$ to $-CF_2$ by means of sulfur tetrafluoride (20). For example, α,β,β-trifluorostyrene presumably could be synthesized as follows:

$$C_6H_5COCF_2Cl \xrightarrow{SF_4} C_6H_5CF_2CF_2Cl \xrightarrow[\text{solvent}]{Zn} C_6H_5CF=CF_2$$

III. SYNTHETIC METHODS

In addition to the classic multistep type of syntheses already described, there are several more direct preparative methods which are of importance from the standpoint of both synthetic utility and theoretical interest.

A. Aryllithium-Fluoroolefin Reaction

A general direct synthesis of vinyl fluorinated styrene is that employing the reaction of aryllithiums with certain fluoroolefins (21,22). For example, α,β,β-trifluorostyrene can be prepared in one step by the condensation of phenyllithium with tetrafluoroethylene (TFE):

$$\underset{\text{(excess)}}{C_6H_5Li + CF_2=CF_2} \xrightarrow[-80 \text{ to } 25°C]{(C_2H_5)_2O} \underset{(30\%)}{C_6H_5CF=CF_2} + \underset{(50\%)}{C_6H_5CF=CFC_6H_5}$$

As can be seen from the relative yields of products, 1,2-difluorostilbene is the preponderant product. Presumably this occurs from a secondary attack of the aryllithium on the initially formed trifluorostyrene, and it seems to suggest that the α,β,β-trifluorostyrene is more susceptible to nucleophilic attack than TFE. In fact, it is possible to synthesize in good yields various β-substituted

α,β-difluorostyrenes by the reaction of various organolithium compounds with the appropriate trifluorostyrene. Thus

$$C_6H_5CF=CF_2 + RLi \longrightarrow C_6H_5CF=CFR + LiF$$

where $R = C_6H_5$, FC_6H_4, CH_3, etc.

The mechanism postulated for this type of reaction involves attack by the nucleophilic phenyl moiety at the more positive carbon of the fluoroolefin (i.e., the carbon carrying the greater number of fluorine atoms) to give an unstable addition intermediate which spontaneously loses lithium fluoride to give the styrene (21):

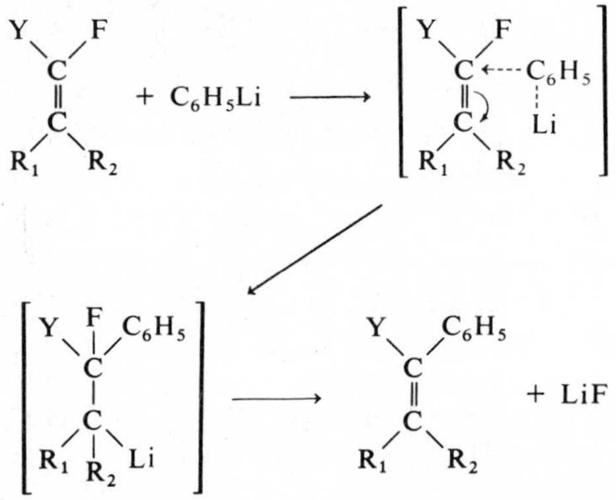

where $Y = F$, or fluoroalkyl; $R_1 =$ electronegative atom or group; $R_2 = R_1$ or H or alkyl group. The nature and distribution of products are dependent on the reaction conditions, such as molar ratio of reactants, temperature, and solvent system.

The scope of this reaction makes it possible to synthesize not only styrenes with fluorinated side chains but also styrene having nuclear substituents. Some typical examples are shown below [for reaction 16, see (21 and 23); 17, (21); 18, (24); 19, (26 and 27); 20, (28)]:

$$C_6H_5Li + CF_2=CFCl \longrightarrow C_6H_5CF=CFCl \quad (16)$$
$$(50\text{–}90\%)$$

$$C_6H_5Li + CF_2=CFCF_3 \longrightarrow C_6H_5CF=CFCF_3 \quad (17)$$
$$(50\%)$$

$$CH_3C_6H_4Li + CF_2=CF_2 \longrightarrow CH_3C_6H_4CF=CF_2 \quad (18)$$
$$o, m, p$$
$$(55\%, 46\%, 40\%)$$

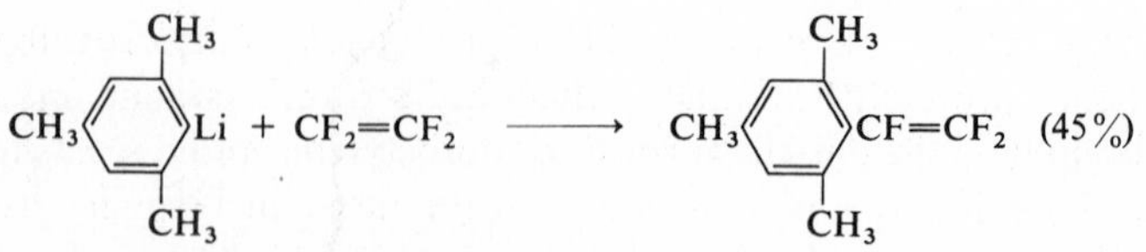

(19)

$$C_6H_5Li + CF_2=CH_2 \xrightarrow{0°C} C_6H_5CF=CH_2 + C_6H_5C\equiv CH \quad (31\%) \quad (20)$$

In reaction 20, the phenylacetylene arises from a secondary reaction of the α-fluorostyrene with phenyllithium,

$$C_6H_5Li + C_6H_5CF=CH_2 \longrightarrow C_6H_6 + C_6H_5C\equiv CH + LiF$$

Phenyllithium will even attack vinyl fluoride to give low yields of styrene, but again the reaction is complicated by the dehydrofluorination of some of the vinyl fluoride to acetylene and by polymer formation (28):

$$C_6H_5Li + CH_2=CHF \xrightarrow{-78-+25°C} \underset{(5\%)}{C_6H_5CH=CH_2} + CH\equiv CH + polymer$$

Not only does this method afford a convenient route to many fluorinated styrenes, it also may be used to synthesize divinylbenzenes with fluorinated side chains. Thus p-bis(β-chloro-α,β-difluorovinyl)benzene is prepared as follows (29):

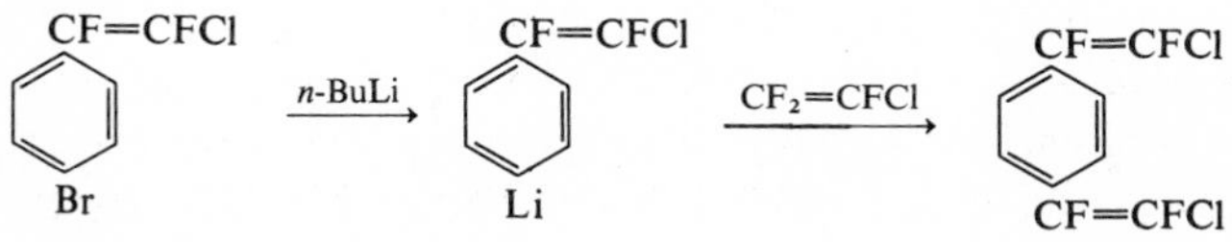

A more direct preparation starts with p-dibromobenzene (30):

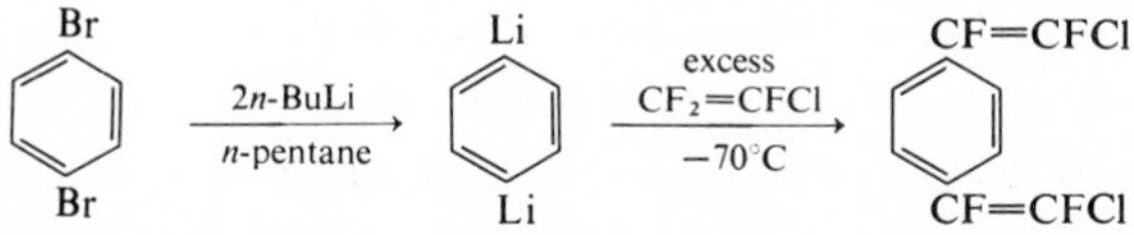

Para-substituted divinylbenzene having one fluorinated and one non-fluorinated vinyl group may be prepared as follows (30):

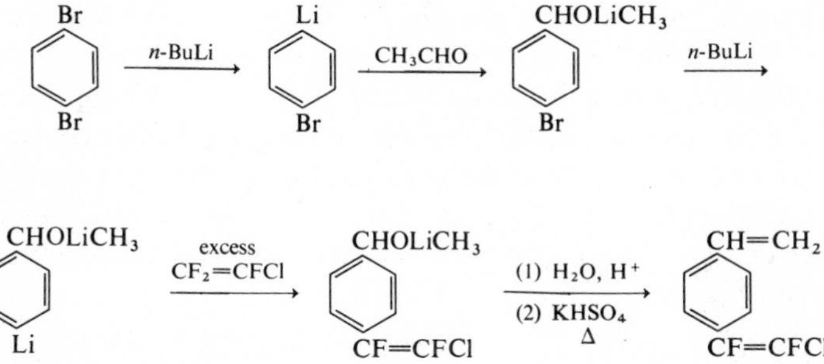

Phenyl Grignard reagents also react with some fluoroolefins (31) to give side-chain fluorinated styrenes:

$$C_6H_5MgBr + CF_2=CCl_2 \longrightarrow \underset{(63\%)}{C_6H_5CF=CCl_2} \qquad (21)$$

$$C_6H_5MgBr + CF_2=CFCl \longrightarrow \underset{(17\%)}{C_6H_5CF=CFCl} \qquad (22)$$

Other arylmetallic compounds, such as phenylsodium, presumably also may be employed, but they may present experimental difficulties (21). In general, it appears that the organolithium compounds are the most versatile reagents for this type of synthesis.

B. Wittig Reaction

In the field of hydrocarbon chemistry, a useful direct synthesis of olefins is the Wittig synthesis, which converts a carbonyl group to a terminal methylene group in one step:

$$\diagdown C=O + R_3P=CH_2 \longrightarrow \diagdown C=CH_2 + R_3PO$$

A recent modification of the Wittig olefin synthesis now makes it possible to prepare terminal difluoroolefins from the corresponding aldehyde or ketone by a similar one-step reaction (32–39). The effective reagent in these syntheses is usually generated by heating a mixture of a salt of chlorodifluoroacetic acid and a tertiary phosphine (usually triphenylphosphine) in the presence of the carbonyl compound in a suitable solvent. Thus β,β-difluorostyrene can be prepared in 74–75% yield (34, 35) by heating a mixture of sodium chloro-difluoroacetate, triphenylphosphine, and benzaldehyde in an inert anhydrous solvent such as 1,2-dimethoxyethane (monoglyme), 2,2'-dimethoxydiethyl ether (diglyme), N-methyl-2-pyrolidone, or dimethylformamide:

$$CClF_2CO_2Na + (C_6H_5)_3P + C_6H_5CHO \xrightarrow{\Delta} [(C_6H_5)_3P=CF_2 + C_6H_5CHO]$$

$$C_6H_5CH=CF_2 + (C_6H_5)_3PO + NaCl + CO_2$$

According to some workers the reaction proceeds by the *in situ* formation of difluoromethylene, which is immediately trapped by the triphenylphosphine to form an ylid. This can then undergo the usual Wittig reaction with the carbonyl compound (36,37). However, others dispute this contention and suggest that the mechanism of the reaction does not involve the prior trapping of the difluorocarbene by the tertiary phosphine but rather involves the decomposition of the unstable intermediate phosphobetaine salt I to the difluoromethylene ylid (38):

$$(C_6H_5)_3P + CClF_2CO_2Na \longrightarrow [(C_6H_5)_3P^+CF_2CO_2{}^-] + NaCl$$

$$\mathbf{I}$$

$$(C_6H_5)_3P=CF_2 + CO_2$$

As previously noted, this synthetic method usually offers a facile one-step, high-yield route to many fluorinated styrenes. A further advantage of this attractive synthesis is that, unlike the Wittig hydrocarbon olefin synthesis, which usually takes place in a basic medium, the fluoroolefin synthesis proceeds in an essentially nonbasic medium.

Some examples of the fluorinated styrenes that have been prepared by this synthesis and their yields are listed in Table 1.

TABLE 1

Fluorostyrenes Prepared by the Wittig Synthesis

Styrene	Yield (%)	Ref.
β,β-Difluorostyrene	75	33
p, β,β-Trifluorostyrene	65	35
p-Methoxy-β,β-difluorostyrene	60	35
β,β-Difluoro-α-methylstyrene	35	32, 34
p-Nitro-β,β-difluorostyrene	2	35
β,β-Difluoro-α-trifluoromethylstyrene	68, 77	38, 39
p-Chloro-β,β-difluoro-α-trifluoromethylstyrene	64	38
p, β,β-Trifluoro-α-trifluoromethylstyrene	66	38
p-Methyl-β,β-difluoro-α-trifluoromethylstyrene	72	38
p-Methoxy-β,β-difluoro-α-trifluoromethylstyrene	78	38
p-Dimethylamino-β,β-difluoro-α-trifluoromethylstyrene	16	38
β,β-Difluoro-α-pentafluoroethylstyrene	48	38
β,β-Difluoro-α,n-heptafluoropropylstyrene	42	38

C. Pyrolytic Reactions

An interesting direct but low-yield synthesis of α,β,β-trifluorostyrene is the high-temperature condensation reaction between benzene and chloro-trifluoroethylene in the presence of certain catalysts such as BF_3 (40,41).

A better pyrolytic synthesis of this monomer involves the copyrolysis of chlorodifluoromethane and α-chloro-α-fluorotoluene in the presence of steam (42):

$$C_6H_5CHClF + CHClF_2 \xrightarrow[H_2O]{600-800°C} C_6H_5CF{=}CF_2$$
$$(42\%)$$
$$+ C_6H_5CF{=}CFC_6H_5 + CF_2{=}CF_2 + C_6H_5CHCl_2$$
$$(25\%) \qquad (8\%) \qquad (16\%)$$

The reaction undoubtedly involves the cross-coupling of the C_6H_5CF: and CF_2: carbenes and should be applicable to the synthesis of other fluorostyrenes.

A somewhat unconventional synthesis of a vinyl fluorinated styrene is that illustrated below for β,β-difluorostyrene (43–45):

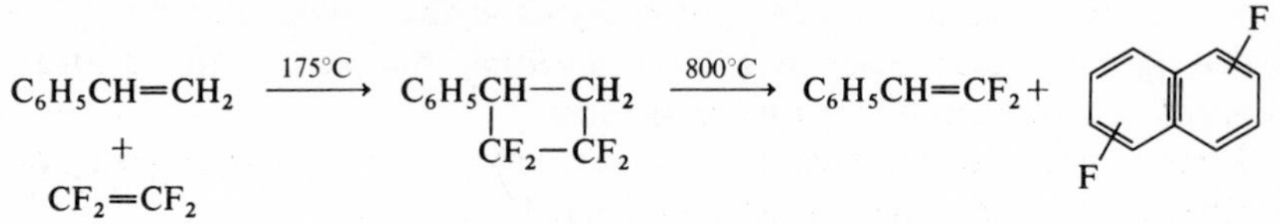

$$C_6H_5CH=CH_2 \xrightarrow{175°C} C_6H_5CH-CH_2 \xrightarrow{800°C} C_6H_5CH=CF_2 +$$
$$+ \qquad\qquad\qquad CF_2-CF_2$$
$$CF_2=CF_2$$

Presumably this method would be applicable to the synthesis of other fluorinated styrenes by the pyrolysis of the appropriate phenyl-substituted fluorinated cyclobutane.

Another thermolytic synthesis of a fluorinated styrene is that described for p,β,β-trifluorostyrene from $\alpha,\alpha,\alpha',\alpha'$-tetrafluoro-$p$-xylene (32):

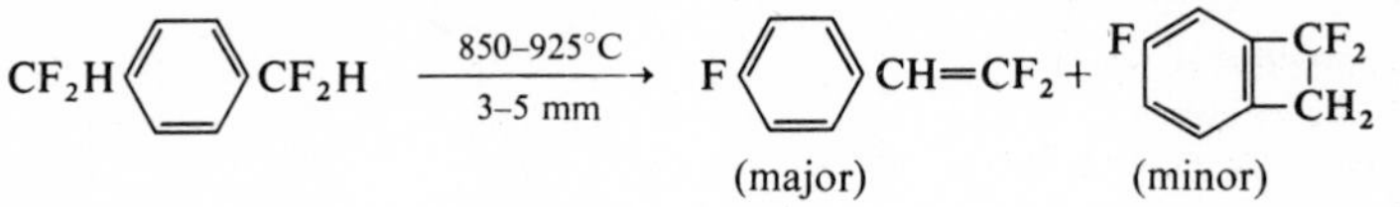

$$CF_2H\bigcirc CF_2H \xrightarrow[3-5\ mm]{850-925°C} F\bigcirc CH=CF_2 +$$

(major)   (minor)

The mechanism for the pyrolytic rearrangement of $\alpha,\alpha,\alpha',\alpha'$-tetrafluoro-$p$-xylene to p,β,β-trifluorostyrene is not known, although it is surmised that it may proceed through a cyclooctatetraene intermediate since both cyclooctetraene and styrene have been found to be formed in the pyrolysis of p-xylene (46). This type of mechanism may explain the conversion of a side-chain fluorine to a nuclear fluorine.

IV. STYRENES WITH FLUORINATED PHENYL GROUPS

Many of the methods used in the preparation of styrenes containing fluoro substituents in the vinyl group also can be used in the synthesis of styrenes with nuclear fluoro substituents. The examples in this section are for the most part of general utility.

A. Monofluorostyrenes

The first reported synthesis of ortho-, meta-, and para-fluorostyrene involved the addition of the methyl Grignard reagent to the corresponding fluorobenzaldehyde followed by dehydration (47, 48):

$$FC_6H_4CHO \xrightarrow[ether]{CH_3MgBr} FC_6H_4CHOHCH_3 \xrightarrow[-H_2O]{\substack{fused\ KHSO_4 \\ +\ hydroquinone}} FC_6H_4CH=CH_2$$

A variation of this method employs the Knoevenagel condensation of the fluorobenzaldehyde with malonic acid followed by the usual decarboxylation of the intermediate cinnamic acid (49,50):

$$FC_6H_4CHO + CH_2(CO_2H)_2 \xrightarrow{\text{pyridine}} FC_6H_4CH=CHCO_2H$$

$$\xrightarrow[200-210°C]{\substack{CuSO_4 \\ \text{quinoline}}}$$

$$FC_6H_4CH=CH_2$$

A similar synthesis makes it possible to prepare ring-fluorinated β-nitro styrenes. Thus o-fluoro-β-nitro styrene has been prepared by both of the following routes (51,52, respectively):

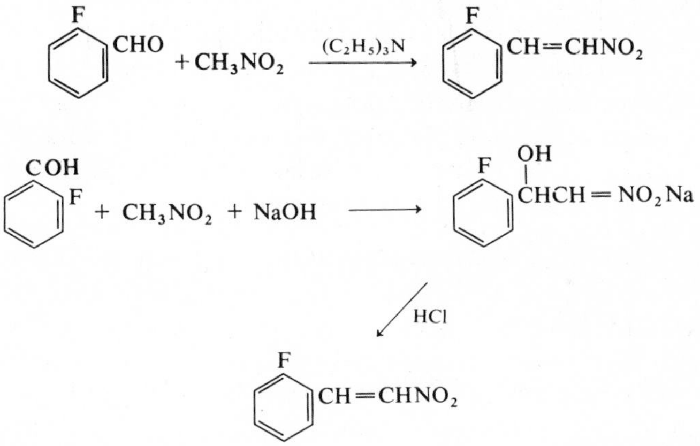

Still another technique of introducing a vinyl group into a fluorobenzene is the Friedel-Crafts acetylation reaction followed by reduction and dehydration of the appropriate intermediates. Thus with fluorobenzene and acetic anhydride in the presence of aluminum chloride a good yield of p-fluoroacetophenone is obtained. Reduction of the ketone to the alcohol followed by the usual dehydration procedure gives a good synthesis of p-fluorostyrene (53):

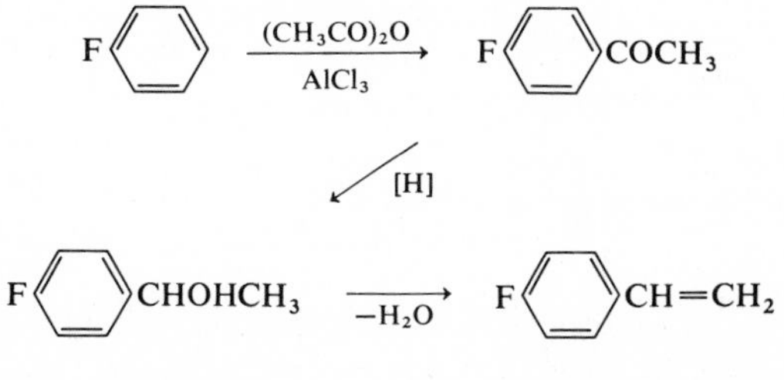

Ring-fluorinated α-methylstyrenes have been synthesized by a Friedel-Crafts reaction employing a propylene halohydrin as the alkylating agent (54). Thus 3-fluoro-4-methyl-α-methylstyrene is prepared as follows:

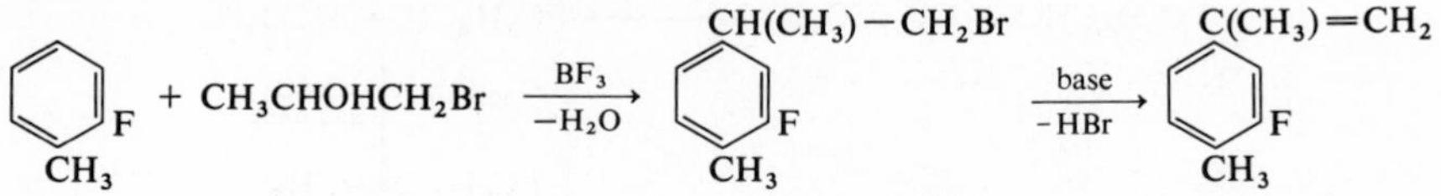

Some of the isomeric 3-fluoro-2-methyl-α-methylstyrene is also produced.

B. Trifluoromethylstyrenes

The Friedel-Crafts approach has not been successful in the preparation of styrenes with nuclear trifluoromethyl groups. For example, the attempted acetylation of benzotrifluoride with aluminum chloride leads to a halogen exchange side reaction involving the trifluoromethyl group (55). The use of boron trifluoride may obviate this difficulty, but forcing conditions may be required to effect acetylation. However, styrenes with trifluoromethyl groups and/or fluorines on the phenyl ring can be prepared in good yield by the following general organometallic syntheses (56–64):

$$CF_3C_6H_4M \xrightarrow{CH_3CHO} CF_3C_6H_4CHOHCH_3 \xrightarrow[-H_2O]{KHSO_4} CF_3C_6H_4CH=CH_2$$
$$(F) \qquad\qquad\qquad (F) \qquad\qquad\qquad\qquad\qquad (F)$$

where M = Li, MgX, etc.

C. Hydroxydi(polyfluoroalkyl)methylstyrenes

An elegant synthesis of styrenes with a hydroxydi(polyfluoroalkyl)methyl substituent involves a Friedel-Crafts condensation of a fluorinated ketone with ethylbenzene as shown below (65–68):

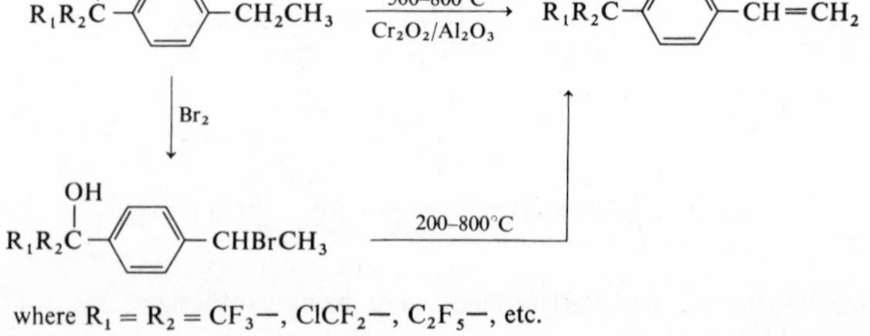

where $R_1 = R_2 = CF_3-$, $ClCF_2-$, C_2F_5-, etc.

where $R_1 = R_2 = CF_3-$, $ClCF_2-$, C_2F_3-, etc.

As already noted, the presence of one fluorine atom in the aromatic nucleus does not deactivate greatly the phenyl ring to the Friedel-Crafts acetylation reaction. It has been reported that even with as many as three fluorine atoms in the benzene ring the Friedel-Crafts acetylation reaction is feasible (69):

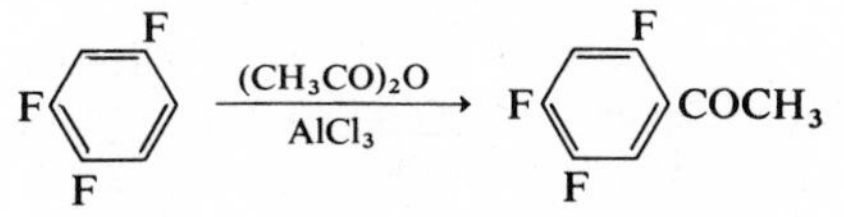

Presumably the corresponding fluorinated styrene can be prepared by the usual reduction-dehydration sequence.

The deactivating presence of even five fluoro substituents has been overcome in some Friedel-Craft alkylations of pentafluorobenzene by employing rather forcing conditions (70). However, the successful acylation of pentafluorobenzene by Friedel-Craft catalysis has not yet been reported.

As in the case of the trifluoromethyl-substituted benzenes, the alternate organometallic condensation type reaction may be employed to synthesize a large variety of ring-fluorinated phenethyl alcohols, which can then be dehydrated to the desired styrene. In general, the reaction sequence is as follows:

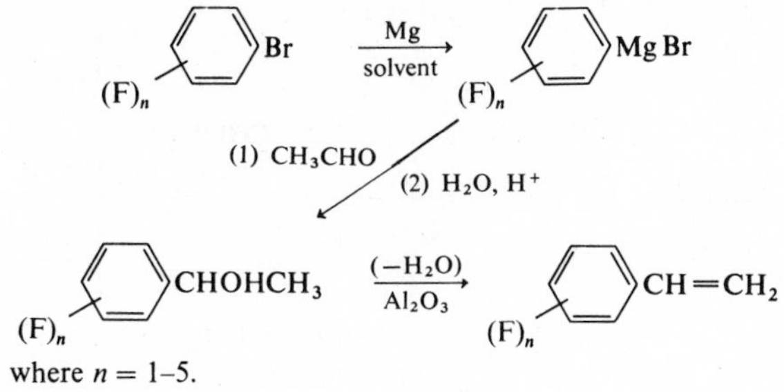

where $n = 1$–5.

where $n = 1$–5.

D. 2,4,6-Tris-(trifluoromethyl)styrene

In those cases where the halofluorobenzene fails to form a Grignard reagent or aryllithium by direct metallation, it is usually possible to effect an exchange reaction and thereby obtain the desired organometallic reagent. For example, in the synthesis of the highly hindered 2,4,6-tris-(trifluoromethyl)styrene an exchange reaction of 2,4,6-tris-(trifluoromethyl)chlorobenzene with butyllithium is necessary in order to prepare the aryllithium (63):

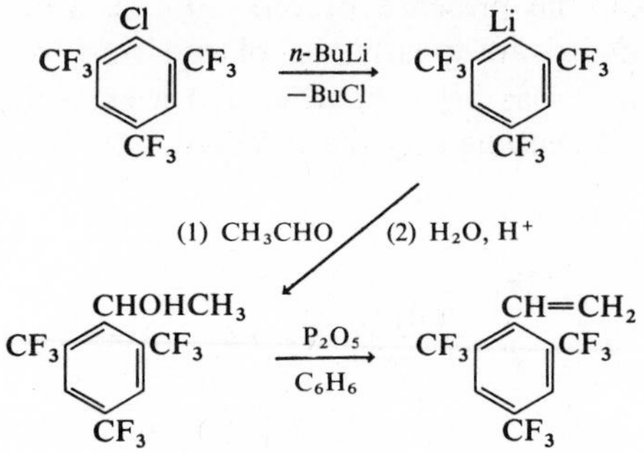

E. Difluorostyrenes

Although the methods of synthesis for nuclear polyfluorinated styrene are essentially the same as those described for nuclear monofluorinated styrenes, relatively few such monomers have been reported. The early patent literature (71) refers briefly to the preparation 2,6-difluorostyrene. Later workers (72) described in some detail the synthesis of 2,4-difluorostyrene by several methods.

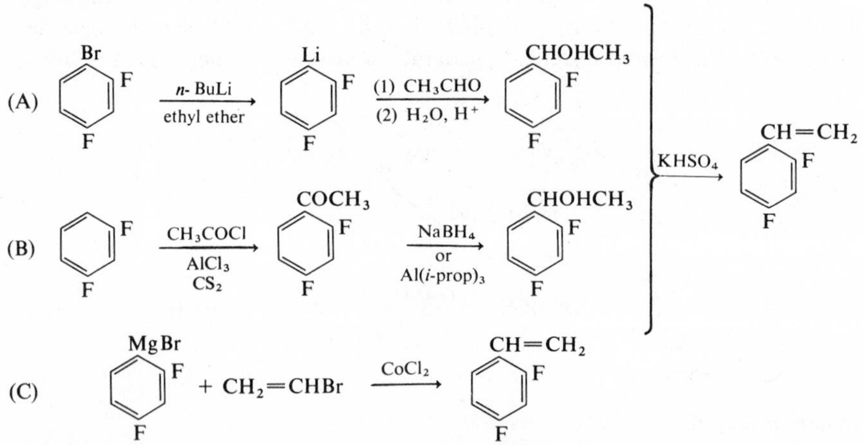

Methods A and B proved to be satisfactory preparative routes to the styrene. The difluorophenyl Grignard reagent could be substituted equally well in Method A for the difluorophenyllithium. Method C provides a direct route to the styrene but the yield is low. Beta-substituted styrenes may be prepared from allyl halides via Method C if an isomerization step is added (73):

$$4-CF_3C_6H_4MgBr + CH_2=CHCH_2Cl \xrightarrow[\text{reflux}]{\text{ether}} 4-CF_3C_6H_4CH_2CH=CH_2$$

$$\downarrow \text{KOH} | \text{CH}_3\text{OH}$$

$$4-CF_3C_6H_4CH=CHCH_3$$

F. 2,3,4,5,6-Pentafluorostyrene

Recently the synthesis of the totally ring-fluorinated styrene 2,3,4,5,6-pentafluorostyrene was reported (74–76). Perhaps the best method developed to date involves the standard condensation of the pentafluorophenyl Grignard reagent with acetaldehyde to give the intermediate secondary carbinol which is subsequently dehydrated to the pentafluorostyrene (74,75):

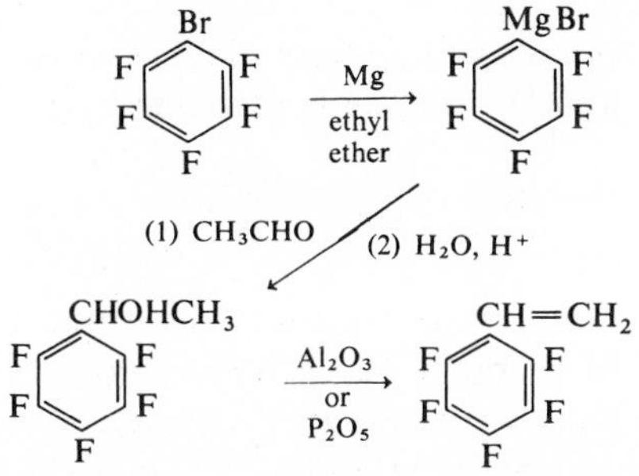

A one-step synthesis of this interesting monomer is also available and simply involves the nucleophilic displacement of a fluorine atom from hexafluorobenzene by vinyllithium (76):

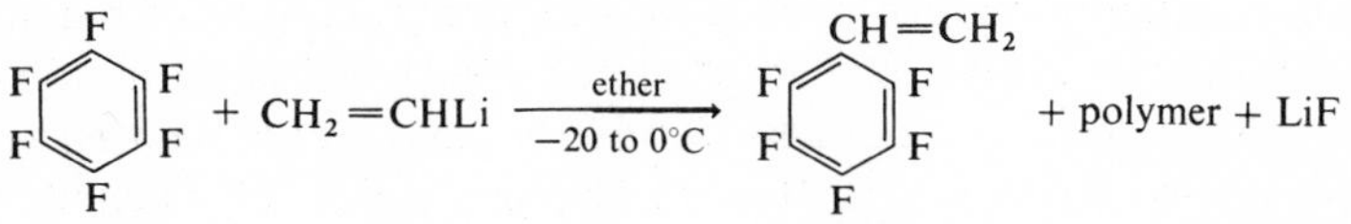

The yield of monomer from the last reaction is usually of the order of 20–25%, which compares favorably with the overall yield of pentafluorostyrene prepared via the Grignard synthesis. However, the vinyllithium reaction does seem to give considerable polymeric products, which may be due to the anionic polymerization of the initially formed styrene monomer. Probably the reaction conditions can be modified to increase the yield of monomer.

G. 2,3,4,5,6-Pentafluoro-α-methylstyrene

In an analogous fashion, 2,3,4,5,6-pentafluoro-α-methylstyrene can be prepared via the pentafluorophenyl Grignard synthesis with acetone (77) or by the nucleophilic reaction of isopropenyllithium with hexafluorobenzene (76):

$$C_6F_5MgBr \xrightarrow[\text{(2) H}_2\text{O, H}^+]{\text{(1) CH}_3\text{COCH}_3} C_6F_5\overset{\overset{\text{CH}_3}{|}}{C}OHCH_3 \xrightarrow[-\text{H}_2\text{O}]{\text{KHSO}_4} C_6F_5\overset{\overset{\text{CH}_3}{|}}{C}=CH_2 \qquad (23)$$

$$\underset{\text{CH}_3}{\overset{|}{\text{CH}_2{=}\text{CLi}}} + \text{C}_6\text{F}_6 \longrightarrow \underset{\text{CH}_3}{\overset{|}{\text{C}_6\text{F}_6\text{C}{=}\text{CH}_2}} + \text{polymer} + \text{LiF} \qquad (24)$$

The overall yield of monomer by reaction sequence 23 is about 58%. Reaction 24 gives a comparable yield (50%) in one step.

H. 2,3,4,5,6-Pentafluoro-β-methylstyrene

The nucleophilic synthesis with alkenyllithiums seems to be of wide scope. 2,3,4,5,6-pentafluoro-β-methylstyrene is prepared by a similar route (78):

$$\text{CH}_3\text{CH}{=}\text{CHLi} + \text{C}_6\text{F}_6 \xrightarrow[-15°\text{C}]{\text{ether}} \underset{(87\%)}{\text{C}_6\text{F}_5{=}\text{CHCH}_3}$$

Either cis, trans, or a mixture of cis and trans isomers of the pentafluoro-propenylbenzene could be prepared depending on the isomeric nature of the propenyllithium reagent. Moreover, by the use of excess alkenyllithium reagent, it is possible to obtain para dipropenyltetrafluorobenzene.

The general reaction scheme for this type of synthesis is

$$\text{C}_6\text{F}_6 + \text{R}_1\text{R}_2\text{C}{=}\text{CR}_3\text{Li} \longrightarrow \text{C}_6\text{F}_5\text{CR}_3{=}\text{CR}_1\text{R}_2$$

R_1, R_2, and R_3 may be any substituents that do not prevent nucleophilic attack on hexafluorobenzene or lead to prior decomposition of the alkenyl-lithium. Probably other polyfluorobenzenes, such as pentafluorobenzene and the various isomeric tetrafluoro-, trifluoro-, and difluorobenzenes, could serve as suitable substrates for the nucleophilic attack of alkenyllithiums, such as vinyllithium. Therefore it would seem that a general method is available for the synthesis of nuclear fluorinated styrenes of varying fluorine content and orientation by a one-step reaction.

Another possible general method for the synthesis of nuclear fluorinated styrenes of varying fluorine content and orientation is the application of Wittig olefin synthesis to polyfluorobenzaldehydes. Thus the generalized reaction would be

$$\text{C}_6\text{F}_x\text{H}_y\text{CHO} + \text{R}_3\text{P}{=}\text{CH}_2 \longrightarrow \text{C}_6\text{F}_x\text{H}_y\text{CH}{=}\text{CH}_2 + \text{R}_3\text{PO}$$

where $x = 1\text{–}5$, $y = 5 - x$.

In this connection it may be significant to note that the application of difluoromethylene phosphorane reagents previously discussed to poly-fluorobenzaldehydes theoretically should provide a general, one-step synthesis of highly fluorinated styrenes such as α-hydroheptafluorostyrene. Presumably the reaction would be

$$\text{C}_6\text{F}_5\text{CHO} + \text{R}_3\text{P} + \text{ClCF}_2\text{CO}_2\text{Na} \xrightarrow[\substack{\text{solvent} \\ \Delta}]{} \text{C}_6\text{F}_5\text{CH}{=}\text{CF}_2$$

Some actual syntheses of totally or highly fluorinated styrenes are discussed in Section V.

V. STYRENES WITH VINYL AND PHENYL GROUPS FLUORINATED

A. Di-, Tri-, and Tetrafluorostyrenes

Relatively few styrenes fluorinated both in the side chain and in the phenyl ring have been prepared. p,β-Difluorostyrene can be prepared by the same method used to prepare β-fluorostyrene discussed earlier. The synthesis of p,β,β-trifluorostyrene, p,β,β-trifluoro-α-trifluoromethylstyrene, and various similar styrenes has already been mentioned. This class of styrenes also includes $\beta,2,4$-trifluoro- and $\beta,\beta,2,4$-tetrafluorostyrenes, which are prepared by the following standard synthetic schemes (79):

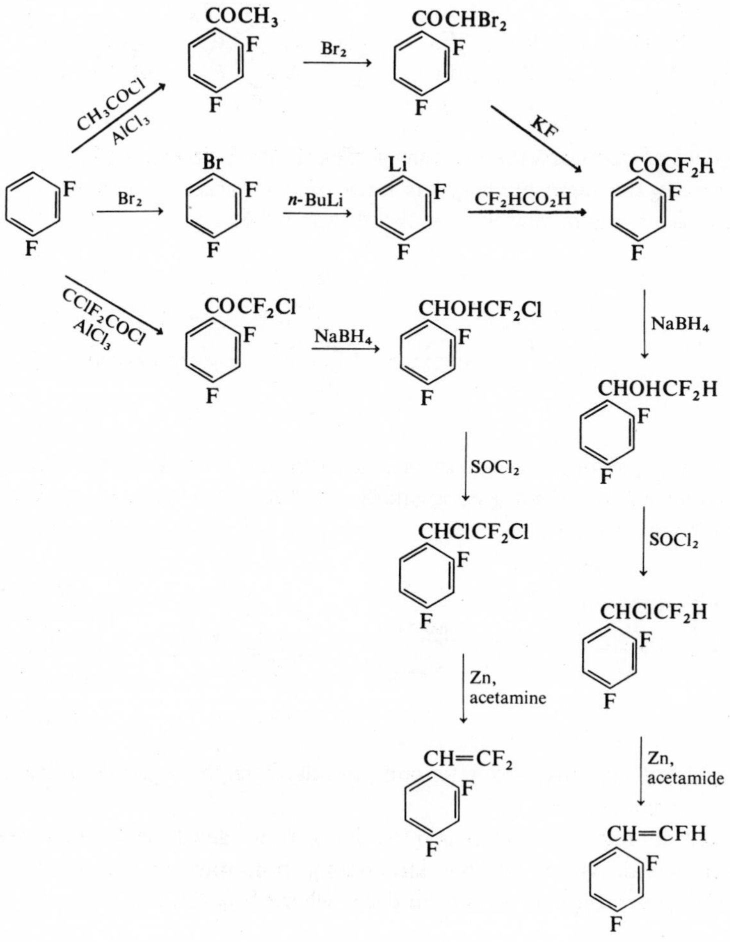

B. Perfluorostyrene, Perfluoro-α-methylstyrene and other Polyfluorostyrenes

Perhaps the most interesting development to occur in this field has been the recent synthesis of perfluorostyrene. This unique monomer is now available by several synthetic methods (80–84).

Tatlow and co-workers have synthesized perfluorostyrene by the same method they used to prepare other highly or totally fluorinated aromatic compounds. The aromatic hydrocarbon is exhaustively fluorinated with cobalt trifluoride to give a highly or totally fluorinated alicyclic derivative. Aromaticity is then reintroduced by thermal cracking over a metal such as iron or nickel (85). In the case of perfluorostyrene, a second cracking step was required to introduce unsaturation in the side chain (80). In outline, the overall method is

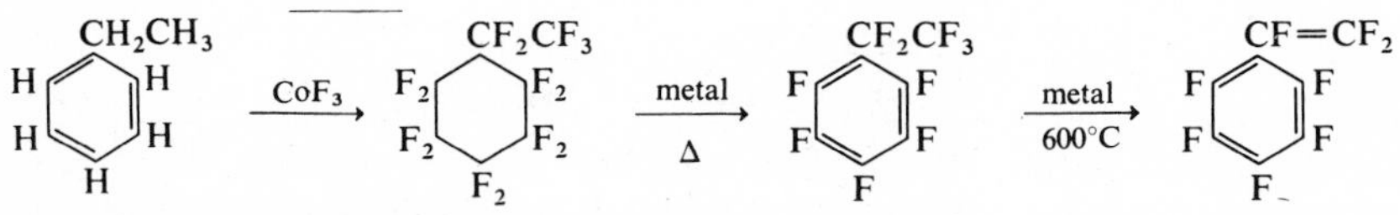

The yield of perfluorostyrene from perfluoroethylbenzene is 15%.

Recently the same method has been used to prepare perfluoro-α-methyl-styrene from perfluoroisopropylcyclohexane (86):

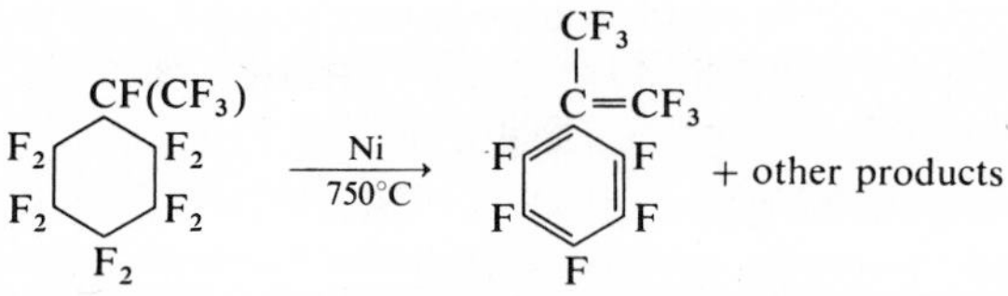

A new pyrolytic synthesis of perfluorostyrene involves the thermolysis of the dimer of hexafluorocyclopentadiene. Perfluoroindane, however, is the major product (83):

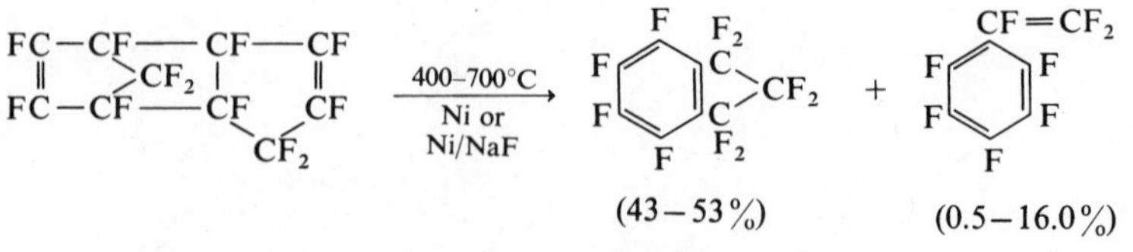

$$(43-53\%) \qquad (0.5-16.0\%)$$

The yield of the vinyl aromatic compounds from these pyrolytic defluorinations is generally low.

Other routes to perfluorostyrene have been developed which result in higher overall yields of this interesting monomer (81,82,84). One such route is shown by reactions 25 and 26, where X = Cl, F.

$$C_6F_5MgBr \xrightarrow[\text{(2) } H_2O, H^+]{\text{(1) } CF_2XCHO} C_6F_5CHOHCF_2X \xrightarrow{SF_4} C_6F_5CFHCF_2X \qquad (25)$$
$$(53-78\%) \qquad\qquad (90\%)$$

$$C_6F_5CFHCF_3 \xrightarrow[500-600°C]{Br_2} C_6F_5CFBrCF_3 \xrightarrow[\substack{0:1 \text{ mm,}\\ \text{steel wool}}]{600-650°C} C_6F_5CF=CF_2 \qquad (26)$$
$$(95\%)$$

Thermal cracking of the intermediate α-hydrononafluoroethylbenzene or α-hydro-β-chlorooctafluoroethylbenzene results in the formation of the two styrenes shown in reaction 27 (81):

$$C_6F_5CHFCF_2X \xrightarrow{500-900°C} \underset{\text{major product}}{C_6F_5CH=CF_2} + \underset{\text{minor product}}{C_6F_5CF=CF_2} \qquad (27)$$

where X = F, Cl. The styrene formed in greater quantity is the α-hydrohepta-fluorostyrene rather than perfluorostyrene. Apparently dehalogenation predominates over dehydrohalogenation in these thermal elimination reactions. Recently α-hydroheptafluorostyrene has been synthesized in good yield by the following scheme (87):

$$C_6F_5MgBr + CClF_2CHO \longrightarrow C_6F_5CHOHCF_2Cl \xrightarrow{PCl_5}$$

$$C_6F_5CHClCF_2Cl \xrightarrow[ETOH]{Zn} C_6F_5CH=CF_2$$

Sometimes it is possible to effect dehydrohalogenation of these highly fluorinated ethylbenzenes to perfluorostyrene by the use of vigorous basic conditions. Thus α-hydro-β-chlorooctafluoroethylbenzene is dehydro-chlorinated in good yield to perfluorostyrene (81):

$$C_6F_5CFHCF_2Cl \xrightarrow[\substack{\text{molten KOH}\\ (250°C) \text{ or}\\ \text{KOH on C}\\ \text{at } 400°C}]{} C_6F_5CF=CF_2$$

In contrast, dehydrofluorination of the intermediate α-hydrononafluoro-ethylbenzene, $C_6F_5CFHCF_3$, by the usual basic reagents even under forcing conditions proved to be unsuccessful. However, the use of very strong bases, such as organolithiums, can provide an excellent means of effecting the dehydrofluorination of highly fluorinated hydrocarbons (88). This novel means of dehydrofluorination involves the metallation of a hydropoly-fluorocarbon at the C—H bond via an exchange reaction with an organol-ithium reagent followed by the subsequent decomposition of the intermediate fluoroorganolithium to the fluoroolefin and lithium fluoride. We have found that this method also works quite well in the case of α-hydrononafluoro-ethylbenzene (82):

$$C_6F_5CFHCF_3 \xrightarrow[\substack{\text{ether}\\ -80°C}]{CH_3Li} [C_6F_5CFLiCF_3] \xrightarrow[-LiF]{-80-+25°C} C_6F_5CF=CF_2$$
$$(85-90\%)$$

Presumably the dehydrofluorination reaction occurs via a prior exchange reaction involving the highly acidic α-hydrogen of the nonafluoroethylbenzene and the highly basic methyllithium. Low enough temperatures should be maintained during the exchange reaction to prevent or at least minimize any nucleophilic displacement of nuclear fluorine by the methyllithium. The presence of any excess methyllithium also is to be avoided since the fluorinated vinyl group is also subject to nucleophilic attack by the organolithium reagent.

Perfluorostyrene also may be prepared directly by the organolithium-fluoroolefin synthesis (cf. Section III-A). Thus pentafluorophenyllithium in the presence of a very large excess of tetrafluoroethylene will give about a 10 % yield of perfluorostyrene in a one-step process (82):

$$C_6F_5Li + CF_2{=}CF_2 \xrightarrow[-20°C]{(C_2H_5)_2O} C_6F_5CF{=}CF_2 + C_6F_5CF{=}CFC_6F_5 + LiF$$
$$\text{(excess) } -20°C \qquad\qquad\qquad (10\%)$$

As is the case in the reaction of C_6H_5Li with tetrafluoroethylene, a considerable quantity of disubstitution product is formed; i.e., it appears that perfluorostyrene is more susceptible to attack by C_6F_5Li than is tetrafluoroethylene.

The pentafluorophenyllithium may be prepared by direct metallation of iodopentafluorobenzene with lithium in the form of a dispersion or from bromopentafluorobenzene or pentafluorobenzene by an exchange reaction with an organolithium reagent such as methyl- or butyllithium (82,89,90).

Pentafluorophenyllithium may also be added to a wide variety of fluoroolefins, thus providing a general method for the synthesis of totally and highly fluorinated styrene derivatives. For example, β-hydroheptafluorostyrene ($C_6F_5CF{=}CFH$), β-chloroheptafluorostyrene ($C_6F_5CF{=}CFCl$), and β-trifluoromethylheptafluorostyrene ($C_6F_5CF{=}CFCF_3$) have been prepared by the reaction of pentafluorophenyllithium with trifluoroethylene, chlorotrifluoroethylene, and perfluoropropylene, respectively (82).

A much improved synthesis of perfluorostyrene and similar polyfluorostyrenes utilizes the unique coupling reaction of pentafluorophenylcopper and iodotrifluoroethylene (84):

$$C_6F_5H \xrightarrow[THF,\,-70°C]{n-C_4H_9Li} C_6F_5Li \xrightarrow[-70°C]{CuI} C_6F_5Cu \xrightarrow{CF_2{=}CFI} C_6F_5CF{=}CF_2$$
$$\qquad\qquad\qquad\qquad\qquad\qquad\qquad\qquad (55\%)$$

In contrast, pentafluorophenyllithium and iodotrifluoroethylene react as follows:

$$C_6F_5Li + CF_2{=}CFI \longrightarrow C_6F_5CF{=}CFI \quad \text{(trans)}$$

The further reaction of pentafluorophenyllithium and β-iodoheptafluorostyrene, surprisingly, results in nucleophilic displacement of a nuclear fluorine rather than a vinylic fluorine:

The high-temperature reaction would seem to follow the standard addition-elimination mechanism, whereas at low temperatures an addition-protonation mechanism prevails. In the case of the preparation of 1,1-difluoro-2,2-dichloroethylphenyl ether by the apparent condensation reaction (reaction 28) it would appear that prior dehydrochlorination of 1,1,2-trichloro-2,2-difluoroethane had occurred to yield 1,1-dichloro-2,2-difluoroethylene. The formation of the saturated ether, then, is a logical consequence of the low-temperature addition of phenol across the olefin. The same mechanism can be adduced to explain the formation of the fluorinated vinyl phenyl ether, $C_6H_5OCF=CClCF_3$, as one of the products resulting from the direct reaction of sodium phenoxide and 2-chloro-1,1,3,3,3-hexafluoropropane $(CF_3CClHCF_3)$ (14). This reaction obviously proceeds by a prior dehydrofluorination step followed by the addition-elimination sequence:

$$C_6H_5ONa + CF_3CClHCF_3 \longrightarrow C_6H_5OH + CF_2=CClCF_3 + NaF \qquad (30)$$

$$C_6H_5ONa + CF_2=CClCF_3 \longrightarrow [C_6H_5OCF_2CClCF_3]^-Na^+] \qquad (31)$$

$$\downarrow$$

$$C_6H_5OCF=CClCF_3$$

C. 2-Chloro-1,2-difluorovinyl Phenyl Ether

Similarly, 2-chloro-1,2-difluorovinyl phenyl ether can be obtained from the base-catalyzed addition of phenol to chlorotrifluoroethylene, but the yield is low (92,94);

$$C_6H_5OH + CF_2=CFCl \xrightarrow[180°C]{C_6H_5OK} \underset{(75-80\%)}{C_6H_5OCF_2CFClH} + \underset{(5-7\%)}{C_6H_5OCF=CFCl}$$

However, the saturated ether (2-chloro-1,1,2-trifluoroethylphenyl ether) may be dehydrofluorinated in good yield with solid potassium hydroxide to the vinyl phenyl ether:

$$C_6H_5OCF_2CFClH \xrightarrow[\text{reflux}]{\substack{\text{solid}\\ \text{KOH}}} \underset{(53\%)}{C_6F_5OCF=CFCl}$$

Recently other workers (95,96) have extended the scope of the nucleophilic addition-elimination reactions involving phenolic salts and fluoro-olefins by employing such potent solvents dimethyl formamide, dimethoxyethane, tetrahydrofuran, dioxane, and various mixed solvents. Thus the previously difficult nucleophilic addition of sodium phenoxide across tetrafluoroethylene occurs readily in dimethylformamide at 80°C to give both 1,1,2,2-tetrafluoroethyl phenyl ether (III) and 1,2-difluoro-1,2-diphenoxyethylene (IV). The latter product presumably arises from the further addition of sodium phenoxide to the transient intermediate 1,2,2-trifluorovinylphenyl ether (II). The formation of these products is visualized as follows:

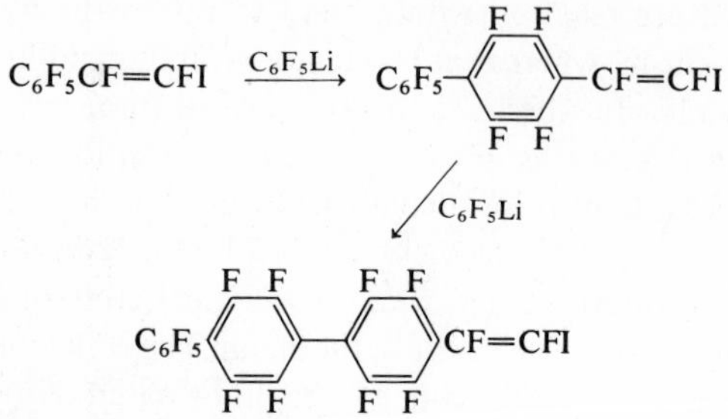

$$C_6F_5CF=CFI \xrightarrow{C_6F_5Li} C_6F_5\text{—}\underset{F\ F}{\overset{F\ F}{\bigcirc}}\text{—}CF=CFI$$

$$\downarrow C_6F_5Li$$

$$C_6F_5\underset{F\ F}{\overset{F\ F}{\bigcirc}}\underset{F\ F}{\overset{F\ F}{\bigcirc}}CF=CFI$$

The lack of stilbene byproducts may be attributed to the increased steric hindrance offered by β-iodo substituent.

A somewhat similar synthesis of polyfluorostyrenes is illustrated below (91):

$$C_6F_5I + CFI=CFCl \xrightarrow{\text{Cu powder}} C_6F_5CF=CFCl$$

VI. SYNTHESIS OF FLUORINATED VINYL PHENYL ETHERS

A. 1-Fluoro-2-chlorovinyl Phenyl Ether

As noted earlier, 1-fluoro-2-chlorovinyl phenyl ether can be prepared by the dechlorofluorination of 1,1-difluoro-2,2-dichloroethyl phenyl ether (14). The latter compound was prepared by an apparent Williamson type of reaction involving sodium phenoxide and 1,2,2-trichloro-1,1-difluoroethane.

B. 1-Fluoro-2,2-dichlorovinyl Phenyl Ether

Dehydrofluorination of the saturated ether results in the formation of 1-fluoro-2,2-dichlorovinylphenyl ether. The appropriate reactions are,

$$C_6H_5ONa + CClF_2CCl_2H \xrightarrow{\text{acetone}} C_6H_5OCF_2CHCl_2 \qquad (28)$$

$$C_6F_5OCF_2CHCl_2 \xrightarrow[\text{KOH, reflux}]{C_6H_5CH_2N^+(CH_3)_3OH^-} \underset{(29\%)}{C_6H_5OCF=CCl_2} \qquad (29)$$

Later it was shown that the same ether could be obtained in one step from the base-catalyzed addition of phenol to 1,1-dichloro-2,2-difluoroethylene if the proper temperature were employed (92,93). At low temperature the saturated ether is obtained; at higher temperatures the vinyl ether is formed:

$$C_6H_5OH + CF_2=CCl_2 \xrightarrow[\text{acetone}]{\text{KOH}} \begin{cases} \xrightarrow{10^\circ C} \underset{(60\%)}{C_6H_5OCF_2CCl_2H} \\ \\ \xrightarrow{20^\circ C} \underset{(62\%)}{C_6H_5OCF=CCl_2} \end{cases}$$

$$C_6H_5O^-Na^+ + CF_2=CF_2 \longrightarrow [(C_6H_5OCF_2CF_2)^-Na^+] \xrightarrow[-NaF]{} C_6H_5OCF=CF_2$$
$$\underset{I}{} \qquad\qquad \underset{II}{}$$

$$\Big\downarrow \begin{array}{l} H^+ \\ (\text{from } C_6H_5OH \\ \text{or DMF or } H_2O) \end{array} \qquad -NaF \Big\downarrow C_6H_5O^-Na^+$$

$$\underset{III}{C_6H_5OCF_2CF_2H} \qquad\qquad \underset{IV}{C_6H_5OCF=CFOC_6H_5}$$

$$(32)$$

D. 1,2,2-Trifluorovinyl Phenyl Ether

Recently the interesting intermediate II, 1,2,2-trifluorovinyl phenyl ether, has been isolated from reaction 32 and also has been derived from III (96).

One of the methods employed involved the dehydrofluorination of 1,1,2,2-tetrafluoroethylphenyl ether (III):

$$C_6H_5OCF_2CF_2H \xrightarrow[-HF]{\begin{array}{c} \text{NaOH, KOH} \\ \text{etc.} \end{array}} \underset{(5\%)}{C_6H_5OCF=CF_2}$$

Dehydrofluorination occurred to only a very limited extent (about 5% or less) with such reagents as 10% platinum on charcoal pellets at 300°C and molten sodium hydroxide alone or admixed with potassium hydroxide at 340–375°C.

Improved yields of this monomer are obtained by modifying the conditions of the addition reaction of phenolic salt to tetrafluoroethylene. It appears that the nature of both the solvent and the cation employed in the preparation of phenolic salt may play a significant role in the determination of the products of the reaction. By maintaining a solvent system that is aprotic and by employing the proper anhydrous phenolic salt, it is possible to obtain appreciable yields of 1,2,2-trifluorovinyl phenyl ether. Table 2 gives a summary of the results of this study. As can be seen from Table 2, the use of mixed solvent systems composed of benzene-dioxane or benzene-tetrahydrofuran seem to offer the best reaction media for the synthesis of 1,2,2-trifluoro-vinylphenyl ether. The potassium salt of phenol, which can be dried quite well by azeotropic distillation, seems to give somewhat higher yields of the vinyl ether than the corresponding sodium phenolate. Side products such as 1,1,2,2-tetrafluoroethyl phenyl ether and 1,2-difluoro-1,2-diphenoxyethylene have not yet been completely eliminated. Traces of moisture, phenol, or other protonic sources still present in the reaction medium would lead to the formation of the saturated ether. The 1,2-disubstituted product probably arises from the nucleophilic attack of the phenoxide anion on the vinyl ether. This reaction will occur even in the presence of a large excess of tetrafluoro-ethylene, which seems to indicate that trifluorovinyl phenyl ether is more susceptible to nucleophilic attack than tetrafluoroethylene. The situation is reminiscent of that prevailing in the reaction of C_6H_5Li and C_6F_5Li with tetrafluoroethylene discussed earlier.

TABLE 2

Reaction of Phenolic Salts with TFE[a] (96)

C_6H_5OM (g)	Metal	TFE (g)	Solvent[d] Type	Solvent[d] Amt (ml)	Temp. (°C)	Pressure (psi)	Yield (%)	Product(s)
94[b]	Na	50	DMF	300	80	114(8kg/cm²)	88	$C_6H_5OCF_2—CF_2H$
118[c]	Na	117	DMF	400	80	200(14kg/cm²)	23.5	$C_6H_5—O—CF_2—CF_2H$
								Mostly higher MW products
25	Na	45	DMF	60	25	400(28kg/cm²)	30	$C_6H_5O—CF{=}CF—O—C_6H_5$
								Multicondensation products
25	Na	25	THF	250	−50	14(.98kg/cm²)	—	Little or no reaction
			DMF	100				
110	Na	110	Ethyl ether	400	100	310(21.8kg/cm²)	—	No reaction
110	Na	110	Benzene	400	140	250(17.5kg/cm²)	16.7	$C_6H_5—O—CF_2—CF_2H$
			Dioxane	100			21.1	$C_6H_5—O—CF{=}CF_2$
							4.6	$C_6H_5—O—CF{=}CF—O—C_6H_5$
132	K	128	Benzene	500	120	208(14.5kg/cm²)	6.0	$C_6H_5—O—CF_2—CF_2H$
			THF	125			34.3	$C_6H_5—O—CF{=}CF_2$
							4.5	$C_6H_5—O—CF{=}CF—OC_6H_5$
							10.0	High MW products

[a] Tetrafluoroethylene.
[b] Salt made *in situ*.
[c] Hydrated salt.
[d] DMF = dimethylformamide; THF = tetrahydrofuran.

E. Vinyl Pentafluorophenyl Ether

Vinyl phenyl ether and vinyl pentachlorophenyl ether have been prepared by the base-catalyzed addition of the phenol to acetylene (97–102). This reaction has been extended to the preparation of vinyl pentafluorophenyl ether by the successful addition of pentafluorophenol to acetylene (103):

$$C_6F_5OH + CH \equiv CH \xrightarrow[\substack{CH_3CON(CH_3)_2 \\ 200°C,\ 1.5\ hr}]{KOH} C_6F_5OCH=CH_2 + (C_6F_5O)_2CHCH_3$$
$$\text{(excess)} \qquad\qquad\qquad \text{(50\%)} \qquad\qquad \text{(10\%)}$$

A fair excess of acetylene is required to control the formation of by-products such as 1,1-bis(pentafluorophenoxy)ethane. Another approach to ring-fluorinated monomers is illustrated by the synthesis of vinyl p-fluorophenyl ether by the chemical dehydrobromination of 2(p-fluorophenoxy)-ethyl bromide (104):

$$\substack{p-FC_6H_4ONa \\ + BrCH_2CH_2Br} \xrightarrow[-NaBr]{} p-FC_6H_4OCH_2CH_2Br \xrightarrow[-HBr]{base} p-FC_6H_4OCH=CH_2$$

Vinyl pentafluorophenyl ether also has been prepared, with some success, by the thermal and chemical decomposition of certain β-substituted ethyl pentafluorophenyl ethers (103). In general, it was found that the decomposition of β-substituted ethyl pentafluorophenyl ethers followed the two paths shown below:

$$C_6F_5O-CH_2CH_2R \begin{cases} \longrightarrow C_6F_5OH + CH_2=CHR \\ \longrightarrow C_6F_5OCH=CH_2 + HR \end{cases}$$

where R = Br, Cl, CH_3CO_2, CF_3CO_2, C_6F_5O, OH.

Table 3 summarizes the results of this study on the pyrolytic behavior of various β-substituted ethyl pentafluorophenyl ethers.

It is evident from these results that the pyrolysis of β-substituted ethers does not provide a good preparative route to vinyl pentafluorophenyl ether. However, it is possible to prepare this monomer in excellent yield by the pyrolysis of 1,1-bis(pentafluorophenoxy)ethane. The decomposition proceeds almost quantitatively to give equivalent quantities of pentafluorophenol and the ether:

$$(C_6F_5O)_2CHCH_3 \xrightarrow{\text{Pyrex glass}} C_6F_5OH + C_6F_5OCH=CH_3$$
$$\text{(49\%)} \qquad\quad \text{(49\%)}$$

The other isomeric diphenoxy ethane, 1,2-bis(pentafluorophenoxy)ethane, does not produce any vinyl pentafluorophenyl ether on pyrolysis, although some (14%) pentafluorophenol is formed (103).

Chemical dehydrohalogenation of some β-substituted ethyl pentafluorophenyl ethers can, under the proper conditions, serve as still another route to

TABLE 3

Pyrolysis of $C_6F_5OCH_2CH_2R$ (103)

R	Weight (g)	Packing[a]	Temp. (°C)	Press (mm)	Weight recovered (g)	Products (VPC)[b]
Br	3	A	500	760	2.5	37% $C_6F_5OCHBrCH_3$; 19% C_6F_5OH; 15% $C_6F_5OCH=CH_2$; 4% $C_6F_5OCH_2CH_2Br$; 2% $(C_6F_5O)_2$ $CHCH_3$
Cl	1	A	520	760	0.75	70% $C_6F_5OCH_2CH_2Cl$; 5% C_6F_5OH; 5% $C_6F_5OCH=CH_2$; 1% $C_6F_5OCClHCH_3$
OAc	1.3	A	560	760	.9	35% C_6F_5OH; 20% $C_6F_5OCH=CH_2$; 15% CH_3CO_2H; 20% 7-unknown products
OAc	1.3	A	600	0.8	1.0	75% $C_6F_5OCH_2CH_2OAc$; 10% C_6F_5OH; 3% $C_6F_5OCH=CH_2$; 2% CH_3CO_2H
OCOCF₃	2	A	540	25	1.6	70% $C_6F_5OCH_2CH_2OCOCF_3$; 25% $(C_6F_5OH; CF_3CO_2H; C_6F_5OCH=CH_2)$
OC₆F₅	3	A	480	760	2.6	70% $C_6F_5OCH_2CH_2OC_6F_5$; 14% C_6F_5OH
OH	5	B	420	760	2.0	99% C_6F_5OH
OH	3	B	300	760	1.2	99% C_6F_5OH
OH	2	KHSO₄	220	50	1.7	No reaction

[a] Key; A, glass helices/B, Al_2O_3 pellets.
[b] The identity along with the approximate relative percentages of the various components present in each pyrolysate was established by analytical vapor phase chromatography (VPC).

60

this monomer, as shown in Table 4. For example, dry, nonaqueous basic systems seem to provide the best conditions for dehydrohalogenation of α-bromopentafluorophenetole, $C_6F_5OCH_2CH_2Br$. Aqueous basic systems seem to favor cleavage at the alkyl-oxygen bond of the aryl ether (76).

TABLE 4

Reaction of C_6F_5OR with some Bases (103)

R	Weight (g)	Base (g)	Temp. (°C)	Weight recovered (g)	Products
$-CH_2CH_2Br$	8.5	KOH (5) paraffin oil (5)	180	3.5	86% $C_6F_5OCH_2CH_2Br$, 14% C_6F_5OH
$-CH_2CH_2Br$	3	KOH (5) K_2CO_3 (10)	200	1.2	50% $C_6F_5OCH_2CH_2Br$, 50% $C_6F_5OCH=CH_2$
$-CH_2CH_2Br$	5	KOH (5) H_2O (20)	100	4.0	62% $C_6F_5OCH_2CH_2Br$, 37% C_6F_5OH
$-CH_2CH_2Br$	5	Collidine	160	3.0	83% $C_6F_5OCH_2CH_2Br$, 16% $C_6F_5OCH=CH_2$
OC_6F_5 | $-CHCH_3$	0.4	KOH (1) H_2O (5)	100	0.3	No reaction
$-CH_2CF_3$	8.5	KOH (5) K_2CO_3 (10)	200	8.0	No reaction
$-CH_2CF_3$	1.4	MgO (5)	150	1.3	No reaction
$-CH_2CF_3$	20	20% KOH/ charcoal	480	16	80% $C_6F_5OCH_2CF_3$, 15% $C_6F_5OCH=CF_2$
$-CH_2CF_3$	2	20% KOH/ charcoal	550	—	Decomposed
$-CF_2CF_2H$	2	20% KOH/ charcoal	550	—	Decomposed

The methods outlined here for the synthesis of vinyl pentafluorophenyl ether should be applicable to the preparation of other nuclearly fluorinated phenyl vinyl ethers of varying fluorine content and orientation.

F. 1,2-Difluorovinyl Pentafluorophenyl Ether

The synthesis of this highly fluorinated monomer was accomplished by the reaction of potassium pentafluorophenoxide with excess trifluoroethylene using the solvent pair benzene-tetrafluorohydrofuran as the reaction medium (103):

$$C_6F_5O^-K^+ + CF_2{=}CFH \xrightarrow{\ C_6H_6/THF\ } C_6F_5OCF{=}CFH + C_6F_5OCF_2CFH_2$$
$$\text{(excess)} \qquad\qquad \text{(29\%)} \qquad\quad \text{(4\%)}$$

G. 2,2-Difluorovinyl Pentafluorophenyl Ether

The synthesis of this monomer was reported by the sequence shown in reactions 33 and 34 (see Table 4) (103).

$$C_6F_6 + CF_3CH_2ONa \xrightarrow{\text{THF/PYR}} \underset{(66\%)}{C_6F_5OCH_2CF_3} \tag{33}$$

$$C_6F_5OCH_2CF_3 \xrightarrow[480°C]{20\% \text{ KOH/C}} C_6F_5OCH=CF_2 + HF \tag{34}$$

Reaction 34, however, proceeded in low yield and conversion.

H. 1,2,2-Trifluorovinyl Pentafluorophenyl Ether

The addition-elimination method also was used to synthesize the completely fluorinated analog of vinyl phenyl ether (96):

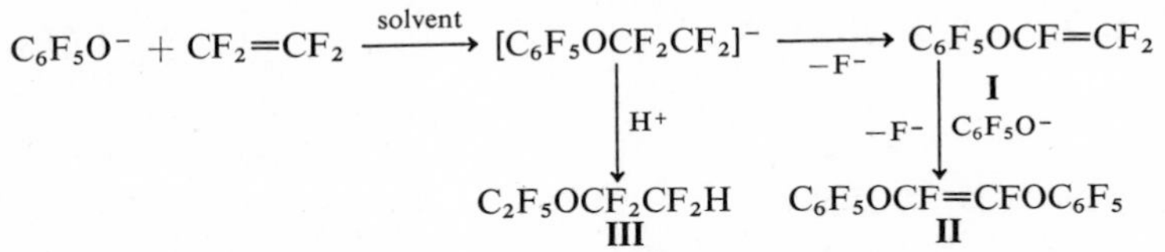

$$C_6F_5O^- + CF_2=CF_2 \xrightarrow{\text{solvent}} [C_6F_5OCF_2CF_2]^- \xrightarrow{-F^-} \underset{\textbf{I}}{C_6F_5OCF=CF_2}$$

with $\Big\downarrow H^+$ and $-F^- \Big| C_6F_5O^-$

$$\underset{\textbf{III}}{C_2F_5OCF_2CF_2H} \qquad \underset{\textbf{II}}{C_6F_5OCF=CFOC_6F_5}$$

Various reaction conditions were explored in an effort to maximize the yield of the perfluoroether. The results of this study, shown in Table 5, indicate that the best yield of unsaturated perfluorinated ether (I) was obtained when anhydrous potassium pentafluorophenoxide was the phenolic salt and benzene-tetrahydrofuran was the solvent system. However, the principal product was the disubstitution product, 1,2-difluoro-1,2-bis (pentafluorophenoxy)ethylene (II). When less stringent anhydrous conditions were employed, good yields of 1,1,2,2-tetrafluoroethyl pentafluorophenyl ether (III) were obtained. This last result indicates that under strictly anhydrous condition the addition-elimination step to the vinyl ether proceeds in good yield. Unfortunately, it appears that the unsaturated perfluoroether is more susceptible to nucleophilic attack by the pentafluorophenoxide anion than is the starting fluoroolefin, tetrafluoroethylene, and, consequently, much of the desired monomer is consumed in going to higher substitution products (cf. Section VI-E).

A seemingly feasible route to 1,2,2-trifluorovinyl pentafluorophenyl ether would be the dehydrofluorination of 1,1,2,2-tetrafluoroethyl pentafluorophenyl ether via an exchange reaction with an organolithium reagent as outlined below:

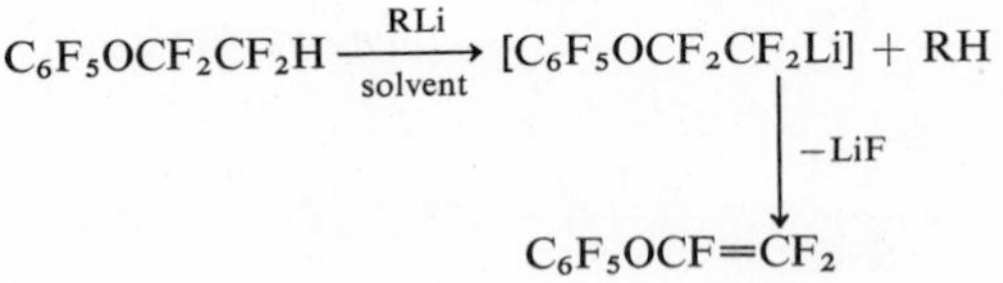

$$C_6F_5OCF_2CF_2H \xrightarrow[\text{solvent}]{\text{RLi}} [C_6F_5OCF_2CF_2Li] + RH$$

$$\Big\downarrow -LiF$$

$$C_6F_5OCF=CF_2$$

TABLE 5

Reaction of Pentafluorophenol Salts with TFE (96)

C_6F_5OM (g)	Metal	TFE (g)	Solvent(s) Type	Amt (ml)	Temp. (°C)	Pressure (psi)	Yield (%)	Product(s)
18	K	30	DMF	40	75	135(9.5kg/cm^2)	83.8	$C_6F_5-O-CF_2-CF_2H$
17	Li	30	DMF	40	80	149(10.1kg/cm^2)	64.1	$C_6F_5-O-CF_2-CF_2H$
18[a]	Na	25	Benzene	100	190	500(35kg/cm^2)	64.2	$C_6F_5-O-CF_2CF_2H$
			THF	34			4.3	$C_6H_5-O-CF=CF_2$
135[a]	K	80	Benzene	350	190	350(24.5kg/cm^2)	6.8	$C_6F_5-O-CF=CF_2$
			THF	100			3.0	$C_6F_5-O-CF_2-CF_2H$
							19.5	$C_6F_5-O-CF=CF-OC_6F_5$

[a] Anhydrous conditions. TFE, tetrafluoroethylene; DMF, dimethylformamide; and THF, tetrahydrofuran.

The same method also should be applicable to the synthesis of 1,2,2-tri-fluorovinyl phenyl ether from the corresponding 1,1,2,2-tetrafluoroethyl phenyl ether.

VII. POLYMERIZATION OF FLUOROSTYRENES

As is well known, styrene is one of the few monomers that will undergo successful polymerization via cationic, anionic, and free radical catalysis. In general this ability to polymerize under a wide variety of conditions is not shared by most fluorinated styrenes, particularly those of high fluorine content. Most of the fluorinated styrenes that have been polymerized were done so under the influence of free radical catalysis.

A. β-Fluorostyrene

In a few instances, cationic catalysis has proved effective. For example, β-fluorostyrene was initially polymerized by treating the monomer with stannic chloride after peroxidic catalysts proved ineffective (6). Later workers, however, did succeed in effecting the polymerization of this monomer by free radical catalysis (105,106). With benzoyl peroxide as the catalyst, bulk polymerization of β-fluorostyrene resulted in 25% yield of polymer. When emulsion polymerization techniques are employed, the yield of polymer is 50–85% (105,106). Attempts at anionic and Ziegler type polymerizations of β-fluorostyrene have thus far proved unsuccessful (105). The softening point of poly (β-fluorostyrene) is reported to be about 200°C (106).

B. α-Fluorostyrene

The isomeric α-fluorostyrene also undergoes polymerization by free radical catalysis. Poly(α-fluorostyrene) of high molecular weight (intrinsic viscosity 0.8–1.2 dl/g) was obtained by employing emulsion polymerization methods (9). Interestingly enough this white polymer, though perfectly stable below its softening point (145–150°C), exhibits the remarkable property of evolving hydrogen fluoride abruptly and quantitatively when heated to 225–235°C. The residual polymer acquires a brick-red hue and has been shown in actuality to be poly(phenylacetylene):

$$\left[\begin{array}{c} CFCH_2 \\ | \\ C_6H_5 \end{array}\right]_n \xrightarrow{\ -nHF\ } \left[\begin{array}{c} C=CH \\ | \\ C_6H_5 \end{array}\right]_n$$

Recently, poly(β-fluorostyrene) has been reported to exhibit a similar type of thermal degradation at a somewhat lower temperature (200–210°C) to give polyphenylacetylene by the quantitative elimination of hydrogen fluoride (106):

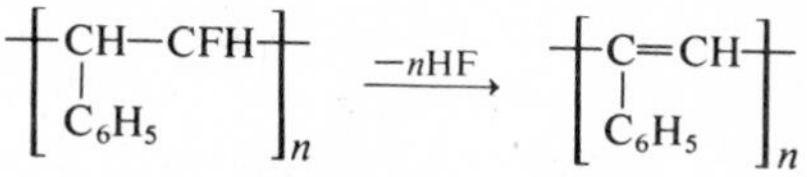

$$\left[\begin{array}{c} CH-CFH \\ | \\ C_6H_5 \end{array}\right]_n \xrightarrow{-nHF} \left[\begin{array}{c} C=CH \\ | \\ C_6H_5 \end{array}\right]_n$$

C. α,β-Difluorostyrene

Another interesting fluorostyrene polymer is poly(α,β-difluorostyrene). Cationic polymerization of this monomer with boron trifluoride proceeds rapidly (12). This polymer also is reported to exhibit thermal instability at moderate temperatures. On heating to 225–230°C, thermal decomposition set in, but neither the nature of the volatile decomposition products nor that of the residue was investigated (12). It might be surmised that, as in the case of poly(α-fluorostyrene) and poly(β-fluorostyrene), poly(α,β-difluorostyrene) could eliminate hydrogen fluoride on heating to give poly(β-fluorophenyl-acetylene):

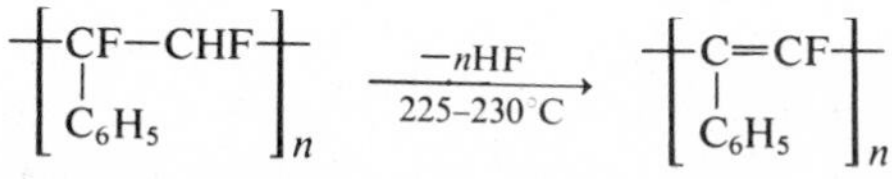

$$\left[\begin{array}{c} CF-CHF \\ | \\ C_6H_5 \end{array}\right]_n \xrightarrow[225-230°C]{-nHF} \left[\begin{array}{c} C=CF \\ | \\ C_6H_5 \end{array}\right]_n$$

The elimination of fluorine (F_2), although a possibility, would not be expected to occur as easily as the loss of hydrogen fluoride. However, the actual mechanism of decomposition still remains to be determined.

D. β,β-Difluorostyrene

Still another fluorostyrene that responds to cationic polymerization is β,β-difluorostyrene, but its yield is low (12). Emulsion polymerization (12, 105) also gives low yields of poly(β,β-difluorostyrene). Unlike the two preceding polymers, no abrupt thermal decomposition occurred on heating this polymer to moderate temperatures (220–225°C).

E. α,β,β-Trifluorostyrene

α,β,β-Trifluorostyrene is ordinarily resistant to the usual free radical polymerization methods. However, it is possible to obtain a high-molecular-weight polymer by emulsion polymerization techniques (12,15,40,105–107). No cationic polymerization of this interesting monomer has been observed, but it is claimed that anionic agents such as sodium, sodium in liquid ammonia, and sodium methoxide induced polymerization (15). However, with ethyllithium in a benzene solution no polymerization was observed (105).

Part of the difficulty encountered in the preparation of high-molecular-weight poly(α,β,β-trifluorostyrene) by free-radical catalysis derives from the fact that the monomer undergoes facile dimerization to a cyclobutane type dimer (12,15). Presumably the dimer could be either the 1,2- or 1,3-diphenyl-hexafluorocyclobutane or a mixture of the two. Since cis-trans stereoisomers of each positional isomer are also possible, the problem of the precise structural determination of the isomer(s) is further complicated.

$$2C_6H_5CF{=}CF_2 \longrightarrow \underset{\underset{\displaystyle C_6H_5CF-CF_2}{|\qquad|}}{C_6H_5CF-CF_2} + \underset{\underset{\displaystyle CF_2-CF-C_6H_5}{|\qquad|}}{C_6H_5CF-CF_2}$$

The relative proportion of products probably is dependent on the temperature of the reaction (108–110).

By analogy, the dimer fraction obtained from the thermolysis of perfluoropropylene at 100°C has been shown to consist primarily of equal amounts of the cis- and trans-isomers of 1,2-perfluoromethylcyclobutane and equal but lesser amounts of the cis- and trans-isomers of the 1,3-perfluorodimethylcyclobutane (108,109). At room temperature, only the cis- and trans-isomers of 1,3-diphenylhexafluorocyclobutane appear to form (110).

The absence of a considerable dimer fraction in the emulsion polymerization of α,β,β-trifluorostyrene is attributable to the lower polymerization temperature made possible by this technique. That is, the rate of dimerization at this temperature is not so competitive with the rate of polymerization.

In contrast to α,β,β-trifluorostyrene, little if any dimerization occurs with β,β-difluorostyrene, α,β-difluorostyrene, and α-chloro-β-fluorostyrene when bulk thermal polymerization is employed (12). The yields of polymer, however, were quite low by free radical techniques, even when emulsion polymerization was employed (12,105).

Table 6 presents a summary of some typical homopolymerization data for some fluorinated styrenes. It is evident from Table 6 that nuclearly fluorinated styrenes (2,4-, 2,5-difluorostyrenes, p-fluorostyrene, m-trifluoromethylstyrene) polymerize quite rapidly, whereas vinylic fluorinated styrenes (β-fluorostyrene, β,β-difluorostyrene, α,β-difluorostyrene, and α,β,β-trifluorostyrene) polymerize at a significantly lesser rate than ordinary styrene.

The last three runs in Table 6 give some typical results of the emulsion polymerization of several vinylic fluorinated styrenes.

As can be seen from examination of the data for the polymerization of α,β,β-trifluorostyrene, the rate of polymerization increases quite sharply in going from ordinary bulk thermal polymerization with a typical peroxidic catalyst to emulsion polymerization.

F. 2,3,4,5,6-Pentafluorostyrene

As already noted, styrenes with fluorine substituents on the nucleus only are generally amenable to polymerization by the usual free-radical methods. This fact has been further underscored since the relatively facile polymerization of 2,3,4,5,6-pentafluorostyrene (111,112). Initial attempts to polymerize

TABLE 6. Homopolymerization of Fluorinated Styrenes (105)

Monomer	Amount of catalyst (% of monomer weight)	Duration of polymerization (hr)	Experimental conditions	Yield (wt. %)
2,4-Difluorostyrene	Benzoyl peroxide 0.5	10.5	Liquid phase polymerization, 60°C	58
2,5-Difluorostyrene	Benzoyl peroxide 0.5	5.5	Liquid phase polymerization, 60°C	92
p-Fluorostyrene	Benzoyl peroxide 0.5	24	Liquid phase polymerization, 60°C	94.9
m-Trifluoromethylstyrene	Benzoyl peroxide 0.5	18	Liquid phase polymerization, 60°C	65
β-Fluorostyrene	Benzoyl peroxide 2.	14	Liquid phase polymerization, 60°C	25
β-Fluorostyrene	0.05 M solution of ethyl lithium in benzene	24	In benzene	Polymerization did not occur
β-Fluorostyrene	Ziegler combined catalyst	40	In octane	Polymerization did not occur
β,β-Difluorostyrene	Benzoyl peroxide 0.5	40	Liquid phase polymerization, 60°C	Traces of polymer
α,β,β-Trifluorostyrene	Benzoyl peroxide 0.5	21	Liquid phase polymerization, 60°C	15
α,β,β-Trifluorostyrene	0.05 M solution of ethyl lithium	50	In benzene	Polymerization did not occur
α,β-Difluorostyrene	Benzoyl peroxide 1	32	In benzene	Traces of polymer
α,β-Difluorostyrene	SnCl₄, 5	20	Liquid phase, 0°	63
α,β-Diflourostyrene	50% solution of AlCl₃ in ethyl chloride	5	Liquid phase, 0°	17
α-chloro-β-fluorostyrene	Benzoyl peroxide 1	15	Liquid phase, 60°	Traces of polymer
α-chloro-β-fluorostyrene	SnCl₄ 4	10	Liquid phase, 20°	25.3
α-Trifluoromethyl-p-fluorostyrene	Benzoyl peroxide 1	21	Liquid phase, 60°	Polymerization did not occur
β-fluorostyrene	0.5% potassium persulfate in 4% "Mersulfate" solution	28	Emulsion polymerization	50
β,β-difluorostyrene		40		Traces of polymer
α,β,β-trifluorostyrene		12		80

this monomer met with little success; either the polymer failed to form, or it formed only slowly, producing low-molecular-weight products. However, it was found that if the monomer is first purified by preparative gas-liquid chromatography, its thermal bulk polymerization proceeds quite nicely at 60°C. Compared to the rate of polymerization of ordinary styrene, the rate for pentafluorostyrene is somewhat slower. Polymerization also occurs at room temperature but again at a slower rate than that observed for styrene.

2,3,4,5,6-Pentafluorostyrene also may be polymerized by the use of anionic agents. Copious polymer formation was observed in the synthesis of 2,3,4,5,6-pentafluorostyrene via the reaction of vinyllithium and hexafluorobenzene (cf. Section IV-F) (76). Similarly, 2,3,4,5,6-pentafluoro-α-methylstyrene gave a polymer under anionic conditions (76). Other attempts at anionic polymerization have thus far been unsuccessful (113). The cationic polymerization of these monomers has not yet been reported.

The mechanism for the thermal initiated polymerization of styrene most widely accepted involves a reversible Diels-Alder dimerization to form the adduct AH, which then transfers a hydrogen to another styrene molecule:

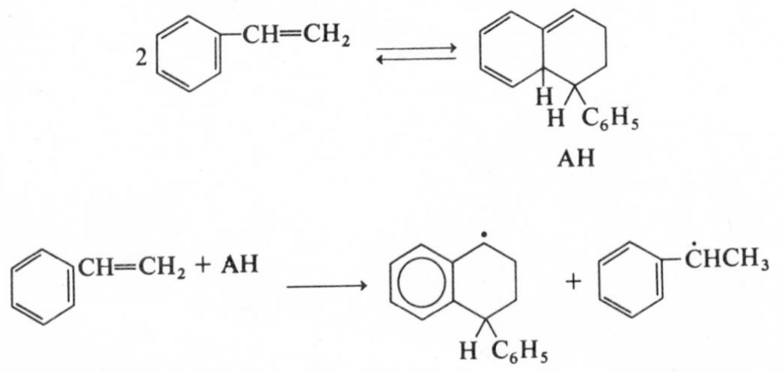

There is some question concerning the application of this mechanism to the thermal polymerization of pentafluorostyrene since this would necessitate the transfer of a fluorine atom from the analogous adduct species to monomer (112). The molecular weights for polypentafluorostyrene ranged downward from 6.8 million for polymer prepared by thermal polymerization at 60°C.

The relationship between the number-average molecular weight $\overline{M}_n$ and the intrinsic viscosity is shown in Fig. 1. A comparison of molecular weights measured osmometrically with those obtained using tritium-labeled azobisisobutyronitrile indicates that termination occurs predominantly by combination. From the line in Fig. 1, the following equation can be derived:

$$[\eta] = 4.37 \times 10^{-5}\overline{M}_n^{0.736}$$
$$\overline{M}_n = 8.50 \times 10^5[\eta]^{1.36}$$

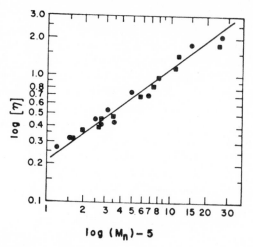

Fig. 1. Intrinsic viscosity in MIBK at 30°C for poly-PFS versus molecular weight (●) data obtained by osmometry; (■) data obtained using tritiated AIBN and assuming that termination is entirely by combination (112). (Reprinted by permission of copyright owner, The American Chemical Society.)

G. Perfluorostyrene

In contrast to 2,3,4,5,6-pentafluorostyrene, initial attempts to polymerize perfluorostyrene by ordinary thermal bulk polymerization methods failed to yield any polymer of significant molecular weight. The main products from the thermal bulk polymerization at 100°C after a period of about two years were two high-boiling products which had rather close retention volumes when examined by gas-liquid chromatography. Preparative gas-liquid chromatography was used to isolate each component, which was then analyzed mass spectrometrically. Both revealed parent mass peaks of 496, indicating that each was a dimer of perfluorostyrene (114,115). The two dimers presumably have structures analogous to that reported for the dimers of α,β,β-trifluorostyrene. That is, the dimers may have cyclobutane structures of the type illustrated below (cf. Section VII-E):

$$
\begin{array}{ll}
\mathrm{C_6F_5-CF-CF_2} & \mathrm{C_6F_5\text{-}CF\ -CF_2} \\
\quad\ \ |\qquad\ | & \qquad\ |\qquad\ | \\
\mathrm{C_6F_5-CF-CF_2} & \mathrm{CF_2-CF-C_6F_5}
\end{array}
$$

In increasing order of their retention volumes, the dimers are present in about a 7 to 6 ratio.

The rate of dimerization of perfluorostryene is far slower than that reported for α,β,β-trifluorostyrene (15). After two years at 100°C, approximately one-third of the perfluorostyrene is recovered unchanged. Attempts at photopolymerization of the monomer using ultraviolet radiation also failed to yield any polymer. Again two dimers were detected having about the same retention volumes as the dimers produced thermally.

Another sample of perfluorostyrene, however, was successfully polymerized by the use of γ-irradiation and the high-pressure techniques developed at the National Bureau of Standards (114–116). In a typical run, the monomer was subjected to a pressure of 11,500 atm at 155°C for 19.7 hr while being irradiated with gamma rays from a Cobalt-60 source at a dose rate of 0.045 Mrad/hr. The conversion of the monomer to polymer was approximately 50–60%. The polymer is a white, glassy solid of high softening point. Unlike poly(2,3,4,5,6-pentafluorostyrene), polyperfluorostyrene is not soluble in the usual ketonic solvents but is only swelled by such solvents as acetone or methyl ethyl ketone. The polymer is soluble in hexafluorobenzene and in the monomer. The intrinsic viscosity of the polymer in hexafluorobenzene is 0.12 dl/g at 29.8°C. The number average molecular weight of the polymer as determined by vapor pressure osmometry is 25,000. Fractionation of the polymer can be accomplished by the use of petroleum ether as the precipitant and hexafluorobenzene as the solvent. In this manner, a fraction having an intrinsic viscosity of 0.22 dl/g at 29.8°C and a softening point of about 255°C was obtained. High-molecular-weight samples (approximately 50,000) have been obtained (115).

H. α-Hydroheptafluorostyrene

Recently α-hydroheptafluorostyrene has been reported to give a solid white opaque polymer simply by moderate heating at atmospheric pressure (87). No other characteristics of the polymer, however, were given. The polymerization of β-hydroheptafluorostyrene has not yet been reported.

I. Effect of Trifluoromethyl Substituents

As noted earlier, styrenes having only one trifluoromethyl substituent, even in the ortho position, are polymerizable by ordinary free-radical catalysis (63,105). However, the presence of two ortho trifluoromethyl substituents, most likely because of steric effects, seems to suppress completely the polymerizability of this class of styrenes (64,117).

Apparently the presence of a trifluoromethyl on the vinyl group also hinders polymerization. The polymerization of such monomers as α-trifluoromethyl-styrene and perfluoro-α-methylstyrene has not been reported (17–19,86). Attempts to polymerize α-trifluoromethyl-β,β-difluorostyrene by the usual free radical catalytic agents have failed (39).

It may be that the free-radical polymerization of styrenes of this type might be feasible if high-pressure techniques were employed.

VIII. COPOLYMERIZATION OF FLUOROSTYRENES

While homopolymers of fluorinated styrenes often have such desirable physical and chemical properties as good thermal stability, good dielectric properties, and low flammability, they also tend to have high softening points. In addition, some of the fluorinated homopolymers such as poly-α,β,β-trifluorostyrene are brittle and difficult to polymerize by the more commercial methods of polymerization.

Copolymerization frequently offers a mode by which the properties of polyfluorostyrenes may be modified. For example, poly(α,β,β-trifluorostyrene)

TABLE 7

Time of Copolymerization, Yield of Copolymers, Their Compositions and Characteristic Viscosities (118)

Time of experiment (hr)	Yield of copolymer (%)	Composition of the copolymer (moles-%)		Characteristic viscosity of benzene solutions of the copolymers at 20°C
		M_1	M_2	
Copolymers of α,β,β-trifluorostyrene (M_1) with styrene (M_2)				
8	74	28.53	71.47	2.7
7	70	39.77	60.23	2.8
8	60	48.53	51.47	1.8
8	40	53.75	46.24	0.9
Copolymers of α,β,β-trifluorostyrene (M_1) with 2,5-dimethylstyrene (M_2)				
4	90	50.54	49.16	1.1
4	50	60.77	39.23	0.6
Copolymers of α,β,β-trifluorostyrene (M_1) with methyl methacrylate (M_2)				
6	—	43.29	56.71	0.4
11	41	53.88	46.12	0.2
Copolymers of m-methyl-α,β,β-trifluorostyrene (M_1) with styrene (M_2)				
2.5	40	46.26	53.73	0.5
Copolymer of o-methyl-α,β,β-trifluorostyrene (M_1) with styrene (M_2)				
33	11	50.41	49.59	0.5
Copolymer of p-methyl-α,β,β-trifluorostyrene (M_1) with styrene (M_2)				
7	55	46.97	53.03	2.0
7	64	36.77	63.23	1.4
Copolymer of α,β-difluoro-β-chlorostyrene (M_1) with styrene (M_2)				
20	30	16.00	84.00	0.05
Copolymer of α,β-difluoro-β-chlorostyrene (M_1) with 2,5-difluorostyrene (M_2)				
20	14	—	—	0.5

TABLE 8

Method of Preparation, Composition and Properties of Copolymers (119)

Polymerization method	Reaction time (hr.)	Conversion (%)	Composition of polymer (mole %)		[η] (20°, benzene)	IFP softening point (°C)
			M_1	M_2		
			α,β,β-Trifluorostyrene-styrene			
III	15	22	32.88	67.12	0.23	135
			α,β,β-Trifluorostyrene-2,5-dimethylstyrene			
I	1	40.7	40.82	59.18	0.59	181
I	0.6	62.3	51.69	48.31	0.68	185
I	0.75	61.6	59.01	40.99	0.47	187
I	2.5	40	74.39	25.60	0.39	190
III	0.5	34.6	59.91	40.09	0.50	171
			α,β,β-Trifluorostyrene-2,5-difluorostyrene			
I	10	60	—	—	0.52	164
			3-Methyl-α,β,β-trifluorostyrene-styrene			
I	8	55	29.89	70.11	0.84	123
I	7	60	48.66	51.34	0.64	135
II	95	40	25.70	74.30	0.39	108
II	0.5	25.2	70.53	29.47	—	131.5
			3-Methyl-α,β,β-trifluorostyrene-2,5-dimethylstyrene			
III	22.5	26.2	72.43	27.57	—	144.5
			4-Methyl-α,β,β-trifluorostyrene-styrene			
I	1.3	68	49.53	50.47	0.86	147
I	8	80	72.97	27.03	0.54	164
III	31.7	26	59.03	40.97	0.08	141.5

TABLE 8 (119) (continued)

Polymerization method	Reaction time (hr)	Conversion (%)	Composition of polymer (mole %)		$[\eta]$ (20°, benzene)	IFP softening point (°C)
			M_1	M_2		
4-Methyl-α,β,β-trifluorostyrene-2,5-dimethylstyrene						
I	12	40	64.55	35.44	0.3	144
III	22.7	26	62.03	37.97	0.1	120
β-Fluorostyrene-styrene						
I	7	65	33.31	66.69	0.28	109
II	95	83	16.47	83.52	1.8	120
3-Trifluoromethylstyrene-styrene						
III	6.75	9.4	76.79	28.21	—	89
Dimethylfluorostyrene-styrene						
I	23	33	39.36	60.63	0.02	116
II	95	66	37.85	62.14	0.42	112
III	70	14.3	58.75	41.25	—	119.5

Three general methods of polymerization were employed to prepare the copolymers shown in Table 8: (I) Emulsion polymerization at 60°C in the presence of azobisisobutyronitrile or potassium persulfate; (II) liquid phase polymerization at 50–170°C; and (III) liquid phase polymerization at 60°C with the copolymerization being terminated at low conversions. Azobisisobutyronitrile (0.5%) was used as the initiator in the liquid phase polymerizations.

73

has a softening point of 240°C ,whereas a copolymer with ordinary styrene containing 0.397 mole fraction of the latter softens at about 160°C (15).

In a study of the copolymerization of side-chain fluorinated styrenes with various other vinyl monomer, it was shown that of all the monomer pairs studied, the copolymerization of α,β,β-trifluorostyrene with 2,5-dimethylstyrene proceeded the most rapidly (118). As may be seen from Table 7, this copolymer is formed in 4 hr in 90% conversion with α,β,β-trifluorostyrene comprising 50.54 mole percent of the copolymer. The molecular weight apparently is quite high ($[\eta] = 1.1$). On the other hand, the copolymerization of the monomer pair α,β,β-trifluorostyrene/styrene proceeds at a slower rate to form a copolymer containing 48.53 mole % of the fluorinated styrene; the conversion is only 60% after 8 hr.

From Table 7 it is apparent that as the composition of the fluorinated styrene increases, the yield and the characteristic viscosity of the copolymer decrease. In contrast to this behavior, the softening point of the copolymer increases with increasing content of the fluorinated styrene. These relationships are further illustrated in Table 8.

The dielectric properties of these copolymers are markedly independent on the copolymer composition. All are dielectric materials with a high dielectric permeability ($\varepsilon = 2.7$–3.5). Their loss factor is of the order of tan $\delta = 1$–3×10^{-3} and their specific volume resistance is given by $p_v = 1$–3×10^{17} ohm/cm (see Table 9).

TABLE 9.   Electrical Properties of Copolymers

Copolymer	Fluorostyrene content (mole %)	ϵ	tan δ $\times 10^3$ (50–5000 kc/sec, 20°)	$p_v \times 10^{-17}$ (20) (ohm · cm)
α,β,β-Trifluorostyrene with 2,5-dimethylstyrene	40.82 51.69 59.01	3.7	1.6	2.3
The same with 2,5-difluorostyrene	—	3.0	1	0.9
3-Methyl-α,β,β-trifluorostyrene with styrene	48.66 29.89	— —	1.3 0.5–2.6	— —
3-Methyl-α,β,β-trifluorostyrene with styrene	63.5	3.4	0.5–2	1.3

A typical copolymer composition curve is shown for the system, α,β,β-trifluorostyrene-styrene, in Fig. 2. The curve conforms to the usual copolymerization equation where the reactivity ratio $r_1 = 0.070$ and $r_2 = 0.66$ are for α,β,β-trifluorostyrene and styrene, respectively. If the Price-Alfrey

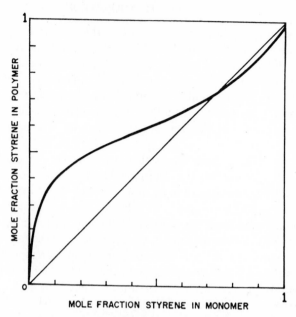

Fig. 2. Copolymerization of α,β,β-trifluorostyrene with styrene; composition of copolymer as a function of the composition of the monomer charged (15).

TABLE 10

Copolymerization Parameters of Some Chloro- and Fluorostyrenes with Styrene

Halostyrene	Copolymerization constants		Ref.
	r_1	r_2	
p-Fluorostyrene	0.7	0.9	105
m-Trifluoromethylstyrene	0.62	0.75	105
2,4-Difluorostyrene	1.05	0.75	105
β-Fluorostyrene	5.96	0.07	105
β,β-Difluorostyrene	10.4	0.	105
β,β-Difluoro-2,4-difluorostyrene	12.	0.	105
α,β-Difluorostyrene	2.42	0.04	105
α-Fluoro-β-chlorostyrene	3.7	0.	105
α-Chloro-β-fluorostyrene	2.1	0.55	105
α,β,β-Trifluorostyrene	3.5	0.15	105
α,β,β-Trichlorostyrene	6.	0.	105
α,β,β-Trifluoromethylstyrene	0.7	0.	105
m-Trifluoromethyl-α-methylstyrene	0.74	0.	105
α-Trifluoromethyl-p-fluorostyrene	0.56	0.	105
2,3,4,5,6-Pentafluorostyrene	0.43	0.22	112
2,3,4,5,6-Pentachlorostyrene	1.31	0.35	112

treatment is used, Q_1, the reactivity of α,β,β-trifluorostyrene, is calculated to be 0.37 and e_1, the polarity, is $+0.95$ on the basis of the revised Q_2 and e_2 values of styrene, which are 1.0 and -0.8, respectively. The copolymerization curve exhibits a crossover at about 0.7 mole fraction of styrene. In this region there also is found a maximum rate of polymerization and a strong tendency to alternate. Copolymerization constants also have been obtained for various other fluorinated styrenes with styrene as shown in Table 10.

It would appear from these results that the influence of polar and steric effects plays a major role in the determination of the composition of the copolymers. The copolymerization constants of such fluorinated styrenes as β-fluorostyrene, α,β-difluoro-, and β,β-difluorostyrenes, and α,β,β-trifluorostyrene are quite small and the product of the copolymerization constants $(r_1 \times r_2)$ is significantly less than unity. Thus it seems that the rates of reaction of these side-chain fluorinated styrenes with their corresponding styryl radicals are drastically low compared with their rates of reaction with the nonfluorinated styryl radical. This observation is in keeping with the expected repulsive interaction set up between the fluorines of the styryl radical and the fluorines of the corresponding styrene monomer.

IX. POLYMERIZATION OF FLUORINATED VINYL PHENYL ETHERS

It has been reported that the homopolymerization of vinyl phenyl ether proceeds with considerable difficulty (120–123). During the attempted polymerization of this monomer, side reactions frequently occurred, usually involving cleavage at the ether linkage; (e.g., Claisen-type rearrangements). Only polymers of low molecular weight and doubtful chemical structure have been obtained. Both ionic and free radical methods of polymerization have been employed.

Only recently has the polymerization of any of the fluorinated analogs of vinyl phenyl ether been attempted (96,103,123). The preliminary data obtained by this study indicate that side-chain fluorinated vinyl phenyl ethers are more difficult to polymerize than monomers having only nuclear fluorines. This observation seems to parallel the situation observed in the case of fluorinated styrenes where the side-chain fluorinated monomers usually are more resistant to polymerization than are the ring fluorinated monomers. The results obtained from polymerization studies with several fluorinated vinyl phenyl ethers are discussed in this section.

A. 1,2,2-Trifluorovinyl Phenyl Ether

Initial attempts to polymerize the monomer, 1,2,2-trifluorovinyl phenyl ether by the usual free-radical and cationic methods failed to yield any polymer (96). Thus ordinary bulk thermal or photoinitiated polymerization techniques either in the presence or absence of such catalysts as benzoyl peroxide and azobisisobutyronitrile (ABIN) proved of no avail. Cationic polymerization with boron trifluoride in liquid propane gave only a clear viscous oil of low molecular weight.

A more successful polymerization of this monomer resulted from the use of high-pressure and gamma irradiation. A tan, powdery solid was produced at a pressure of 6400 atm and a dose rate 1.3 Mr/hr for 63 hr at 100°C. The polymer was soluble in benzene, $[\eta] = 0.03$ dl/g, and had a calculated molecular weight of 4500. Efforts to increase the molecular weight of the polymer by increasing both the applied pressure (10,000 atm) and the temperature of polymerization (191°C) but lowering the dose rate (0.0033 Mr/hr for 51 hr) were unsuccessful. Only small quantities of low-molecular-weight polymer were formed along with a considerable quantity dimeric material. The dimer presumably had the basic cyclobutane structure associated with the dimeric products formed in the thermal polymerization of α,β,β-trifluorostyrene and perfluorostyrene. Thus a likely structural representation for the dimeric fraction would be that of 1,2,3,3,4,4-hexafluoro-1,2,diphenoxycyclobutane and perhaps also the 1,3-diphenoxy isomer:

$$2C_6H_5OCF{=}CF_2 \xrightarrow[\Delta]{\text{high press}} \begin{array}{c} C_6H_5O{-}CF{-}CF_2 \\ | \quad \cdot \quad | \\ C_6H_5O{-}CF{-}CF_2 \end{array}$$

When both the pressure (14,000 atm) and reaction time (111 hr) were increased while the temperature (103°C) and dose rate (0.0037 Mr/hr) were kept about the same, a similar polymer of about 4500 molecular weight was obtained.

It would appear from the foregoing experiments that serious transfer reactions may be occurring to limit the molecular weight of the growing polymer chain.

B. Vinyl Pentafluorophenyl Ether

Unlike the preceding 1,2,2-trifluorovinyl phenyl ether, vinyl pentafluorophenyl ether is a monomer more amenable to polymerization (103). It may be polymerized cationically at -78°C in n-pentane with boron trifluoride gas as the catalyst. A hard, white, solid polymer, soluble in benzene, acetone, and hexafluorobenzene, was obtained. Vapor-phase osmometry (VPO) gave number average molecular weights ranging from 10,000 to 17,000. Other cationic catalysts such as aluminum chloride were not as effective as polymerization agents. With aluminum chloride in benzene a polymer of only 2300 (VPO) molecular weight was obtained. In contrast to its action on vinyl phenyl ether, where it produces a resinous material after a slight induction period, hydrobromic acid had no effect on pentafluorophenyl vinyl ether. This is not surprising since the fluorinated phenyl ring of the latter monomer would be expected to render difficult, if not impossible, side reactions involving attack at nuclear positions (Claisen-type rearrangements).

High-pressure techniques with gamma irradiation gave the highest molecular weight poly(vinyl pentafluorophenyl ether). At a pressure of 9500 atm and a dose rate of 0.009 Mr/hr for 68 hr at 105°C, a polymer of 30,000 molecular weight (VPO) was formed in low yield. The polymer was a fluffy pink solid softening at 95°C and melting below 150°C (103).

C. 1,2-Difluorovinyl Pentafluorophenyl Ether

As in the case of 1,2,2-trifluorovinyl phenyl ether, both cationic and peroxidic catalysis failed to give any significant polymer of 1,2-difluorovinyl pentafluorophenyl ether (96,103). With azobisisobutyronitrile (ABIN) no polymer at all was obtained. Thermal bulk polymerization at 110°C for 4 days gave a black, viscous liquid with a little solid. Dimeric products appeared to be the main products. High-pressure polymerization of this monomer has not been attempted.

D. 1,2,2-Trifluorovinyl Pentafluorophenyl Ether

The polymerization behavior of 1,2,2-trifluorovinyl pentafluorophenyl ether was in keeping with that observed for both 1,2,2-trifluorovinyl phenyl ether and 1,2-difluorovinyl pentafluorophenyl ether. Attempts at thermal and photoinitiated bulk polymerizations gave a dimeric product, assigned the structure 1,2,3,3,4,4-hexafluoro-1,2-bis(pentafluorophenoxy)cyclobutane (96). Rudimentary polymerization of this monomer, however, was achieved by the use of high pressure (12,000 atm) and gamma irradiation (0.0032 Mr/hr) at 100°C for 5 days. The crude polymer was a tacky substance which, when precipitated by perfluoro-n-heptane from a solution in hexafluorobenzene, became a white, powdery solid, apparently of low molecular weight.

References

1. P. J. Plunket, U.S. Patent 2,230,654 (1941).
2. W. E. Hanford and R. M. Joyce, *J. Am. Chem. Soc.*, **68**: 2082 (1946).
3. F. Swarts, *Bull. Soc. Chem.*, **25**(4):145 (1919).
4. F. Swarts, *Bull. Soc. Chem.*, **25**(4): 325 (1919).
5. E. Elkik, *Bull. Soc. Chim. France*, **5**:1569 (1967).
6. F. Bergmann, A. Kalmus, and E. Breuer, *J. Am. Chem. Soc.*, **80**: 4540 (1958).
7. N. M. Nad', T. V. Talalaeva, G. V. Kazennikova, and K. A. Kocheshkov, *Izv. Akad. Nauk SSSR*, **1959**: 250.
8. T. Ando, F. Namigata, M. Kataoka, K. Yachida, and W. Funasaka, *Bull. Chem. Soc. (Japan)*, **40**(5):1275 (1967).
9. K. Matsuda, J. A. Sedlak, J. S. Noland, and G. C. Gleckler, *J. Org. Chem.*, **27**: 4015 (1962).

10. S. G. Cohen, H. T. Wolosinski, and P. J. Sheuer, *J. Am. Chem. Soc.*, **71**:3439 (1949).
11. S. G. Cohen, H. T. Wolosinski, and P. J. Sheuer, *J. Am. Chem. Soc.*, **72**:3952 (1950).
12. M. Prober, *J. Amer. Chem. Soc.*, **75**:968 (1953).
13. E. T. McBee, W. B. Ligett, and V. V. Lingren, U.S. Patent 2,586,364 (1952).
14. E. T. McBee and R. B. Bolt, *Ind. Eng. Chem.*, **39**:412 (1947).
15. D. I. Livingston, P. M. Kamath, and R. S. Corley, *J. Polymer Sci.*, **20**:485 (1956).
16. J. M. Antonucci and L. A. Wall, *J. Research NBS*, **70A**(6):473 (1966).
17. J. B. Dickey and T. E. Stanin, U.S. Patent 2,475,423 (1949).
18. P. Tarrant and R. E. Taylor, *J. Org. Chem.*, **24**:238 (1959).
19. G. V. Kazennikova, T. V. Talalaeva, A. V. Zimin, and K. A. Kocheshkov, *Izv. Akad. Nauk SSSR*, **1961**:1066.
20. V. A. Englehardt, W. R. Hasek, and W. C. Smith. *J. Am. Chem. Soc.*, **82**:543 (1960).
21. S. Dixon, *J. Org. Chem.*, **21**:400 (1956); U.S. Patent 2,874,197 (1955).
22. T. P. McGrath and R. Levine, *J. Am. Chem. Soc.*, **77**:4168 (1955).
23. R. Meier and F. Bohler, *Chem. Ber.*, **90**:2344 (1957).
24. G. V. Kazennikova, T. V. Talalaeva, A. V. Zimin, A. P. Simonov, and K. A. Kocheshkov, *Izv. Akad. Nauk SSSR*, **1961**: 1063.
25. T. V. Talalaeva, O. P. Petrii, G. V. Timofeyuk, A. V. Zimin, and K. A. Kocheshkov *Doklady Akad. Nauk SSSR*, **154**(2): 398 (1964).
26. N. M. Segree, N. N. Shapet'ko, and G. V. Timofeyuk, *Zh. Strukturn. Khim.*, **6**:300 (1965).
27. O. P. Petrii, A. A. Makhina, T. V. Talalaeva, and K. A. Kocheshkov, *Dokl. Akad. Nauk SSSR*, **167**:594 (1966).
28. R. Fontanelli and D. Sianesi, *Ann. Chim. (Rome)*, **55**(8-9): 862 (1965).
29. E. M. Panov, R. S. Sorokina, A. V. Zimin, and K. A. Kocheshkov, *Doklady Akad. Nauk SSSR*, **145**: 1068 (1962).
30. E. M. Panov, R. S. Sorokina, and K. A. Kocheshkov, *Zhur. Obschei Khim.*, **35**:1426 (1965).
31. P. Tarrant and D. A. Warner, *J. Am. Chem. Soc.*, **76**:1624 (1954); U.S. Patent 2,804,484 (1957).
32. S. A. Fuqua, R. M. Parkhurst, and R. M. Silverstein, *Tetrahedron*, **20**:1625 (1964).
33. S. A. Fuqua, W. G. Duncan, and R. M. Silverstein, *Tetrahedron Letters*, **1963**(9): 521.
34. S. A. Fuqua, W. G. Duncan, and R. M. Silverstein, *Tetrahedron Letters*, **1964**(23): 1461.
35. S. A. Fuqua, W. G. Duncan, and R. M. Silverstein, *J. Org. Chem.*, **30**:1027 (1965).
36. S. A. Fuqua, W. G. Duncan, and R. M. Silverstein, *J. Org. Chem.*, **30**:2543 (1965).
37. D. J. Burton and F. E. Herkes, *Tetrahedron Letters*, **1965**(23):1883.
38. D. J. Burton and F. E. Herkes, *J. Org. Chem.*, **32**:1311 (1967).
39. P. M. Barna, *Chem. Ind. (London)*, **49**:2054 (1966).
40. M. Prober, U.S. Patent 2,651,627 (1953).
41. M. Prober, U.S. Patent 2,752,400 (1956).
42. H. Shingu and M. Hisazumi, U.S. Patent 3,489,807 (1970).
43. P. L. Barrick, U.S. Patent 2,462,346 (1949).
44. J. L. Anderson, U.S. Patent 2,773,278 (1956).
45. J. J. Drysdale and W. D. Phillip, *J. Am. Chem. Soc.*, **79**:319 (1957).
46. L. A. Errede and F. DeMaria, *J. Phys. Chem.*, **66**:2664 (1962).
47. L. A. Brooks, *J. Am. Chem. Soc.*, **66**:1295 (1944).
48. L. A. Brooks and M. Nazzewski, U.S. Patent 2,406,319 (1946); *Chem. Abstr.*, **41**: 315 (1947).
49. C. S. Marvel and D. W. Hein, *J. Am. Chem. Soc.*, **70**:1895 (1948).

50. M. M. Koton, E. P. Moskvina, and F. S. Florinskii, *Zhur Obschei. Klim.*, **21**:1843 (1951).
51. D. E. Worrall and H. T. Wolosinski, *J. Am. Chem. Soc.*, **62**:2449 (1940).
52. A. V. Topchiev, V. P. Alaniya, and M. V. Vagin, *Doklady Akad. Nauk SSSR*, **151**(1): 114 (1963).
53. M. W. Renol, *J. Am. Chem. Soc.*, **68**:1159 (1946).
54. G. B. Bachman and H. M. Hellman, *J. Am. Chem. Soc.*, **70**:1772 (1948).
55. A. L. Henne and M. S. Newman, *J. Am. Chem. Soc.*, **60**:1697 (1938).
56. C. S. Marvel, C. G. Overberger, R. E. Allen, and J. H. Saunders, *J. Am. Chem. Soc.*, **68**:736 (1946).
57. F. Bergmann and D. J. Szmuszkowizc. *J. Am. Chem. Soc.*, **70**:2748 (1948).
58. G. B. Bachman and L. L. Lewis, U.S. Patent 2,580,504 (1952).
59. C. G. Overberger, J. H. Saunders, R. E. Allen, and R. Gander, *Org. Syn. Coll.*, **3**:200 (1955).
60. C. G. Overberger and J. H. Saunders, *Org. Syn. Coll.*, **3**:204 (1955).
61. D. Seymour and K. B. Wolfstirn, *J. Am. Chem. Soc.*, **70**:1177 (1948).
62. G. B. Bachman and L. L. Lewis, *J. Am. Chem. Soc.*, **69**:2022 (1947); U.S. Patent 2,414,330 (1947).
63. E. T. McBee and R. A. Sanford, *J. Am. Chem. Soc.*, **72**:4053, 5574 (1950).
64. C. A. Kraus and A. B. Conciatori, *J. Am. Chem. Soc.*, **72**:2283 (1950).
65. D. C. England, French Patent 1,325,204 (1963).
66. B. S. Faral, E. E. Gilbert and J. P. Sibilia, *J. Org. Chem.*, **30**:998 (1965).
67. W. A. Sheppard, *J. Am. Chem. Soc.*, **87**:2410 (1965).
68. W. J. Middleton, U.S. Patent 3,179,640 (1965).
69. A. E. Pavlath and A. J. Leffler, "Aromatic Fluorine Compounds," Reinhold, New York, 1962, p. 144.
70. W. F. Beckert and J. V. Lowe, Jr., *J. Org. Chem.* **32**:582 (1967).
71. L. A. Brooks and M. Mazzerski. U.S. Patent 2,406,319 (1946).
72. N. M. Nad', T. V. Talalaeva, G. V. Kazennikova, and K. A. Kocheshkov, *Izv. Akad. Nauk SSSR*, **1959**:58.
73. A. Rocca, French Patent 1,457,450 (1966).
74. E. Nield, R. Stephen, and J. C. Tatlow, *J. Chem. Soc.*, **1959**:166.
75. W. J. Pummer and L. A. Wall. *J. Research NBS*, **63A**:167 (1959).
76. L. A. Wall, W. J. Pummer, J. E. Fearn, and J. M. Antonucci, *J. Research NBS*, **67A**: 481 (1963).
77. A. K. Barbour, M. W. Buxton, P. L. Coe, R. Stephens, and J. C. Tatlow, *J. Chem. Soc.*, **1961**:808.
78. J. M. Birchall, T. Clarke, and R. N. Haszeldine, *J. Chem. Soc.*, **1962**:4977.
79. M. M. Nad', T. V. Talalaeva, G. V. Kazennikova, and K. A. Kocheshkov, *Izv. Akad. Nauk SSSR*, **1959**:64.
80. B. R. Letchford, C. R. Patrick, M. Stacey, and J. C. Tatlow, *Chem. Ind.* (*London*), **32**:1472 (1962); C. R. Patrick, M. Stacey, and J. C. Tatlow, U.S. Patent 3,187,058 (1965).
81. J. M. Antonucci and L. A. Wall, *SPE Trans.*, **3**:225 (1963); U.S. Patent 3,265,746 (1966).
82. J. M. Antonucci, D. W. Brown, and L. A. Wall, Abstract of papers, 156th Natl. Am. Chem. Soc. Meeting, Div. of Fluorine Chem., paper No. 20, Atlantic City, N.J. (1968).
83. A Bergomi, J. Burdon, and J. C. Tatlow, French Patent 1,544,617 (1968); *Chem. Abstr.*, **71**:112,689d (1969).

84. R. J. DePasquale, E. J. Solski, and C. Tamborski, *J. Organometal*, **15**:494 (1968).
85. M. Stacey and J. C. Tatlow, "Advances in Fluorine Chemistry," Vol. I, Buttersworth, London, 1960, p. 166.
86. B. B. Letchford, C. R. Patrick, and J. C. Tatlow, *Tetrahedron*, **20**:1381 (1964).
87. P. L. Coe, R. C. Pleney, and J. C. Tatlow, *J. Chem. Soc.*, **1966**:597.
88. S. Andreade, *J. Am. Chem. Soc.*, **86**:2003 (1964).
89. P. L. Coe, R. Stephens, and J. C. Tatlow, *J. Chem. Soc.*, **1962**:3327.
90. R. J. Harper, E. J. Soloski, and C. Tamborski, *J. Org. Chem.*, **29**:2385 (1964).
91. G. Camaggi, Italian Patent 758,250 (1967); *Chem. Abstr.*, **68**:77,946 (1968).
92. P. Tarrant and H. C. Brown, *J. Am. Chem. Soc.*, **73**:5831 (1951).
93. L. S. Croix and A. J. Buselli, U.S. Patent 2,799,712 (1957).
94. R. S. Corley, J. Lal, and M. W. Kane, *J. Am. Chem. Soc.*, **78**:3489 (1956).
95. D. C. England, L. R. Melby, M. A. Dietrich, and R. V. Lindsey, *J. Am. Chem. Soc.*, **82**:5116 (1960).
96. W. J. Pummer and L. A. Wall, *SPE Trans.*, **3**:220 (1963).
97. W. Reppe, U.S. Patent 1,959,927 (1934).
98. W. M. Lauer and M. A. Spielman, *J. Am. Chem. Soc.*, **55**:1572 (1933).
99. H. V. R. Iengar and P. D. Ritchie, *J. Chem. Soc.*, **1957**:2562.
100. W. Reppe, *Ann.*, **601**:81 (1956).
101. A. K. H. Filippora, E. S. Domnino, T. I. Ermolova, M. L. Navtanovich, and G. V. Dmitrieva, *Izv. Sibir Otdel. Akad. Nauk SSSR*, **11**:9 (1958).
102. D. J. Foster and E. Tobler, Abstracts Papers Am. Chem. Soc. Meeting, **138** (1960).
103. W. J. Pummer and L. A. Wall, *J. Research NBS*, **70A** (3):233 (1966).
104. M. Julia, *Bull. Soc. Chim. (France)*, **1956**:185.
105. A. R. Gantmahher, Yn. L. Spirin, and S. S. Medvedev, *Vysokomol. soed.* **1**:1526 (1959).
106. J. Parrod and C. Hugelin, *C. R. Acad. Sci. (Paris)*, Ser. **C**(6):267 (1968).
107. M. M. Koton, A. F. Dokukina, and E. I. Egorova, *Doklady Akad. Nauk SSSR* **155**(1): 139 (1964).
108. M. Hauptschein, A. H. Fainberg, and M. Braid, *J. Am. Chem. Soc.*, **80**:842 (1958).
109. D. I. McCane and I. M. Robinson, U.S. Patent 3,316,312 (1967).
110. M. P. Voti, V. A. Kosobutskii, and A. F. Dokukina, *Vysokomol. soyed.*, **A10**(5): 1137 (1968).
111. J. M. Antonucci, S. Straus, M. Tryon, and L. A. Wall, *Proc. Sym. Polymer Degradation, Soc. Chem. Ind. Monograph No. 13*, 205 (1961).
112. W. A. Pryor and Tzu-lee Huang, *Macromolecules* **2**(1):70 (1969).
113. L. J. Fetters, unpublished results.
114. J. M. Antonucci and W. J. Pummer, *Am. Chem. Soc. Polymer Preprints*, **7**(2):107 (Sept. 1966).
115. J. M. Antonucci, D. W. Brown, and L. A. Wall, unpublished results.
116. L. A. Wall, D. W. Brown, and R. E. Florin, *Am. Chem. Soc. Polymer Preprints*, **2**(2): 366 (1961).
117. V. V. Korshak and N. G. Matveeve, *Doklady Akad. Nauk SSSR*, **85**:797 (1952).
118. A. F. Dokukina, Ye. I. Yegorova, G. V. Kazennikov, M. M. Koton, K. A. Kocheshkov, Z. A. Smirnova, and T. V. Talalaeva, *Polymer Science, USSR*, **4**(6):270 (1962).
119. Ye. I. Yegorova, Z. A. Smirnova, and A. F. Dokukina, *Polymer Science, USSR*, **6**(7):1306 (1964).
120. C. E. Schildnecht, "Vinyl and Related Polymers," John Wiley and Sons, New York, 1952, pp. 622–625.

121. A V. Kalabina, N. A. Tyukavkina, G. P. Mantsivoda, and R. V. Krasovskii, *Vysokomol. soedin.*, **3**:1150 (1961).
122. A. V. Kalabina, N. A. Tyukavkina, and V. A. Kruglova, *Vysokomol. soedin.*, **3**:1155 (1961).
123. A. V. Kalabina, A.Kh. Fillipova, G. V. Smitrieva, and L. Ya. Tsarik, *Vysokomol. soedin.*, **3**:1120 (1961).

3. AROMATIC FLUOROCARBON POLYMERS

WALTER J. PUMMER, *Polymer Chemistry Section, National Bureau of Standards, Washington, D.C.*

Contents

I. INTRODUCTION

Before 1955, one of the few heat-resistant polymers known was polytetrafluoroethylene, a wholly aliphatic polymer. Since then, many new polymers have been prepared; most possess exceptional mechanical properties and exhibit excellent thermal stabilities. Some of these new types of polymers which contain aromatic structures are based on the following ring systems: (1) simple benzenoid; (2) aromatic heterocycles; (3) heteroaromatic; and (4) to a lesser extent the perfluoroaromatics. This chapter deals primarily with the research and development of the perfluoroaromatics.

II. SYNTHESES AND REACTIONS OF PERFLUOROARYL COMPOUNDS

Aromatic compounds that contain several fluorine atoms have been known for some time (1–3). The synthesis of a fully fluorinated aryl compound, hexafluorobenzene, was first reported in 1947 (4). More recent methods for the synthesis of hexafluorobenzene and other highly fluorinated derivatives in reasonably good yields and larger quantities have greatly accelerated progress in this area of research since 1947. In general, three basic processes have been employed: (1) pyrolysis of fluorohalogenomethanes (5–7), ethanes (8,9), and ethylenes (10) at high temperatures ($> 500°C$); (2) fluorination of an aromatic hydrocarbon with cobalt trifluoride (11,12) and subsequent dehalogenation of the perfluoroalicyclic intermediates (13–16); and (3) exchange fluorination of highly chlorinated (17–20) or brominated (21-23) aromatics with potassium fluoride with (17,23) or without solvent (24-26). A wide variety of aromatic fluorocarbons and their derivatives are now

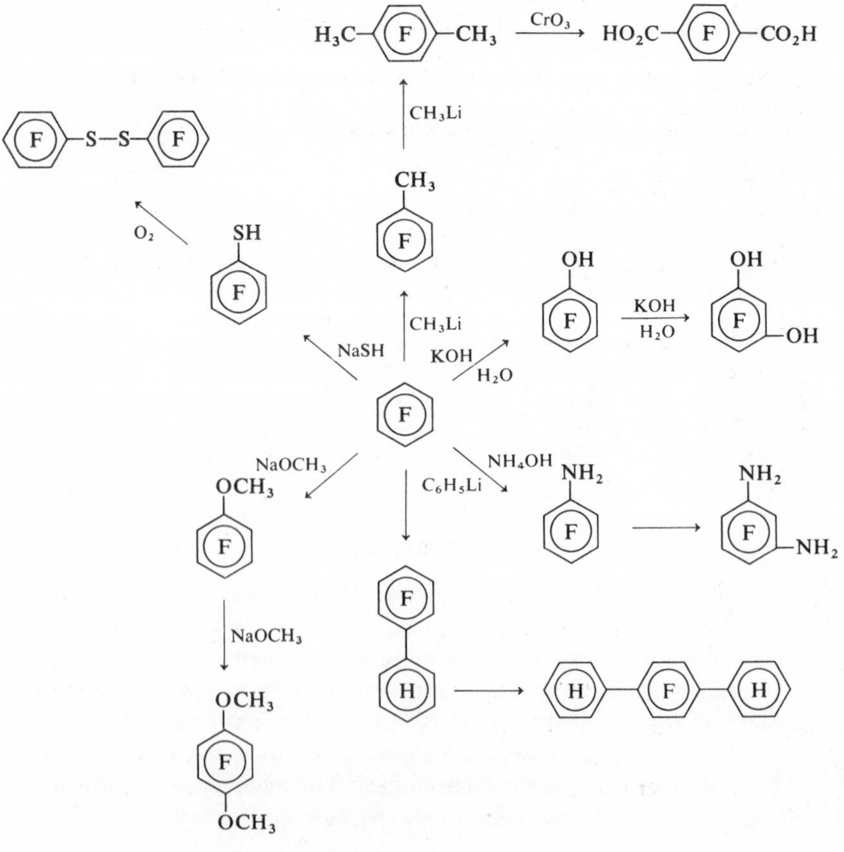

(1)

commercially available. Highly fluorinated derivatives are also available which contain functional groups such as

$$\text{H, Br, Cl, I, OH, NH}_2, \text{CO}_2\text{H}, \ -\overset{\overset{\text{O}}{\|}}{\text{C}}-, \ -\overset{\overset{\text{O}}{\|}}{\text{C}}\text{H, CN, and S.}$$

Difunctional derivatives of these types can also be found.

In general, hexafluorobenzene is attacked by nucleophiles (anions) (12,27), whereas benzene reacts readily with electrophiles (cations). Both species, perfluoro (28,29) and hydro (30,31), are susceptible to reactions with free radicals and undergo radiolysis after exposure to gamma irradiations (32–34). Some replacement reactions (35–38) of hexafluorobenzene by typical nucleophiles are shown in reaction 1.

Disubstitution reactions can also occur, and in some cases isomers are also formed. The position occupied by the second entering group depends to a large extent upon the nature of first functional group (39–41) already present in the molecule, as shown by the outer circle of compounds in scheme 1. A great deal of effort was expended in elucidating the directive effects of various functional groups toward attacking nucleophiles (35–43). This knowledge of the orienting position taken by various nucleophiles in these small molecules is extremely useful as an aid to deduce the structures of polymers prepared from highly fluorinated material via nucleophilic reaction processes (condensation polymers).

Pentafluorobenzene and other monosubstituted derivatives are also valuable intermediates in perfluoroaromatic chemistry. In these compounds, several reaction paths are possible depending upon the conditions employed in the reaction. Advantage can be taken of the normal behavior of the functional group to prepare other monosubstituted derivatives, such as pentafluorostyrene (44–46), perfluorobenzonitrile (47), and perfluorobiphenyl (44,45). Under nucleophilic conditions, replacement of a fluorine atom (35,40,48) occurs and a new difunctional derivative is formed. Using pentafluorobenzene and pentafluorobromobenzene as examples, these reactions are summarized in scheme 2.

Each of these new species can, in turn, be converted to still other perfluoroaryl derivatives. Thus many of these highly fluorinated derivatives can be used in polymer-forming reactions as either a monomer or a comonomer.

The perfluoroaryl polymers discussed in this chapter fall into two structural classes, polyphenyl types and polymers containing an atom or group of atoms between the perfluoroaryl rings. Fluoropolymers, which contain a pendant pentafluorophenyl group attached to an aliphatic side chain, are discussed elsewhere in this book. In most cases, the perfluoroaromatic polymers, except for their fluorine content, are structurally similar to their hydrogenic counterparts, i.e., perfluoropolyphenylenes versus polyphenylene.

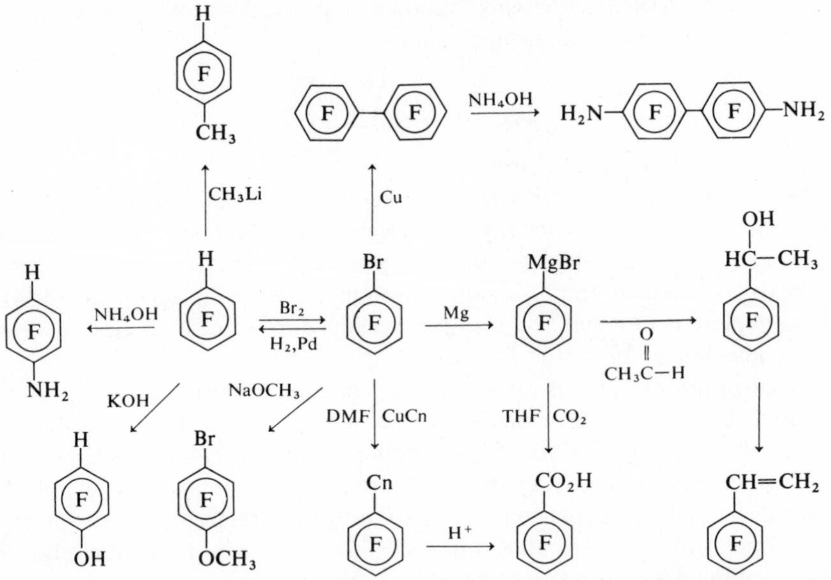

(2)

Therefore, whenever possible, comparisons are made between the fluorinated and hydrocarbon polymers with respect to preparative method, molecular weights, and properties.

III. PERFLUOROPOLYPHENYLS (CLASS I)

In this class of polymers, the fluoroaryl rings are joined together by a single carbon–carbon bond. The bonding sites of the aryl group used most often are the meta (1,3-) and para (1,4-) positions. In some cases, both types of linkage are utilized. The ortho (1,2-) position is seldom employed in polymer formations of this type. Occasionally, an ortho site may be used to introduce a center of dissymmetry (kink) into a predominantly meta or para polymer. The general structures of these polymers are shown in scheme 3.

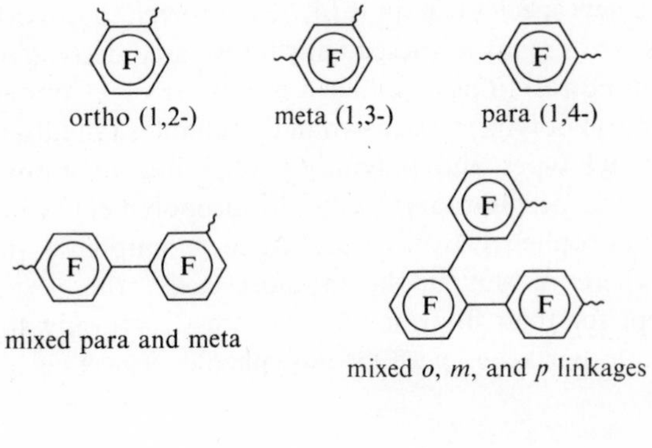

ortho (1,2-)   meta (1,3-)   para (1,4-)

(3)

mixed para and meta

mixed *o*, *m*, and *p* linkages

A. Monomer Synthesis

The compounds generally used as monomers for the preparation of perfluoropolyphenylenes are the mono- and dihalo-fluorobenzenes. Pentafluorobenzene (13,45,49,50) and the

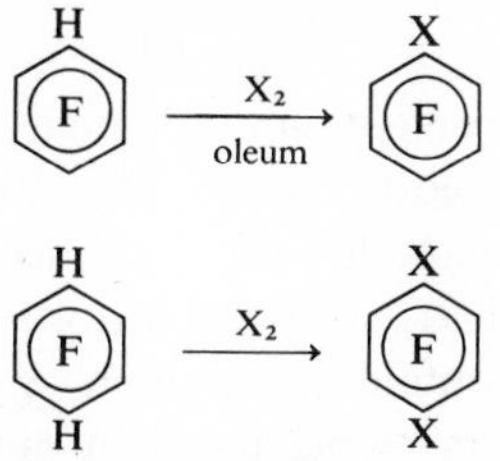

$$(4)$$

$$(X = Br, I)$$

isomeric tetrafluorobenzenes (13,51–53) are easily brominated (53,54) or iodinated (55) in 65% oleum to yield their respective halogenated analogs (Eq. 4). The chlorinated derivatives (55) cannot be prepared in oleum by the use of sulfuryl chloride and the hydrofluorobenzenes. 1,4-Dichlorotetrafluorobenzene is conveniently prepared by the reaction of either chloropentafluorobenzene and sodium chloride (56) in sulfolane at 220°C or from 1,4-dibromotetrafluorobenzene and cuprous chloride (47) in refluxing dimethylformamide. Chloro- and bromopentafluorobenzenes (57,58), along with 1,3-dichloro- and dibromotetrafluorobenzenes, are also obtained as by-products from the many syntheses now available for hexafluorobenzene (18–26).

B. Perfluoro-*p*-Polyphenylenes

1. ULLMANN METHOD

Early attempts to effect the condensation of 1,4-dibromo- or diiodotetrafluorobenzenes with reactive metals or organometallic reagents failed to yield any polymers (55). The reaction with sodium metal in ether or dioxane led to degradation, while complex mixtures were produced by Grignard reagents or organolithium reagents. A coupling technique used successfully for perfluoroalkyl bromides (59) involving zinc and acetic anhydride also failed to yield polymers with either the dibromo- or the more reactive diiodotetrafluorobenzene.

Perfluoro-*p*-polyphenylenes of low molecular weights (55) were formed by an Ullmann reaction in which activated copper powder and the dihalotetrafluorobenzene derivative were heated to 220°C as shown in Eq. 5. The partial solubility of the

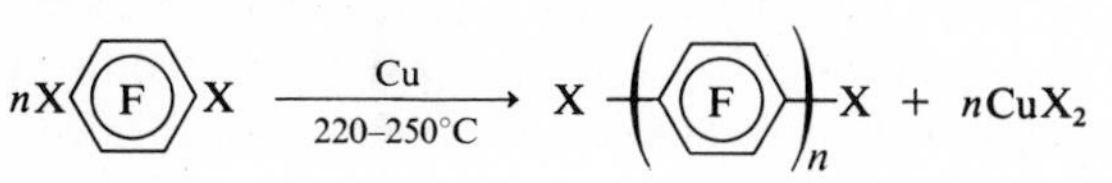

$$(X = Cl, Br, I) \quad (5)$$

products in benzene afforded a means of polymer fractionation. The number average molecular weight of the soluble polymer was 750–1000 ($n = 4$–5), while the insoluble portion showed a number average molecular weight of 1300–1700 ($n = 8$–10). The molecular weights of these polymers were determined by end-group analysis assuming two bromine or iodine atoms per polymer chain.

Recently, perfluoro-*p*-polyphenylenes containing chlorine end groups were prepared from 1,4-dichlorotetrafluorobenzene (56) and activated copper powder in refluxing dimethylformamide (47). The incorporation of a polar solvent in this type of reaction is a useful extension of the work reported earlier. However, these chlorine-containing polymers appear to have the same physical properties as those polymers produced in bulk from the bromine or iodine derivatives.

2. GRIGNARD METHOD

The polymers prepared by the Ullmann method from 1,4-dihalotetrafluorobenzenes all contain labile atoms, such as Cl, Br, or I, as end groups. Atoms of this type are undesirable and often yield weak links, which adversely affect the thermal stability of the polymer. More recently, a procedure was developed for preparing perfluoro-*p*-polyphenylenes in which these labile end-group atoms were reduced considerably and were replaced by the more stable pentafluorophenyl group. In refluxing tetrahydrofuran, the Grignard reagents, pentafluorophenylmagnesium bromide (60–62) and pentafluorophenylmagnesium chloride (63) decomposed to form perfluoro-*p*-polyphenylenes according to scheme 6.

$$n \langle F \rangle - Br \xrightarrow[\text{THF}]{+Mg} \left[\langle F \rangle MgBr \right] \xrightarrow{\Delta} F - \left(\langle F \rangle \right)_n Br + MgBrF \qquad (6)$$

In this reaction, the solvent plays an important role, and to date tetrahydrofuran was the only solvent employed in the polymerization reactions. In ethyl ether, the usual solvent for Grignard reagents, pentafluorophenylmagnesium bromide (44,45) or C_6F_5MgCl (63) forms readily, but propagation to polymer does not occur.

Solutions of pentafluorophenylmagnesium bromide in tetrahydrofuran (64) are quite stable at 25°C, for after 5 days, little or no decomposition of the Grignard reagent was observed. After 8 hr at reflux temperatures, mainly soluble (THF) perfluoropolyphenylenes (60) of high molecular weight (DP = 100, VPO) were produced from pentafluorophenylmagnesium bromide in tetrahydrofuran. Extending the reflux time period to 24 hr, 70% of the Grignard reagent was decomposed (64) and resulted in the formation of insoluble and intractable polymers. Similar polymers were formed spontaneously from pentafluorophenylmagnesium chloride (63) in tetrahydrofuran at 25°C. These polymers still contain residual amounts of halogen other than fluorine, but in smaller quantities ($>5\%$) than was observed ($\sim 18\%$) from the Ullmann method. Hydrolysis of noncoupled Grignard-type chain ends is accompanied by the introduction of hydrogen into the polymer as shown in reaction 7.

$$F \left[\langle F \rangle \right]_n MgBr \xrightarrow{H^+} F \left[\langle F \rangle \right]_n H \qquad (7)$$

Completely fluorinated, low-molecular-weight polyphenylenes (DP = 25) were synthesized by the reaction of pentafluorophenylmagnesium bromide and perfluorobiphenyl (61) in tetrahydrofuran. The addition of perfluoro-biphenyl to

$$n \langle F \rangle MgBr + \langle F \rangle - \langle F \rangle \xrightarrow[\Delta]{THF} \langle F \rangle \left[\langle F \rangle \right]_n \langle F \rangle \qquad (8)$$

the Grignard solution causes the reaction to proceed at room temperature. Perfluorobiphenyl is a more effective initiator and chain-capping reagent than is hexafluorobenzene (due to the more reactive fluorine atoms at the 4,4'- positions of the former compound). When two moles of perfluoro-biphenyl are added to the Grignard solution (C_6F_5MgBr), the chain lengths of the materials formed are severely limited, and result in the formation of oligomers (62).

The isolation and identification of perfluoro derivatives (62) of p-terphenyl, p-quaterphenyl, and p-quinquephenyl strongly suggest that the higher poly-mers (DP = 25) obtained by the addition of perfluorobiphenyl (0.1 mole) are also para linked. On this basis the perfluoro-polyphenylenes prepared from the decomposition of pentafluorophenylmagnesium bromide alone also contain a high percentage of para linkages. This concept is further supported by the failure of the Grignard reagents from 1,4-dibromotetrafluorobenzene and 4-bromotetrafluorobenzene to propagate into polymers (60). In these compounds, the position para to the Grignard groups does not contain fluorine atoms, which appear to be required for the polymerization

$$Br \langle F \rangle MgBr \xrightarrow{\Delta} \text{no propagation} \qquad (9)$$

$$H \langle F \rangle MgBr \xrightarrow{\Delta} \text{no propagation} \qquad (10)$$

reaction at these temperatures. In contrast, both the 1,2- and 1,3-dibromo-tetrafluorobenzenes (65) (compounds which contain a fluorine atom para to the Grignard group) yield deep red, infusible, insoluble materials after refluxing in tetrahydrofuran with magnesium.

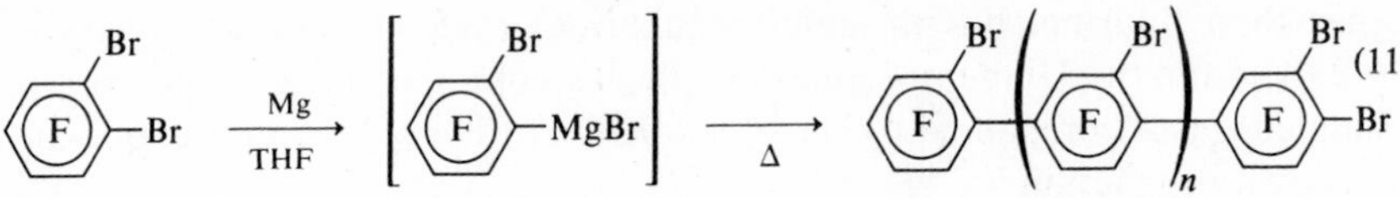

(11)

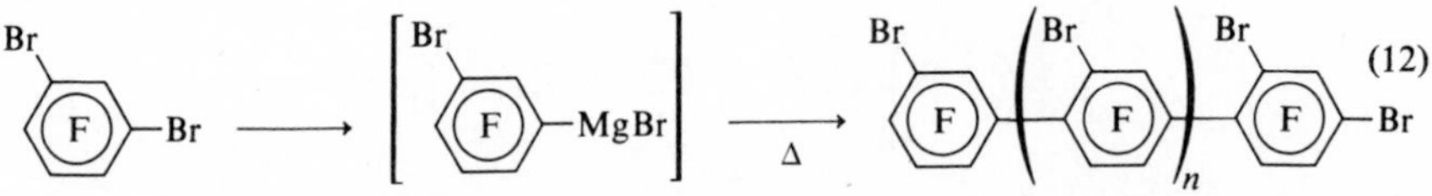

(12)

The near identity of the infrared spectra of these two polymers is readily explicable from an examination of the proposed structures (reactions 11 and 12.) Under the same conditions, 1,3-dichlorotetrafluorobenzene (65) yields a similar insoluble but cream-colored polymer having characteristic infrared bands (1500 cm^{-1}) of the polyphenyl structure.

Although the polymerization reactions of the various fluorinated Grignard reagents appear to be a simple intermolecular nucleophilic displacement of a fluorine atom para to the Grignard group, the mechanism of the reaction is more intricate and complex in nature. The possibility of pentafluorophenyl magnesium bromide or chloride to form tetrafluorobenzyne (66,67), by the intramolecular elimination of MgBrF, was suggested earlier (60) to explain some of the irregularities (e.g., solubility, molecular weight) observed in the perfluoropolyphenylenes. For example, once the tetrafluorobenzyne is formed, many reaction paths become available (scheme 13)

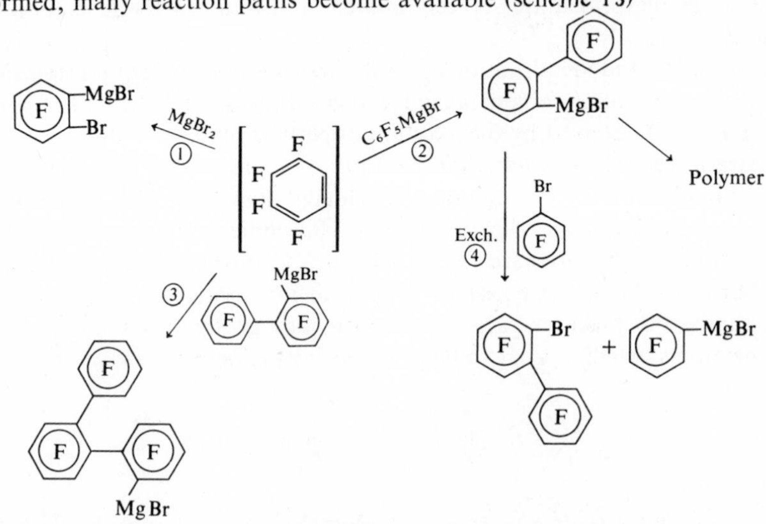

(13)

Metal halides (route 1) or other Grignard reagents (routes 2 and 3) can add to the benzyne intermediate to form new species which are capable of entering into the polymerization reaction. Such intermediates may be responsible for the introduction of other types of linkages into the predominantly para polymers. They may also provide a route for the undesirable distribution of bromine and hydrogen along the polymer chain length.

Perfluoro-*p*-phenylene oligomers (68), which contain a hydrogen atom at one end, were also obtained as by-products from the reactions of pentafluorophenylmagnesium bromide and some group II metallic halides, i.e., tetrachlorides of silicon, tin, and germanium.

3. PERFLUOROARYLLITHIUM REAGENTS

The development of highly fluorinated aryllithium derivatives (69) has added another potent synthetic tool for use in aromatic fluorocarbon chemistry. Thus far, methods for the preparations and reactions of these fluoroaryllithio intermediates have received the most attention of interested investigators (70–79). In general, the fluoroaryllithium compounds are more reactive than the corresponding Grignard reagents (70), but they undergo similar reactions, as shown in Eq. 14.

$$
\begin{array}{c}
\xrightarrow{\text{RMgX}} \quad C_6F_5MgX \quad \xrightarrow{\hspace{2cm}} \\[0.2cm]
\Big\downarrow \textcircled{1}CO_2, \textcircled{2}H^+ \\[0.2cm]
C_6F_5H \qquad\qquad\qquad C_6F_5CO_2H \qquad\qquad (14) \\[0.2cm]
\Big\downarrow \textcircled{1}CO_2, \textcircled{2}H^+ \\[0.2cm]
\text{THF} \xrightarrow{\text{RLi}} \quad C_5F_5Li \quad \xrightarrow{\hspace{2cm}}
\end{array}
$$

Only a few reactions have been reported which deal with the polymerization of pentafluorophenyllithium. Pentafluorophenyllithium decomposes in ethyl ether (69) (above −20°C) and in tetrahydrofuran (71) at 0°C to yield intractable materials. The insolubility of these products made characterization difficult.

Like Grignard reagents, pentafluorophenyllithium can act as a nucleophile or as a source of tetrafluorobenzyne. The predominance of one type of reaction over the other depends considerably upon the reagents and conditions employed in the reactions. Usually, the maintenance of low temperatures, −40° and below, for the preparation and reactions of the fluoroaryllithio intermediate favors the formation of normal reaction products (nonbenzyne types). Thus pentafluorophenyllithium (75) reacts with a reactive substrate, such as perfluorobiphenyl, to form a mixture of low-molecular-weight perfluoropolyphenyls in ether as solvent. This reaction (Eq. 15) is similar to those

$$
\text{F}\!\!-\!\!\text{Li} \;+\; \text{F}\!\!-\!\!\text{F} \longrightarrow \text{F}\!\!\left(\!\!-\!\!\text{F}\!\!-\!\!\right)_{\!\!n}\!\!\text{F} \qquad (n = 1\text{–}3)^{(15)}
$$

reactions reported earlier for the Grignard reagents. Confirmation of the para-linked structure was obtained by comparisons with authentic samples prepared by mixed Ullman reactions. Other polyfluorobiphenyl and terphenyl derivatives were also prepared in this fashion.

The reaction of pentafluorophenyllithium with pentafluorobromobenzene at $-40°C$, beside forming the perfluoropolyphenyl bromides, gave small quantities of 1,2-dibromotetrafluorobenzene, 2-bromononafluorobiphenyl (both products attributed to benzyne intermediates), and 4,H-nonafluoro-biphenyl (a nucleophilic product) in the ratio of $1:1:8$. If the same reaction was performed at $15°C$, the proportions of the compounds arising from the benzyne intermediates were increased and the product ratios were $40:10:1$, respectively. The formation of the polymers at the higher reaction temperatures was negligible, whereas at the lower temperature it was the major product. These reactions are summarized in the following series of Eqs. 16 (nucleophilic) and 17 (benzyne).

Nucleophilic type:

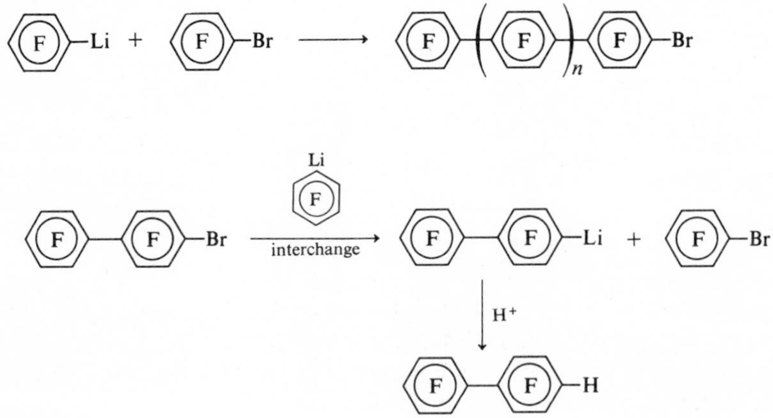

Benzyne type:

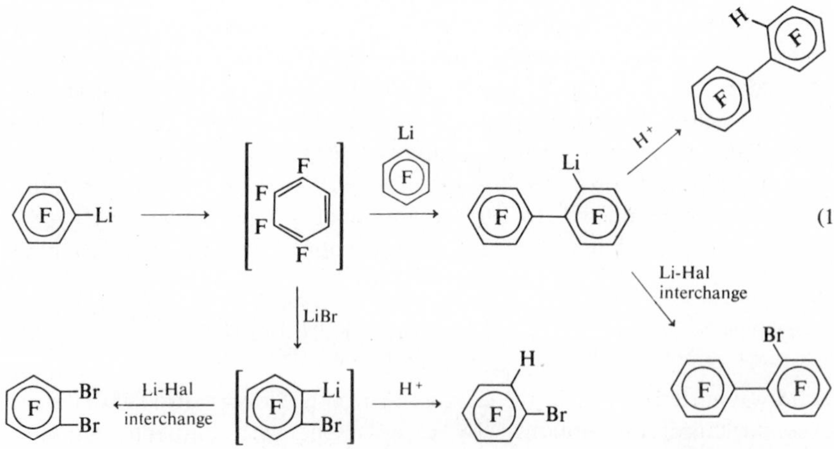

The multiplicity of products possible from the fluoroaryllithio derivatives via the benzyne intermediates often present formidable problems in separation and characterization. The complexity of the problem may be best shown by the reaction scheme (Eq. 18) envisaged for pentafluorophenyllithium (78) at 25°C in ether-hexane mixture.

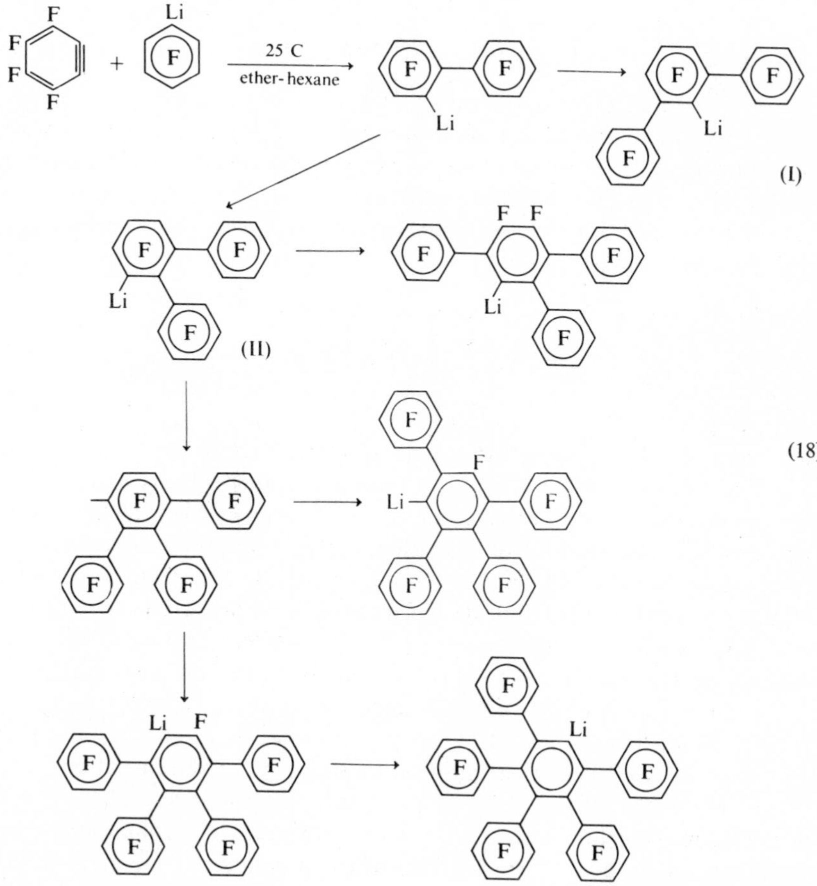

(18)

The reaction products are formed from a series of benzyne formations followed by the addition of pentafluorophenyllithium. When this addition of pentafluorophenyllithium to the intermediate benzyne takes place in such a manner so that the lithio atom is isolated by ortho pentafluorophenyl groups (right side of scheme), benzyne formation by the intramolecular elimination of lithium fluoride is no longer possible. Although these lithio derivatives can lose lithium fluoride by an intermolecular (nucleophilic) process, no perfluoropolyphenylenes were detected. To date, only compounds I and II have been isolated and characterized.

Although the fluoroaryllithium reagents are capable of forming polymers, their main function appears to be as a synthetic tool for the preparation of valuable fluorinated phenyl or biphenyl derivatives, not available by other processes, which can be used to prepare new and novel polymeric systems.

C. Perfluoro-*m*-Polyphenylenes

Pure meta-linked perfluoropolyphenylenes are prepared from 1,3-dichloro- (65,80) or dibromotetrafluorobenzenes (65,81) and copper powder in an Ullmann reaction in bulk (65) or refluxing dimethylformamide (80,81). The bromine derivatives are more reactive than the chloro compounds since the former materials yield higher-molecular-weight polymers at lower temperatures than do the latter materials in the

$$ n \bigcirc \underset{X}{\overset{X}{F}} \quad \xrightarrow{\text{Cu}} \quad \bigcirc\overset{X}{F} \left(\bigcirc F \right)_n \bigcirc\overset{X}{F} \qquad (X = Cl, Br) \qquad (19)$$

bulk reactions. In refluxing dimethylformamide, slightly higher-molecular-weight species are produced from 1,3-dibromotetrafluorobenzene in shorter time periods than was observed from the bulk methods. Like the para polymers produced by the Ullmann reaction, these perfluoro-*m*-polyphenylenes also contain bromine of chlorine end groups (Eq. 19). Unlike the para polymers the meta-linked fluoropolymers are soluble in acetone, tetrahydrofuran (65,80), and hexafluorobenzene (65). Low-molecular-weight polymers, (>2200) are also soluble in ether (80,81) and benzene (80). Thus fractionations and molecular-weight determinations are performed without difficulty.

Some of the conditions used for the preparation of the perfluoro-*m*-polyphenylenes as well as their physical properties are listed in Table 1. The molecular weights (M_n) of some of the fractions range from 875 to 16,000 for the bromofluoropolymers, while those from the chlorofluoropolymer range from 600 to 3400. In general, the softening points of these meta-linked polymers increased as the molecular weights increased, but the softening points were still lower than the para polymers.

Several attempts were made to obtain higher-molecular-weight polymers by conducting the condensation reaction of 1,3-dibromotetrafluorobenzene and copper in decafluorobiphenyl as the solvent (65) at 230°C; only low-molecular-weight species were formed (~4000). No materials of higher molecular weights were formed when the preformed polymers (~4000) were recycled in the process.

TABLE 1.

Polymerization Conditions and Properties of the Perfluoro-m-Polyphenylenes from m-$C_6F_4X_2$

X	Wt. (g)	Cu (g)	Temp. (°C)	Time (hr)	Wt %	Products M.W.	S.P.	Ref.
Cl	20	25	260	15	1	dimer	oil	65
Cl	10	10	{300 / 340}	{96 / 24}	{16 / 10 / 50}	{>600 / 770 / 1,500}	{oil / 65–70 / 125–140}	65
Cl	20	7	350	24	40	dimer	oil	80
Cl[a]	40	40	154	24	{37 / 20}	{3,410 / 2,086}	{240 / 160}	80
Br	30	32	200	15	{41 / 58}	{2,100 / 875}	{150–170 / 55–60}	65
Br	15	20	280	36	{10 / 73 / 10}	{13,600 / 4,800 / >4,000}	{260–280 / 220–235 / 190–208}	65
Br[a]	40	50	Reflux DMF	24	{45 / 46}	{2,200 / 6,000 / 9,000 / 16,000}	{220–240 / >272 / >290 / >330}	81
Br[a,b]	40	50	Reflux DMF	24	94	1,860	150–200	81

[a] In refluxing dimethylformamide, B.P. 152°C (all others in bulk).
[b] Pentafluorobromobenzene (3.58 g, 0.0145 moles) added.

The addition of bromopentafluorobenzene to the reaction of 1,3-dibromo-tetrafluorobenzene and copper powder severely limits the molecular weight of the polymer (81) formed (Table 1, last line) but the polymer still contains traces of bromine. Small molecules, such as perfluoro-m-ter- and quaterphenyl, were also prepared by varying the mole ratios of the reactants in the preceding reaction (81). To date, no high-molecular-weight perfluoro-m-polyphenylene is available containing only carbon and fluorine atoms. The Ullmann method is the only synthetic method suitable at present by which perfluoro-m-polyphenylenes of relatively high molecular weights can be prepared in good yields.

D. Perfluoropolyphenylenes Containing Meta and Para Linkages

Polymers were prepared by two different methods which presumably contained both meta and para linkages in varying quantities. In the first procedure, mixtures of 1,3- and 1,4-dichlorotetrafluorobenzenes were heated with copper powder in dimethylformamide in a mixed Ullmann reaction as seen in Eq. 20 (82).

$$Cl\text{—}(F)\text{—}Cl + (F)_{Cl}^{Cl} \xrightarrow[\text{DMF}]{\text{Cu}} Cl\text{—}\left(\!(F)\text{—}(F)\!\right)_n^{Cl} \tag{20}$$

When the weight ratio of 1,4-dichlorotetrafluorobenzene to 1,3-dichlorotetrafluorobenzene is equal to 1, as assumed in Eq. 20, 62% of the bulk polymer is soluble in benzene ($M_n \sim 5000$, M.P. 250-2°C) and 37% of the polymer is insoluble ($M_n \sim 2400$, M.P. >350°C). As the weight ratio of 1,4-/1,3- increases, the melting point of the bulk polymer also increases and approaches the melting points of the pure para polymers.

In a similar reaction, some perfluoropolyphenylenes of low molecular weight and having mixed linkages were prepared from chloropentafluorobenzene and a mixture of dichlorotetrafluorobenzene isomers (o-30%, m-43%, p-27%) after heating with copper powder at 352°C (9).

In the foregoing reactions, cross-coupling of unlike monomer units would be the ideal reaction course to produce polymers that have alternating meta and para linkages. As in most mixed Ullmann reactions, self-condensations of like units may also occur before cross-coupling reactions are realized. Depending upon the reactivity of each monomer unit, the actual polymer may contain segments in which similar monomer units are linked together as shown in Eq. 21. Thus the

$$(F)_F^{Cl}\text{—}\!\left(\!(F)\text{—}(F)\!\right)_x\!\left(\!(F)\text{—}(F)\text{—}(F)\!\right)_y\!(F)\text{—}(F)\text{—}Cl \tag{21}$$

properties of the polymer may depend upon the quantity and sequential ratio (x/y) of each monomer unit finally incorporated into the polymer chain length.

Perfluoropolyphenylenes having mixed meta and para linkages were also prepared by the coupling (27,44,45) of the Grignard reagent, 3-bromotetrafluorophenylmagnesium bromide (65), with perfluorobiphenyl in tetrahydrofuran (83) as solvent. Since the fluorine atoms in the 4,4'-positions of perfluorobiphenyl are susceptible to attack by nucleophiles, perfluorobiphenyl may serve as both an initiator (60–62) and comonomer. However, in order to prevent the self-condensation of the Grignard reagent, 3-bromotetrafluorophenylmagnesium bromide (as shown in Eqs. 11 and 12), perfluorobiphenyl was placed in the system before the initiation of the Grignard reagent. In this reaction, the solubility of the polymers formed depends upon the mole ratio of the reactants as shown in Table 2. When

TABLE 2

Reaction Conditions and Properties of Meta-Para Polymers from
3-Bromotetrafluorophenylmagnesium Bromide and Perfluorobiphenyl (65)

$m - C_6F_4Br_2$ (mol)	$(C_6F_5)_2$ (mol)	Mg (g-atom)	Sol.	Time (hr)	Products
0.05	0.06	0.055	Et$_2$O and THF	24	low conversion $C_6F_4H_2$, C_6F_4BrH $C_6F_5-C_6F_4-C_6F_4H$
0.05	0.1	0.055	THF	40	soluble polymers M.W. 1400 S.P. 205–215
0.2	0.1	0.25	THF	1	insoluble gel S.P. > 300°C

the mole ratio of the Grignard reagent to perfluorobiphenyl was 1:2, a soluble low-molecular-weight polymer ($M_n \sim 1400$) was obtained after 40 hr of reflux time. An insoluble gel was formed in less than 1 hr at reflux temperatures when the mole ratios were reversed. In ether as the solvent, the preparations of the Grignard reagents (60,65,69) from the dihalotetrafluoro-benzenes are sluggish even after 15 hr of reflux times. Upon the addition of tetrahydrofuran to the reaction medium, the Grignard reagent will form, but simple coupling reactions rather than polymerization is the main path in the presence of perfluorobiphenyl and mixed solvents. The isolation and identification of the coupling product, 3,H-perfluoroterphenyl, suggests that the structure of these polymers may be as shown in Eq. 22.

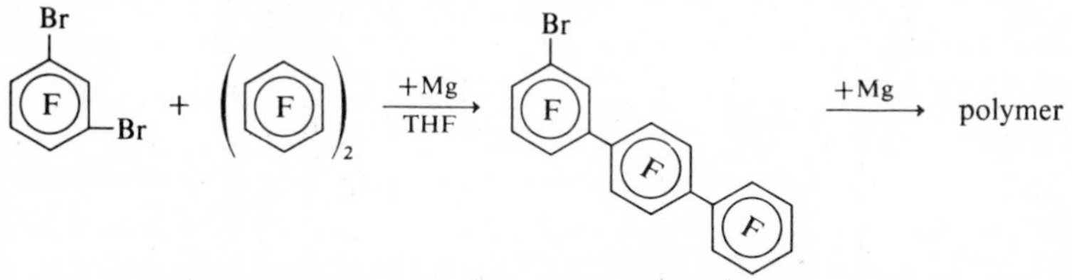

$$(22)$$

For perfluoropolyphenylenes of comparable molecular weights ($M_n \sim 1500$), the softening points (205–215°C) of the mixed meta- and para-linked poly-mers are intermediate between the pure para- (>300°C) and meta-linked polymers (S.P. 150–170°C).

TABLE 3

Polymerization of 2-Chloroheptafluorotoluene with Magnesium in Tetrahydrofuran (65)

Exp. no.	C$_6$F$_4$ClCF$_3$ g	C$_6$F$_4$ClCF$_3$ moles	Mg g	Mg g-atom	Solvent (ml)	Preparation temp. (°C)	Reflux time (hr)	Crude polymer yield (g)	Purified fractions wt %	Purified fractions M.W.·t$_d$	Purified fractions [η]^f
1	25.3	0.1	2.4	0.1	100	0	4[a]	23.2	{20 / 16	35,600[e]	— / —
2	5	0.02	0.75	0.03	30	0	19[b]	4.0	{15 / 13	88,000[e] / 20,000	0.1979 / 0.0842
3	5	0.02	0.75	0.03	100	−30	19[b]	4.0	27.5	—	0.0825
4	5	0.02	0.6	0.025	100	0	120[c]	2.5	29.2	—	0.32

[a] 1 ml of C$_6$F$_6$ added 1 hr before termination.
[b] 0.1 g decafluorobiphenyl added 2 hr before termination.
[c] 0.1 g anhydrous cobalt chloride added 48 hr before termination and 0.1 g decafluorobiphenyl added 24 hr before termination.
[d] By vapor pressure osmometry.
[e] Contains 4.8% Cl.
[f] In tetrahydrofuran.

E. Perfluoropolytolylenes

Perfluoropolytolylene (65,84) is a polyphenyl type polymer with a pendant trifluoromethyl group. These polymers were prepared from the decomposition of the Grignard reagent of 2-chloroheptafluorotoluene in tetrahydrofuran as the solvent (Eq. 23).

The reaction performed in tetrahydrofuran proceeds in the same fashion as for chloropentafluorobenzene (63) in that only polymers were isolated from attempts to carbonate the Grignard reagent at low temperatures. The conditions and the results of the polymerization of 2-chloroheptafluorotoluene with magnesium are shown in Table 3. The perfluoropolytolylenes all contain about 5% of residual chlorine. The capping of chain ends by the addition of hexafluorobenzene or perfluorobiphenyl, which proved effective in reducing the bromine and hydrogen content of the perfluoro-p-polyphenylenes, had little effect on the chlorine content of the perfluoropolytolylenes. In general, the conversion to polymer by this method was 70% or better. Molecular weights of the polymers increased with longer reaction times.

Some of the physical properties of the perfluoropolytolylenes are listed in Table 4. The most striking differences between the perfluoropolytolylenes and the perfluoropolyphenylenes are the increased molecular weights and the greater solubility of the former polymers. Only the crude bulk and low-molecular-weight polymers ($M_n > 20,000$) soften below 300°C. Most of the polymers were brown powders or glasses. The crude polymers were fractionated by using acetone or tetrahydrofuran as the solvent and benzene as the precipitant. The highest-molecular-weight fraction found ($\eta = 0.32$) was soluble in acetone, tetrahydrofuran, and hot xylene but insoluble in hot benzene.

Most of the perfluoropolytolylene polymers when heated to 300°C in either air or a vacuum for a short period of time (~5 min) lose some of their solubility in tetrahydrofuran. The extent of insolubility depends upon the molecular weight of the polymer as shown in Table 4. The low-molecular-weight sample remains soluble, but the high-molecular-weight fraction is completely insoluble. Many other types of aromatic polymers, whether fluorinated (85, 86) or not (87,88), have shown this behavior.

The structures of the perfluoropolytolylene polymers are more difficult to determine than are the perfluoropolyphenylenes. Due to the presence of the trifluoromethyl group, numerous reactive sites become available during

TABLE 4

Properties of Perfluoropolytolylene (65)

M.W. (VPO)	Color appearance	S.P.	Other solubility[a]	% Sol[b]
Bulk	Brown powder	170–265	Most organic	—
5,100	Brown powder	235–255	Most organic	75
20,000–36,000	Cream to buff powders	>300	Most organic	25
88,000	Dark brown glass	>300	C_6F_6, hot benzene	>5
200,000[c]	Dark brown glass	Some softening ~310	C_6F_6 (partial) hot xylene insol. benzene	0

[a] All fractions soluble in acetone and tetrahydrofuran (THF).
[b] Solubility of small samples in THF after being heated in air at 300°C for 5 min.
[c] Extrapolated mol wt from plot of log $[\eta]$ versus log $[M_n]$ for other polymers.

polymerization. The choice between multiple reaction paths becomes speculative at best in the absence of experimental evidence. Infrared spectroscopy was of little value for structure determinations of the perfluoropolytolylenes, since the spectra of the low- and high-molecular-weight fractions as well as the thermally cross-linked samples were all identical. In a similar attempt to determine structure, the infrared spectrum of a perfluoropolytolylene was compared with the spectrum of perfluoropolybenzylene (89,90). The monomers of these polymers are structure isomers. Although

$$\sim CF_3-\left(\!\!\!\!\bigcirc\!\!\!\!\!\!\!\!F\right)\!\!-M^+ \xrightarrow{\text{THF}} \left\{\!CF_2-\left(\!\!\!\!\bigcirc\!\!\!\!\!\!\!\!F\right)\!\!\right\}_n\!\!\sim \qquad (M = \text{Li or MgCl}) \quad (24)$$

the spectra of the two polymers were slightly different, the interpretation of these differences due to the CF_2 and CF_3 bonds (91) were inconclusive.

The high molecular weights obtained for the perfluoropolytolylenes and their solubility behavior also suggest that these polymers are highly branched and probably contain sequences of meta and para linkages along the polymer chain length.

More recently, 3-chloroheptafluorotoluene and 3,5-dichlorohexafluorotoluene (92) were prepared by exchange methods, but no polymers from these compounds were reported at this time.

F. Thermal Stability of Perfluoropolyphenylenes and Perfluoropolytoiylenes

The thermal stabilities of the perfluoropolyphenylenes were determined by thermogravimetric analyses. The TGA curves (86) resulting from the thermal decompositions of several samples of perfluoro-p-polyphenylenes are shown in Fig. 1. Sample A ($DP_n = 100$) contains carbon, fluorine, and 2.6%

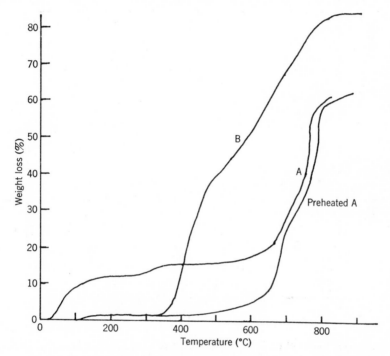

Fig. 1.   TGA curves on various perfluoropolyphenylenes (86). (Reprinted by permission of copyright owner, The Am. Chem. Soc.).

bromine, while sample B ($DP_n = 25$) contains only carbon and fluorine, but the molecular weight of this polymer is significantly lower than that of polymer A. A portion of sample A was also pretreated at 400°C for 1 hr before thermal stability determinations were performed and the difference in thermal stability between sample A and the pretreated (heat) sample A is also shown in Fig. 1.

The TGA curves (81) for perfluoropoly-m-phenylene ($M_n = 16,000$) are shown in Fig. 2. The results show that there is little difference in the thermal stabilities between meta- and para-linked polymers of comparable molecular weight.

The results, from the TGA curves in Figs. 1 and 2, show that the perfluoro-

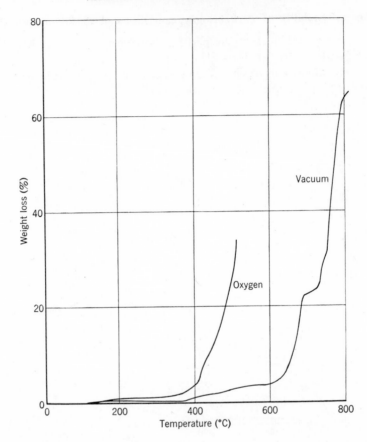

Fig. 2. Loss in weight of perfluoro-*m*-polyphenylene in vacuum and in oxygen (200 mm) (81). (Reprinted by permission of the copyright owner The Am. Chem. Soc.).

polyphenylenes have good thermal stability in a vacuum up to 600°C. In an oxygen atmosphere (Fig. 2) decomposition begins below 400°C and increases rapidly above this temperature. Similar results were reported for perfluoro-*p*-polyphenylene in air (86). In contrast, the hydrocarbon analog, poly-*p*-phenylene (93), is markedly more resistant to oxidation at temperatures up to 500°C. However, if the perfluoropoly-*p*-phenylene, as in sample A, is pretreated at 400°C for 1 hr (as noted above, Fig. 1) the oxidative stability increases considerably and compares favorably with the stability of poly-*p*-phenylene in air (86).

The thermal stability of the perfluoropolytolylenes (84) is essentially the same as for the perfluoropolyphenylenes in a vacuum up to 600°C. The presence of the trifluoromethyl group on the polymer chain does not adversely affect the thermal properties of the polymers.

The mechanism(s) by which these perfluoropolyphenylenes degrade thermally is not clearly understood at present. The evidence indicates that long perfluorophenylene chains do not degrade markedly into smaller volatile compounds, but instead, the residue becomes carbonized (86).

IV. PERFLUOROARYL POLYMERS CONTAINING A HETEROATOM IN THE POLYMER CHAIN (CLASS II)

The syntheses of perfluoroaryl polymers, as well as the required monomers, which contain an atom or group of atoms (X) between the fluorinated aryl rings are described in this section. The role of the heteroatom was to intro-

$$-\left[\bigotimes_{F}\right]-X-\left[\bigotimes_{F}\right]-X-\left[\bigotimes_{F}\right]- \quad (X = O, S, CF_2O, CO_2-, -\overset{R}{\underset{}{N}}- \text{ etc.})^. \quad (25)$$

duce some flexibility into the rigid polyphenyl structure. Since these heteroatoms all form relatively strong chemical bonds with carbon and have bond angles different from 180°, the derived polymers were expected to have good thermal stability and lower softening points than the polyphenyl types. Some of the polymers in this group may also possess the desirable property of having elastomeric (rubberlike) characteristics if polymers of sufficiently high molecular weight could be synthesized.

A. Perfluoropolyphenylene Ethers

1. MONOMER SYNTHESIS

Most of the synthetic preparations of perfluoropolyphenylene ethers involve the use of pentafluorophenol or its salts. The syntheses, in many ways, are similar to those used for the preparations of hydrocarbon (94) as well as some partially fluorinated polyaryl ethers (95–97) in an Ullmann reaction.

Pentafluorophenol is usually prepared directly from hexafluorobenzene by the nucleophilic replacement of a fluorine atom by hydroxide ion (Eq. 26) in such solvents as pyridine (27,98), t-butyl alcohol (99), and water (closed pressure vessel) (35). Demethylation reactions of pentafluoroanisole by

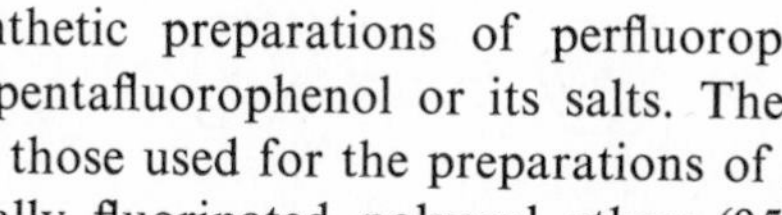

$$\bigotimes_{F} \xrightarrow[\underset{②H^+}{Solvent}]{①KOH} \bigotimes_{F}-OH + KF \qquad (26)$$

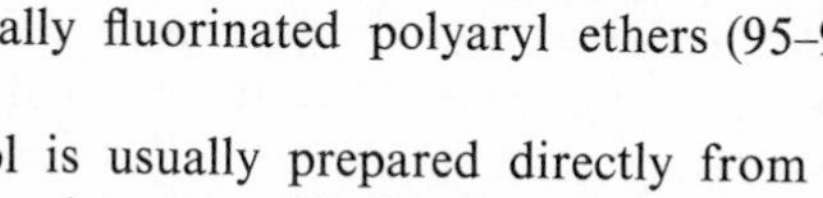

$$\left(\bigotimes_{F}\right)-OCH_3 \xrightarrow[②H^+]{①AlCl_3} \left(\bigotimes_{F}\right)-OH \qquad (27)$$

both acidic reagents (HBr, HI, or $AlCl_3$) and aqueous nitrogen bases (39) such as ammonia or hydrazine hydrate also result in the formation of pentafluorophenol (Eq. 27). In the absence of bases, hexafluorobenzene is extremely resistant to hydrolysis up to 300°C in a closed pressure vessel (85).

The three isomeric fluorinated dihydroxy derivatives are also potentially useful for the preparation of polyfluoroaryl ethers. These compounds were prepared; 3,4,5,6-tetrafluorocatechol (109) (Eq. 28), 2,4,5,6-tetrafluoro-

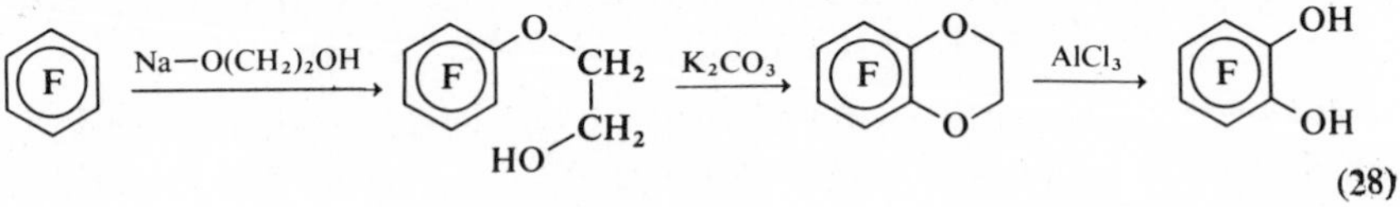

$$\text{(28)}$$

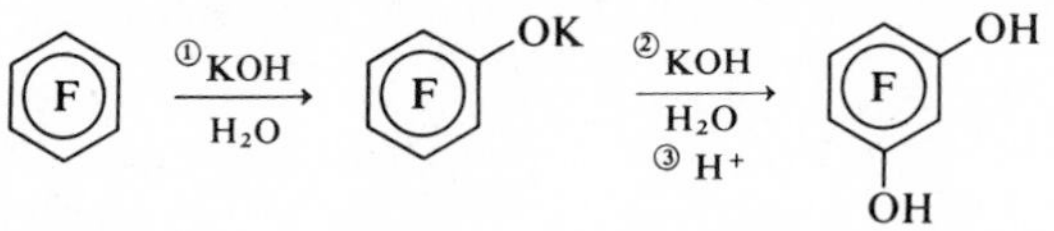

$$\text{(29)}$$

resorcinol (35,85) (Eq. 29), and 2,3,5,6-tetrafluorohydroquinone (24,101) (Eq. 30).

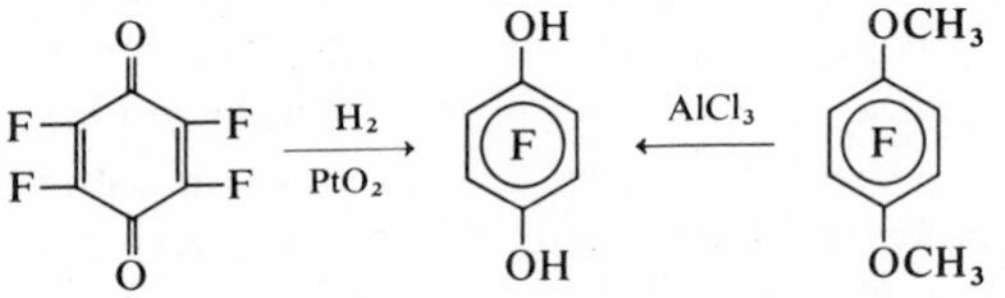

$$\text{(30)}$$

The simple perfluorophenyl ether (35,85) was first prepared in a low-yield (16%) reaction between hexafluorobenzene and potassium pentafluorophenolate in refluxing dimethylformamide (Eq. 31). More recently, excellent

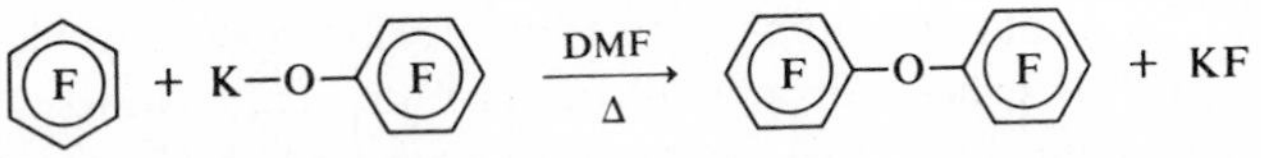

$$\text{(31)}$$

yields of perfluorophenyl ether and, in general, other perfluoroaryl ether derivatives (102–104) have been realized by the use of dimethylacetamide as the reaction medium rather than the lower-boiling solvents (Eq. 32). The perfluoroaryl ether derivatives contain

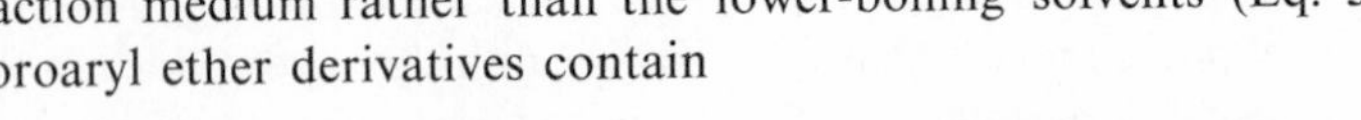

$$\text{(32)}$$

(X and R = Br, Cl, CN, etc.)

reactive Br. CN, CO_2H, and chlorine (105) end groups in various positions which can be utilized to prepare other polymeric systems such as polyetheresters. A number of hydrofluoroaryl ethers (106), having 2–6 rings, were also synthesized. These were used as model compounds for thermal degradation studies.

2. POLYMERIZATION OF FLUOROPHENOLS

Early attempts to prepare perfluoropolyphenylene ethers involved the thermal decomposition of sodium or silver pentafluorophenoxide in a vacuum at 360°C (107–109). Although small quantities of polymer ($M_n \sim 1500$) were formed in this uncontrolled decomposition reaction, the main product was the simple ortho dimer (Eq. 33). Perfluorophenylene ethers (85) of similar

$$n \left[\begin{array}{c} F \end{array} \right] \text{—ONa} \quad \xrightarrow[0.1 \text{ mm}]{360\,°C} \quad \left(F \right) \overset{O}{\underset{O}{\bigcirc}} \left(F \right) + \left(C_6 F_4 O \right)_n \qquad (33)$$

molecular weights ($M_n \sim 1700$) were synthesized under less drastic conditions by the reaction of potassium pentafluorophenoxide in excess pentafluorophenol. The pentafluorophenol served as both the substrate and reaction solvent. The perfluoropolyethers obtained by this method in pentafluoro-

$$n \left(F \right) \text{—OK} + \left(F \right) \text{—OH} \quad \xrightarrow[150°C]{\Delta} \quad F \left[\left(F \right) \text{—O} \right]_n \left(F \right) \text{—OH} \qquad (34)$$

phenol all contain a hydroxyl end group. Significantly higher-molecular-weight materials ($M_n \sim 4300$ and 6500) were realized by heating the low-molecular-weight species (~ 1700) in the form of potassium salts in a melt polymerization process (110). Polymer I is formed first and by recycling of polymer I in the melt process, polymer II is obtained.

$$F \left[\left(F \right) \text{—O} \right]_n \left(F \right) \text{—OK} \quad \xrightarrow{\Delta} \quad \text{polymer I} \quad \xrightarrow{\Delta} \quad \text{polymer II} \qquad (35)$$

$$\text{MW 1700} \qquad\qquad\qquad \text{MW 4300} \qquad \text{MW 6500}$$

A perfluoropolyphenylene ether of high molecular weight ($M_n \sim 12{,}500$) was also prepared directly from excess potassium pentafluorophenoxide and pentafluorophenol. The stoichiometry is the reverse of those quantities which gave the 1700-MW polymer. The resulting perfluoropolyphenylene ether can be cross-linked, as can other polyaryl ethers (87,88), simply by heating at 260°C for 18 hr. The cross-linked polymer is a hard, brittle, insoluble glass

$$F \left[\left(F \right) \text{—O} \right]_n \left(F \right) \text{—OK} \quad \xrightarrow{\Delta} \quad \text{X-linked polymer} \qquad (36)$$

$$\text{MW 12,500}$$

at 25°C, but it softens at 90°C and exhibits rudimetry elastomeric characteristics up to 300°C.

The physical properties and the solubility of these perfluoropolyphenylene ethers (PPPE) are shown in Table 5. In general, the softening points increase

TABLE 5

Some Physical Properties of PPPE

M.W.	Appearance	S.P. (°C)	Sol.	Insol.
1,700	Amber gum	40–60	Most organic	H_2O
3,700–6,500	Amber, brittle glass	60–80	Ether, CCl_3H, benzene	CH_3OH, H_2O
12,500	Brown, brittle glass	80–100	CCl_3H, benzene	Ether, CH_3OH
×-link	Brown, brittle glass	90[a]	None	

[a] Temperature at which rubbery behavior appears.

as the molecular weights increase. Also, a change in solubility of these polymers in various solvents accompanies this increase in molecular weights.

The spectral characteristics exhibited by these fluorinated polyaryl ethers are shown in Table 6. The perfluoroaryl ether band for these compounds (C to G) appears at 1170 cm^{-1}, shifted considerably from the usual 1270–1230 cm^{-1} region for nonfluorinated aryl ethers (111). The presence of the hydroxyl groups in these polymers was determined from the infrared spectra by the presence of two bands, one at 3570 cm^{-1} (35) and the other at 1235 cm^{-1}. These bands disappear when the hydroxyl groups are converted to salts.

3. THERMAL STABILITY OF THE PERFLUOROPOLYPHENYLENE ETHERS

The thermal stabilities of the perfluoropolyphenylene ethers are shown in Fig. 3. Various molecular weight fractions were heated isothermally for $\frac{1}{2}$ hr at 400°C. The cross-linked (insoluble) polymer is also included, even though its molecular weight is not known. The leveling off of the decomposition (at approximately 50%) curve as the molecular weights increased is interpreted as the breaking of the fluorinated phenyl-to-oxygen linkages. Thus the thermal stability of these perfluoropolyphenylene ethers is not exceptional at 400°C. This is in accord with data recently reported for the thermal stabilities of some perfluorinated and partially fluorinated, small, aryl ether molecules (106). In all cases, the fluorinated species had lower thermal decomposition points than the corresponding hydrogenic analog. The mixture of hydrogen and fluorine atoms distributed along the aromatic polymer chain is not conducive

TABLE 6

Infrared Spectral Bands of Some Fluoroaromatic Ethers and Phenols in the 1300 cm^{-1} to 1100 cm^{-1} Region

Compounds	C_6F_5—O—R	C_6H_5—O—R	OH	CF	Misc.
			Bands (cm^{-1})		
		Phenols			
A. C_6F_5OH			1,242(S)	1,330(S)	1,105(M)
B. $C_6F_4(OH)_2$			1,235(S)	1,300(S)	1,138(M)
C. $C_6F_5OC_6F_4OH$	1,171(S)		1,237(S)	1,315(S)	1,116(M)
D. $C_6F_5O(C_6F_4O)_xC_6F_4OH$	1,170(VS)		1,237(S)	1,315(S)	1,117(M)
		Ethers			
E. $C_6F_5OC_6F_5$	1,170(S)			1,310(S)	1,125(W)
F. $C_6F_5OC_6H_5$	1,169(S)	1,205(VS)		1,315(S)	1,130(W)
G. $(C_6H_5O)_2C_6F_4$	1,168(S)	1,195(VS)		1,310(S)	1,110(W)
		Fluorocarbons			
H. C_6F_6				1,300(M)	1,157(M)
I. $(C_6F_5)_2$				1,292(M)	1,150(M)

S = strong.   M = medium.   W = weak.

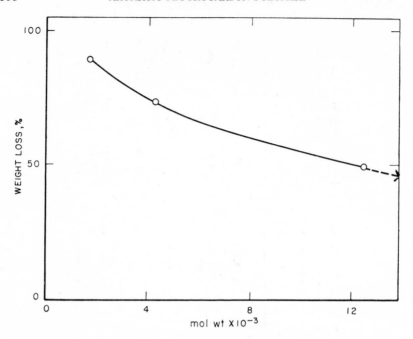

Fig. 3. Pyrolysis of poly(perfluorophenylene ethers) percentage weight loss upon heating for 1/2 hr at 400°C in a vacuum (85).

to high thermal resistance. Hydrogen fluoride is easily stripped from the polymer by thermal excitation. Although model compounds are useful entities for the study of the relationship(s) that may exist between structure and thermal stability, certain useful data such as elasticity, glass transition temperatures, and film-forming properties can be assessed only from the polymers. In this respect, the perfluoropolyphenylene ethers have began to show rubberlike characteristic and may find application where temperatures do not exceed 400°C.

The thermal stabilities of a wide range of polyphenylene ethers were determined by thermogravimetric analyses (112). The TGA curves resulting from the decomposition of various polyaryl ethers are shown in Figs. 4 and 5. It can be seen that the thermal stability tends to decrease as the degree of substitution in the aromatic nuclei increases. Thus the order of stability in these polymers is polyphenylene oxide > monosubstituted > disubstituted > trisubstituted > tetrasubstituted polyphenylene oxides. Hence the perfluoropolyphenylene ethers, based on the results in Fig. 3, appear to fall into this latter category of tetrasubstituted polymers.

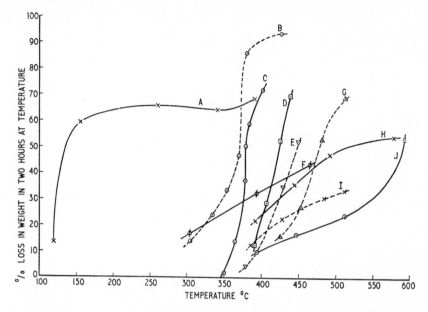

Fig. 4.   Relative thermal stabilities of poly(phenylene oxides) prepared by decomposition of metal halogenophenoxides: (A) poly(p-2,5-dichlorophenylene oxide); (B) poly(p-2,3,5,6-tetrachlorophenylene   oxide);   (C)   poly(p-2,3,5,6-tetrabromophenylene   oxide);   (D) poly(p-2,3,6-trichlorophenylene oxide); (E) poly(p-2,6-dibromophenylene oxide); (F) poly(p,p-diphenylene oxide); (G) poly(p-2,6-dichlorophenylene oxide); (H) poly(p-2-bromophenylene oxide); (I) poly(p-phenylene oxide); (J) poly(p-2-chlorophenylene oxide) (112).

B. Other Fluoroaryl Ether Polymers

A perfluoropolyphenylene ether derivative was recently reported from the decomposition of p-pentafluorophenoxytetrafluorophenyllithium or the corresponding Grignard reagent (113).

$$\langle F \rangle{-}O{-}\langle F \rangle{-}M^+ \xrightarrow[\substack{65°C \\ 3\ hr}]{THF} \text{polymer}\quad (M = Li\ or\ MgBr) \qquad (37)$$

$$MW\ 4000$$
$$S.\ P.\ 120°C$$

At 25°C, in refluxing tetrahydrofuran as the solvent, only polymer was formed. These fluoroorganometallic intermediates are also capable of undergoing benzyne formation, similar to those already mentioned for pentafluorophenyllithium. The polymers obtained from the decomposition of p-pentafluorophenoxytetrafluorophenylmagnesium bromide were described as glassy and tractable. The softening point for this polymer was higher than

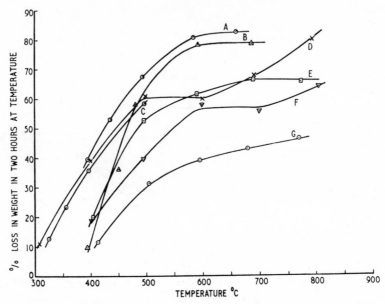

Fig. 5. Relative thermal stabilities of polyphenylene oxides prepared by oxidation of halogenophenols or Ullmann condensations: (A) poly[(*p*-2,6-dibromophenylene oxide) (*o*-2,6-dibromophenylene oxide)]; (B) poly*o*-[6-bromo-4-*p*-chlorophenylphenylene oxide]; (C) poly[*o*-4-fluorophenylene oxide)(*p*-phenylene oxide)]; (D) poly(*o*-6-bromo-4-phenylphenylene oxide); (E) poly(*p*-phenylene oxide); (F) poly[(*o*-4-bromophenylene oxide) (*p*-phenylene oxide)]; (G) poly(*m*-phenylene oxide) (112).

that of a comparable molecular weight fraction of perfluoropolyphenylene ether (~ 4300).

The thermal decomposition of potassium 4-chlorotetrafluorophenoxide (105) in benzophenone in the presence of a small amount of iodine yields a polyether which still contains chlorine in the repeating polymer unit as shown in Eq. 38.

$$Cl\text{---}\underset{}{\underset{}{\bigcirc}}\!\!(F)\text{---}OK \xrightarrow[\substack{\Phi_2C=O \\ \Delta}]{I_2} \text{+}C_6ClF_3O\text{+}_n \qquad (38)$$

The polymer ($M_n \sim 1000$ to 2500) was described as a brown powder (M.P. 180–190°C). Since chlorine was still in the polymer, the polymerization apparently occurred at positions other than para. This is in contrast to the failure of 4-bromotetrafluorophenylmagnesium bromide (60) to propagate into polymer at other positions.

Several other types of fluorinated polyaryl ethers were reported recently. In one case, the unstable tetrafluoro-*o*-quinone (114) simply polymerizes to a high-melting solid (M.P. $>200°C$) on repetitive recrystallization. No structure was assigned to the solid polymer.

In the other case, a polymer was formed when 2,5-dihydroxyoctafluoro-

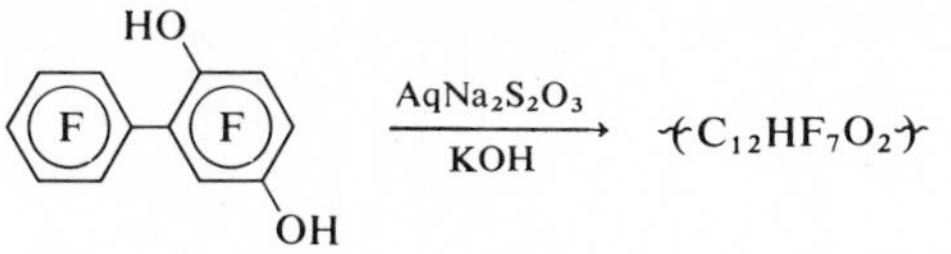

$$\xrightarrow[\text{KOH}]{\text{AqNa}_2\text{S}_2\text{O}_3} \quad -(\text{C}_{12}\text{HF}_7\text{O}_2)- \qquad (39)$$

biphenyl (115) was treated with an aqueous solution of sodium thiosulfate in the presence of potassium hydroxide.

C. Perfluorobenzylene Oxide Polymers

1. MONOMER SYNTHESIS

Several perfluoropolybenzylene ethers were prepared from heptafluoro-*p*-cresol. The preparation of the monomer requires special attention. The usual treatment of octafluorotoluene (116) with excess base (OH⁻) leads to a complex mixture of solids and no heptafluoro-*p*-cresol is formed. Several methods are now available by which heptafluoro-*p*-cresol is synthesized in excellent yields (70%). These method are summarized in reaction 40.

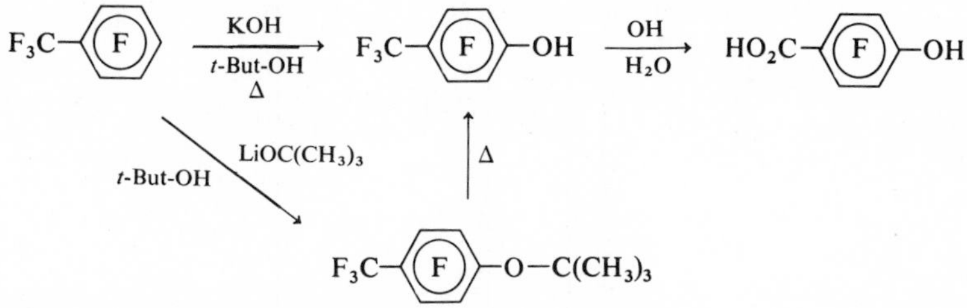

$$(40)$$

In the first method (117) the heptafluoro-*p*-cresol can be prepared directly from octafluorotoluene by excess potassium hydroxide in tertiary butanol at $20°C$ or by use of only stoichiometric amounts of potassium hydroxide in *t*-butanol at reflux temperatures ($83°C$) for a short period of time. In the second method the *t*-butyl ether of heptafluoro-*p*-cresol (118) is first prepared and isolated. Only the para isomer is obtained in this reaction because of the bulky *t*-butyl group. The *t*-butyl heptafluoro-*p*-cresyl ether can be pyrolyzed in either the liquid or vapor state to give excellent yields of pure heptafluoro-*p*-cresol and gaseous isobutylene.

2. POLYMERIZATION AND THERMAL STABILITY

The relative ease by which the heptafluoro-*p*-cresol can be polymerized under mild, anionic conditions is shown in Table 7. In contrast, in aqueous

TABLE 7

Polymerization Conditions of Heptafluoro-*p*-Cresol

Reagent	Temp. (°C)	Time (hr)	M_n (VPO)	Color	S.P.	Ref.
1. 5% aq. NaHCO₃	100	17	—	White	70–75	118
2. Aq. KOH	25	672⎫	1,500–3,000	Amber	90–110	117
3. Aq. KOH	70–100	2⎭				
4. 1% Aq. KF	100	3	4,000–5,000	White	70–80	118
5. Pyrolysis	950/1.5 mm	—	—	Yellow	110–120	118

systems, temperatures in excess of 175°C are required to effect the self-condensation of pentafluorophenoxide. In the polymerization of hepta-fluoro-*p*-cresol in the presence of base, the quinoidal intermediate is postulated to be involved as shown in reaction 41.

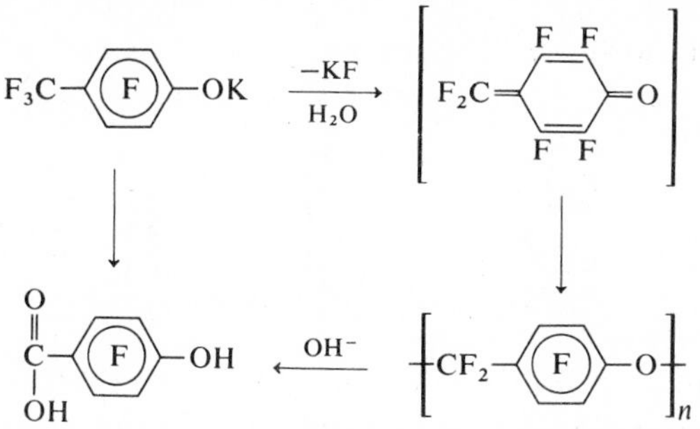

$$(41)$$

A similar intermediate was proposed to account for the low-temperature polymerization of 4-trifluoromethylphenol (119). Pyrolytic polymerization (118) of heptafluoro-*p*-cresol at 950°C and 1–5 mm pressures, believed to involve an identical intermediate, resulted in lower conversions (20–30%) to polymer, but higher softening points of the products were reported. After treatment with cold, dilute alkali, these perfluoropolybenzylene oxide polymers (MW 1500–3000) react further on heating above 300°C *in vacuo* to produce a polymer which shows crude elastomeric properties above 120°C. This effect is similar to those reported for the perfluoropolyphenylene ethers (85). The perfluoropolybenzylene oxide polymers lose weight rapidly above 300°C *in vacuo* (118). Thus their thermal stability is somewhat lower than the perfluoropolyphenylene ethers.

The para orientation assumed for the structure of these polymers was confirmed by degradation with strong alkali to *p*-hydroxytetrafluorobenzoic acid (117). This same compound can also be obtained directly from heptafluoro-*p*-cresol. The *p*-hydroxytetrafluorobenzoic acid may also serve as a monomer for the syntheses of fluorinated polyaryl esters.

D. Other Fluoropolyethers Containing Both Aromatic and Aliphatic Groups in the Chain

A trimer was reported which is related to the perfluorobenzylene oxide polymers but contains hydrogen (X = CH_2O) instead of fluorine in the side chain. In an attempt to prepare pentafluorobenzyl alcohol from pentafluoro-benzyl bromide and aqueous alkali, only the trimeric product, as shown in Eq. 42, was isolated (120).

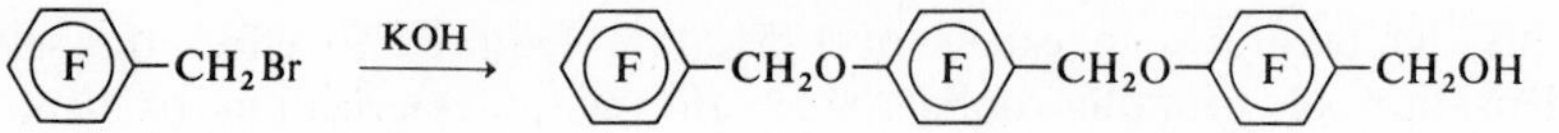

$$\text{(F)}-CH_2Br \xrightarrow{\text{KOH}} \text{(F)}-CH_2O-\text{(F)}-CH_2O-\text{(F)}-CH_2OH \quad (42)$$

A fluorocarbon polyether (121) was also formed by heating a mixture of hexafluoropropylene and fluoranil (24) at temperatures of 200–250°C and pressures of 1500–5000 atm. The copolymers are believed to form according to the scheme shown in Eq. 43, where n is the number of hexafluoropropylene

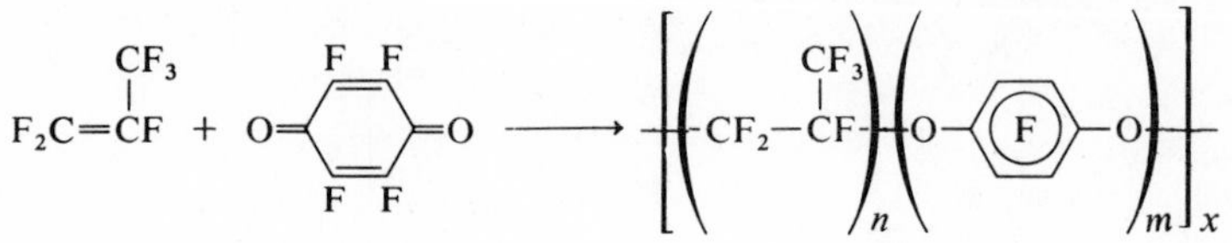

$$\underset{F_2C=CF}{\overset{CF_3}{|}} + O=\text{(ring)}=O \longrightarrow \left[\left(CF_2-\underset{|}{\overset{CF_3}{CF}}\right)_n\left(O-\text{(F)}-O\right)_m\right]_x \quad (43)$$

units, m is the number of fluoranil units in the polymer chain section having one hexafluoropropylene-fluouranil bond, and x represents the number of polymer chain sections in the copolymer. The molecular weight of the copolymers can be varied to form greases and waxes or high-molecular-weight resins.

E. Perfluoropoly Aromatic Esters

Completely fluorinated polyaryl esters (122) were prepared by the interfacial polymerization of tetrafluoroisophthaloyl chloride or tetrafluoroterepthaloyl chloride and the disodium salt of octafluoro-4,4'-dihydroxybiphenyl in trichloroethylene-water media at 5–30°C (Eq. 44).

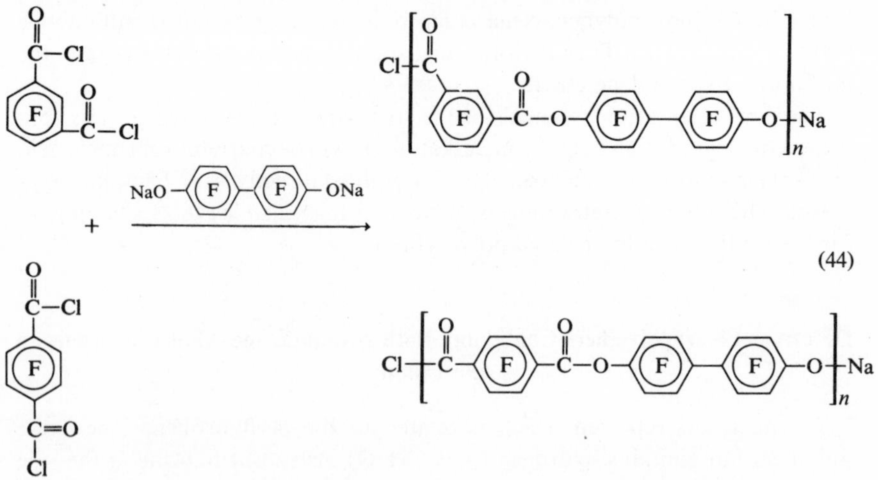

$$(44)$$

Only the softening points were reported for these polymers. The meta-linked, poly(octafluoro-4,4'-biphenylenetetrafluoroisophthalate) melted at about 300°C but was not completely fluid below 400°C. The para-linked polymer melted above 400°C. No other polymer properties were disclosed. This method of polymerization is generally applicable for the syntheses of a wide variety of perfluorinated or partially fluorinated polyaromatic esters.

F. Perfluoropolyphenylene Sulphides

1. MONOMER SYNTHESIS

Perfluoroaromatic thiols were used as the intermediates to prepare several perfluoropolyphenylene sulfide polymers. The thiols were synthesized by the treatment of hexafluorobenzene (123,124), octafluorotoluene (116,124), *o*- and *p*-perfluoroxylenes (125) with sodium hydrogen sulfide in pyridine or ethylene glycol mixture as shown in Eq. 45.

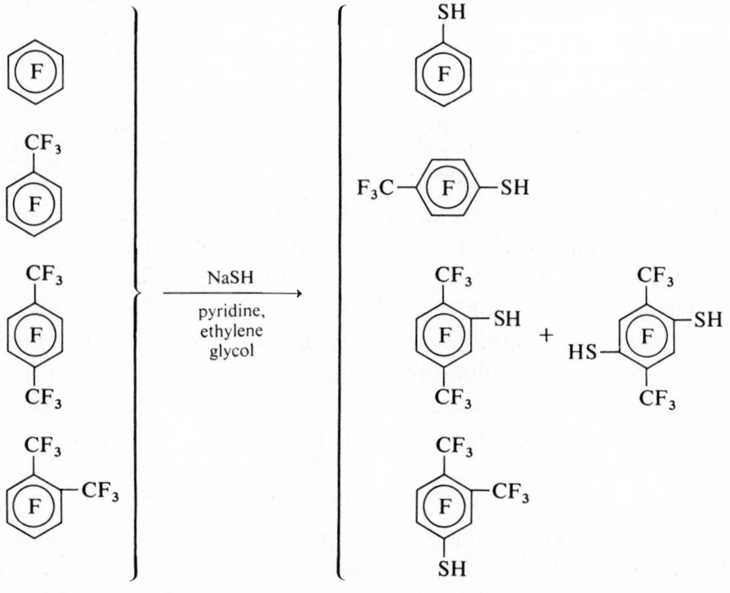

(45)

Whereas the treatment of the perfluoroxylenes (125) with sodium hydrogen sulfide produced a dithiol as well as the monothiol, the same reaction with pentafluorothiophenoxide does not yield a tetrafluorobenzenedithiol (123). Tetrafluorobenzene-1,4-dithio was prepared by the action of elementary sulfur on the 1,4-dilithiotetrafluorobenzene in tetrahydrofuran at $-70°C$ (Eq. 46) (123).

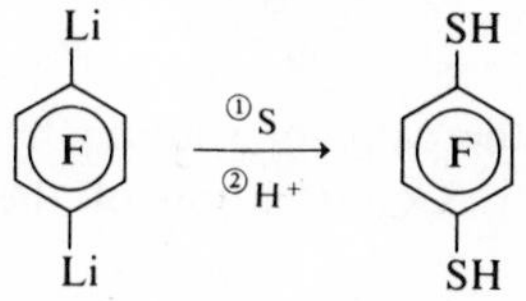

(46)

More recently a number of three- and four-ring perfluorothioethers were prepared by the reaction of cuprous pentafluorothiophenoxide and penta-fluorohalobenzenes in dimethylformamide as shown in reactions 47 (47). The

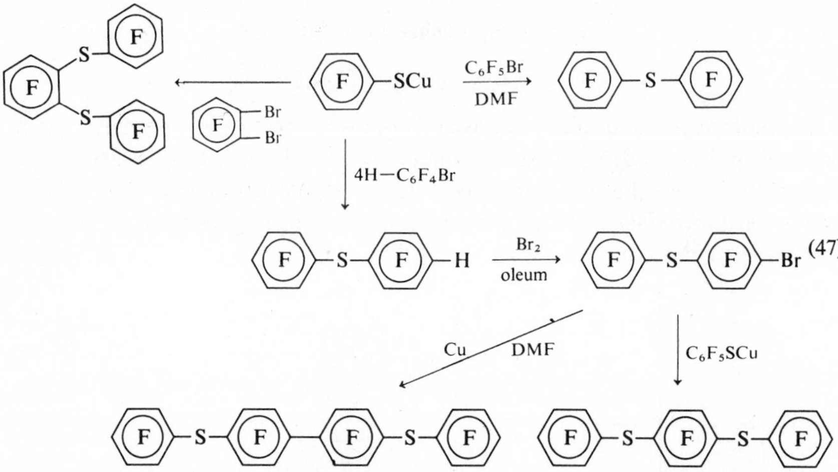

reaction is particularly interesting since the cuprous salts are highly specific for bromine, iodine, and chlorine. Fluorine is not replaced under these conditions. The dihalo derivatives such as 1,2-dibromotetrafluorobenzene can also be effectively used as intermediates. It is not yet known whether the dicuprous salts of the dithiols can be employed to produce polymers by reactions with dihalotetrafluorobenzenes. The simple fluoroaryl sulfide, pentafluorophenyl sulfide, was also prepared from the reaction of penta-fluorophenyllithium and sulfur dichloride (127).

2. POLYMERIZATION OF FLUOROARYLTHIOLS

Perfluoropolyphenylene sulfides are prepared by the reaction of potassium pentafluorothiophenoxides alone (pyrolysis) or with hexafluorobenzene in pyridine as solvent. In pyridine, an amorphous solid perfluoropolyphenylene

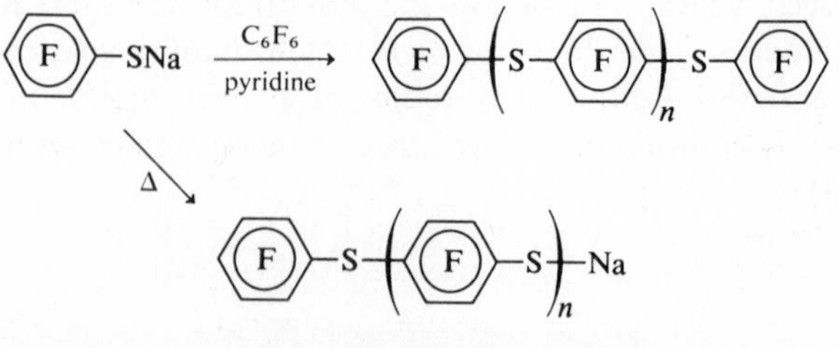

sulfide was formed when the ratio of sodium hydrosulfide (123) to hexa-fluorobenzene was 1.5 to 1. Confirmation of the para orientation of the polymer was obtained by desulfurization with Raney nickel in butanol-1. The products, 1,2,4,5-tetrafluorobenzene and pentafluorobenzene, were obtained in a 4 : 1 ratio.

The pyrolysis of potassium pentafluorothiophenoxide at 250°C (123) also yields a perfluoropolyphenylene sulfide, but this polysulfide may contain ortho and meta linkages as well as para bonds based on differences in the infrared spectra of the two polymers.

The potassium salt of 2,3,5,6-tetrafluorothiophenoxide was also decomposed thermally at 200°C, but since the salt lacks a fluorine atom in the para (4-) position, the polymer is probably linked in positions other than para.

The pyrolysis of the potassium 1,2,4-trifluoro-3,6-bis (trifluoromethyl) thiophenoxide (125) also resulted in a fluorinated polysulfide.

3. Thermal Stability

The perfluoropolyphenylene sulfides, in contrast to the perfluorophenylene ethers (85), are high-melting (300°C), >intractable solids and show a low order of solubility in organic solvents. Thus molecular-weight determinations are more difficult to perform by usual methods. Some characterization studies accomplished by desulfurization yielded the component linkages as various

$$(49)$$

fluorinated benzenes (123). Analysis of this volatile fraction by vapor-phase chromatography can reveal not only the relative quantities of each component but also the type of bonding in the original fluoropolysulfide. For example, pentafluorobenzene from end groups, 1,4-tetrafluorobenzene from para-linked phenyl rings, 1,3- and 1,2-tetrafluorobenzenes from meta- and ortho-linked rings.

The thermal stabilities of two perfluoropolyphenylene sulfide polymers were determined by thermogravimetric analyses (128). One sample was prepared from the reaction of hexafluorobenzene and sodium hydrosulfide in pyridine; the other was the pyrolytic product from the thermolyis of potassium pentafluorothiophenoxide. In Fig. 6, the relative thermal stabilities of the fluoropolysulfides (samples B and C) and a hydrocarbon polyphenylene sulfide (A) in a vacuum and oxygen atmosphere are shown. The results indicate that neither type, hydro versus fluorinated, is adversely affected by oxidizing conditions. Nevertheless, the perfluoropolyphenylene sulfides are again less stable thermally than the hydrocarbon analog. Similar differences in

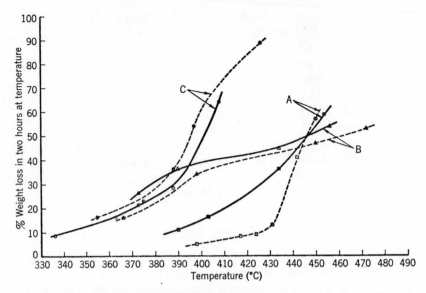

Fig. 6. Comparison of thermal stabilities of polymers (——) in vacuum and (– – –) in oxygen: (A) poly(phenylene sulfide); (B) perfluoropoly(phenylene sulfide), Sample A; (C) perfluoropoly(phenylene sulfide), Sample B (128).

thermal stability between small ring hydrogenic and fluorinated analogs were recorded from isoteniscope data (129).

G. Perfluoroarylamines

Heptafluoro-*p*-toluidine (116,118) decomposes at temperatures above its boiling point ($\geq 186°C/760$ mm) to yield poly-*p*-aminoperfluorobenzylene (118) according to the route shown in reaction 50. The initial polymer

$$
F_3C-\underset{}{\bigcirc}(F)-NH_2 \xrightarrow[\Delta]{-HF} \left(F_2C=\underset{F\ F}{\overset{F\ F}{\bigcirc}}=NH \right)
$$

$$
\left(\underset{}{\bigcirc}(F)-\underset{F}{\overset{}{C}}=N \right)_n \xleftarrow{-HF} \left(F_2C-\underset{}{\bigcirc}(F)-\underset{}{\overset{H}{N}} \right)_n
$$

(50)

(M.P. 80–135°C) is soluble in most organic solvents. However, on standing, a second evolution of hydrogen fluoride ensues, which is believed to have resulted in the formation of some azomethine linkages ($-\overset{\text{F}}{\underset{}{\text{C}}}=\text{N}-$) in the polymer (118). Although this secondary emission of hydrogen fluoride may occur intermolecularly as well as intramolecularly, the bulk of the polymer was still soluble.

In a similar manner, the distillation of pentafluorobenzylamine (120) at atmospheric pressures also results in a polymer believed to be a poly-*p*-aminotetrafluorobenzylene. Unlike the polybenzylene just mentioned this solid was water soluble.

$$
\overset{\text{F}}{\underset{}{\bigotimes}}-\text{CH}_2-\text{NH}_2 \xrightarrow[\Delta]{} \left(\overset{\text{F}}{\underset{}{\bigotimes}}-\text{CH}_2-\overset{\text{H}}{\underset{}{\text{N}}}\right)_n \tag{51}
$$

A fluoropolyaryl amine was also prepared by the self-condensation of the sodium salt of *N*-methyl, pentafluoroaniline (130) in the presence of hexafluorobenzene in a two-step process (Eq. 52). After the salt was prepared, the

$$
\overset{\text{F}}{\underset{}{\bigotimes}}-\overset{}{\underset{\text{CH}_3}{\overset{|}{\text{NH}}}} \xrightarrow[\substack{\text{NH}_3 \\ -33°\text{C}}]{\text{NaNH}_2} \left(\overset{\text{F}}{\underset{}{\bigotimes}}-\overset{}{\underset{\text{CH}_3}{\overset{|}{\text{N}^-}}}\right)\text{Na}^+ \xrightarrow[\Phi\text{H/THF}]{C_6F_6} \left(\overset{\text{F}}{\underset{}{\bigotimes}}-\overset{}{\underset{\text{CH}_3}{\overset{|}{\text{N}}}}\right)_n \tag{52}
$$

liquid ammonia was slowly replaced as solvent by a benzene-tetrahydrofuran (1:1) mixture and the temperature gradually raised to reflux. The isolated polymer was insoluble in water and most common organic solvents including formic and acetic acids, acetonitrile, triethylamine, and pyridine. The poly(*N*, methylaminotetrafluorophenylene) was soluble in concentrated sulfuric acid. The polymer was precipitated with water from the acid solution as a grey solid, M.P. 220–250°C. No other polymer properties are available for this class of polymers.

H. Perfluoropolyarylimides

The syntheses of a number of fluorine-containing polyarylimides (131) were disclosed recently. In general, the fluoroarylimides are prepared by reactions similar to their hydrogenic analogs (132–134). For example, poly [(tetrafluoro-*m*-phenylene)pyromellitimide] was obtained from tetrafluoro-*m*-phenylene diamine and pyromellitic dianhydride as shown in Eq. 53. The

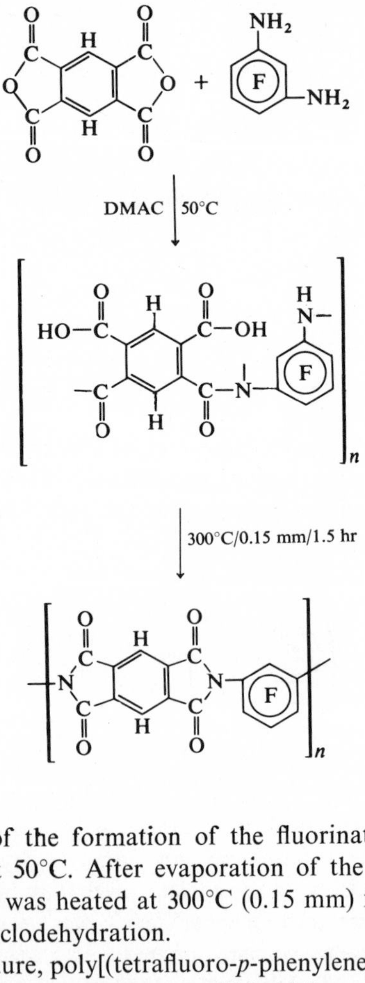

(53)

first step consisted of the formation of the fluorinated polyamic acid in dimethylacetamide at 50°C. After evaporation of the solvent, the residual solid (polyamic acid) was heated at 300°C (0.15 mm) for 1.5 hr to form the fluoroarylimide by cyclodehydration.

By the same procedure, poly[(tetrafluoro-*p*-phenylene)pyromellitimide] and poly[(octafluoro-4,4'-biphenylene)pyromellitimide] were also prepared. No physical data are available for these polymers. Although a number of perfluorinated aromatic dianhydrides and diamines are mentioned as possible comonomers, no examples were cited.

A variety of three- and four-ring perfluorinated arylimides (135,136), amides (137), and hydrazides (138, 139) have been synthesized. The latter compounds, fluoroarylhydrazides, are useful for the preparation of the perfluorinated oxa- and thiadiazole derivatives (138,139).

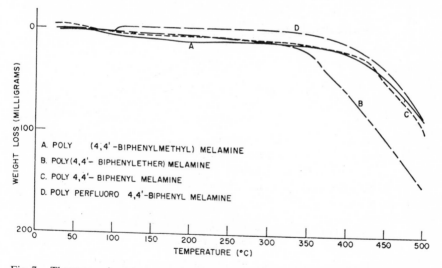

$$(54)$$

I. Perfluoropolypyridyls

Some perfluoropolypyridyls were prepared by the Ullmann reaction of 3,5-dichlorotrifluoropyridine $(X = Cl)$ and copper powder in refluxing dimethylformamide (Eq. 55) (140). When the reaction is performed in

$$X-\underset{F}{\overset{N}{\bigcirc}}-Cl \xrightarrow[DMF]{Cu} X\left[\underset{F}{\overset{N}{\bigcirc}}\right]_n Cl + Cu_2Cl_2 \qquad (X = F \text{ or } Cl) \quad (55)$$

the presence of 3-chlorotetrafluoropyridine $(X = F)$, perfluorobipyridyl (140,141) and other low-molecular-weight materials are formed. The polymers ranged from viscous liquids $(X = F,\ n = 2\text{--}5)$ to solids $(X = Cl,\ M.P.\ 180\text{--}190^\circ C,\ n = 15\text{--}16)$. The molecular weights of these perfluoro-polypyridyls are comparable to the molecular weights of perfluoro-p-polyphenylenes prepared by the same technique.

A. POLY   (4,4'-BIPHENYLMETHYL) MELAMINE

B. POLY(4,4'- BIPHENYLETHER) MELAMINE

C. POLY 4,4'- BIPHENYL MELAMINE

D. POLY PERFLUORO 4,4'-BIPHENYL MELAMINE

Fig. 7.   Thermogravimetric curves of melamine polymers to 500°C at 3°C/min in air (142).

J. Perfluoropolymelamines

Polyoctafluoro(4,4'-biphenyl)melamine (142) was prepared in 81% yield from cyanuric chloride and 4,4'-diaminooctafluorobiphenyl as shown in Eq. 56.

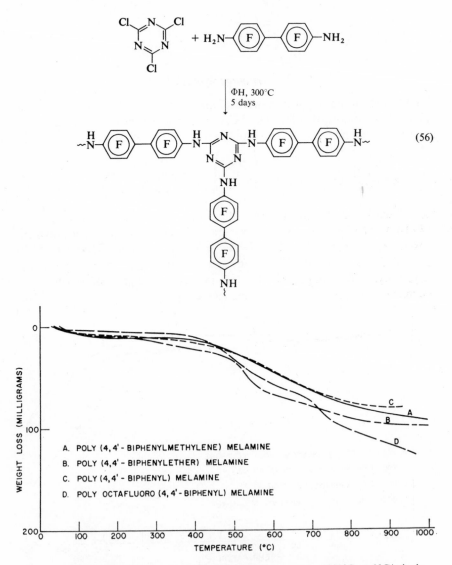

Fig. 8. Thermogravimetric curves of melamine polymers to 1000°C at 3°C/min in nitrogen (142).

The thermogravimetric curves for this perfluoroarylmelamine as well as for some hydrogenic analogs are shown in Figs. 7 and 8. In air (Fig. 7), and up to 500°C, the fluoromelamine polymer was the most stable, whereas above 700°C (Fig. 8) in a nitrogen atmosphere, it was the least stable thermally (losing about 63% of its weight above 700°C). This perfluoroarylmelamine polymer is one of the few perfluoropolymers which is more stable in air than its hydrogenic analog.

Discussion of other nitrogen-containing perfluoroaryl polymers, such as the urethanes and nitroso derivatives, can be found in other chapters of this book.

References

1. G. Balz and G. Shiemann, *Ber. dtsch. chem. Ges.*, **60**:1186 (1927).
2. A. Roe, "Organic Reactions," Vol. V, John Wiley and Sons, New York, 1949, p. 193.
3. A. E. Pavlath and A. J. Leffler, "Aromatic Fluorine Compounds," Rheinhold Publishing Corp., New York, 1962.
4. E. T. McBee, V. V. Lindgren, and W. B. Ligett, *Ind. Eng. Chem.*, **39**:378 (1947).
5. Y. Desirant, *Bull. Classe Sci. Acad. Roy. Belg.*, **41**(5):759 (1955).
6. Y. Desirant, *Bull. Soc. Chim. Belg.*, **67**:676 (1958).
7. M. Hellmann, E. Peters, W. J. Pummer, and L. A. Wall, *J. Am. Chem. Soc.*, **79**:5654 (1957).
8. R. E. Banks, J. M. Birchall, R. N. Haszeldine, R. N. Simon, J. H. Sutcliffe, and H. Umfreville, *Proc. Chem. Soc.*, **1962**:281.
9. R. A. Falk, *Sperry Eng. Rev.*, **16**:24 (1963).
10. J. M. Birchall, R. N. Haszeldine, and A. R. Parkinson, *J. Chem. Soc.*, **1961**:2204.
11. M. Stacey and J. C. Tatlow, "Advances in Fluorine Chemistry," Vol. 1, Butterworths, London, 1960, p. 166.
12. J. A. Godsell, M. Stacey, and J. C. Tatlow, *Nature*, **178**:199 (1956).
13. J. A. Godsell, M. Stacey, and J. C. Tatlow, *Tetrahedron*, **2**:193 (1958).
14. E. Nield, R. Stephens, and J. C. Tatlow, *J. Chem. Soc.*, **1958**:159.
15. B. Gething, C. R. Patrick, M. Stacey, and J. C. Tatlow, *Nature*, **183**:588 (1959).
16. W. K. R. Musgrave, British Patent 960,279 (1964).
17. G. C. Finger and C. W. Kruse, *J. Am. Chem. Soc.*, **78**:6034 (1956).
18. J. T. Maynard, *J. Org. Chem.*, **28**:112 (1963).
19. Imperial Smelting Corp., French Patent 1,360,917 (1964).
20. Imperial Chemical Industries, British Patent 970,746 (1964).
21. G. G. Yakobson, N. E. Mironova, A. K. Petrov, and N. N. Vorozhtsov, Jr., *Zh Obshch.Khim.*, **36**(1):147 (1966).
22. H. W. Halbrook, L. A. Loree, and O. R. Pierce, *J. Org. Chem.*, **31**(4):1259 (1966).
23. G. Fuller, *J. Chem. Soc.*, **1965**:6264.
24. K. Wallenfels and K. Draber, *Chem. Ber.*, **90**:2819 (1957).
25. Imperial Smelting Corp., British Patent 1,393,516 (1965).
26. N. N. Vorozhtsov, V. E. Platonov, G. G. Yakobson, *Izv. Akad. Nauk. SSSR, Ser. Khim.*, **1963**:1524.
27. W. J. Pummer and L. A. Wall, *Science*, **127**:643 (1958).
28. S. W. Charles, J. T. Pearson, and E. Whittle, *Trans. Faraday Soc.*, **59**:1156 (1963).
29. P. A. Claret, G. H. Williams, and J. Caulson, *J. Chem. Soc. (C)*, **1968**:341.

30. C. Walling, "Free Radicals in Solution," John Wiley and Sons, New York, 1957.
31. G. H. Williams, "Homolytic Aromatic Substitution," Pergamon Press, New York, 1960.
32. R. E. Florin, D. W. Brown, and L. A. Wall, *J. Research NBS*. **64A**: 269 (1960).
33. V. A. Khramchenkov, *At. Energ.*, **21**(5): 375 (1966).
34. D. R. MacKensie, F. W. Bloch, and R. H. Wiswall, Jr., *J. Phys. Chem.*, **69**: 2526 (1965).
35. L. A. Wall, W. J. Pummer, J. E. Fearn, and J. M. Antonucci, *J. Research NBS*, **67A**(5): 481 (1963).
36. J. C. Tatlow, *Endeavor*, **22**: 80 (1963).
37. A. K. Barbour and P. Thomas, *Ind. Eng. Chem.*, **58**(1): 52 (1966).
38. R. E. Banks, "Fluorocarbons and Their Derivatives," Oldbourne Press, London, 1964, p. 139.
39. E. J. Forbes, R. D. Richardson, M. Stacey, and J. C. Tatlow, *J. Chem. Soc.*, **1959**: 2019.
40. G. B. Brooke, J. Burdon, M. Stacey, and J. C. Tatlow, *J. Chem. Soc.*, **1960**: 1768.
41. B. Gething, C. R. Patrick, and J. C. Tatlow, *J. Chem. Soc.*, **1961**: 1574.
42. K. C. Ho and J. Miller, *Aust. J. Chem.*, **19**: 423 (1966).
43. J. Burdon, W. B. Hollyhead, C. R. Patrick, *J. Chem. Soc.*, **1964**: 4663.
44. E. Nield, R. Stephens, and J. C. Tatlow, *J. Chem. Soc.*, **1959**: 166.
45. W. J. Pummer and L. A. Wall, *J. Research NBS*, **63A**: 167 (1959).
46. L. A. Wall, J. M. Antonucci, S. Straus, and M. Tryon, *Soc. Chem. Ind. Monograph*, **13**: 295 (1961).
47. L. Belf, J. Buxton, M. W. Fuller, *J. Chem. Soc.*, **1965**: 3372.
48. J. Burdon, *Tetrahedron*, **21**(12): 3373 (1965).
49. R. Stephens and J. C. Tatlow, *Chem. and Ind.*, (London), **1957**: 821.
50. R. E. Florin, W. J. Pummer, and L. A. Wall, *J. Research NBS*, **62**(3): 122 (1959).
51. G. C. Finger, F. H. Reed, D. M. Burness, D. N. Fort, and R. R. Blough, *J. Am. Chem. Soc.*, **73**: 146 (1951).
52. G. C. Finger, F. H. Reed, and R. E. Oesterling, *J. Am. Chem. Soc.*, **73**: 153 (1951).
53. W. J. Pummer, R. E. Florin, and L. A. Wall, *J. Research NBS*, **62**(3): 113 (1959).
54. M. Hellmann and A. J. Bilbo, *J. Am. Chem. Soc.*, **75**: 4590 (1953).
55. M. Hellmann, A. J. Bilbo, and W. J. Pummer, *J. Am. Chem. Soc.*, **77**: 3650 (1955).
56. Imperial Smelting Corp., Neth. Appl. 6,500,322 (1965).
57. J. M. Antonucci and L. A. Wall, *J. Research NBS*, **70A**(6): 475 (1966).
58. R. E. Florin, W. J. Pummer, and L. A. Wall, *J. Research NBS*, **62**(3): 107 (1959).
59. A. L. Henne, *J. Am. Chem. Soc.*, **75**: 5750 (1953).
60. E. J. P. Fear, J. Thrower, and M. A. White, Paper presented at XIX Interntl. Congress of Pure and Applied Chemistry, London, 1963, p. 103.
61. E. J. P. Fear and J. Thrower, British Patent 1,100,261 (1968).
62. J. Thrower and M. A. White, *Abstracts Papers, Am. Chem. Soc. Meeting*, **148**: 19K (1964).
63. G. M. Brooke, R. D. Chambers, J. Heyes, and W. K. R. Musgrave, *J. Chem. Soc.*, **1964**: 731.
64. W. J. Respess and C. Tamborski, *J. Organometal. Chem.* **11**: 620 (1968).
65. W. J. Pummer and J. M. Antonucci, *Am. Chem. Soc. Polymer Preprints*, **7**(2): 1072 (1966).
66. J. P. N. Brewer and H. Heaney, *Tetrahedron Letters*, **1965**(51): 4709.
67. N. N. Vorozhtsov, N. G. Ivanova, V. A. Barkhash, *Zh. Org. Khim.*, **3**(1): 220 (1967).
68. C. Tamborski, E. J. Soloski, and J. P. Ward, *J. Org. Chem.*, **31**(12): 4230 (1966).

69. P. L. Coe, R. Stephens, and J. C. Tatlow, *J. Chem. Soc.*, **1962**: 3227.
70. R. J. Harper, E. J. Soloski, and C. Tamborski, *J. Org. Chem.*, **29**: 2385 (1964).
71. C. Tamborski and E. J. Soloski, *J. Org. Chem.*, **31**: 743 (1966).
72. D. E. Fenton, A. J. Park, D. Shaw, and A. G. Massey, *J. Organometal. Chem.*, **2**: 437 (1964).
73. D. E. Fenton and A. G. Massey, *Tetrahedron*, **21**: 3009 (1965).
74. C. Tamborski, E. J. Soloski, and S. M. Dec, *J. Organometal. Chem.*, **4**: 446 (1965).
75. D. D. Callander, P. L. Coe, and J. C. Tatlow, *Tetrahedron*, **22**: 419 (1966).
76. S. C. Cohen, D. E. Fenton, D. Shaw, and A. G. Massey, *J. Organometal. Chem.*, **8**: 1 (1967).
77. C. Tamborski and E. J. Soloski, *J. Organometal. Chem.*, **10**(3): 385 (1967).
78. S. C. Cohen, A. J. Tomlinson, M. R. Wiles, and A. G. Massey, *J. Organometal. Chem.*, **11**(3): 385 (1968).
79. R. D. Chambers, J. A. Cunningham, and D. J. Spring. *J. Chem. Soc.*, (*C*), **1968**: 1560.
80. Imperial Chemical Industries Ltd., Belgian Patent 640,491 (1964).
81. J. Thrower and M. A. White, *Am. Chem. Soc. Polymer Preprints*, **7**(2): 1077 (1966).
82. Imperial Chemical Industries, Ltd., Neth. Appl. 6,505,550 (1965).
83. R. J. Harper and C. Tamborski, *Chem. and Ind.*, **1962**: 1824.
84. W. J. Pummer and L. A. Wall, *Abstracts Papers Am. Chem. Soc. Meeting*, **150**: 10K (1965).
85. W. J. Pummer and L. A. Wall, *J. Research NBS*, **68A**: 281 (1964).
86. J. L. Cotter, J. M. Lancaster, and W. W. Wright, *Am. Chem. Soc. Polymer Preprints*, **5**(2): 475 (1964).
87. G. P. Brown and A. Goldman, *Am. Chem. Soc. Polymer Preprints*, **4**(2): 45 (1963).
88. K. C. Tsou, H. E. Hoyt, and B. D. Halpern, *Am. Chem. Soc. Polymer Preprints*, **4**(2): 583 (1963).
89. J. M. Antonucci and L. A. Wall, *Abstracts Papers, Am. Chem. Soc. Meeting*, **150**: 11K (1965).
90. D. G. Holland, Wright-Patterson A. F. Base, Dayton, Ohio, unpublished results.
91. J. K. Erown and K. J. Morgan, "Advances in Fluorine Chemistry," Vol. 4, Butterworths, Washington, D.C., 1965, p. 279.
92. Imperial Smelting Corp., Belgian Patent 659,239 (1965).
93. P. Kovacic and A. Kryiakis, *J. Am. Chem. Soc.*, **85**: 456 (1963).
94. F. Ullmann and P. Sponagel, *Justus Liebigs' Ann. Chem.*, **350**: 83 (1906).
95. S. Aftergut, R. J. Blackington, and G. P. Brown, *Chem. Ind.*, **1959**: 1090.
96. E. S. Blake, W. C. Hammond, J. W. Edwards, T. E. Reichard, and M. E. Ort, *J. Chem. Eng. Data*, **6**: 87 (1961).
97. J. M. Cox, B. A. Wright, and W. W. Wright, *J. Appl. Polymer Sci.*, **9**: 513 (1965).
98. A. L. Rocklin, *J. Org. Chem.*, **21**: 1478 (1956).
99. J. M. Birchall and R. N. Haszeldine, *J. Chem. Soc.*, **1961**: 3221.
100. J. Burdon, V. A. Damodaran, and J. C. Tatlow, *J. Chem. Soc.*, **1964**: 763.
101. E. Nield and J. C. Tatlow, *Tetrahedron*, **8**: 39 (1960).
102. R. J. DePasquale and C. Tamborski, *J. Org. Chem.*, **32**: 3163 (1967).
103. R. J. DePasquale and C. Tamborski, *J. Org. Chem.*, **33**(2): 830 (1968).
104. R. J. DePasquale and C. Tamborski, *J. Org. Chem.*, **33**(4): 1658 (1968).
105. N. Ya. Tel'pina and S. V. Solokov, *Zh. Obshch. Khim.*, **37**(10): 2175 (1967).
106. G. A. Richardson and E. S. Blake, *I and EC Product Research and Development*, **7**(1): 22 (1968).
107. L. A. Wall, *NBS Tech. News Bull.*, **43**: 78 (1959).
108. W. J. Pummer and L. A. Wall, *J. Chem. Eng. Data*, **6**(1): 78 (1961).

109. National Polychemicals, Inc., British Patent 887,691 (1962).
110. W. R. Sorenson and T. W. Campbell, "Preparative Methods of Polymer Chemistry," Interscience Publishers, New York, 1961, pp. 61–63.
111. L. J. Bellamy, "Infrared Spectra of Complex Molecules," John Wiley and Sons, New York, 1954, p. 102.
112. J. M. Cox, B. A. Wright, and W. W. Wright, *J. Appl. Polymer Sci.*, **9**: 520 (1965).
113. R. J. DePasquale and C. Tamborski, *J. Organometal. Chem.*, **13**: 274 (1968).
114. V. D. Shteingarts and A. G. Budnik, *Izv. Sib. Otd. Akad. Nauk SSSR, Ser. Khim. Nauk*, **3**:124 (1967).
115. V. D. Shteingarts, N. G. Kostina, G. G. Yakobson, and N. N. Vorozhtsov, *Izv. Sib. Otd. Akad. Nauk. SSSR, Ser. Khim. Nauk*, **3**: 117 (1967).
116. D. Alsop, J. Burdon, and J. C. Tatlow, *J. Chem. Soc.*, **1962**: 1801.
117. V. C. R. McLoughlin and J. Thrower, *Chem. and Ind. (London)*, **1964**:1557.
118. J. M. Antonucci and L. A. Wall, *J. Research NBS*, **71A**: 35 (1967).
119. R. G. Jones, *J. Am. Chem. Soc.*, **69**: 2346 (1947).
120. J. M. Birchall and R. N. Haszeldine, *J. Chem. Soc.*, **1961**: 3725.
121. W. J. Brehm and A. S. Williams, E. I. DuPont de Nemours Co., U.S. Patent 3,053,823 (1962).
122. W. Cummings and E. R. Lynch, British Patent 1,079,516 (1967).
123. P. Robson, M. Stacey, R. Stephens, and J. C. Tatlow, *J. Chem. Soc.*, **1960**: 4754.
124. Imperial Smelters Corp., Neth. Appl. 6,605,036 (1966).
125. E. V. Aroskar, M. T. Chaudry, R. Stephens, and J. C. Tatlow, *J. Chem. Soc.*, **1964**: 2976.
126. R. H. Mobbs, *Chem. and Ind.*, **1965**:1562.
127. R. S. Chambers, J. A. Cunningham, and D. A. Pyke, *Tetrahedron*, **24**(6): 2783 (1968).
128. N. S. J. Christopher, J. L. Cotter, G. J. Knight, and W. W. Wright, *J. Appl. Polymer Sci.*, **12**(4): 867 (1968).
129. W. Cummings, E. R. Lynch, and E. B. McCall, Paper presented at the 3rd International Symposium Fluorine Chemistry, Munich, Germany, 1965.
130. W. J. Pummer, National Bureau of Standards, unpublished results.
131. W. Cummings and E. R. Lynch, British Patent 1,077,243 (1967).
132. C. E. Scroog, S. V. Abramo, C. E. Berr, W. M. Edwards, S. L. Endrey, and K. E. Olivier, *Am. Chem. Soc. Polymer Preprints*, **5**(1):132 (1964).
133. J. F. Heacock and C. E. Berr, *Soc. Plastics Engrs. Reg. Tech. Conf. Preprints, Stability of Plastics*, Washington, D.C., June 4, 1964, p. 345.
134. A. H. Frazer, "High Temperature Resistant Polymers," Interscience Publishers, New York, 1968, p. 160.
135. E. R. Lynch and W. Cummings, British Patent 1,099,096 (1968).
136. D. G. Holland and C. Tamborski, *J. Org. Chem.*, **31**: 281 (1966).
137. W. Cummings, British Patent 1,075,166 (1967).
138. E. R. Lynch and W. Cummings, British Patent 1,096,600 (1967).
139. A. T. Prudchenko, S. A. Vereshihagina, V. A. Barkhash, and N. N. Vorozhtsov, *Zh. Obshch. Khim.*, **37**(10): 2195 (1967).
140. H. C. Fielding, Imperial Chemical Industries Ltd., British Patent 1,085,882 (1967).
141. R. D. Chambers, D. Lomas, and W. K. R. Musgrave, *J. Chem. Soc. (C)*, **1968**(6): 625.
142. D. R. Anderson and J. M. Holova, *J. Polymer Sci.*, Part A-1, **4**(7): 1967 (1966).

4. HIGH-PRESSURE POLYMERIZATION

Leo A. Wall, *Polymer Chemistry Section, National Bureau of Standards, Washington, D.C.*

Contents

I. INTRODUCTION

In previous chapters of this book the polymerization of fluoroethylenes is discussed and reviewed. These and a few other fluoromonomers may be polymerized at pressures of several thousand pounds per square inch. Such pressures are adequate for obtaining the desired rates and products. In general, for low-boiling monomers that easily polymerize to high polymer, such as tetrafluoroethylene, the effect of pressure is chiefly to increase the concentration of the reagent. On the other hand, for numerous vinylic or olefinic compounds, both fluorinated and nonfluorinated, attempts at polymerization have not succeeded at low pressures. On structural grounds many of these monomers would be considered to be unpolymerizable by common procedures. Ultra high pressures—pressures above 1000 atm—have shown promise for converting these monomers into useful polymeric materials. In recent years much work in this direction has been begun. Fully fluorinated olefins having more than two carbon atoms, such as hexafluoropropylene and perfluoroheptene, require pressures above 1000 atm for the production of homopolymers of high molecular weight.

The polymerization of tetrafluoroethylene and the other fluoroethylenes are known to proceed via free-radical mechanisms. Cationic catalysts are ineffective (1) and the usual anionic catalysts, e.g., the lithium alkyls, are very reactive in processes that strip off fluorine and compete with polymerization. Other catalysts such as the Ziegler and alfin may also be expected to strip off fluorine and participate in competing processes that are inimical to the polymerization reaction. On the basis of our present knowledge of the reactivity of fluoroolefins, however, it is anticipated that anionic processes involving the fluoride ion will be developed for homopolymerization.

High pressure in the order of 20,000–30,000 atm, on the other hand, can be expected to convert many as yet unpolymerized monomeric compounds to high polymers via free radical and possibly anionic intermediates. At present there are at least three books (2–4) that review in varying degrees the early literature on the effects of pressure on organic reactions and polymerization. Although the effect of pressure on addition polymerization is generally predictable by the principle of Le Chatelier, the process, because of its complex mechanism, may not respond to pressure, in the desired or anticipated fashion. For instance, a given rate may be achieved but not at the same time as adequate molecular weight or the preferred molecular structure. Assuming that the existence of the desired polymer is thermodynamically feasible under the experimental conditions, we can on elementary grounds separate the effects of pressure on polymerization into several categories:

1. In the pressure range of 1–2000 atm, the homogeneous polymerization of gaseous monomers is favored by the enhancement of the monomer concentration.

2. In the pressure range of 10^3–10^4 or more atm, homogeneous liquid bulk polymerization is favored by:

a. The acceleration of the bimolecular propagation or growth step.
b. In some situations, retardation of the termination step as a consequence of increased viscosity.
c. The elevation of the ceiling temperature.
d. Elimination of inhibition by impurities.

3. In the pressure range, particularly above 10^4 atm, phase changes may determine the polymerization process through diffusion and morphological factors.

These categories by no means completely cover the various and intricate possibilities for pressure to alter the course of polymerization processes. This is particularly true since very little research of a definitive nature has yet been carried out. High pressure has, however, been successfully used in the preparation of numerous polymers and copolymers and is a technique of some generality.

II. MECHANISM OF BULK POLYMERIZATION

In this section and throughout most of this chapter we discuss the mechanisms of addition polymerization in terms of a free radical chain process. For peroxide and thermally initiated systems this is not open to serious question. For initiation with γ-rays, however, there exists the possibility that ionic and other mechanisms occur. Most of the polymerization reviewed herein were initiated by γ-rays.

The following elementary processes constitute the usual mechanism of free radical polymerization:

	Reaction	Rate expression
Initiation	$\left.\begin{array}{l} C \\ \text{photo, } M \\ \text{radio, } M \\ \text{thermal, } M \end{array}\right\} \longrightarrow R_1$	$\mathscr{I}$
Propagation	$R_i + M \xrightarrow{k_2} R_{i+1}$	$k_2 M \sum\limits_1^\infty R_i$
Transfer	$R_i + M \xrightarrow{k_3} P_i + R_1$	$k_3 M \sum\limits_1^\infty R_i$
Termination	$R_i + R_j \xrightarrow{k_4} P_i + P_j$	$2k_4 \left(\sum\limits_1^\infty R_i\right)^2$
	or	
	P_{i+j}	

Here it is indicated that radicals R_1 are formed by the initiation process which can be the result of the dissociation of an added chemical catalyst C, or by the action of radiation on the monomer M. It is assumed that the catalyst is 100% efficient. The symbol $\mathscr{I}$ stands for the total rate of radical formation. Each propagation or growth act adds a monomer unit to the radical. The number of units per radical or polymer molecule P_i is indicated by the subscript. The transfer process is shown with monomer and competes with propagation along with termination.

A. Role of Pressure

Pressure on bulk liquid monomer will then influence the polymerization process only insofar as it affects the individual processes just discussed. Processes thus far neglected would include polymer transfer and transfer with diluents or impurities. Changes in the concentration of the species as a result of the compression of the liquid state are disregarded here since such changes are minor over the usual experimental pressure ranges.

A rate theory that provides an interpretation of the effect of pressure on homogeneous elementary processes is the transition state theory. If the volume of the transition state is smaller than that of the reactants, the rate is accelerated. The quantitative relation is

$$\left(\frac{\partial \ln k}{\partial P}\right)_T = \frac{-\Delta V\dagger}{RT} \tag{1}$$

where $\Delta V\dagger$ is the volume change upon the formation of the transition state from the reactants.

All bimolecular processes are expected to be favored by pressure since the formation of the transition state requires to varying degrees a close approach of the reacting molecules. On this premise pressure will certainly increase the consumption of monomer by propagation and transfer. Unimolecular or dissociative reaction rates would be expected to be decreased by pressure since a dissociative transition state, unless there were solvation effects, would necessarily be an expanded one. Since in polymerization processes the molecular weight and molecular structure are of greater importance than is rate of formation alone, it is best to discuss the problem by presenting the relationships for the rate and number average degree of polymerization. The fractional rate r of polymerization based on the mechanism just outlined is

$$r = \frac{1}{M}\frac{dM}{dt} = k_2\left(\frac{\mathscr{I}}{2k_4}\right)^{1/2} \tag{2}$$

The number average degree of polymerization $\bar{P}_n$ is given by

$$\frac{1}{\overline{P}_n} = \frac{k_3}{k_2} + \frac{(2k_4\mathscr{I})^{1/2}}{k_2 M} \tag{3}$$

or

$$\frac{1}{\overline{P}_n} = \frac{k_3}{k_2} + \frac{2k_4}{k_2{}^2 M}r \tag{4}$$

From Eq. 2 it may be anticipated that for pure monomer pressure will greatly accelerate the rate of polymerization since it contains the propagation rate constant in the numerator. The termination process, being bimolecular, would in the absence of diffusion control also be accelerated and hence produce an opposite effect. However, in actuality the termination process is prone to be controlled by diffusion and is likely to be very much retarded by high pressure. In fact, since pressure greatly increases the viscosity of fluids, with high-energy radiation a relatively large concentration of free radicals might be achieved, particularly at pressures in the region of the glass transition.

Equation 4 reveals that the molecular size, $\overline{P}_n$, is controlled by both the transfer and termination process. The former will be increased with pressure but often not to as great an extent as propagation. An important point to be made is that the initiation process chosen for the polymerization is critical in high-pressure polymerization.

B. The Initiation Process

Table 1 compares on the basis of the preceding mechanism the different results possible with three different initiation processes. The chemical and thermal initiations are the type most often presented, i.e., first-order catalyst decomposition and thermal initiation dependent on the second power of the monomer concentration. In table I the γ-ray initiation is given by the relation

$$\mathscr{I} = \frac{10 \, G(R) \, I\rho}{N} \tag{5}$$

where $G(R)$ is radicals produced per $100\,\mathrm{eV}$ (electron volts) of radiation energy absorbed, I the intensity of radiation in electron volts per gram-second, ρ the density in grams per cubic centimeter, and N Avogadro's number.

The decomposition of chemical free-radical catalyst of the dissociating type will be retarded by pressure. This process is restricted to a characteristic temperature range. Thus both the temperatures and pressures feasible for the initiation of polymerization may not be those favorable for propagation.

TABLE 1

Comparison of Rate, Number Average Degree of Polymerization, and Activation Volume Relations for Different Modes of Initiation

$\mathscr{I}$ (ml^{-1}sec^{-1})	r	$\Delta V_r^\dagger$	No transfer		Transfer	
			$\bar{P}_n$	$\Delta V_{P_n}^\dagger$	$\bar{P}_n$	$\Delta V_{P_n}^\dagger$
Chemical, $2k_1C$	$k_2\left(\dfrac{k_1C}{k_4}\right)^{1/2}$	$\dfrac{\Delta V_1^\dagger}{2} + \Delta V_2^\dagger - \dfrac{\Delta V_4^\dagger}{2}$	$\dfrac{k_2M}{(4k_4k_1C)^{1/2}}$	$\Delta V_2^\dagger - \dfrac{\Delta V_1^\dagger}{2} - \dfrac{\Delta V_4^\dagger}{2}$	$\dfrac{k_2}{k_3}$	$\Delta V_2^\dagger - \Delta V_3^\dagger$
Thermal, $2k_1M^2$	$k_2M\left(\dfrac{k_1}{k_4}\right)^{1/2}$	$\dfrac{\Delta V_1^\dagger}{2} + \Delta V_2^\dagger - \dfrac{\Delta V_4^\dagger}{2}$	$\dfrac{k_2}{2(k_4k_1)^{1/2}}$	$\Delta V_2^\dagger - \dfrac{\Delta V_1^\dagger}{2} - \dfrac{\Delta V_4^\dagger}{2}$	$\dfrac{k_2}{k_3}$	$\Delta V_2^\dagger - \Delta V_3^\dagger$
γ-radiation $\dfrac{10G(R)I\rho}{N}$	$k_2\left(\dfrac{5G(R)I\rho}{k_4N}\right)^{1/2}$	$\Delta V_2^\dagger - \dfrac{\Delta V_4^\dagger}{2}$	$\dfrac{k_2M}{2}\left(\dfrac{N}{5k_4G(R)I\rho}\right)^{1/2}$	$\Delta V_2^\dagger - \dfrac{\Delta V_4^\dagger}{2}$	$\dfrac{k_2}{k_3}$	$\Delta V_2^\dagger - \Delta V_3^\dagger$

Thermal initiation is also restricted by conditions, but since it will probably be bimolecular as shown or be of a higher order in monomer concentration, pressure will favor it. On the other hand, gamma radiation is anticipated to be nearly pressure and temperature independent and hence permits experimentation over a wide range of conditions. Since the gamma rays readily penetrate pressure vessels, the technique is relatively simple. Large accelerations in the rates result when the activation volume changes are large negative quantities.

The only unfavorable activation volume terms are $\Delta V_1^{\ddagger}$ for chemical initiation and $\Delta V_3^{\ddagger}$ for transfer. The $\Delta V_4^{\ddagger}$ term, although expected to be positive because of diffusion control, always occurs in Table 1 with a negative sign. The degree of polymerization Eqs. 3 and 4 is a linear composite of two terms that can be separated graphically by plotting $\bar{P}_n^{-1}$ against the fractional rate r. The slope from such a plot is

$$s = \frac{2k_4}{k_2^2}M$$

If a finite slope is measured the rate of initiation is obtainable since $\mathscr{I} = r^2 sM$. This is particularly useful for obtaining the rate of initiation in gamma-radiation studies.

C. Dissociation of Initiators

The decompositions of typical free-radical catalysts, benzoyl peroxide, azoisobutyronitrile, and t-butyl peroxide have been shown to decrease with pressure (5–8). The results varied with different solvents and temperatures. The activation volumes reported ranged from ~ 4–13 cc/m. In a study (6) of azoisobutyronitrile the effect of pressure was greater when the extent of reaction was followed by an iodine-scavenger method than by the direct measurement of the dissociating compound. The solvent was toluene. The results appear to indicate that the decomposed radicals were also more effectively trapped with increasing pressure and mutually terminated before diffusing. This plus the effect of pressure on the termination rate in styrene polymerization calculated from rotating sector and emulsion polymerization measurements supports the concept that the bimolecular reactions requiring diffusion are diminished by pressure. The activation volume change is estimated as $\Delta V_4^{\ddagger} \geq 4$ cc/m (9). Using the viscosity pressure dependence of toluene to approximate that of the monomer styrene, it is shown that the product of the termination constant with the square root of the viscosity is quite constant (10). Thus it seems that the termination constant varies in the fashion $k_4 \sim \eta^{-1/2}$ over a pressure range of 1–3000 atm.

III. STYRENE

A. Propagation

Styrene polymerization under pressures to 6000 atm has been studied extensively (9–13). Emulsion polymerization (9) and rotating sector measurements (10) have been utilized to obtain the effect of pressure on the elementary propagation reaction. These measurements give, respectively, values of (4,10)

$$\Delta V_2^\ddagger = -13.4$$

and (9)

$$\Delta V_2^\ddagger = -11.5$$

For the overall rate of catalyzed styrene polymerization

$$\Delta V_0^\ddagger \cong -17 \ \text{cc/m}$$

Thus most of the effect of pressure on the overall rates is the result of a greatly accelerated propagation rate. For bulk sytrene polymerization under pressure, monomer transfer appears to be absent and the molecular weight appears dependent entirely on the termination process (10).

B. Solvent Transfer

For styrene polymerization chain transfer with the active solvent CCl_4 is increased in almost the same amount as propagation, thus $\Delta V\dagger = -11$ cc/m (9). Also at 10,000 atm the rate of reaction between diphenylpicrylhydrazyl and mercaptans is greatly increased (14). Four mercaptans, n-butyl, t-butyl, n-hexyl, and t-octyl, gave, respectively, $-\Delta V\dagger = 14.5, 29.5, 16.8$, and 20.7 cc/m for their reaction (14) with diphenylpicrylhydrazyl between 1 and 2700 atm. In contrast, studies of the transfer between the styryl radical and triethylamine gives $-\Delta V\dagger \sim 2$ cc/m (15).

IV. PERFLUOROSTYRENE

A. Thermal Initiation

Unlike styrene, perfluorostyrene is extremely difficult to polymerize. The polymerization of partially fluorinated styrenes has been reviewed in an earlier chapter. Of these the α, β, β-trifluoro monomer tends to form cyclic dimers which interfere with the producton of high polymer. Thus it is not surprising that under autogenous pressures attempted thermal polymerization of perfluorostyrene at 100°C gives after two years 60 % conversion to a mixture of two presumably cyclic dimers in the ratio 7/6 based on chromatographic analysis. Under pressures of $\sim 10^4$ atm thermal polymerization produces high polymer (17) with intrinsic viscosities measured in hexa-fluorobenzene at 29.7°C, up to 0.4 dl/g. Both the rate of polymerization and the intrinsic viscosities go through a maximum with temperature and pressure.

B. Effect of Crystalline and Glass State

The experimental results were apparently erratic and uninterpretable until a phase study of the liquid monomer indicated the underlying reasons for the observed behavior. Crystalline phases are relatively easily detected during the compression of a liquid since an isobaric volume change occurs at the crystallization point. However, a glass transition is much more difficult and tedious to detect since it is not a well-defined phenomenon. In the study of perfluorostyrene under pressure, crystallization did not occur readily and a superpressed condition above the crystallization point is easily attained. When highly superpressed, the viscosity of a fluid can become sufficiently large that pressure gradients can develop in the sample and creep behavior is seen. For perfluorostyrene this happens in the region of twice the crystallization pressure. In this range the monomer behaves in the manner depicted in Fig. 1, which shows typical creep behavior of the superpressed monomer. In this particular plot outward creep is shown.

Consider the upper curve, in this case the load, i.e., the pounds force on the ram is reduced from a higher value to 27.5 klb (kilopounds) and the time for the outward movement of the piston measured, and is shown in the plot

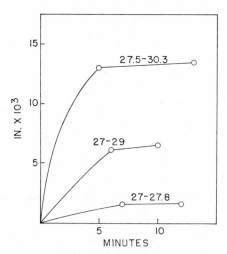

Fig. 1. Perfluorostyrene in the region of 10,000 atm, 22°C. Characteristic behavior in the vicinity of the glassy state. Outward piston creep (in.) after lowering load to the initial values in kilopounds shown on curves, as function of the time required for the system to equilibrate, at which time the load rises to the final values indicated on curves.

as inches. At the end of the curve the load value had risen to 30.3 klb. Below the highly superpressed region the creep is too fast to measure. At higher pressures it is too slow to measure or even detect. The onset of this behavior is taken here to be the onset of the transition into a glassy state. The pressure for this behavior is possibly quite far below that at which the "true" glass transition point occurs. It is, however, a lower boundary of a region where the viscosity is $> 10^7$ poises, and hence where the bimolecular termination should be slow.

Figure 2 shows a plot of piston position as a function of load, which shows the behavior seen as the system is pressed. Both crystalline and the glass phenomena are detected. The longer curves show the behavior of the system before crystallization is induced. The open points were obtained as the pressure was applied, the closed ones as the pressure was released. At the

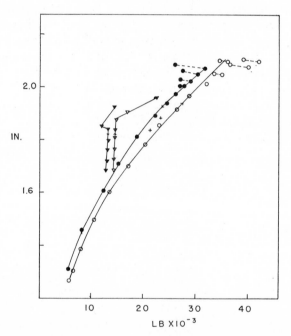

Fig. 2.   Perfluorostyrenes in the range of 2–15 Katm at 22°C. Piston position as function of load.

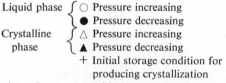

Dashed lines connect points where measureable creep occurred; see Fig. 1.

upper end of the curves there are pairs of points connected by dashed lines. These pairs of points come from measurements discussed in connection with Fig. 1. The conditions first set on the apparatus resulted in the outer points. After a time during which observable creep of the piston occurred as the system equilibrated, the points on the curves were obtained. The pressure where this phenomenon is first detected, we call for convenience the glass transition.

The other two curves show the behavior of the perfluorostyrene when it crystallizes. Crystallizing systems were induced by storage conditions shown in the figure as crosses. Crystallization was extremely difficult to induce and was first detected after overnight storage at 21° and at a pressure subsequently found to be 2 katm higher than the equilibrium value. Once crystals were present the freezing was rapid and determination of the equilibrium pressure

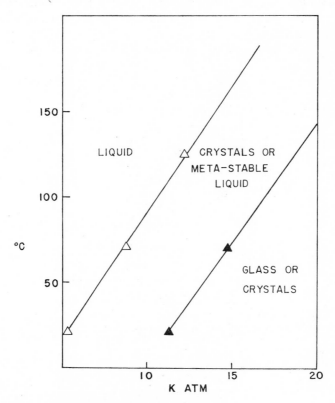

Fig. 3.   Phases of Perfluorostyrene (16):
  △ Melting temperature at an indicated pressure
  ▲ Viscosity becomes high, near or approaching the
      glassy state

easy. During this phase-transition study the discovery of polymer in the sample suggested that thermal initiation may be more effective than gamma-ray initiation. Gamma-ray experiments had been tried first but produced polymer of relatively low intrinsic viscosities.

Figure 3 shows the melting temperatures and the glass temperatures as a function of pressure. Proceeding horizontally across the figure, say at 100°C, we first pass through the liquid region into a region where we can have either a crystal phase of metastable liquid or both. Finally we enter the glass or high-viscosity region or crystalline region. Since the polymerization behavior may depend on the phase and the pressure can drop off as polymerization or crystallization occurs, it is clear that especially in the vicinity of the transitions rates may not be simple functions of the temperature and pressure. This situation could be rendered more ambiguous by chemical effects resulting from inhibitory, catalytic, or transferring impurities.

Figure 4 summarizes data (16) on the polymerization rates with an Ar-

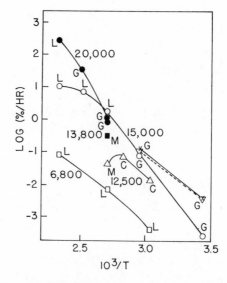

Fig. 4. Polymerization of perfluorostyrene, thermally initiated rates (16):

      ☐   6,800 atm
      △ 12,500
      ■ 13,800
      ○ 15,000
      × 15,000 γ-ray
      ▽ 15,000 γ-ray corrected for thermal
      ● 20,000 atm

Letter associated with each point indicates phase of monomer at termination of polymerization. C = crystalline, M = metastable liquid, L = liquid.

rhenius plot. Most of the data are for thermal polymerization. The lines are not straight, but where sufficient data were obtained they curve downward at higher temperatures. The letters near the points indicate whether the monomer was in the liquid (L), metastable liquid (M), or the crystalline (C) phase. It seems clear that the reaction rate in a given phase condition increases with pressure but that the rate tends to go through maximum. A better insight to the effects of phase can be seen by considering Figure 5. In

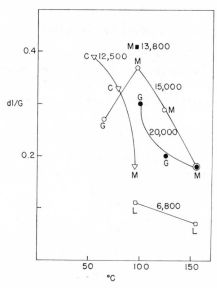

Fig. 5. Perfluorostyrene, intrinsic viscosity of polymer as a function of temperature polymerization at different pressure. Letters near points indicate phase at termination of polymerization:

$\square$ 6,800 atm
$\triangledown$ 12,500
$\blacksquare$ 13,800
$\bigcirc$ 15,000
$\bullet$ 20,000

this figure the influences of temperature and pressure on the intrinsic viscosity of the polyperfluorostyrene produced are shown. At 15,000 atm the intrinsic viscosity shows a maximum as the temperature is raised. It seems evident that at a constant temperature a maximum also exists in the intrinsic viscosity as the pressure is raised. The complexities introduced by the phase problem and the limited data obscure the interpretation considerably. Still it seems clear that the optimum condition for high-polymer formation exists in the metastable liquid region.

Referring back to Figure 3, to the left and above the melting line there also exists a line for the ceiling temperature as a function of pressure. The ceiling temperature is that temperature where at given pressure the rate of propagation equals that for depropagation. Above this critical temperature, polymer cannot form. An approach to this condition can in theory produce maxima in the rates of polymerization as a function of temperature but not of pressure. This phenomenon is not important to results obtained with perfluorostyrene so far.

C.　Gamma-Ray Initiation

Figure 4 and Table 2 show some results for gamma-ray-induced polymerization of perfluorostyrene. The intrinsic viscosities are lower, as would be expected with the greater initiation rate produced by adding gamma rays to a thermally initiating system. In the last two runs listed in Table 2, glassy and

TABLE 2

Radiation-Induced Polymerization of Perfluorostyrene

Pressure (atm. × 10^{-3})	Temp. (°C)	Time (hr)	Dose rate (rad/hr)	Conv. (%)	R_p (10^3 %/hr)	$[\eta]$ (dl/g)
15	17[a]	141	1,200	0.74	5.2	0.15
15	65[a]	65.6	1,200	8.11	129	0.24
15	17[a]	2.4[b]	150,000	5.43	45[e]	0.23
15	17[c]	3.2[d]	150,000	0.065	0.39[e]	—

[a] Irradiated as a glass.
[b] In source, stored 122 hr.
[c] Irradiated as a crystalline solid.
[d] In source, stored 166 hr.
[e] Overall rate

crystalline solids were irradiated and then stored at high pressure outside the radiation field. The conversion was higher with the glassy sample and the intrinsic viscosity is higher than that of the continuously irradiated sample even though the stored sample had a hundredfold higher dose rate. This implies that considerable polymerization occurred during storage.

V.　PROPYLENE

Since a considerable number of results on various fluoropropylenes will be reviewed, we first briefly review work on the high-pressure polymerization of propylene. Unlike styrene, which is readily polymerized to high-molecular-weight polymer by free-radical initiators, propylene at 1 atm is not. In recent years propylene has been polymerized to amorphous polymer by acid catalysts and to crystallizable isotactic polymer by the Ziegler-Natta systems. The failure of propylene to polymerize with free-radical initiators is explained on the basis of the rapidity of allylic hydrogen abstraction from propylene. The resultant allyl radicals are resonance stabilized and incapable of propagation.

Under high pressures up to 16,000 atm and γ-radiation extremely high rates of polymerization can be achieved (17). Degrees of polymerization obtained, however, were relatively low, in the range of 50–100. They were independent of the rate of polymerization, which was varied by altering the radiation intensity. In the absence of radiation, there was essentially no thermal polymerization; see Table 3 which shows the effect of radiation

TABLE 3

Polymerization of Propylene Induced by Gamma Rays (Variation of rate and number average degree of polymerization as a function of radiation intensity at 21°C and 14,600 atm pressure)

r	$\bar{P}_n^{-1}$	$I \times 10^2$ (Mrad/hr)	$I^{1/2} \times 10^2$ (Mrad/hr)
0.015	—	0.0	0.0
0.32	0.15	0.31	5.6
0.68	0.16	1.08	10.4
1.08	0.14	13.0	36.0

intensity on rate and degree of polymerization $\bar{P}_n$. The degree of polymerization is unchanged as intensity is increased, whereas the rate increases. The increase in the rate is at low dose rates proportional to the square root of the intensity but drops off at high dose rates. This behavior is seen with many of the fluoromonomers polymerized in the liquid state. Such behavior is common also in ordinary radiation-induced polymerization and has been attributed (18) to the termination of a substantial portion of the initiating radicals by polymer radicals or other initiating radicals.

Figure 6 shows both the variation in the rate and the degree of polymerization as a function of pressure, at several temperatures. The rate accelerates strongly with both temperature and pressure. The number of monomer molecules consumed by 100 eV of radiation energy absorbed, $G(-M)$, varies from 570 to 108,000. Since the $G(R)$ for the primary radicals produced by γ-rays probably is in the range $1 < G(R) < 10$, this indicates very long kinetic chain lengths. Because the degree of polymerization is independent

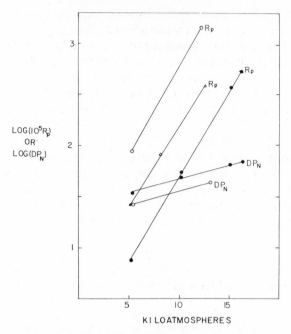

Fig. 6. Gamma-ray polymerization of propylene, rate and number average degree of polymerization as a function of pressure (17). Dose rate 0.0031 Mrad/hr. (○) 83°C; (△) 48°C; (●) 21°C. (Reprinted by permission of copyright owner The Am. Chem. Soc.).

of rate, no estimate of the actual $G(R)$ can be made from the experimental data obtained on the rate of propylene polymerization. This independence plus the large $G(-M)$ values and low $\bar{P}_n$ indicate that transfer with monomer controls the molecular weight, i.e., $\bar{P}_n = k_2/k_3$. Referring to Fig. 6, it is seen that $\bar{P}_n$ is only slightly increased with pressure, and unlike the rate it decreases with temperature. Thus it can be concluded that both $\Delta V_2^{\ddagger}$ and $\Delta V_3^{\ddagger}$ have appreciable negative values with $\Delta V_2^{\ddagger}$ only slightly more so. From the temperature behavior of $\bar{P}_n$ it can be concluded that $\Delta H_2^{\ddagger} > \Delta H_3^{\ddagger}$ by about 800 cal/mol (see Table 4).

Table 4 lists some estimated changes in the values of the composite thermodynamic quantities on going into the transition state, deduced from the rate and molecular weight results. To calculate the free energy and hence the entropy change, a value of $G(R)$ is needed. For the estimated $\Delta H\dagger$ and $\Delta S\dagger$ values listed in Table 4, a $G(R)$ of one was assumed. In the original publication (17) an error involving radiation units resulted in the listing of erroneous $\Delta S_{\gamma}^{\ddagger}$ values for the rate of polymerization. The values in Table 4 have been corrected. Calculations using a value of 10 for $G(R)$ increase the listed values

TABLE 4

Thermodynamic Quantities of the Transition State for Rate and Degree of Polymerization of the Gamma-Ray Initiated Polymerization of Propylene [G(R) assumed to be 1.0]

	Pressure (Katm)	$\Delta H_y^{\dagger}$ (cal/mol)	$-\Delta S_y^{\dagger}$ (cal/mol °C)
Rate	5.0	7,800	13.8
	12.0	8,160	7.5
	16.4	8,400	3.3
Degree of	5.0	−855	4.1
polymerization	12.0	−760	5.3
	16.4	−700	6.1

by a fraction of a kilocalorie. Since the actual $G(R)$ value is probably nearer to 1 than to 10, the values in the table are probably quite reliable. For the rate, $\Delta V_y^{\dagger} = \Delta V_2^{\dagger} - \frac{1}{2}\Delta V_4^{\dagger} = -9.62$ cc/mole at 21°C, −9.64 at 48°C, and −12.22 at 83°C. For the degree of polymerization $\Delta V_{Pn}^{\dagger} = \Delta V_2^{\dagger} - \Delta V_3^{\dagger}$ = −1.57 at 21°C and −1.67 cc/mol at 83°C.

VI. HEXAFLUOROPROPYLENE

Under autogenous pressures the polymerization of hexafluoropropylene produces high-boiling liquids (19,20). High polymers can be formed by either catalytic (21,22) or gamma-ray initiation (23) and pressures of 3–15 kbar and temperatures of 100–230°C. Polymers of hexafluoropropylene have also been reported to have been prepared by the use of Ziegler-Natta catalyst (24) and these have been claimed to be isotactic and crystallizable. They were also reported to be soluble in organic solvents such as acetone. Since the polymers formed at high pressure are amorphous and soluble in FC-75, a technical perfluoro liquid composed of a mixture of cyclic perfluoroethers manufactured by the 3M Co., and in hexafluorobenzene but not in acetone, it can be concluded that the product produced by the Ziegler-Natta catalysts is not a pure perfluoro polymer. Since polymers such as those of trifluoroethylene and 3,3,3-trifluoropropylene are soluble in acetone, it can tentatively be concluded that the Ziegler-Natta catalyst facilitaties fluoride exchange with some proton donor substances in the polymerization system.

In Fig. 7 results on the polymerization of hexafluoropropylene (23) are shown as Arrhenius plots at three pressures, 4.5, 10, and 15 katm. Two samples of perfluoropropylene were studied and erratic results were obtained with the second sample. Mass-spectrometric analysis showed both samples to be pure, but small peaks were present in the mass spectra of the second sample.

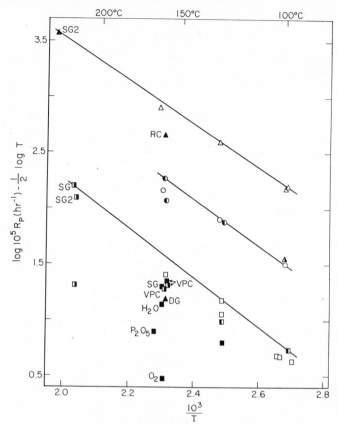

Fig. 7. Polymerization rates at high pressure of perfluoropropylene induced by γ-rays (23):

For 15,000 atm pressure

△ Tank I monomer, 47,000 rad/hr

▲ Tank II monomer, 47,000 rad/hr

◮ Tank II monomer, calculated for 47,000 rad/hr from other dose rates

For 10,000 atm pressure

○ Tank I monomer, 47,000 rad/hr

◑ Tank I monomer, calculated for 47,000 rad/hr from other dose rates

For 4500 atm pressure

□ Tank I monomer 47,000 rad/hr

◪ Tank I monomer calculated for 47,000 rad/hr from other dose rates

▣ Tank II monomer, 47,000 rad/hr

■ Tank II monomer calculated for 47,000 rad/hr from other dose rates

O_2 13 ppm Oxygen added

H_2O Saturated with water at $-40°C$

P_2O_5 Condensed on P_2O_5 before final transfer

DG More thoroughly outgassed than normal

RC Recovered monomer

SG Passed over activated silica gel at $-15°C$

SG2 Passed twice over activated silica gel at $-15°C$

VPC Purified on a preparative-scale vapor-phase chromatograph

These peaks were too small for specific identification of the compounds responsible for them. The compounds appear to be unsaturated hydrocarbons with three or less carbon atoms and their total concentration is approximately 0.1 mole % or less. Various methods of purification were tried and rates were obtained (see Fig. 7) which were substantially comparable to those obtained with the first sample.

The lines in Fig. 7 are drawn through the highest rate data obtained at specific temperatures and tentatively assumed to be representative of pure perfluoropropylene. The rate varied as the square root of the radiation intensity. The polyfluoropropylene formed was soluble in fluorocarbon solvents and intrinsic viscosities were measured in FC-75. Number average molecular weights were measured with a vapor pressure osmometer using hexafluorobenzene solutions. The largest intrinsic viscosity measured was 2.0 dl/g. From this value a number average molecular weight of 5.7×10^6 is estimated. Table 5 presents characterization and polymerization data for polymers formed at rates on the isobars of Fig. 7. The rate and degree of polymerization increase with pressure and temperature. It appears that in the

TABLE 5

Polyperfluoropropylene Formed at Rates on the Isobars of Figure 7

Conditions							
Pressure (atm $\times 10^{-3}$)	Temp. (°C)	Dose rate (rad/hr $\times 10^{-3}$)	R_p (hr^{-1})	$DP_{n_a}{}^a$	$[\eta]$ (dl/g)	$DP_{n_b}{}^b$	$10^{10} R_p$ $4.7\ I$
4.5	100	140	0.0018	18	0.03	20	27
4.5	130	47	0.0031	83	0.07	94	103
4.5	220	47	0.035	100	0.19	560	1,600
10	100	47	0.0061	62	0.10	180	270
10	130	47	0.016	46	0.12	250	730
10	128	140	0.027		0.14	320	410
10	160	140	0.066		0.24	850	1,020
15	100	47	0.030		0.33	1,500	1,400
15	130	47	0.079		0.59	4,400	3,600
15	160	47	0.173		0.72	6,300	7,900
15	230	1.5	0.152		2.0	38,000	220,000

^{a} DP_{n_a} is from measurements with a vapor pressure osmometer.

^{b} DP_{n_b} is calculated from a relation determined for fractions of poly-3,3,3-trifluoropropene in acetone, $[\eta] = 3.8 \times 10^{-4} (96\ DP_v)^{0.55}$, where $[\eta]$ is the intrinsic viscosity at 29.6°C and DP_v is the viscosity-average degree of polymerization. It is assumed that DP_v is 1.5 times DP_n.

range of 4.5–15 katm, 100–230°C, and dose rates used that the molecular weight is controlled by both transfer and termination. It may easily be that slight amounts of impurities are involved. The activation energy or the enthalpy change, $\Delta H_\gamma^\ddagger$, for the rate expression for the γ-ray initiated process, is 10 kcal/mol at 4.5 katm, 9.6 at 10 katm, and 9 at 15 katm, while the $\Delta V_\gamma^\ddagger$ is -10 cc/mol.

VII. 3,3,3-TRIFLUOROPROPYLENE

In general, fluorohydropropylenes polymerize under pressure more readily than either propylene or perfluoropropylene. Polymers have been produced from 3,3,3-trifluoropropylene (25), 2,3,3,3-tetrafluoropropylene, and 1,3,3,3-tetrafluoropropylene.

The polymerization of the monomer 3,3,3-trifluoropropylene has been reported in several patents (26,27). Also, the use of coordination catalysts has been reported to produce low-molecular-weight polymer (28). The polymerization of 3,3,3-trifluoropropylene under pressure has been studied in detail (25). Table 6 lists some representative results. As seen in the variation

TABLE 6

Polymerization of 3,3,3,-Trifluoropropylene, Variation of Rate and Intrinsic Viscosity with Square Root of Dose Rate

Pressure (atm)	Temp. (°C)	$I^{1/2}$[rad/hr)$^{1/2}$]	$10^3 R_{p-1}$ (10^3hr)	$[\eta]$ (dl/g)	$\bar{M}_v \times 10^{-6}$
1,820	25	37	1.04	0.35	0.25
		90	3.0	0.23	0.12
5,060	25	39	11.4	1.28	2.8
		90	27	0.88	1.4
5,060	68	39	35	1.18	2.4
		90	80	1.06	2.0

of intrinsic viscosity of the polymers, the molecular weight increases with pressure but not with temperature. The rate of polymerization varies with the square root of the γ-radiation intensity.

The polymer is soluble in acetone and hexafluorobenzene. Two samples for instance, gave intrinsic viscosity values in acetone of 0.15 and 3.5 dl/g compared, respectively, to values of 0.25 and 8.5 dl/g in hexafluorobenzene. A sample of the polymer was fractionated from mixtures of acetone and benzene. On solutions of four fractions osmotic measurements gave number average molecular weights from 1.42×10^5 to 6.9×10^5. Based on these data and intrinsic viscosity values for the same samples, the following relationship between intrinsic viscosity of acetone solution at 29.6°C and the viscosity average molecular weight is obtained:

$$[\eta]_{acetone} = 4.2 \times 10^{-4} \, \overline{M}_v^{0.54} \tag{6}$$

The low value of 0.54 for the exponent of $\overline{M}_v$ indicates that acetone is a poor solvent. At the somewhat lower temperature of 25°C a polymer of $[\eta]_{acetone}$ = 5.2 dl/g separates from acetone.

By using Eq. 6, only the viscosity average molecular weight can be calculated. However, it is very likely that the polymers studied have the "most probable" molecular weight distribution and we may assume that $\overline{M}_v/\overline{M}_n = 1.8$. This presumes that the polymer radicals are terminated by transfer, disproportionation or both. If termination were purely by recombination, this ratio would be 1.4. Application of Eq. 4 to experimental data at different intensities enables us to make estimates of the ratio of monomer transfer to propagation, k_3/k_2 and also k_4/k_2^2. From a given rate of polymerization this last ratio permits the calculation of the overall rate of initiation from which, when radiation initiation is used, follows the radiation yield $G(R)$. Table 7 lists values of

TABLE 7

Transfer Constants in the Polymerization of 3,3,3-Trifluoropropylene

Pressure (atm)	Temp. (4C)	$10^3 k_3/k_2$	$G(R)$ (radical/100 eV)
1,800	25	30	5
	99	340	(4.0)
5,000	25	2	4
	69	7	3
	99	40	(4.0)
8,000	25	0.4	(4.0)
	68	0.8	(4.0)
10,000	25	0.3	(4.0)

$G(R)$ and k_3/k_2 for 3,3,3-trifluoropropylene. Experimental values of $G(R)$ were obtained for only three sets of conditions. The values are relatively constant, as theoretical considerations suggest, and hence values at other conditions were assumed to be 4.0. This permits other k_3/k_4 ratios to be calculated from a minimum of experimental data. In table 7 these assumed values are shown in parentheses. The relative importance of the transfer step varies considerably with reaction conditions. At 1800 atm and 99°C, 98% of the growing radicals terminate by transfer, whereas at 5000 atm and

25°C only 25% do so. Thus the molecular weight decreases with temperature at constant pressure and eventually becomes completely controlled by transfer. At constant temperature the molecular weight increases with pressure since pressure increases the rate and decreases k_3/k_2.

It is possible that the observed transfer is not with monomer but with some impurity. If the impurity to monomer ratio were 10^{-3}, then its reactivity ratios to propagation are 10^3 times the values shown in Table 7. From the variations of k_3/k_2 with pressure and temperature it is estimated that $\Delta V_3^\ddagger - \Delta V_2^\ddagger$ and $\Delta H_3^\ddagger - \Delta H_2^\ddagger$ are, respectively, 20 cc/mol and 10 kcal/mol.

The thermal decomposition of the poly-3,3,3-trifluoropropylene varied considerably with the conditions of polymerization (29). A sample prepared at 10 katm and $-80°C$ gave the slowest rate of thermal decomposition. It was

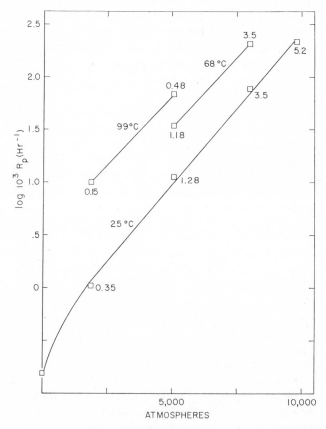

Fig. 8. Gamma-ray induced polymerization of 3,3,3-trifluoropropylene as a function of pressure. Dose rate 1.5 krad/hr. The intrinsic viscosities in acetone (dl/g) of polymer produced at each point are given by the numbers adjacent to the points (25). (Reprinted by permission of copyright owner The Am. Chem. Soc.).

concluded that this variation in thermal stability was the result of variation in the polymer transfer process during the polymerization reaction. By analogy with the behavior of various polyethylenes, the more branched polymers would give initially higher rates of decomposition.

The effect of pressure on the rate 3,3,3-trifluoropropylene polymerization is shown in Fig. 8. The radiation intensity used was 1.5 krad/hr. The temperatures of the three isotherms are listed to the left of the lines. The number adjacent to each point is the intrinsic viscosity of the polymer formed. The slopes of these lines indicate the activation volume change, presumably $\Delta V_2^{\ddagger} - 0.5 \, \Delta V_4^{\ddagger}$. It can be seen from Fig. 8 that these values are not entirely constant with pressure. In the low-pressure range from 5 to 1800 atm the activation volume change at 25°C is 26 cc/mol. At the same temperature the value goes from 19 to 16 cc/mol as the pressure is increased to 8000 atm. At 5100 atm the activation enthalpy term, $\Delta H_2^{\ddagger} - 0.5 \, \Delta H_4^{\ddagger}$, is 5 kcal/mol and the entropy term, $\Delta S_2^{\ddagger} - 0.5 \, \Delta S_4^{\ddagger}$, is 14 cal/mol°C. These values decrease slightly at higher pressures.

VIII. 2,3,3,3-TETRAFLUOROPROPYLENE

The polymerization of 2,3,3,3-tetrafluoropropylene has been studied at high pressure using monomer prepared by the deiodofluorination of 1,1,1,2,2-pentafluoroiodopropane (30). Deiodofluorination using zinc in acetic acid

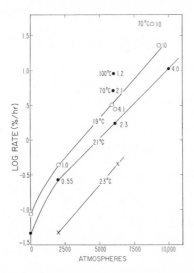

Fig. 9.   Gamma-ray induced polymerization of 2,3,3,3-tetrafluoropropylene as a function of pressure. Dose rate 1.4 krad/hr. The numbers adjacent and to the right of the points are the intrinsic viscosities (dl/g) in acetone (30). Monomer prepared by deiodofluorination with (○) magnesium, (●) zinc, (×) lithium methyl.

gave monomer containing about 5% 1,1,1,2,2-pentafluoropropane, which, however, polymerized well to give high-molecular-weight polymer. To avoid the formation of the pentafluoropropane, the defluorination was then carried out using magnesium in dry ether. The monomer, although now >99.8% pure, polymerized at a slow rate until purified by vapor-phase chromatography. This purified monomer gave faster rates and higher molecular weight polymer than the zinc preparation.

Lithium methyl was used for the defluorination in order to obtain a better yield. The purity of this product was better than 99.9%, but its rate of polymerization was low. Purification of this preparation improved the rate of polymerization, but rates as high as the highest of previous preparations were not achieved. Some results of this study are shown in Fig. 9. The isotherms for the differently synthesized products are similar, giving an overall activation volume change of between -11 and -12.5 cc/mol. The temperature dependence of the rates were also insensitive to its synthetic history and overall activation enthalpies of 4 kcal/mol were obtained.

IX. 1,3,3,3-TETRAFLUOROPROPYLENE

This monomer has been studied only to a very limited extent (31). At 13 kbar and 70°C, the rate polymerization for the liquid state was quite low, 0.5%/hr. The resulting polymer has low molecular weight. Experiments at higher pressures would put the material in the crystalline state. At 22°C the monomer freezes at 9 kbar, at 70°C the freezing pressure is 13 kbar.

X. THE EFFECT OF STRUCTURE, FLUOROPROPYLENES

Comparison of the high-pressure polymerization of the fluoropropylene monomers can best be made by considering Table 8. For the indicated conditions there are listed the overall rates and activation volumes. It is seen that

TABLE 8

Polymerization Rates of Fluoropropylenes at 5 katm Pressure, 25°C, and 3 krad/hr

Monomer	Rate (% hr $\times 10^4$)	$-(\Delta V_2^{\dagger} - 0.5 \Delta V_4^{\dagger})$ (cc/mol)
$CH_2 = CHCH_3$	89	9.6
$CH_2 = CHCF_3$	14,000	17.0
$CH_2 = CFCF_3$	16,000	11
$CHF = CHCF_3$	330	—
$CF_2 = CFCF_3$	3.2	12.3

the highest rates are found with the partially fluorinated monomers, particularly the 2,3,3,3-tetrafluoropropylene and the 3,3,3-trifluoropropylene. These monomers also generally give higher-molecular-weight polymers. Their polymerizability may be enhanced by polar effects, similar to those in vinyl and vinylidene chloride. With these fluoromonomers pressures as low as 2 kbar can produce polymer with molecular weight adequate for good physical properties.

Since the activation volume change is for the gamma-induced process, it is a composite of two terms, the propagation term and the termination term. It is likely that termination is diffusion controlled and hence retarded by pressure. The $\Delta V_4^{\ddagger}$ term would thus be positive and the values for $\Delta V_2^{\ddagger}$ would be less negative than the values listed in Table 8. The data in Table 8 suggest that the viscosity of the trifluoropropylene increases most rapidly with pressure of all the five monomers listed. This monomer very likely has the highest permanent dipole moment.

For the five monomers listed, the activation energy, $\Delta H_2^{\ddagger} - 0.5 \ \Delta H_4^{\ddagger}$, varied from 4 to 10 kcal/mol and the activation entropy changes, $\Delta S_2^{\ddagger} - 0.5 \ \Delta S_4^{\ddagger}$, were between 11 and 16 cal/degree-mol. Again, interpretation of any trend in these values is difficult in view of the strong possibility that termination is diffusion controlled. If $\Delta H_4^{\ddagger} \gg 0$, we would deduce that the $\Delta H_2^{\ddagger}$ values are perhaps 20 or so kcal/mol. The lack of polymerizability at low autogenous pressures is then explicable.

XI. VINYL *n*-PENTAFLUOROETHYL AND VINYL *n*-HEPTAFLUORO-PROPYL

By comparison with vinyl trifluoromethyl (3,3,3-trifluoropropylene), the monomers

$$CH_2=CH \qquad \text{and} \qquad CH_2=CH$$
$$| \qquad\qquad\qquad\qquad |$$
$$CF_2CF_3 \qquad\qquad\qquad CF_2CF_2CF_3$$

would be expected to polymerize with difficulty. High polymers of these monomers can have desirable properties, whose study can greatly clarify the role of molecular structure on physical properties. With gamma-ray initiation, these monomers require pressures of 10–15 kbar to produce polymers with appreciable intrinsic viscosities (25). These polymers, unlike vinyl trifluoromethyl, are not soluble in acetone, but they are soluble in hexafluorobenzene, in which their intrinsic viscosities were determined. In Fig. 10, the variation of the log of the rate polymerization with pressure is presented. Some single point measurements also are indicated in the figure. Maximum polymerization rates were from 10 to 30%/hr, using the lowest dose rate convenient, $\sim 2\,\text{krad/hr}$. Over the pressure range studied the vinyl perfluoroethyl does not freeze. However, at 22°C the vinyl perfluoropropyl freezes in the vicinity of 13–14 kbar and the maximum in the rate observed at 22°C corresponds to the freezing of the monomer. Thus for this monomer, polymerization is prevented by the solid state. For these conditions the $\Delta V_2^{\ddagger}$ values quoted in the tables are for the liquid state, i.e., for pressures below 13 kbar.

As the size of the pendant fluorocarbon groups is increased, markedly smaller intrinsic viscosities for polymers prepared under similar conditions are found. The effect is large enough to imply that the molecular weights change in the same direction, regardless of conceivable variations in the molecular weight-intrinsic viscosity relationships for the polymers. Upper limits for the number average molecular weights $\overline{M}_n$ of the polymers can be estimated by dividing the rate of polymerization by an initiation rate calculated by assuming the radiation yield of radicals $G_i(R)$ is 4. Values of $\overline{M}_n$ so calculated are, respectively, 3×10^6 and 3×10^5 for the polypentafluorobutene and the polyheptafluoropentene formed at 21°C and 15 kbar. The values are unreasonably large for polymers with intrinsic viscosities of 0.67 and 0.12 dl/g. Transfer of some type must therefore be involved in these polymerizations.

At still higher temperatures transfer must become increasingly important because as the rate increases, $[\eta]$ decreases. At higher pressure, temperature remaining constant, $[\eta]$ and the rate increase. Table 9 compares the apparent transition state volume, enthalpy, and entropy change for the gamma-ray induced polymerization of the three vinyl perfluoroalkyl monomers. We see that enthalpies or activation energies are significantly lower for the 3,3,3-trifluoropropene. The volume and entropy vary considerably and are not independent of pressure and temperature. Since the radical termination process, which is diffusion controlled, contributes to these overall values any interpretation is difficult. It may be that the viscosity of the medium, by determining the $\Delta V_4^{\ddagger}$, $\Delta H_4^{\ddagger}$, and $\Delta S_4^{\ddagger}$ terms, plays the most significant role in these reactions. The activation energies for $\Delta H_2^{\ddagger}$ are undoubtedly somewhat higher than the composite values listed.

In this connection recent experiments (32) have compared the rates of polymerization and intrinsic viscosities, $[\eta]$, for polyvinyl-n-heptafluoropropyl with the bulk viscosity η of the initial monomer at the pressures of the polymerizations. A postirradiation effect was observed at 10 katm and above. Table 10 summarizes these results. The second column of Table 10 lists the viscosity of the initial monomer at the pressure indicated in the first column. The third column gives the hours the sample was exposed to gamma rays of 38 kr/hr intensity while the fourth column indicates hours of subsequent storage under pressure. At 5 katm it is seen that the conversion is proportional

TABLE 9

Polymerization of Vinyl Perfluoroalkyls

Monomer	$-(\Delta V_2^\ddagger - 0.5 \Delta V_4^\ddagger)$ (cc/mol)	Pressure range (katm)	Temp. (°C)	$\Delta H_2^\ddagger - 0.5 \Delta H_4^\ddagger$ (kcal/mol)	$-(\Delta S_2^\ddagger - 0.5 \Delta S_4^\ddagger)$ (cal/mol°)	Pressure (katm)
CH_2CHCF_3	16	5–8	25	5	15	5
	17	5–8	68	4	13	8
	12	2–10	22	4	14	10
CH_2CH | CF_2CF_3	8	10–15	22	10	6	5
	9	10–15	70	7	11	10
	10	5–10	126	7	8	15
CH_2CH | $CF_2CF_2CF_3$	19	2–5	22			
	11	5–10	22	7	16	5
	5	10–15	126	6	13	10

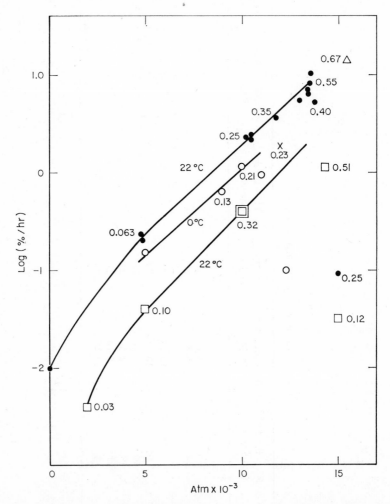

Fig. 10. Variation of polymerization rate for vinyl *n*-perfluoropropyl with pressure. Numbers adjacent to points are the intrinsic viscosities of the polymer. Full lines end at freezing pressures.

● 22°C, 38 krad/hr
△ 22°C, 38 krad/hr, 20 mol % CH₃CHF₂ diluent
○ 0°C, 38 krad/hr
× 0°C, 38 krad/hr, 20 mol % CH₃CHF₂ diluent
□ 22°C, 2 krad/hr

TABLE 10

Polymerization of Vinyl *n*-Heptafluoropropyl at 22°C and 38 krad/hr

Pressure (katm)	η (poise)	Hours irradiated	Hours stored	Rate (% /hr)	[η] (dl/g)
5	0.005	38.3	0	0.23	0.063
		75.0	0	0.20	0.064
		36.5	50	0.25	0.065
		43.8	120	0.22	0.065
10	0.2	3.5	0	2.1	0.25
		6.0	0	2.3	0.25
		3.5	17	5.3	0.29
14	$>10^5$	1.0	0	3.7	0.46
	solid state	1.0	3	30.0	0.57

to hours irradiated. At 10 and especially 15 katm it is seen that appreciable polymerization occurs during the storage period.

Further experiments have shown that the post-polymerization is not the result of high viscosity and of long-lived free radicals, since irradiation of the monomer at low autogenous pressures produces the catalytic agent, which is not distillable. The catalytic substance decays faster at low than at high pressure but is too long-lived to be a radical active enough for polymerization. The substance is believed to be an unstable peroxide (33).

In addition to the complication of the radiation-created catalyst, the pentene polymerization is complicated, as shown in Fig. 10, by the freezing of the monomer.

XII. PERFLUOROHEPTENE

Perfluoroheptene polymerization requires severe conditions to produce appreciable rates of polymerization and even moderate molecular weights (34). Table 11 lists results of some experiments on this monomer. It is seen that even at 17.1 katm pressure quite high temperatures are required. The degrees of polymerization $\bar{P}_n$ are low and in many cases adequate amounts of polymer for $\bar{P}_n$ determinations were not produced. It should be noted that the rates of polymerization R_p go through a maximum with temperature at a given pressure. They increase markedly with pressure, however. Monomer transfer constants are listed in Table 11 in three instances. The rate at a given pressure and temperature was varied by varying the dose rate. Figure 11 shows a plot of the reciprocal degrees of polymerization as a function of rate at 17.1 katm and 139°C.

Equation 4 was applied and a value of 0.028 was obtained for the ratio

TABLE 11

Polymerization of Perfluoroheptene-1 Induced by Gamma Rays, 2.7 krad/hr

Pressure (katm)	Temp. (°C)	$10^6 R_p$ (hr^{-1})	$\bar{p}_n$	k_3/k_2
8.1	97	23.5		
	125	54.5		
	146	73.9		
	168	40.9		
11.9	93	88		
	140	378	42	<0.005
	166	521		
	183	386		
	237	2.5		
17.1	94	109	19	
	139	1047	26	0.28
	189	3760	76	
	211	6113	52	0.17
	267	7240	17	

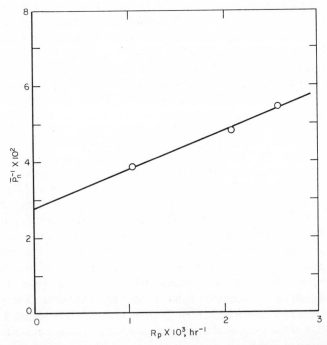

Fig. 11. Polymerization of perfluoroheptene-1 induced by gamma rays at 17.1 katm and 139°C. Variation in the reciprocal number average degree of polymerization with rate.

k_3/k_2 from the intercept. From the slope the rate of initiation is obtained, which permits an estimate of the radiation yield of primary radicals $G(R)$. Assuming pure disproportionation, we can calculate a value of 11.5, assuming combination gives the value of 23. These results are reasonable since a value of 20 was determined for perfluoroheptane at low pressure (35).

The transfer constant could not be determined for all situations. At 94°C and 17.1 katm the pressure-volume behavior indicates that the monomer has a change of state. The rate constants are abnormal and the estimated k_3/k_2 is found to be negative. At 267°C the rate falls off in its variation with temperature and again k_3/k_2 cannot be calculated. This fall off is attributed to an increasingly important reverse polymerization process as the temperature is raised. At other conditions the isolation of representative polymer is difficult since the molecular weights are low and some dimer and trimer are evidently present.

The maximum in the rates with increasing pressure can be attributed to a back reaction in the radical growth process. When the data are placed on an Arrhenius plot the anomalous maximum is still very pronounced. With a proper theory curves can be much more informative than straight lines. The curves in Fig. 12 were evaluated by introducing explicitly the depropagation reaction

$$R_{i+1} \xrightarrow{\ k_{-2}\ } R_i + M$$

into the previously discussed polymerization mechanisms. Straight lines through the low-temperature portions of the curves were assumed to be entirely determined by the forward polymerization process alone. The deviation of the high-temperature portion of the curves from these straight lines (dotted lines in Fig. 12) were taken to be a measure of the depolymerization rate. These deviations are plotted as the points with primes in Fig. 12. Straight lines through these points intersect the dotted lines. At these points of intersection the forward and back reactions are equal in magnitude. From such an analysis of the data we obtained the thermodynamic quantities for the composite transition state terms for both the polymerization and depolymerization rates (Table 12) and the overall thermodynamics for the polymerization (Table 13).

In contrast to the effect of the perfluoroalkyl group seen in Table 9, the perfluoropentyl has little effect on the activation enthalpy of the fluorinated vinyl polymerization but makes the activation entropy twice as negative. The activation volume change, $\Delta V_2^\ddagger - 0.5\,\Delta V_4^\ddagger$, varied with pressure at 139°C; in the vicinity of 11.9 katm it was -9.4 cc/mol. The activation volume for depolymerization is expected to be positive. The dashed lines in Fig. 12 indicate a somewhat negative value. The dashed lines are very close to one another

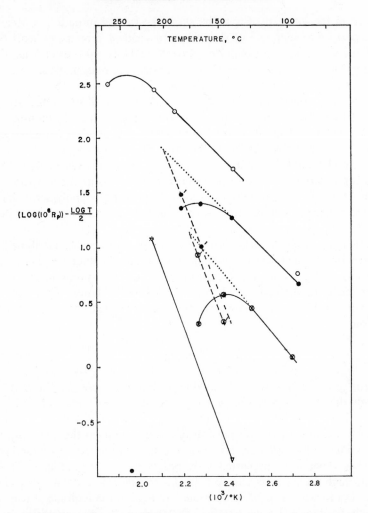

Fig. 12. Arrhenius plots for the gamma-ray, 2.7 kr/hr-induced polymerization of per-fluoroheptene-1 (34). (○) 17.1 katm, (●) 11.9 katm, (⊗) 8.1 katm, (▽) 17.1 katm, thermal. Points with primes are calculated for back reaction.

and a small error in the measured rates could easily reverse the relative position of the lines. Therefore any value estimated would be highly suspect. Although the perfluoroheptene is believed to be in a liquid state for all measurements shown in Fig. 12, some mechanism brought about by abrupt changes in physical properties rather than the back reaction may be operative. It is also possible that the behavior may be the result of impurities, auto-catalysts or autoinhibitors.

TABLE 12

Transition State Thermodynamics for the Polymerization and Depolymerization of Perfluoroheptene-1

Pressure (katm)	Temp. (°C)	$\Delta H_2^{\ddagger} - 0.5\,\Delta H_4^{\ddagger}$ (kcal/mol)	$-(\Delta S_2^{\ddagger} - 0.5\,\Delta S_4^{\ddagger})$ (cal/mole °C)
Polymerization			
8.1	139	10.5	19.8
11.9	139	9.1	20.5
17.1	139	9.6	17.1
17.1	189	9.6	17.2
Depolymerization		$\Delta H_{-2}^{\ddagger} - 0.5\,\Delta H_4^{\ddagger}$	$-(\Delta S_{-2}^{\ddagger} - 0.5\,\Delta S_4^{\ddagger})$
8.1	139	21.9	−5.8
11.9	139	21.5	−5.5

TABLE 13

Thermodynamics of Polymerization and Ceiling Temperatures for Perfluoroheptene-1

Pressure (katm)	$-\Delta H$ (kcal/mol)	$-\Delta S$ (cal/mol °C)	T_c
8.1	11.4	25.6	172
11.9	12.4	26.0	205

The thermodynamics and ceiling temperatures for the overall polymerization of perfluoroheptene shown in Table 13 are reasonable and in accord with the molecular structure and the chemical properties of the polymer. At 1 atm, ΔH_p very likely has a low value, ~ 9 kcal mol, with $\Delta S_p \sim 25$ e.u. and therefore $T \sim 100°C$.

XIII. n-PERFLUORO-α,ω-DIENES

The polymerization of hydrocarbon α, ω-dienes has been shown to give in some cases soluble polymer (36,37). This observation leads to the concept of cyclo or alternating intra-intermolecular polymerization. The first highly fluorinated monomer of this type reported was 4-chloroperfluoroheptadiene-1,6, which requires the high pressure of 13.6 katm to produce high polymer (38). The polymer so prepared had an intrinsic viscosity of 0.60 dl/g, and it forms transparent, strong, flexible film with good thermal and oxidative stability. The synthesis of n-perfluorodienes of the general structure

$$CF_2{=}CF(CF_2)_n{-}\,CF{=}CF_2$$

with n varying from 1 to 4 has been reported (39–41). On general grounds these monomers should polymerize, at least to some degree, through the propagation mechanism illustrated below. If this process or some other

$$R\cdot + CF_2{=}CF \underset{(CF_2)_n}{\overset{F_2C}{\diagup\diagdown}} CF \longrightarrow R{-}CF_2{-}CF \underset{(CF_2)_n}{\overset{CF_2}{\diagup\diagdown}} CF\cdot$$

similar ring-forming polymerization occurs, exclusively soluble polymer will be produced. However, if quite high polymer is produced, only a few 1,2-additions will lead to the formation of a gel or network structure.

XIV. n-PERFLUORO-1,4-PENTADIENE

The polymerization of the perfluoropentadiene by gamma rays and high pressure was complicated by a very facile isomerization of the monomer to the perfluoro-1,3-pentadiene. This monomer gave polymers with rubbery properties, whereas the former monomer gave a more brittle, plastic product. The polymers produced (39,40) were copolymers of the two pentadienes and infrared spectroscopy enabled the investigators to determine the ratio of isomers in the polymeric structure. Table 14 lists the overall thermodynamic

TABLE 14

Transition State Thermodynamics for the Gamma-Ray Induced Polymerization of Perfluoro-1,4-pentadiene

Pressure (katm)	$\Delta H_2^{\ddagger} - 0.5\,\Delta H_4^{\ddagger}$	$-(\Delta S_2^{\ddagger} - 0.5\,\Delta S_4^{\ddagger})$	Temp. (°C)	$-(\Delta V_2^{\ddagger} - 0.5\,\Delta V_4^{\ddagger})$
8.0	15	3	106	10
11.4	13	6	140	10
14.9	11	7	167	10

quantities of the transition state. The activation volume is quite constant with temperature and pressure. On the other hand, the activation energy decreases with pressure and the activation entropy becomes more negative.

All the perfluorodienes have rates of polymerization of 4–10%/hr at 14 katm, 150°C, and dose rate of 45 kr/hr (31). High purity and low conversions are required in order to isolate soluble polymer. With the octadiene-1,7 gel formation is difficult to avoid. Table 15 lists the intrinsic viscosities in hexafluorobenzene and glass transition temperatures for some of the better polymer samples that have been prepared. All form amorphous polymers with the

TABLE 15

Intrinsic Viscosities and Glass Transitions of Polyperfluorodienes
Prepared with High Pressures and Gamma Radiation

Presumed polymer structure	$[\eta]$ (dl/g)	T_g (°C)
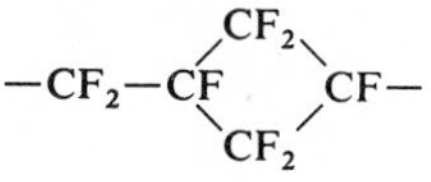 $-CF_2-CF{\overset{CF_2}{\underset{CF_2}{\diagdown\diagup}}}CF-$	0.03	105
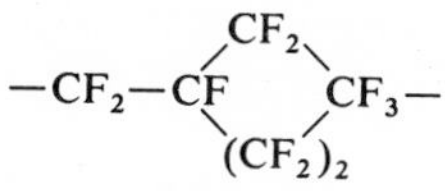 $-CF_2-CF{\overset{CF_2}{\underset{(CF_2)_2}{\diagdown\diagup}}}CF_3-$	0.70	122
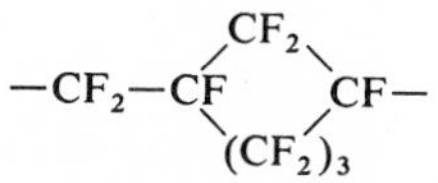 $-CF_2-CF{\overset{CF_2}{\underset{(CF_2)_3}{\diagdown\diagup}}}CF-$	insol below 80°C	101
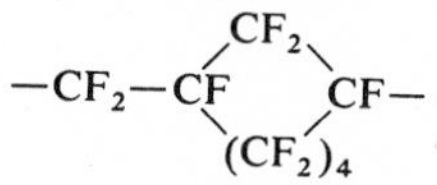 $-CF_2-CF{\overset{CF_2}{\underset{(CF_2)_4}{\diagdown\diagup}}}CF-$	gel	198
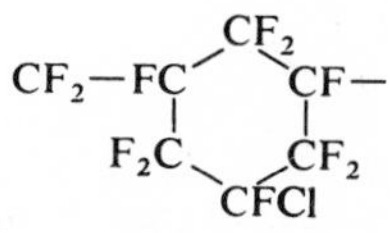 $CF_2-FC{\overset{CF_2}{\underset{\underset{CFCl}{F_2C\diagdown\diagup CF_2}}{\diagdown\diagup}}}CF-$	0.60	147

exception of the perfluoroheptadiene, which is crystalline and insoluble below
60°C.

XV. SOLID-STATE POLYMERIZATION

On four monomers, perfluorocyclobutene, perfluoropentadiene-1,4,
perfluorooctadiene-1,7, and vinyl perfluoropropyl measurements of the
gamma-ray induced polymerization have been made above and below the
freezing points (32,42). The results on the last monomer were shown in
Fig. 10, where it is seen that the rate of polymerization drops precipitously at
pressures above the freezing pressure. Figure 13 shows the effect of pressure
(42) on the melting temperature for the first three monomers, as determined
by isobaric volume changes at constant temperature. The influence of
pressure on the melting temperature can be very marked. For example, the
normal freezing temperature is -60°C. At 14.7 katm it freezes at 197°C. High

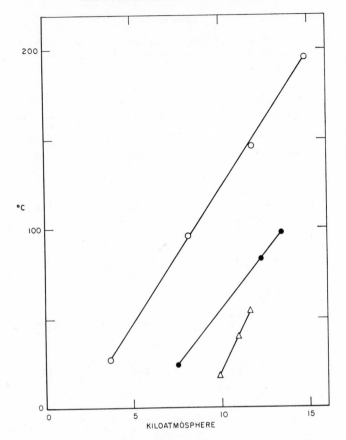

Fig. 13. Variation of melting temperature with pressure. (42) (○) perfluorocyclobutene, (●) n-perfluorooctadiene-1,7, (△) n-perfluoropentadiene-1,3. (Reprinted by permission of copyright owner The Am. Chem. Soc.).

pressure thus will permit the study of such processes over a wide temperature range. Pressure may increase the rate of solid-state polymerization if reorganization of the solid state is not required, i.e., by diffusion, migration, orientation, or rotation of monomer. If such rearrangement is necessary, then pressure should prevent polymerization.

XVI. PERFLUOROCYCLOBUTENE

Perfluorocyclobutene has been homopolymerized in both the solid (42) and liquid state and copolymerized with hexafluoroacetone (43). An acceleration of sixfold in the gamma-ray induced rate is observed in the melting region

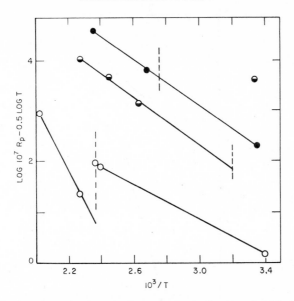

Fig. 14. Arrhenius plots of polymerization rates in the liquid and solid phases. Vertical dashed lines indicate the freezing temperature. (42) (○) perfluorocyclobutene, 11.4 katm; (●) n-perfluoroocatadiene-1,7, 12.4 katm; (◐) n-perfluoropentadiene-1,4, 10.8 katm. (Reprinted by permission of copyright owner, The Am. Chem. Soc.).

(see Fig. 14) for the cyclobutene. At 168°C the rate of polymerization appears to decrease with pressure below the freezing condition and then to increase drastically with pressure above the freezing condition, where it is in the solid state (see Fig. 15).

For cyclobutene, polymerization conditions were very critical. When in the solid state at moderately high temperature and pressures the monomer polymerizes explosively. Several pressure vessels were ruptured in studying the process (42). It is evident from Figs. 14 and 15 that with the cyclobutene, polymerization in the solid state is very much favored and that a discontinuity occurs in the polymerization mechanism at the freezing condition. Impurities will add even greater complications to such solid-state polymerizations since they will concentrate in the melt during the freezing process. On complete freezing, the composition may then be heterogeneous, varying from pure crystals with impurities in the grain boundaries to that of some eutectic.

The polyperfluorocyclobutene appears to have the structure

$$\left(\begin{array}{c} FC - CF \\ | \qquad | \\ F_2C - CF_2 \end{array}\right)$$

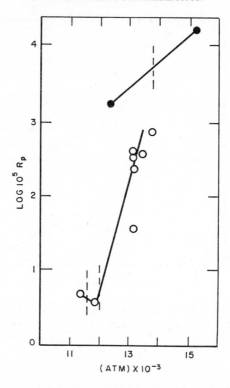

Fig. 15. Influence of pressure on the polymerization rate in the liquid and solid phases. Vertical dashed lines indicate freezing (42). (○) perfluorocyclobutene, 168°C; (●) n-perfluorooctadiene-1,7, 107°C. (Reprinted by permission of the copyright owner, The Am. Chem. Soc.).

which would seem to have the possibility of isomerization to the polybutadiene structure. The monomer can presumably isomerize to perfluorobutadiene-1,3.

The perfluorooctadiene-1,7 monomer, shown in Figs, 14 and 15, displays no discontinuities in its polymerization rate at the freezing condition. One point at the lowest temperature for the perfluoropentadiene-1,4 (Fig. 14) suggests that this monomer has a discontinuity in the rate. However, due to the ease of isomerization of the pentadiene, this point is not especially significant. More studies of high-pressure polymerization in the solid state would seem to be very desirable. However, it is anticipated that more difficulties and complexities will be encountered in the study of polymerization in the solid state than in the liquid state.

XVII. FLUORINATED PHENOXYETHYLENES

Perfluorophenoxyethylene, phenoxy trifluoroethylene, and pentafluorophenoxyethylene have been synthesized (44) and their polymerization investigated (34,44). These compounds were all difficult to polymerize by normal free-radical or cationic methods. It is known also that the hydrocarbon phenoxyethylene is quite difficult to polymerize (45). Thus far there has been only a cursory investigation of the polymerization of these monomers at high pressures. In Table 16 the best results (34,44) of the high-pressure

TABLE 16

Gamma-Ray Polymerization of Fluorinated Phenoxyethylenes, Dose Rate 3 krad/hr

Monomer	Conditions		Polymer solid, sol C_6F_6
	Temp. (°C)	Pressure (katm)	
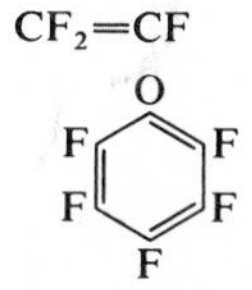 $CF_2=CF$ O F⌬F F⌬F F	105	12	$\bar{M}_n \sim 2,000$
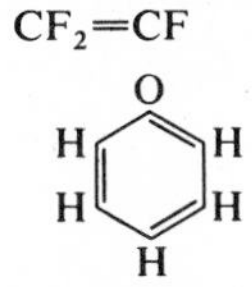 $CF_2=CF$ O H⌬H H⌬H H	103	14	Solid $[\eta]C_6H_6 = 0.03$ $\bar{M}_n \sim 20,000$
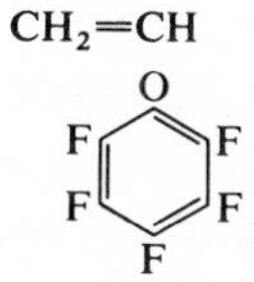 $CH_2=CH$ O F⌬F F⌬F F	110	9.5	Solid $[\eta]C_6F_6 = 0.12$ $\bar{M}_n \sim 30,000$

studies are summarized. Only one monomer gave reasonably high-molecular-weight polymer. In the other two cases pressures above 10 katm failed to produce molecular weights above a few thousand. It is interesting to note than the polymers with perfluorophenyl group were soluble in hexafluorobenzene but not in benzene. The reverse is true if the phenyl group is unsubstituted.

XVIII. COPOLYMERS OF TETRAFLUOROETHYLENE

A. Comonomers

Tetrafluoroethylene copolymerizes with hexafluoropropylene and perfluoroheptene-1 at autogenous pressures (46). Hexafluoropropylene is incorporated in the copolymers relatively efficiently. Perfluoroheptene-1 appears to enter copolymers reluctantly. Approximately equal molar charges of the two monomers produce copolymers containing at most about 5% of the perfluoroheptene. However, this small amount in the polymer structure lowers the melt viscosity over a millionfold.

Vinyl perfluoroalkyls are essentially retarders or inhibitors for the polymerization of tetrafluoroethylene at autogenous pressures. Primary perfluororadicals add apparently rapidly to these partially fluorinated olefins, whereas the radicals formed presumably of the structure

$$\sim CH_2C \cdot \overset{H}{\underset{R_F}{}}$$

fail to propagate effectively. Apparently, high pressure can accelerate this copropagation step since at sufficiently high pressure very high rates and molecular weights are obtained while the copolymer compositions are essentially unchanged.

Copolymerization studies of seemingly poorly polymerizing new monomers are also a useful tool for their study, since copolymer compositions are affected only in a limited way by impurities with inhibiting properties. Copolymerization of tetrafluoroethylene with vinyl perfluoromethyl (47) and with vinyl perfluoropropyl (32) have been investigated at high pressures using gamma-ray initiation.

B. Copolymerization Kinetics

Provided the kinetic chain length of the copolymerization process is large, the copolymer composition depends, according to the most successful theory, on the monomer charge composition and two reactivity ratios. These are defined as

$$r_a = \frac{k_{aa}}{k_{ab}} \quad \text{and} \quad r_b = \frac{k_{bb}}{k_{ba}} \tag{7}$$

where the k's are the rate constants for four specific propagation steps. For instance, k_{aa} is the constant for the rate a radical terminating with an A monomer adds another A unit. The meaning of the other constants is then self-evident.

Numerous expressions have been developed for relating instantaneous polymer composition to that for monomer. The one utilized for the copolymerization now being discussed is

$$F_A = \frac{r_a f_a^2 + f_a f_b}{r_a f_a^2 + 2 f_a f_b + r_b f_b^2} \tag{8}$$

where F_A and f_a are the mole fractions of A in polymer and monomer, respectively. Figure 16 is a plot of F_A as a function of f_a where monomer A is tetrafluoroethylene and monomer B is vinyl perfluoromethyl. The points are

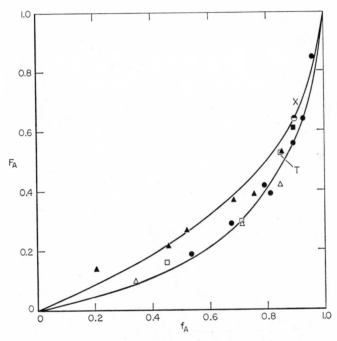

Fig. 16. Copolymerization of tetrafluoroethylene with vinyl perfluoromethyl. Variation of mole fraction CF_2CF_2 in polymer as a function of the mole fraction in monomer charged. (47) Curves given by Eq. 7:

Upper curve	$r_a = 0.15$	$r_b = 2.5$
Lower curve	$r_a = 0.12$	$r_b = 5.0$

Experimental points:

▲ 21°C, autogenous pressure
× 21°C, 1. katm
■ 21°C, 1.8 katm
◓ 21°C, 3.5 katm
● 21°C, 5 katm
△ 21°C, 8 katm
□ 100°C, 5 katm

the results of experiments under various conditions of pressure, temperature, and dose rates. In most of the experiments a dose rate of 1.4 krad/hr was employed. In a few cases a dose rate of 250 krad/hr was tried with no great effect on the polymer composition. The lines are theoretical from Eq. 8 and the indicated values of r_a and r_b. Values $r_a = 0.12$ and $r_b = 5.0$ lower curves fit most the results well. The upper curve is a better fit to the autogenous pressure results but the values of r_a and r_b are only minimally shifted. Since r_a is less and r_b is greater than one, both radicals under all conditions studied add the vinyl perfluoromethyl preferentially. The point indicated by T is the result of a thermal polymerization. The position of this point suggests the operation of

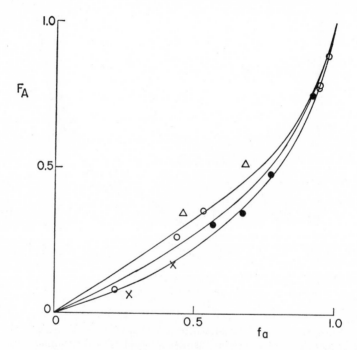

Fig. 17. Copolymerization of tetrafluoroethylene with vinyl-n perfluoropropyl.(48) Variation of mole fraction CF_2CF_2 in polymer with mole fraction in monomer charge.
Points experimental: 22°C   100°C
                                         ○ 5 katm   △ 5. katm
                                         ● 10 katm
                                         ✕ 14 katm
Full lines theoretical (Eq. 7):

	r_a	r_b
Upper curve	0.21	1.5
Middle curve	0.21	2.3
Lower curve	0.21	3.2

the same mechanism as in the radiation-induced process, presumably a free-radical mechanism.

Figure 17 is a plot of F_A as a function of f_a for the copolymerization of tetrafluoroethylene (A) with vinyl perfluoropropyl (B). The results and conclusion are closely similar to those seen in Fig. 16. The r_a and r_b values for the two copolymerizations are shown in Table 17. The reciprocals of the r_a values

TABLE 17

Reactivity Ratios in Copolymerizations with Tetrafluoroethylene (Monomer A) and Relative Reactivity ($1/r_a$) of B monomers with Radicals Ending in A Units

Monomer B	r_a	r_b	$1/r_a$
$CH_2{=}CHCF_3$	0.12	5.0	8.3
$CH_2{=}CHCF_2CF_2CF_3$	0.21	2.3	4.8
$CH_2{=}CFCF_3$	0.37	5.4	2.7
$CH_2{=}C(CF_3)_2$	0.58	${\sim}0.05$	1.7

give a measure of the relative reactivity of the monomers toward a tetrafluoroethylene radical, $-CF_2CF_2\cdot$. The relative rate this radical reacts with CF_2CF_2, $CH_2{=}CHCF_3$, and $CH_2{=}CHCF_2CF_2CF_3$ is then in the order 1, 8.3, and 4.8, respectively.

C. Rates of Copolymerization

In radiation-initiated copolymerization a theoretical expression can be readily developed for the overall rate provided termination and initiation rate constants are not dependent on monomer composition. This equation is

$$\frac{R_p}{R_B} = \frac{R_A/R_B(r_a f_a^2 + 2f_a f_b + r_b f_b^2)}{\{r_a^2 f_a^2 + 2\phi f_a f_b r_a r_b (R_A/R_B) + [r_b f_b (R_A/R_B)]^2\}^{1/2}} \quad (9)$$

R_p, R_A, R_B are the rates for the copolymerization, pure monomer A and pure monomer B, respectively. The parameter ϕ is the ratio $k_{4_{ab}}/(2k_{4_{aa}}^{1/2}k_{4b_b}^{1/2})$ where the ks are the rates of termination for the three ways the A and B radical can mutually terminate. The larger the ϕ the more the unlike radicals prefer to react with each other and the more the net polymerization rate is decreased. In general, R_A, R_B, r_a, and r_b are known and plots of experimental results according to Eq. (9) permit a determination of the ϕ value. For these complex systems, the insensitivity in many cases of the plots to the ϕ values, as well as the tediousness of the calculations, introduce considerable uncertainty into the estimates.

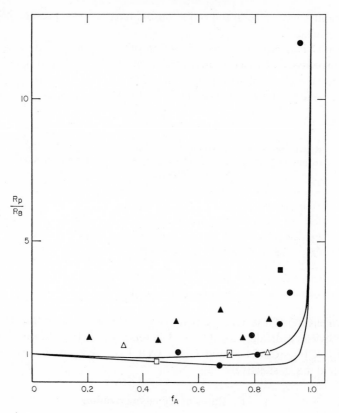

Fig. 18.   Variation in the copolymerization rate of tetrafluoroethylene with vinyl perfluoromethyl with fraction of tetrafluoroethylene in monomer charge (47). Curves given by Eq. 8

For both curves R_A/R_B	ϕ		r_a	r_b
10	1	lower curve	0.15	2.5
10^3	1	lower curve	0.12	5.0
10^3	10^2			

For the particular copolymerizations under discussion R_A, the homopolymerization rate of tetrafluoroethylene, is unknown but is extremely high compared to R_B. This requires the calculation of several curves by varying
numerical values of both the ratio R_A/R_B and ϕ and comparing the theoretical
curves to the experimental points. Figures 18 and 19 show such comparisons
for two copolymerization. For the vinyl perfluoromethyl copolymerization,
many of the points are far above the theoretical curves. Better agreement is
seen with the vinyl perfluoropropyl system. Both systems behave qualitatively
in the manner of the theoretical curves. In these copolymerizations the ratio
R_p/R_B is not appreciably dependent on pressure and temperature. Therefore

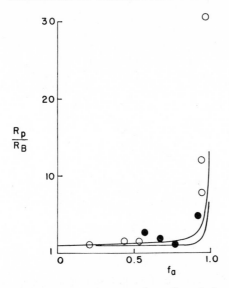

Fig. 19. Variation in the copolymerization rate of tetrafluorethylene with vinyl perfluoro *n*-propyl. (48)
Theoretical curves give by Eq. 8.
Both curves $r_1 = 0.21$, ϕ 1 to 10 R_A/R_B 10^2 to 10^3
Upper curve $r_2 = 1.5$
Lower curve $r_2 = 3.2$
Experimental points: ($\bigcirc$) 5 katm, ($\bullet$) 10 katm.

R_p should have approximately the same activation volume and energy parameters as R_B. Estimated values of the parameters for R_p with the vinyl perfluoromethyl system is -15 cc/mol and 5 kcal/mol. For the vinyl perfluoropropyl system the respective estimated values are -10 cc/mol and 7 kcal/mol.

D. Intrinsic Viscosities

In the homopolymerization of the vinyl perfluoroalkyl monomers, molecular weights depend to some appreciable degree on monomer transfer. The introduction of tetrafluoroethylene should tend to decrease or eliminate the monomer transfer process. In the complete absence of such transfer, the kinetic chain length is the same as the degree of polymerization assuming termination by disproportionation. In Fig. 20 two essentially independent plots are superimposed. One is the intrinsic viscosity in hexafluorobenzene of vinyl perfluoromethyl homopolymer as a function of its viscosity average molecular weight. The second plot is that of the intrinsic viscosities of radiation-produced copolymers and homopolymers as a function of the kinetic

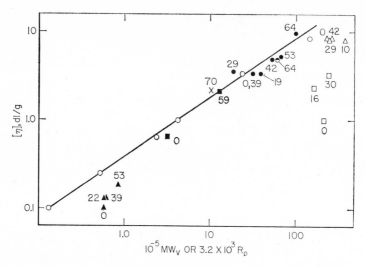

Fig. 20.   Variation of the intrinsic viscosity in hexafluorobenzene of polyvinylperfluoro methyl with viscosity average molecular weight, straight line and (○) experimental points. (47) Also, variation of intrinsic viscosity for all other experimental polymers with kinetic chainlength function at 1.4 kr/hr.

Experimental points at 21°C

    ▲ autogenous pressure
    × 1 katm
    ■ 1.8 katm
    ◓ 3.5 katm
    ● 5 katm
at 100°C: □ 5 katm

The number adjacent to points is the percentage tetrafluoroethylene in polymer.

chain length of the reaction. The kinetic chain length scale is adjusted to be essentially the degree of polymerization for the corresponding MW_v values. Thus the vertical deviation of the experimental points is a measure of the monomer transfer process. The numbers adjacent to the points are the percentages of tetrafluoroethylene in the copolymers. It is noted that the higher the tetrafluoroethylene percentage, the closer the points are to the line. The results are interpreted to signify that all the polymers and copolymers have nearly the same intrinsic viscosity-molecular weight relationship. Thus to determine the viscosity average molecular weight for a given polymer point it is necessary to move horizontally to the line then vertically down to its molecular weight. From this figure the importance of transfer under the conditions of polymerization can be deduced.

References

1. R. E. Florin, L. A. Wall, D. W. Brown, L. A. Hymo, and T. P. Michaelsen, *J. Research NBS*, **53**:121 (1954).
2. P. W. Bridgman, "The Physics of High Pressure," G. Bell and Sons, London, 1952.
3. S. D. Hamman, "Physico-Chemical Effects of Pressure," Academic Press, New York, 1957.
4. K. E. Weale, "Chemical Reactions at High Pressures," E. & F. N. Spon Ltd., London, 1967.
5. A. E. Nicholson and R. G. W. Norrish, *Disc. Faraday Soc.*, **22**:97 (1956).
6. A. H. Ewald, *Disc. Faraday Soc.*, No. **22**:138 (1956).
7. C. Walling and J. Pellon, *J. Am. Chem. Soc.*, **79**:4786 (1957).
8. C. Walling and G. Metzger, *J. Am. Chem. Soc.*, **81**:5365 (1959).
9. C. Walling and J. Pellon, *J. Am. Chem. Soc.*, **79**:4776 (1957).
10. A. E. Nicholson and R. G. W. Norrish, *Disc. Faraday Soc.*, **22**:97 (1956).
11. G. Tamman and A. Pape, *Z. Anorg. Chem.*, **200**:113 (1931).
12. R. C. Gillham, *Trans. Faraday Soc.*, **46**:497 (1950).
13. F. M. Merrett and R. G. W. Norrish, *Proc. Roy. Soc.*, **A206**:309 (1951).
14. A. H. Ewald, *Trans. Faraday Soc.*, **55**:792 (1959).
15. A. C. Tookey and K. E. Weale, *Trans. Faraday Soc.*, **58**:2446 (1962).
16. D. W. Brown, J. M. Antonucci, L. A. Wall, *Am. Chem. Soc. Polymer Preprints*, **9**:1279 (1968).
17. D. W. Brown and L. A. Wall, *J. Phys. Chem.*, **67**:1016 (1963).
18. A. Chapiro and M. Magat, "Actions Chimiques et Biologiques des Radiations," Mason et Cie, Paris, 1958, p. 90.
19. E. V. Volkova and A. I. Skobina, *Vysokomol. Soedin.*, **6**:984 (1964).
20. B. Monowitz, *Nucleonics*, **11**:18 (1953).
21. H. S. Eleuterio, U. S. Patent 2,958,685 (1960).
22. H. S. Eleuterio and E. P. Moore, Second Internat. Sym. on Fluorine Chem., Estes Park, Colorado, 1962.
23. R. E. Lowry, D. W. Brown, and L. A. Wall, *J. Polymer Sci. Al.* **4**:2229 (1966).
24. D. Sianesi and G. Caporiccio, *Makromol. Chem.*, **60**:213 (1965).
25. D. W. Brown, *Am. Chem. Soc. Polymer Preprints*, **6**(2):965 (1965).
26. G. H. Crawford, U.S. Patent 3,084,144 (1963).
27. E. M. Sullivan, E. W. Wise, and F. P. Reding, U.S. Patent 3,110,705 (1963).
28. D. Sianesi and G. Caporiccio, *Makromol. Chem.*, **81**:264 (1965).
29. S. Straus and D. W. Brown, *Am. Chem. Soc. Polymer Preprints*, **7**:1128 (1966).
30. D. W. Brown, R. E. Lowry, and L. A. Wall, unpublished.
31. L. A. Wall, *Am. Chem. Soc. Polymer Preprints*, **7**:1112 (1966).
32. D. W. Brown, R. E. Lowry, and L. A. Wall, *Am. Chem. Soc. Polymer Preprints*, **10**:1472 (1969).
33. D. W. Brown, R. E. Lowry, and L. A. Wall, *J. Polymer Sci.,* **8A-1**:3483 (1970).
34. D. W. Brown and L. A. Wall, *SPE Trans.*, **3**:300 (1963).
35. D. W. Brown and L. A. Wall, *J. Polymer Sci.*, **44**:325 (1960).
36. C. S. Marvel, *J. Polymer Sci.*, **48**:101 (1960).
37. G. B. Butler, *J. Polymer Sci.*, **48**:279 (1960).
38. J. E. Fearn and L. A. Wall, *SPE Trans.*, **3**:231 (1963).
39. D. W. Brown, J. E. Fearn, and R. E. Lowry, *J. Polymer Sci.*, **3A**:1641 (1965).
40. J. E. Fearn, D. W. Brown, and L. A. Wall, *J. Polymer Sci.*, **4**:131 (1966).

41. D. W. Brown and L. A. Wall, *J. Polymer Sci. A*2, **7**: 601 (1969).
42. D. W. Brown and L. A. Wall, *Am. Chem. Soc. Polymer Preprints*, **5**: 907 (1964).
43. E. S. Jones and W. Mears, U.S. Patent 3,383,363 (1968).
44. W. J. Pummer and L. A. Wall, *SPE Trans.*, **3**: 220 (1963).
45. C. E. Schildkrecht, "Vinyl and Related Polymers," John Wiley & Sons, New York, 1952, pp. 622–625.
46. M. I. Bro., U.S. Patent 2,943,080 (1960).
47. D. W. Brown and L. A. Wall, *J. Polymer Sci.* **6A–1**: 1367 (1968).
48. D. W. Brown, R. E. Lowry, and L. A. Wall, *J. Polymer Sci.*, **8A-1**: 2441 (1970).

5. NITROSO FLUOROPOLYMERS

LEWIS J. FETTERS, *Institute of Polymer Science, The University of Akron, Akron, Ohio*

Contents

I. INTRODUCTION

The quest for solvent resistant elastomers capable of retaining rubbery character over a broad temperature range has been under way for some time. Additional sought-after properties have been high-temperature stability and resistance to the degradative influences of ultraviolet radiant energy and ionizing radiation. To fulfill these needs, fluorine-containing polymers are being appraised in hope that they will mimic the chemical and thermal resistance of Teflon (1,2) while simultaneously exhibiting elastomeric properties over as extensive a temperature span as possible. In regard to this latter point, the low-temperature recovery from elastic deformation of highly fluorinated elastomers is generally found to be much slower than that exhibited by conventional hydrocarbon rubbers.

The reaction, with the exclusion of light and air, of trifluoronitrosomethane (CF_3NO) and tetrafluoroethylene (CF_2CF_2) at 20°C yields a dual product consisting of perfluoro-2-methyl-1,2-oxtazetidine

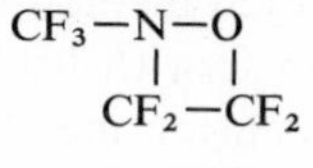

$$CF_3-N-O$$
$$\;\;\;\;\;|\;\;\;\;|$$
$$CF_2-CF_2$$

and a viscous oil (3–8). At temperatures above 100°C the reaction product is almost exclusively the oxazetidine, whereas below 0°C polymeric material is

175

obtained in high yield. This material is an elastomeric copolymer made up, for the most part, of alternating trifluoronitrosomethane and tetrafluoroethylene units. In a sense, the nitroso copolymer constitutes the first true fluorocarbon elastomer since, unlike other fluorine-containing rubbers, hydrocarbon units are absent.

The nitroso copolymer is a clear, transparent rubber which has been found to be completely nonflammable, even in pure oxygen. In this feature it is unique among organic polymers. The rubber is devoid of crystallinity in the extended state as evidenced by the diffuse nature of its X-ray diffraction patterns. The polymer reputedly has a glass transition temperature (T_g) of $-51°C$ (9) and has demonstrated excellent resistance in the presence of corrosive chemical environments such as hydrogen-bearing solvents, acids, oxidizing agents (nitrogen tetroxide; chlorine trifluoride), and ozone (10,11). Solubility is limited to halogenated alkanes, perfluorocyclic ethers, and perfluorotributylamine. The copolymer has a fluorine content of 66.8 wt % and a specific gravity of 1.93 gm/cm^3. Its dielectric constant is 2.41 at 60 Hertz. In the presence of strong bases, e.g., triethylamine, the polymer degrades readily.

The polymer chain structure has been shown by high-resolution nuclear magnetic resonance studies (12) to contain the $[-CF_2CF_2-N(CF_3)-O-]$ unit along with some segments of $[-N(CF_3)-CF_2-CF_2N(CF_3)-]$ and $[-O-CF_2CF_2-O-]$. No evidence was uncovered which indicated the presence of units where the nitrogen is pentavalent and linked coordinately to oxygen; i.e.,

$$
\begin{array}{c}
\text{O} \\
\uparrow \\
[-\text{N}-\text{CF}_2\text{CF}_2-] \\
| \\
\text{CF}_3
\end{array}
$$

II. POLYMERIZATION MECHANISM

The polymerization of CF_3NO and CF_2CF_2 is unusual in that it occurs spontaneously upon the mixing of equimolar quantities of the two monomers. Suspension, bulk, and solution systems have been found to be effective for the preparation of high polymer with bulk polymerizations usually generating the higher-molecular-weight material. Polymer prepared in bulk systems at $-25°C$ was found to contain fractions with viscosity molecular weights in excess of 2×10^6 (13). The favored method for large-scale polymerizations is an aqueous suspension at $-25°C$ with lithium bromide used as a freezing point depressant (14). The suspension method is preferred since this permits the exothermic heat of polymerization to be easily dissipated. Magnesium carbonate is often used as the suspending agent.

Initially an anionic mechanism was advanced (3) for this interpolymerization. This proposal was later amended to include a gas-phase, radical mechanism for the formation of the oxazetidine and polymer (15). Recently it has been shown (12) that an equimolar mixture of CF_3NO and CF_2CF_2 at $-80°C$ exhibits an ESR spectrum with concentrations estimated to be about 10^{14} spins/gm. The free-radical mechanism is further fortified by the fact that polymerization can take place in an aqueous medium and in the presence of such classic carbanion deactivators as CO_2, BF_3 etherate, and $TiCl_4$ (16). These compounds exerted virtually no influence on either the viscosity average molecular weight or on the extent of conversion in bulk systems at $-16°C$. A cationic mechanism appears to be precluded since this would involve electrophilic attack on CF_2CF_2 with the subsequent generation of an intermediate fluorocarbon carbonium ion—an unfavorable species energetically.

Molecular weight observations seem to lend further support to the free radical mechanism. The viscosity average molecular weight is decreased by the presence of radical chain transfer agents and solvents containing labile chlorine or hydrogen atoms. (16) Perfluorinated solvents exert no influence on the molecular weight. The presence of diphenylpicrylhydrazine results in a color alteration (violet to reddish-brown) of the DPPH and its eventual incorporation into the polymer chain, thus indicating the presence of radicals. Also in line with conventional free-radical polymerization behavior, the reaction rate exhibits typical temperature dependence, i.e., decreasing temperature gives decreasing rates.

It must be mentioned that an anionic mechanism retains some adherents (17). It is thought that the first stage of the polymerization is the formation of an ion-radical [reminiscent of the mechanism encountered in some homogeneous anionic polymerization systems (18)] followed by propagation through the carbanion. It has been proposed that the reactive species has the structure

$$CF_3\overset{\cdot}{N}CF_2\overset{\cdot\cdot}{\overset{\cdot\cdot}{C}}F_2.$$

Although intriguing, this proposed mechanism needs additional investigation before adequate comment can be made concerning its validity.

The formation of the proposed free-radical initiating species is believed to involve the combination of CF_2CF_2 with CF_3NO, the latter being (to some extent) in the triplet or diradical state. The existence of the triplet state in nitroso compounds was shown by Lewis and Kasha (19), and this has been used to explain the unusual colors of these compounds (CF_3NO is a dark blue gas, whereas aromatic groups impart a green color). Orgel (20) has presented a theoretical treatment of the spectra of the RNO species. The visible absorption was ascribed to a singlet-singlet $n - \pi^*$ transition; in other words, one electron from the lone pair of the nitrogen atom is elevated to an antibonding π orbital. When the electron participates in bond formation, the color vanishes.

An alternative free-radical producing step involves the decomposition of CF_3NO to $\cdot CF_3$ and $\cdot NO$. It has been shown, however, that CF_3NO does not decompose over long periods of time at ambient temperatures (16). Furthermore, $\cdot NO$ will react rapidly with CF_3NO to yield N_2, NO_2, and CF_3NO_2.

The initiation step has been envisaged as follows:

$$CF_3NO \underset{k_{-1}}{\overset{k_1}{\rightleftharpoons}} CF_3-\overset{\cdot}{N}-O\cdot \tag{1}$$

$$CF_3-\overset{\cdot}{N}-O\cdot + CF_2CF_2 \xrightarrow{k_2} CF_3\overset{\cdot}{N}-O-CF_2CF_2\cdot \tag{2}$$

The diradical nature of this species can cause self-termination by intramolecular combination with the resultant formation of the oxazetidine and perhaps other low-molecular-weight ring structures. The diradical mechanism provides a ready explanation for the formation of the oxazetidines resulting from the reaction of CF_3NO with an unsymmetrical olefin, e.g., CF_2CXY. The oxazetidine generated from this olefin has the following as the predominant structure:

$$\begin{array}{ccc} CF_3-N & \!\!\!\!-\!\!\!\! & O \\ | & & | \\ CF_2 & \!\!\!\!-\!\!\!\! & CXY \end{array}$$

Since free-radical attack will take place on the CF_2 group of the olefin the following diradicals would be formed:

$$\begin{array}{cc} \overset{\displaystyle X}{\underset{\displaystyle Y}{\cdot N-OCF_2\overset{|}{\underset{|}{C}}\cdot}} & \quad\text{and}\quad CF_3-N-CF_2\overset{\cdot}{C}XY \\ | & \qquad\qquad\quad | \\ CF_3 & \qquad\qquad\quad \underset{\cdot}{O} \end{array}$$

with the latter species favoring ring formation. Supposedly this latter structure is formed only at higher ($>0°C$) temperatures.

Propagation then ensues when the diradical (Eq. 2) reacts with CF_3NO. It is assumed (16) that one site of the initiating species, the nitrogen atom, is sufficiently stable so as to be inactive as a propagating species. Chain growth would then occur by means of monoradical chains exclusively. It is, however, difficult to see why the nitrogen bearing the free radical may be reactive in ring formation but not in the propagation reaction leading to linear chains. Peripheral evidence suggests that some diradical chains exist. This will be discussed at a later point.

The propagation reaction may be denoted in the conventional fashion:

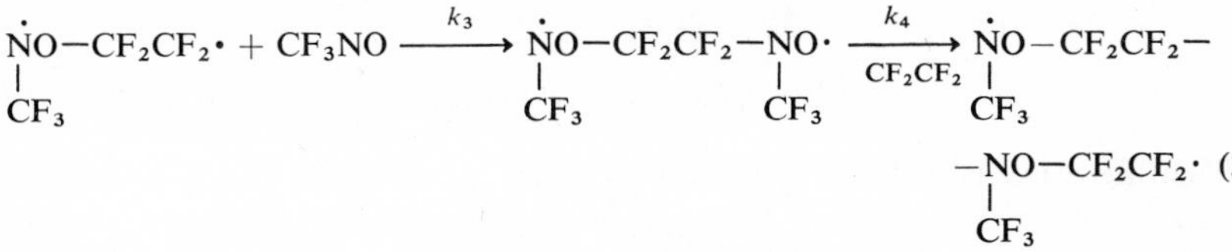

$$\underset{\underset{CF_3}{|}}{\overset{\cdot}{N}O}-CF_2CF_2\cdot + CF_3NO \xrightarrow{k_3} \underset{\underset{CF_3}{|}}{\overset{\cdot}{N}O}-CF_2CF_2-\underset{\underset{CF_3}{|}}{NO}\cdot \xrightarrow[CF_2CF_2]{k_4} \underset{\underset{CF_3}{|}}{\overset{\cdot}{N}O}-CF_2CF_2-$$

$$-\underset{\underset{CF_3}{|}}{NO}-CF_2CF_2\cdot \quad (3)$$

In this reaction scheme the product* r_1r_2 is of the order of 10^{-4}–10^{-5} since an alternating copolymer of the two monomers is obtained. Hence this monomer pair is one of the few that are azeotropic in nature.

The termination step apparently evolves through the usual combination procedures:

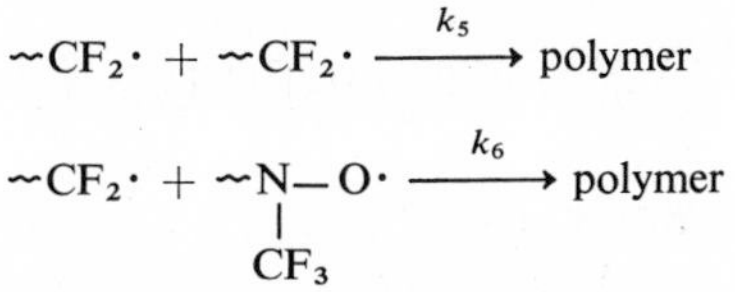

$$\sim CF_2\cdot + \sim CF_2\cdot \xrightarrow{k_5} \text{polymer}$$

$$\sim CF_2\cdot + \sim \underset{\underset{CF_3}{|}}{N}-O\cdot \xrightarrow{k_6} \text{polymer} \qquad (4)$$

Combination by the oxygen atoms would yield a relatively unstable peroxide linkage. This group, if indeed it is formed, would rupture, leaving the final product free of this bond.

This polymerization scheme has been advanced as the most likely and reasonable process for polymer formation. Studies of the initial polymerization rate were conducted in $(C_4F_9)_3N$ solution at $-30°C$. The polymerization rate was found to have the following orders with respect to the two monomers:

$$[CF_3NO]^{1.3} \quad \text{and} \quad [CF_2CF]_2^{0.9}$$

Rate expressions have been derived which seem to accommodate these results. In this presentation M_1 represents CF_3NO, M_2 denotes CF_2CF_2, M_1^* the diradical CF_3N-O, and P_1 and P_2 the chain ends $\sim N(CF_3)-O\cdot$ and $\sim CF_2^\cdot$, respectively.

Hence, under the usual steady-state conditions,

$$\frac{d[M_1^*]}{dt} = 0 \qquad (a)$$

with $$k_1M_1 = k_{-1}[M_1^*] + k_2[M_1^*][M_2] \qquad (b)$$

were $$[M_1^*] = k_1[M_1]/(k_{-1} + k_2[M_2]) \qquad (c)$$

* The monomer reactivity ratios r_1 and r_2 are the ratios of the rate constant for a given radical adding its own monomer to that for the radical adding the other monomer. Hence $r_1 < 1$ and $r_2 < 1$ means that the radical $M_1\cdot$ prefers to add $M_2\cdot$ and the radical $M_2\cdot$ prefers to add $M_1\cdot$.

Since the monomer pair is azeotropic,

$$k_3[P_2][M_1] = k_4[P_1][M_2] \qquad (d)$$

Then for the case where termination occurs as shown in Eq. 5,

$$k_2[M_2][M_1^*] = k_5[P_2]^2 \qquad (e)$$

Rearrangement yields

$$[P_2] = \left(\frac{k_2}{k_5}\right)^{0.5}[M_1^*]^{0.5}[M_2]^{0.5} \qquad (f)$$

Substitution with the use of Eq. c yields

$$[P_2] = \left(\frac{k_1 k_2}{k_5}\right)^{0.5}[M_1]^{0.5}[M_2]^{0.5}/(k_{-1} + k_2[M_2])^{0.5} \qquad (g)$$

R_p, the rate of polymerization, is

$$R_p = k_3[P_2][M_1] \qquad (h)$$

By combination of Eqs. g and h, Eq. i is obtained:

$$R_p = \frac{k_3(k_1 k_2/k_5)^{0.5}[M_1]^{1.5}[M_2]^{0.5}}{(k_{-1} + k_2[M_2]^{0.5}} \qquad (i)$$

Hence for the case of termination of $CF_2 \cdot$ radicals the polymerization rate will be proportional to $[M_1]^{1.5}$ and $[M_2]^{0-0.5}$ depending upon the value of k_1' relative to $k_2[M_2]$. For termination by Eq. 5,

$$R_p = (k_3 k_1 k_2 k_4/k_6)^{0.5}[M_1][M_2]/(k_{-1} + k_2[M_2])^{1/2}$$

where the polymerization rate is proportional to $[M_1]^{1.0}$ and $[M_2]^{0.5}$ to $[M_2]^{1.0}$. Since the simultaneous presence of the two termination reactions is plausible, the resultant polymerization rate should show a dependence on $M_2 > 1$ and on $M_1 > 1$. This is in apparent accord with the observed dependencies.

Various chain transfer agents such as CF_3I, CF_3Br, or CF_2ClCF_2Cl cause only minute alterations in the overall rate of polymerization. Hence the newly formed radicals seemingly are efficient initiators of new chains presumably by attack on a CF_3NO molecule. Trace amounts of $\cdot NO$ and $\cdot NO_2$ increase the polymerization rate while simultaneously decreasing the molecular weight. $\cdot NO_2$ will react rapidly with CF_2CF_2 to yield a variety of products, all of which arise from the intermediate $O_2\dot{N}CF_2CF_2$ (9). Since this species is analogous to a growing chain, an increase in the polymerization rate is understandable. $\dot{N}O$ reacts with CF_3NO to produce, among other species, $\dot{N}O_2$ and $\cdot CF_3$ (16). Both radical species are capable of generating new chains and thus causing a concurrent increase in the polymerization rate.

The importance of monomer purity in this polymerization is amplified by the observed lowering of the viscosity average molecular weight by the presence of trifluoromethyl iodide and trifluoromethyl bromide. Both species are alternative starting materials in the synthesis of CF_3NO.

A fairly large group of nitroso polymers has been prepared in addition to the copolymer prepared from CF_3NO and CF_2CF_2. This has been achieved through the dual expedient of altering the structure of the nitroso monomer or by varying the olefin. Two variations achieved have been polymers where CF_3NO has been replaced by 1-nitroso-2-nitrotetrafluoroethane ($ONCF_2$ CF_2NO_2) and 1-nitroso-2-chlorotetrafluoroethane ($ONCF_2CF_2CL$). Tetra-fluoroethylene has been replaced by olefins such as CF_2CFCF_3, CF_2CFCl, and $CFClCFOCH_3$. According to Lewis and Kasha (19), the reactivity of the nitroso group is not appreciably influenced by the type of substituent it carries, provided the nitroso unit is buttressed by a $-CF_2-$ group. This has been verified by rate studies on a variety of nitroso compounds with CF_2CF_2. All were found to be of nearly the same reactivity. Conversely, the nature of the olefin has a pronounced effect on the rate of polymerization. The reactivity of the olefins decrease with decreasing halogen content. Thus no yield is achieved when CF_2CH_2 is used and low-molecular-weight oils are isolated when styrene or butyl methacrylate are substituted for CF_2CF_2. Some of the more promising systems are indicated in Table 1. It is appropriate to note here that details concerning nitroso monomer synthesis are compiled elsewhere (23,26,28).

As mentioned previously, evidence obtained from a high-resolution, ^{19}F NMR probe (12) indicates that the nitroso polymer (prepared in suspension at $-25°C$) is garnished with structures of the following type: $[-ON(CF_3)-CF_2CF_2N(CF_3)O-]$ and $[-O-CF_2CF_2-O-]$. The NMR examination (Fig. 1) showed a five-line spectrum rather than the three signals anticipated if only head to tail addition occurs. These signals were found at 5570, 4300, 4135, 3755, and 3610 Hertz. The signal at 4135 Hertz was assigned to the difluoromethylene fluorine in the $[-N(CF_3)O-CF_2CF_2-ON(CF_3)-]$ unit; the signal at 3755 Hertz was assigned to the difluoromethylene fluoride attached to nitrogen in the $[-ON(CF_3)-CF_2CF_2-N(CF_3)O-]$ segment. This latter assignment was based on the observation that the 3755 Hertz signal exhibited ^{14}N quadrupole broadening, indicating nitrogen nuclei as near neighbors.

The presence of a detectable amount of these unsymmetrical units raises questions as to the mechanism of their incorporation into the polymer chain. In the foregoing polymerization mechanism, initiation generates a diradical species, whereas propagation progresses through only one end and termination involves only two of the three radical ends present. The proposed low

TABLE 1

Nitroso Copolymer Systems[a]

Comonomer	Time (hr)	Temp. (°C)	Conversion	Polymer	Reference
Copolymers of CF_3NO					
CF_2CFCl	24	-15	100	Elastomer	22
CF_2CFBr	96	-30	81	Elastomer	23
CF_2CFH	6	-20	85	Elastomer	22, 23
CF_2CCl_2	24	-20	100	Plastic	22
$CFClC(F)OCH_3$	12	-15	100	Plastic	22
Copolymers of C_2F_4					
C_2F_5NO	24	-16	100	Elastomer	22
C_3F_7NO	24	-20	100	Elastomer	22
$C_8F_{17}NO$	24	-20	50	Elastomer	22
C_2HF_4NO	24	-50 to -20	90	Elastomer	22
CF_2BrCF_2NO	12	-78	84	Elastomer	23
$CClF_2NO$	192	-65 to -35	82	Elastomer	23
$ONCF_2CF_2NO$	24	-25	90	Elastomer	22, 24
p-FC_6H_4NO	24	-25	50	Resin	25
p-$COOHC_6H_4NO$	24	-25	80	Resin	25
p-$ClCF_2CF_2NO$	20	-20	80	Elastomer	23
p-BrC_6H_4NO	24	-25	42	Resin	25

[a] All systems polymerized in bulk.

reactivity of the $[\sim O-\dot{N}CF_3]$ end was assumed to be sufficient justification for excluding it from both the propagation and the termination reaction. If the initiation reaction generates a monoradical rather than the proposed diradical species, the presence of the unsymmetrical units can be explained by the occasional reverse incorporation of nitroso methane units into the chain. This explanation implies that the $[\sim O-\dot{N}CF_3]$ is reactive in the polymerization.

However, the presence of the $[-ON(CF_3)CF_2CF_2N(CF_3)O-]$ segment can also be interpreted as indicating the presence of diradical chains resulting from the participation of the $[\sim O-\dot{N}CF_3]$ chain end in the polymerization; thus the propagation would be as follows:

$$\sim O\dot{N}CF_3 + CF_2CF_2 \longrightarrow \sim ON(CF_3)CF_2CF_2\cdot$$

$$\sim ON(CF_3)CF_2CF_2\cdot + CF_3NO \longrightarrow \sim ON(CF_3)CF_2CF_2N(CF_3)NO\cdot$$

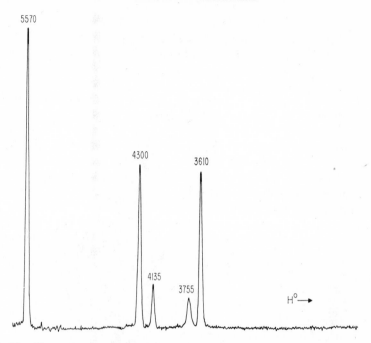

Fig. 1. The ^{19}F NMR spectrum of a nitroso fluoropolymer (~ 10 weight percent in hexafluorobenzene) (12).

An alternative means of generating this structure would be the following:

$$CF_3-\overset{\cdot}{N}-\overset{\cdot}{O} + CF_2CF_2 \longrightarrow CF_3N-CF_2\overset{\cdot}{C}F_2$$
$$\underset{\overset{|}{\underset{\cdot}{O}}}{}$$

$$CF_3N-CF_2CF_2\cdot + CF_3NO \quad \text{and} \quad 2\,CF_2CF_2 \longrightarrow$$
$$\underset{\overset{|}{\underset{\cdot}{O}}}{}$$

$$\cdot CF_2CF_2-ONCF_2CF_2N-OCF_2CF_2\cdot$$
$$\underset{CF_3}{|} \qquad \underset{CF_3}{|}$$

Either reaction would generate a diradical chain and as a consequence exert a far-reaching influence on the ultimate molecular weight and molecular weight distribution. This structure could also owe its existence to a combination step involving a $[\sim ON(CF_3)CF_2CF_2\cdot]$ unit and a $[\sim O\overset{\cdot}{N}(CF_3)]$ chain end. For the

$$[-ONCF_2CF_2NO-]$$
$$\underset{CF_3}{|} \qquad \underset{CF_3}{|}$$

segment to be the result of an occasional reverse addition (certainly a not unfamiliar feature in radical polymerizations) would mean that the $[\sim\text{O}\dot{\text{N}}\text{CF}_3]$ unit is capable of reacting with CF_2CF_2 and would thus open the possibility for the existence of some diradical chains. The $[-N(CF_3)O-CF_2CF_2-ON(CF_3)-]$ segment could result from the combination of the $[\sim N(CF_3)O\cdot]$ and $[\sim N(CF_3)O-CF_2CF_2\cdot]$ chain ends as well as from the reaction of the $\cdot CF_2$ radical with $ONCF_3$ to yield a $\sim O(\dot{N}CF_3)$ chain end.

The presence of some diradical chains during polymerization is fraught with intriguing consequences. Their presence implies that intermolecular termination (by combination) will cause chain extension. In other words, the diradical species would cause a large increase in polymer molecular weight while still retaining chain-end activity. This new, larger chain would in its turn be eligible to add monomer and eventually take part in another combination reaction. Thus as long as at least one chain end remained active, a chain would dramatically increase its molecular weight by simple combination reactions alone.

A somewhat similar mechanism has been advanced (29,30) to explain the propagation step of the spontaneous alternating copolymerization of bicyclo[2,2,1]hept-2-ene and sulfur dioxide. The propagation reaction is believed to progress through a diradical coupling step. The bicyclo[2,2,1]hept-2-ene and sulfur dioxide apparently form a 1:1 molecular charge transfer complex, which in turn spontaneously rearranges to yield a diradical. The subsequent combination of diradicals with each other (and possibly also with the charge transfer complexes) manifests itself as a rapid propagation reaction with the corresponding formation of high molecular weight $(\overline{M} \simeq 2\cdot2 \times 10^6)$ alternating copolymer. In regard to the molecular weight $(\overline{M}_v)$ of this copolymer, it was noted that molecular weights increased with time. This is, of course, contrary to the results obtained from conventional, free-radical polymerizations where molecular weights remain virtually constant throughout the reaction.

The possibility of molecular-weight enhancement by the combination of diradical chains could account for the extremely high-molecular-weight (2×10^6) fraction found in the nitroso polymer (13). A study of the relation between the reciprocal number average degree of polymerization, $1/P_n$, and the polymerization rate, R_p, would reveal whether or not growing diradical chains are present. In this vein, it was pointed out (31,32) some time ago that diradical propagation would be distinguishable from mono-radical propagation from the effect on the relation between $1/P_n$ and R_p. The absence of termination by disproportionation in this polymerization would simplify this task.

As Flory (33) has noted, there are several objections that can be raised against diradical chains generating high polymer. For the most part, at least for the uncatalyzed polymerization of styrene, this concerns the propensity of a diradical chain to annihilate itself through ring formation (34,35). Thus a major product of the diradical species would be expected to be low-molecular-weight rings. Along this line the nitroso polymers have a high (about 26 wt %) proportion of low-molecular weight material ($\overline{M}_v \simeq 7300$).

It remains to be seen if diradical chains play any significant role in this polymerization. The presence of some hitherto unsuspected segments in the chain and the unusual spread of molecular weights found in fractionated material[1] indicates the existence of some peculiarities in the polymerization mechanism which have not been elucidated.

III. SOLUTION PROPERTIES

An examination of the dilute solution properties of a polymer can result in much insight as to the molecular configuration as well as provide information about the solvent resistance of the polymer. The study by Morneau, Roth, and Schultz (13) is the only examination of the solution behavior of the nitroso rubber. From light-scattering measurements on selected fractions (in 1,1,2-trichloro-1,2,2 trifluoroethane) it was possible to relate intrinsic viscosities, $[\eta]$, to the molecular weight (Table 2).

TABLE 2

Mark-Houwink Constants

Solvent	Temp. (°C)	K	α
Trichlorotrifluoroethane	35.0	2.80×10^{-4}	0.51
Perfluorotributylamine	25.0	8.77×10^{-5}	0.66

Note: $[\eta] = KM^{\alpha}$.

The value of the light scattering second virial coefficient, A_2, has the unusually low value of about 2×10^{-5} cm^3-mole/gm^2. This value of A_2 is lower by two orders of magnitude than the usual value encountered in the case where a polymer exhibits Gaussian behavior in a good solvent. The α exponent in trichlorotrifluoroethane is close to the value of 0.50 for a polymer in a θ solvent. The magnitude of these two parameters are in concert with each other and reflect the lack of polymer-solvent interaction. In contrast, the value of α in $(C_4F_9)_3N$ fortifies the observed insolubility of this chain in all but highly fluorinated solvents. A θ temperature of 28.4°C was calculated for trichlorotrifluoromethane by extrapolation of the critical solution temperatures. The intercept/slope ratio of this plot gives $\Psi = 0.975$, where Ψ is the Flory entropy parameter. Hence these solutions of the nitroso rubber in CCl_2FCClF_2 have a small excess entropy of dilution ($+0.94v_2^2$ cal mole^{-1} deg^{-1}) and a small endothermic heat of dilution ($+584v_2^2$ cal mole^{-1}) where v_2 is the volume fraction of polymer.

The solubility parameter δ (34) was estimated (13) to be 5.2 $cal^{1/2}cm^{-3/2}$. The validity of this value seems to be fortified by the apparent low T_g observed for this polymer. The values of δ for other rubbery polymers fall in the range of 7–10. It is valid to recall at this time that one of the molecular features that affects T_g is the magnitude of intermolecular forces. Additional parameters include the amount of chain symmetry, the free-volume contribution of flexible side chains, and the hindrance to rotation of the chain units.

An estimate of the last-mentioned parameter has been obtained from light-scattering measurements. The freely jointed segment length has been deduced (13) from the observed mean-square end-to-end distance, $<\bar{r}^2>$ of the chain, and $<r^2>_0$, which is calculated for the freely jointed chain. The freely jointed chain length (which is defined as the linear end-to-end distance of the number of chain atoms comprising a unit having rigid bonds which may be considered as freely jointed to its neighboring units) was determined to be about 24 chain bonds. This value does not indicate a highly flexible chain. By way of comparison, the freely jointed segment of polystyrene is also 24 units. Thus the low T_g of this polymer would seem, in the main, to arise from a low degree of intermolecular interaction and the lack of chain symmetry rather than from inherent flexibility of the chain. This conclusion is supported by the low value of δ and the observed amorphous nature of the chain.

Fractionation of a sample prepared in bulk at $-25°C$ revealed the presence of a large proportion of low-molecular-weight material. Approximately 26 wt % of the polymer possessed an $\overline{M}_v$ of less than 7500. This is to be expected for a system where the spontaneous generation of active sites occurs with such ease. Apparently the majority of newly formed chains live for only a short time before their growth is arrested by adventitious impurities or by intermolecular or intramolecular coupling. This early annihilation of the chain growth sites is a serious problem, since a sizable proportion of the polymer is incapable of contributing effective network chains in the cross-linked polymer. The low-molecular-weight material will also serve as a plasticizer and thus lead to an erroneous low value for T_g. On the other hand, better than 39 wt % of the fractionated material was polymer with a molecular weight in excess of 10^6. Thus the nitroso rubber possesses an apparent molecular-weight distribution which is far broader than that usually encountered in conventional free-radical polymerizations. As mentioned previously, this high-molecular-weight material could, in part, arise from the repetitive combination steps of diradical chains. The integral molecular-weight distribution curve is shown in Fig. 2. A considerable portion ($\sim 65\%$) of the polymer has a viscosity average molecular weight of either <7300 or $>10^6$. The molecular-weight distribution is certainly dissimilar to that usually encountered in polymer formed in conventional free-radical systems. The sample from which the fractions were obtained was bulk polymerized at $-25°C$.

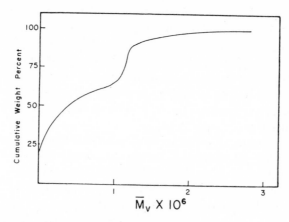

Fig. 2. Integral molecular weight distribution cure for a polymer prepared in bulk at _25°C (13).

IV. THERMAL DEGRADATION

The stability of the linear, uncross-linked nitroso rubber has been surveyed in several environments. An early investigation (37) of the thermal degradation of the CF_3NO/CF_2CF_2 polymer and the oxazetidine identified the decomposition species as COF_2 and $CF_3N=CF_2$ when the degradation was carried out at 400°C under vacuum. More recently, a detailed picture of the stability of this copolymer with respect to heat, gamma radiation, and ultraviolet radiant energy has appeared (38). It was concluded that the mechanism of degradation was similar in all three cases.

Weight-loss measurements on an unfractionated sample (prepared in bulk at about −25°C) at 204.4°C under vacuum revealed a loss within the first 3 hr of about 27 wt %. Concurrently, the viscosity average molecular weight increased from 7.8×10^5 to about 10^6. This strongly indicates the loss of the low-molecular-weight fraction with very little chain scission or depolymerization of the higher-molecular-weight material. Unfortunately it was not possible to ascertain whether the low-molecular-weight fraction possessed some chain defect which prompted its rapid degradation to yield volatile material. Prolonged heating (34 days) at 204°C caused a weight loss of approximately 79% and a drop in the viscosity average molecular weight to about 1.24×10^5. This reflects the onset of chain scission accompanied by incomplete depolymerization of the high-molecular-weight polymer.

Degradation studies run in air showed that the polymer underwent some chain scission at 177°C. The weight loss of high-molecular-weight fractions was about 1.5% after 114 hr. Concurrently the intrinsic viscosities dropped by about 9%. Thus elimination of the low-molecular weight material reveals that a small amount of chain scission (but virtually no unzipping) occurs at this relatively low temperature. The onset of chain scission (*not weight loss*) denotes the maximum temperature at which a polymer may serve and still retain useful physical properties. At 204°C in air, heating for 504 hr caused a weight loss of 57%. It was concluded that degradation of the high-molecular-weight copolymer can be avoided in air at temperatures below 149°C. Between 149 and 177°C the onset of chain scission and depolymerization gradually becomes evident.

Observation of the weight loss of individual fractionated samples during a period of time over a relatively narrow temperature range (250, 260, and 265°C) revealed a first-order dependence on residual copolymer. An activation energy for volatilization of 58 kcal/mole was computed. The first-order rate constant (in units of min^{-1}) can be expressed as

$$\log k = 21.5 - \left(\frac{12.6}{T}\right)$$

A plot of the viscosity average molecular weight versus the weight fraction of residue suggests that the initial molecular-weight decrease is the result of chain scission, which is followed in its turn by an unzipping process that released somewhere between 1000 and 2000 repeat units. This value is roughly equivalent to the zip lengths of poly-α-methylstyrene (1230) (39) and poly-(methyl methacrylate) (3000) (40). The N—O bond with a dissociation energy of 36–40 kcal/mole nominates it as the most likely candidate for the initial chain fracture site. Thus a freshly cleaved chain would have the form

$$\overset{\displaystyle CF_3}{\underset{\displaystyle |}{}}\qquad \overset{\displaystyle CF_3}{\underset{\displaystyle |}{}}\qquad\qquad \overset{\displaystyle CF_3}{\underset{\displaystyle |}{}}$$
$$-CF_2-CF_2-N-O-CF_2-N\cdot\ \ \cdot O-CF_2-CF_2-N-O-CF_2-CF_2-$$

Since the overall degradation process generates COF_2 and CF_3NCF_2, further bond dissociations probably occur in the order of every other chain bond starting with the second bond in from the newly formed chain end. This unraveling procedure implies that four dissimilar bonds are eligible for the secondary cleavage step. Each step generates a radical differing from its predecessor (41). This can be compared to the unzipping of a vinyl monomer (α-methyl styrene) where two dissimilar radicals are involved, depending upon which direction from the fracture site the unzipping reaction takes. However, each new radical is a twin of its predecessor. Unfortunately it has thus far not been ascertained which (if any) of the two dissimilar radicals is the more susceptible to the unzipping reaction.

The production of the inert monomer pair CF_2O and CF_2NCF_3 indicates that they are thermodynamically more stable than the copolymer. This behavior is paralleled by the linear siloxane chain $[-(CH_3)_2SiO-]_n$, which is formed from the cyclic tetramer octamethylcyclotetrasiloxane, which decomposes into the cyclic trimer. These two examples are systems where the polymer is more stable with respect to one monomer but is labile with respect to the other. The heat of reaction (ΔH) for the production of CF_2O and CF_2NCF_3 has been estimated to be between -8.7 and -24.7 kcal/mole (38).

V. RADIOLYSIS

The effects of ionizing radiation on macromolecules are considerably more complicated than pyrolysis since high-energy radiation amplifies the possibilities for the existence of a variety of intermediates, ions, and excited states in addition to the radicals normally present in thermal degradation. In addition to chain degradation, the influence of high-energy particles or electromagnetic radiation upon polymers can cause branching, grafting, and crosslinking. In general it has been observed that polymers with long zip-lengths and high monomer yields in pyrolysis tend to degrade upon irradiation, whereas those with low monomer yields cross-link rather than degrade. The less energetic ultraviolet radiation directly and in a more selective manner brings about effects such as electronic excitation and dissociation into radicals.

The effects of these two forms of energy on the nitroso copolymer revealed a degradative mechanism similar (in the main) to that induced by pyrolysis. From a plot of the reciprocal of the viscosity average molecular weight versus radiation dose (in megarads) under vacuum at 24°C the total energy absorption per primary chain scission was found to be $E_d = 44$ eV/scission. This indicates that the nitroso copolymer is foremost among those polymers with little resistance to ionizing radiation. This is to be expected in view of the presence of electronegative groups in the chain.

From the quantity of gas (as in pyrolysis, CF_2O and CF_2NCF_2 are evolved) isolated at different doses a G (gas) value of 39.4 (molecules/100 eV) was deduced. The primary chain scission efficiency, G (random scissions) was found to equal 3.7 scissions/100 eV. Hence the ratio of G(gas)/G(random scissions) is equal to 10.6 gas molecules generated for each radiation-induced fracture point. This is equivalent to 5.3 $[-CF_2CF_2-N(CF_3)O-]$ units decomposed for each chain scission. As was the case in the pyrolysis study, the primary rupture joint was assumed to be the electronegative N—O bond.

Exposure of the nitroso polymer to 2537 Å radiation (25°C) revealed some surprising similarities with the gamma-ray induced degradation study. Molecular weight $(\bar{M}_v)$ measurements revealed a primary chain scission efficiency of about 0.9×10^{-3} main chain scissions per absorbed 2537 Å photon. An average of 3.6 copolymer repeat units are destroyed for each rupture point. The efficiency of producing chain scissions is greater for gamma radiation than ultraviolet/radiant energy ($E_d = 5400$ eV/scission) on the basis of energy input. However, the nearly identical values of the number of degraded copolymer units per chain scission implies that the surplus energy at the gamma-induced radiolysis site is ineffective in sustaining the zip length beyond 5–6 units.

In both the gamma-radiation and ultraviolet-radiant-energy induced degradation, the presence of water vapor led to the formation of some proton bearing byproducts in addition to the primary decomposition species. The occurrence of these secondary reactions was enhanced at high gas pressures. These reactions (42) are as follows:

$$CF_2O + H_2O \longrightarrow CO_2 + HF$$

$$CF_3NCF_2 + H_2O \longrightarrow CF_3NCO + 2HF$$

$$CF_3NCO + H_2O \longrightarrow CF_3NH_2 + CO_2$$

$$CF_3NCF_2 + HF \longrightarrow (CF_3)_2NH$$

$$(CF_3)_2NH + CF_2O \longrightarrow (CF_3)_2NCOF + HF$$

VI. CURING, FORMULATION, AND CROSS-LINKING

The development of the nitroso copolymer to the level of a useful elastomer has been plagued in part by its instability to heat and electromagnetic radiation. The problem has been complicated further by the difficulty of finding efficient vulcanization systems. The nitroso rubber fails to respond to the usual vulcanization systems employed for hydrocarbon elastomers, e.g., sulfur, peroxides, or metallic oxides. However, it was discovered that triethylene tetramine and hexamethylene diamine could generate a network in the gum nitroso rubber. This behavior is reminiscent of the vulcanization system used for the copolymer of hexafluoropropylene and vinylidene fluoride (Viton A) (43). No detailed knowledge exists concerning the chemistry of the cross-linking reaction or of the structure of the cross-link itself. Whether or not there exist similarities between the nitroso rubber and Viton A cures remains unanswered.

In order to impart improved strength to the cured gum a variety of reinforcing agents were examined. Carbon blacks, regardless of type and acidity, caused severe sponging when used in the amine cures. The most efficient reinforcing filler is a fine-particle silicon oxide. It was eventually determined that a press cure of 60 min at 121°C followed by an oven cure of 100°C for 18 hr yielded stock with the highest tensile strengths. The oven cure is necessary for the development of the optimum physical properties, although the mechanism is not understood. Table 3 lists a typical compound formulation and the physical properties of the vulcanized nitroso copolymer.

In an attempt to circumvent the vulcanization difficulties encountered in

TABLE 3

Compound Formulation for Nitroso Copolymer

	Parts (by wt.)
Recipe	
Polymer	100
$(SiO)_2)_x$	15
Triethylene tetramine	1.25
Properties	
Tensile strength (kg/cm^2)	86.1
Stress at 300% elongation (kg/cm^2)	28.7
Ultimate elongation (%)	640
Hardness, shore A	60
Tensile set at break (%)	34

the nitroso copolymers, polymers have been developed with a small amount of a third monomer containing a reactive group which can serve as a potential site for cross-link formation. The canonical principal in selecting the ter-monomer is that its reactivity be such that it insures its random distribution along the chain. For this reason the third monomer is usually of the nitroso variety. The favored reactive group has been the carboxy species. These terpolymer systems are in Table 4.

The inclusion of carboxyl groups along the chain backbone qualifies the elastomer for cross-linking by several procedures. Thus cross-links may be introduced by metal oxides, metal salts of organic and fluorocarbon acids (chromium trifluoroacetate), and epoxy compounds (44). These methods have proven successful in the curing of carboxy-terminated polybutadiene and polyisobutylene, both of which have been used as binder material in solid rocket motors.

Vulcanizates generated from carboxyl-containing nitroso rubber have proven to be stable to strong acids, nitrogen tetroxide, and chlorine tri-fluoride. By comparison the amine cured stocks, with and without silica reinforcing, revert to the uncross-linked state when exposed to nitrogen tetroxide and chlorine trifluoride. This instability is apparently due to the disintegration of the amine crosslinks.

A further effort to simplify cross-linking has been made by producing terpolymers with dienes such as $CF_2=CFCF_2=CF_2$ and $CF_2=CFCF_2=CH_2$. The pendent vinyl group has proven to be susceptible to cross-linking by dicumyl peroxide cures.

In some respects the nitroso polymer holds a unique combination of properties exhibited by no other elastomer. However, its future applications

TABLE 4

Terpolymer Systems of the Nitroso Elastomers $CF_3NO \mid CF_2CF_2 \mid A$

Monomer A	System	Time (hr) | Temp. (°C)	Molar ratio	Conversion (%)	Product	Ref.
p-FC$_6$H$_4$NO	Solution	24 | −25	0.9 | 1.0 | 0.1	76	Elastomer	24
p-BrC$_6$H$_4$NO	Solution	24 | −25	0.9 | 1.0 | 0.1	82	Elastomer	24
p-HOOCC$_6$H$_4$NO	Solution	24 | −25	0.9 | 1.0 | 0.1	80	Elastomer	24
CF_2CFH	Suspension	96 | −30	1.0 | 0.96 | 0.4	70	Elastomer	22
CF_2CFBr	Bulk	12 | −78	1.0 | 1.0 | 0.25	92	Elastomer	22
		to 25	1.0 | 1.0 | 05.0	84	Elastomer	22
			2 | 1 | 1	89	Elastomer	22
$CF_2CFCFCF_2$	Suspension	48 | −30	0.8 | 0.06 | 0.02	50	Elastomer	22
C_6F_5NO	Solution	—	0.9 | 1 | 0.1	76	Elastomer	22
	Solution	—	0.5 | 1 | 0.5	86	Stiff elastomer	22
$NO(CF_2)_3COOH$	Bulk	—	0.96 | 1 | 0.04	—	Elastomer	22

must be reconciled with its many deficiencies. These include its lack of thermal stability, its feeble resistance to electromagnetic radiation, and the bedevilling problems encountered in generating cross-linked material of high strength and stability. The large fraction of low-molecular-weight material partially vitiates this last endeavor. It would seem that a study of the polymerization with the hope of finding conditions whereby the formation of low-molecular-weight material might be avoided would be a worthwhile goal. Its proposed (45) use as a binder in solid rocket engines is not practical since the formation of an effective network remains a very difficult task. The unique characteristic of the nitroso polymer in failing to burn in pure oxygen has resulted in its extensive use by NASA in the Apollo spacecraft. The use of this material, however, carries the penalty of the extremely toxic nature of the decomposition products, i.e., CF_2O and CF_3NCF_2, which result upon the degradation of the nitroso polymer. Currently, the use of this polymer would appear to be limited to situations where a nonflammable elastomer is required or to ambient temperature conditions where resistance to hydrocarbons or strong oxidizing reagents is necessary. The present high cost [$300–400 per pound (46)] is a potent factor in retarding its use. In summation, the future, if indeed it has one, of this material would seem to be limited to that of a specialty elastomer.

Acknowledgement

The author expresses his appreciation to Mr. D. D. Lawson of the Jet Propulsion Laboratory for his assistance and early access to the results of his NMR study. It is also a pleasure to thank Dr. L. A. Wall and Dr. R. F. Fedors for their counsel and advice.

References

1. S. L. Madorsky, V. E. Hart, S. Straus, and V. A. Sedlak, *J. Research NBS*, **51**:327 (1953).
2. J. C. Siegle, L. T. Muus, T.-P. Lin, and H. A. Larsen, *J. Polymer Sci.*, **2A**:391 (1964).
3. D. A. Barr and R. W. Hazeldine, *J. Chem. Soc.*, **1955**:1881; 2532; **1960**:1151.
4. D. A. Barr and R. W. Hazeldine, *Nature*, **175**:991 (1955).
5. J. B. Rose, U.S. Patent 3,065,214 (1962).
6. D. A. Barr, R. W. Hazeldine, and C. J. Willis, *Proc. Chem. Soc.*, **1959**:230.
7. C. E. Griffen and R. W. Hazeldine, *Proc. Chem. Soc.*, **1959**:369; *J. Chem. Soc.*, **1960**:1398.
8. J. B. Rose, British Patent 789, 254 (1958).
9. G. H. Crawford, *J. Polymer Sci.*, **45**:259 (1960).
10. J. C. Montromoso, C. B. Griffis, A. Wilson, and G. H. Crawford, *Rubber Plastics Age*, **42**:514 (1961).
11. J. Green, N. B. Levine, and W. Sheehan, *Rubber Chem. Tech.*, **39**:1222 (1966).
12. D. D. Lawson and J. D. Ingham, *J. Polymer Sci.*, **6B**:181 (1968).
13. G. A. Morneau, P. I. Roth, and A. R. Shultz, *J. Polymer Sci.*, **55**:609 (1961).

14. C. B. Griffis and M. C. Henry, *Rubber Chem. Tech.*, **39**:481 (1966); *Rubber Plastics Age*, **46**:63 (1965).
15. D. A. Barr and R. W. Hazeldine, *J. Chem. Soc.*, **1956**:3424.
16. G. H. Crawford, D. E. Rice, and B. F. Landrum, *J. Polymer Sci.*, **1A**:565 (1963).
17. V. A. Ginsburg, S. S. Dubov, A. N. Medvedev, L. L. Martynova, B. I. Tetel'baum, M. N. Vasil'eva, and A. Ya. Yakubovich, *Doklady Akad. Nauk SSSR*, **152**:1104 (1963).
18. M. Szwarc, *Proc. Roy. Soc.*, **A279**:260 (1964).
19. G. N. Lewis and M. Kasha, *J. Am. Chem. Soc.*, **67**:994 (1945).
20. L. E. Orgel, *J. Chem. Soc.*, **1953**:1276.
21. I. L. Knunyants and A. V. Fokin, *Bull. Acad. Sci. USSR*, **1957**:1463.
22. J. C. Montromoso, *Rubber Chem. Tech.*, **34**:1521 (1961).
23. M. C. Henry, C. B. Griffis, and E. C. Stump, Chapter in Fluorine Chem. Rev. **1**, P. Tarrant (ed.), M. Dekker, New York, 19, p.1.
24. British Patent 983,486 (1965).
25. J. Green, N. Mayes, and E. Cotrill, *Am. Chem. Soc. Polymer Preprints*, **7**:1084 (1966); *J. Macromolecular Sci.*, **A1**(7):1387 (1967).
26. H. Hai-chun, *HachsuehT'ung-pao* (*Chem. Bull.*) (*Peiping*), No. 7:11 (1964).
27. J. A. Castellano, J. Green, and J. M. Kauffman, *J. Org. Chem.*, **31**:821 (1966).
28. J. D. Park, R. W. Rosser, and J. R. Lacher, *J. Org. Chem.*, **27**:1462 (1962).
29. N. L. Zutty and C. W. Wilson, III, *Tetrahedron Letters*, **30**:2181 (1963).
30. N. L. Zutty, C. W. Wilson, III, G. H. Potter, D. C. Priest, and C. J. Whitworth, *J. Polymer Sci.*, **3A**:2781 (1965).
31. C. H. Bamford and M. S. Dewar, *Proc. Roy. Soc.* (*London*), **192A**:309 (1948).
32. D. H. Johnson and A. V. Tobolsky, *J. Am. Chem. Soc.*, **74**:938 (1952).
33. P. J. Flory, "Principles of Polymer Chemistry," Cornell University Press, Ithaca, N.Y., 1953, Chapter IV.
34. R. N. Haward, *Trans. Faraday Soc.*, **46**:204 (1950).
35. B. H. Zimm and J. K. Bragg, *J. Polymer Sci.*, **9**:476 (1952).
36. J. H. Hildebrand and R. L. Scott, "The Solubility of Nonelectrolytes," 3rd ed., A.C.S. Monograph Series, No. 17, Reinhold, New York, 1956.
37. R. L. Stone, *Anal. Chem.*, **29**:1273 (1957).
38. A. R. Shultz, K. Noll, and G. A. Morneau, *J. Polymer Sci.*, **62**:211 (1962).
39. D. W. Brown and L. A. Wall, *J. Phys. Chem.*, **62**:848 (1958).
40. L. A. Wall and J. H. Flynn, *Rubber Chem. Tech.*, **35**:1157 (1962).
41. L. A. Wall, in "Polymer Degradation Mechanisms," National Bureau of Standards Circular 525, Washington, D.C., 1953, p. 239.
42. D. A. Barr and R. W. Hazeldine, *J. Chem. Soc.*, **1956**:3416.
43. J. F. Smith and G. T. Perkins, *Rubber Plastics Age*, **42**:59 (1961).
44. H. P. Brown, *Rubber Chem. Tech.*, **36**:931 (1963).
45. *Aviation Week and Space Tech.*, p. 81 (July 10, 1967).
46. Conference on Elastomers for Extreme Environments, Oct. 24–25, 1969, Dayton, Ohio; Chem. Eng. News, **47**; (29), 72 (1969)

6. FLUORINATED POLYURETHANES

JEROME HOLLANDER,* *Narmco Research and Development Division, Whittaker Corporation, San Diego, California*

Contents

I. INTRODUCTION

Polyurethanes are polymers which contain the characteristic linkage

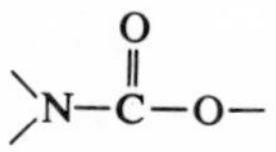

$$\begin{array}{c} \quad\;\; O \\ \quad\;\; \| \\ {>}N{-}C{-}O{-} \end{array}$$

and can be considered esters of the corresponding unstable carbamic acids or amide esters of carbonic acid. The polymers are named polycarbamates, but the term polyurethane is used in referring to this general class of compounds.

In recent years, polyurethanes have become commercially important materials. Homopolymers containing only urethane linkages are used as

*Current address: Burlington Industries Research Center, Greensboro, North Carolina

plastics, fibers, and adhesives. Most of the "polyurethanes" that have found commercial use, however, are copolymers that contain only a small number of urethane linkages. These copolymers, which are prepared from a variety of prepolymers such as polyesters and polyethers, are used as elastomers, foams, and coatings.

Polyurethanes have been discussed thoroughly in a number of books (1–5) and reviews (6–14). In none of these discussions is there any mention of fluorine-containing polyurethanes.

This chapter deals with the preparation of, and to a small extent the properties of, fluorine-containing polyurethanes. Various synthetic routes for the preparation of polyurethanes have been investigated; the most widely used methods are the reaction of diisocyanates with diols (or difunctional hydroxy compounds):

$$OCN-R-NCO + HO-R'-OH \longrightarrow \left[\begin{matrix} O \\ \parallel \\ C-NH-R-NH-C \end{matrix} \underset{\parallel}{\overset{O}{\underset{}{}}} -O-R'-O \right]_n$$

and the reaction of bischloroformates with diamines:

$$\underset{\parallel}{\overset{O}{Cl-C}}-O-R'-O-\underset{\parallel}{\overset{O}{C}}-Cl + H_2N-R-NH_2$$

$$\downarrow$$

$$\left[NH-R-NH-\underset{\parallel}{\overset{O}{C}}-O-R'-O-\underset{\parallel}{\overset{O}{C}} \right]$$

All of the fluorinated polyurethanes reported have been prepared by one of these methods.

Although the chemistry and technology of polyurethanes date back to Bayer's work in 1937, the first reported preparation of a fluorine-containing polyurethane appeared in a patent issued in 1958 (15). This patent and several later reports of work at other laboratories described the synthesis of polyurethanes based on the reactions of nonfluorinated diisocyanates with fluorinated diols, polyesters, and polyethers.

The field of fluorinated polyurethanes was expanded greatly by the work of Narmco Research & Development Division of Whittaker Corporation. This work, begun in 1963, included the synthesis of a large number of more highly fluorinated polyurethanes based on several new fluorinated diisocyanates and a variety of fluorinated diols, polyesters, and polyethers.

II. SYNTHESIS OF INTERMEDIATES

A. Diisocyanates

The preparation of nonfluorinated diisocyanates has been quite adequately described in a number of reviews (13, 16–19). The most commonly used, and the only really commercial method is the phosgenation of primary amines. Other methods have been successfully used for laboratory preparations. These include the Curtius rearrangement of acid azides, the Hofmann rearrangement of acid amides, the Lossen rearrangement of hydroxamic acids, and the reaction of organic halides or sulfates with metal cyanates.

The most commonly used nonfluorinated diisocyanates are the tolylene diisocyanates, hexamethylene diisocyanate, and the bitolylene diisocyanates. The properties and commercial sources of these and other diisocyanates are described in Saunders' book (3) on polyurethanes.

The syntheses of two highly chlorinated aromatic diisocyanates have been reported recently. Tetrachloro-*p*-phenylene diisocyanate was prepared from tetrachloroterephthaloyl chloride by the Curtius reaction (20,21) and by chlorination of *p*-phenylene diisocyanate (22,23):

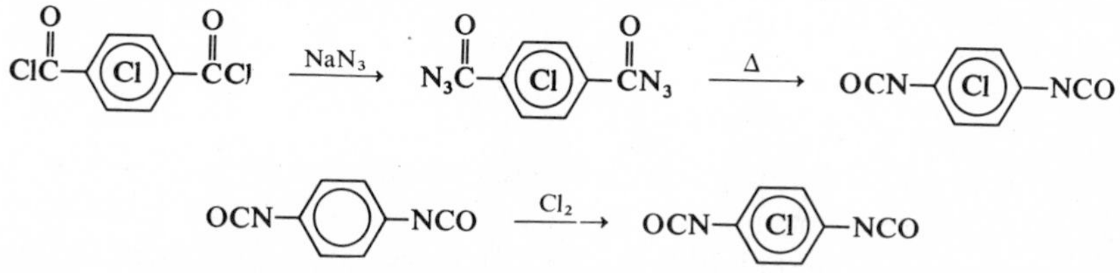

Tetrachloro-*p*-xylylene-α,α'-diisocyanate was prepared by reaction of the corresponding diamine with chlorocarbonyl pyridinium chloride (prepared at low temperature from phosgene and pyridine) (21,24):

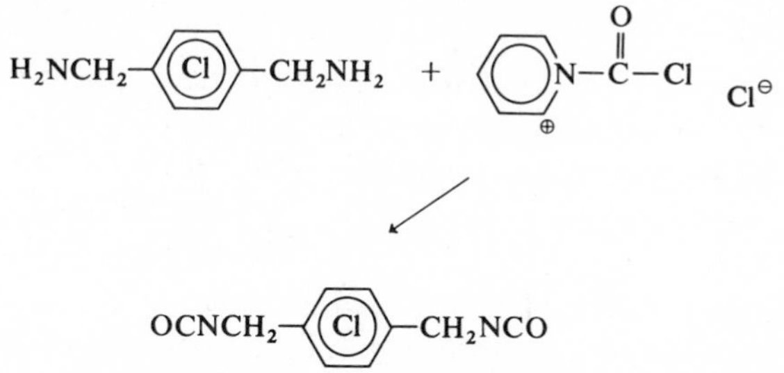

An unusual phosphorus-containing diisocyanate, butyl phosphonic diisocyanate, was prepared by reaction of 1-chlorobutane with phosphorus trichloride and aluminum chloride, followed by hydrolysis and reaction with silver cyanate (25):

$$C_4H_9Cl + PCl_3 \xrightarrow[\text{(2) H}_2\text{O}]{\text{(1) AlCl}_3} C_4H_9-\overset{\overset{\displaystyle O}{\|}}{P}Cl_2$$

$$C_4H_9-\overset{\overset{\displaystyle O}{\|}}{P}Cl_2 + 2\ AgNCO \longrightarrow C_4H_9-\overset{\overset{\displaystyle O}{\|}}{P}(NCO)_2$$

The reported properties of these new diisocyanates are shown in Table 1.

TABLE 1

Properties of Some New Diisocyanates

Diisocyanate	Melting point (°C)	Boiling point (°C)	Characteristic NCO infrared absorption (μ)
Tetrachloro-p-phenylene diisocyanate	91.5–93.5	—	4.5
Tetrachloro-p-xylylene-α,α'-diisocyanate	116–117	162–164/3 mm	4.5
Butyl phosphonic diisocyanate	—	71–73/0.4 mm	—

Two methods of synthesis of fluorinated monoisocyanates have been reported, the Curtius reaction and the direct phosgenation of amines. Perfluoroalkyl monoisocyanates of the general formula $CF_3(CF_2)_nNCO$ have been prepared by reacting perfluoroacyl chlorides with sodium azide, followed by thermal decomposition of the intermediate perfluoroacyl azides (26–30):

$$CF_3(CF_2)_n\overset{\overset{\displaystyle O}{\|}}{C}Cl \xrightarrow{NaN_3} CF_3(CF_2)_n\overset{\overset{\displaystyle O}{\|}}{C}N_3 \xrightarrow{\Delta} CF_3(CF_2)_nNCO$$

α,α'-Dihydroperfluoroalkyl monoisocyanates of the general formula CF_3-$(CF_2)_nCH_2NCO$ have been prepared by treatment of the corresponding α,α-dihydroperfluoroalkyl amines with phosgene (31,32):

$$CF_3(CF_2)_nCH_2NH_2 + COCl_2 \longrightarrow CF_3(CF_2)_nCH_2NCO$$

Only recently have these reactions been extended to the synthesis of perfluoroalkyl diisocyanates, perfluoroaryl diisocyanates, and $\alpha,\alpha,\omega,\omega$-tetrahydroperfluoroalkyl diisocyanates.

Perfluorotrimethylene diisocyanate was prepared by the Curtius degradation of perfluoroglutaryl diazide (20,21,91):

$$\mathrm{ClC(O)(CF_2)_3C(O)Cl} \xrightarrow{\mathrm{NaN_3}} \mathrm{N_3C(O)(CF_2)_3C(O)N_3} \xrightarrow{\Delta} \mathrm{OCN(CF_2)_3NCO}$$

Hexafluoropentamethylene diisocyanate was prepared by reaction of hexafluoropentanediamine with chlorocarbonyl pyridinium chloride (21,24):

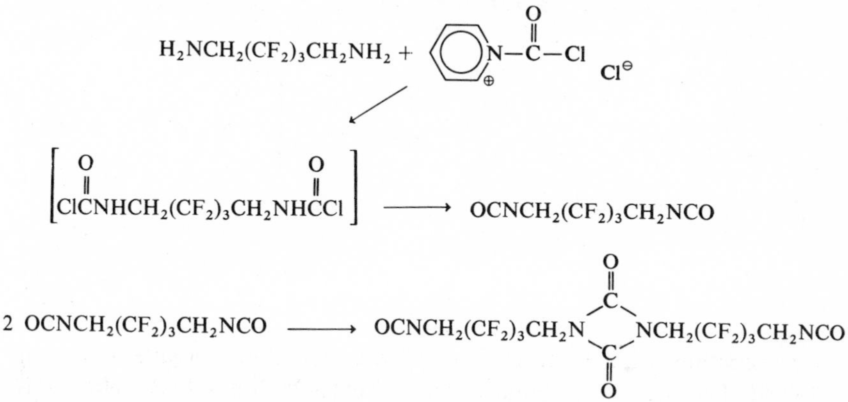

This diisocyanate is believed to exist, at least partially, as the dimer at room temperature.

Perfluoroglutaryl diisocyanate was synthesized by reaction of perfluoroglutaramide with oxalyl chloride and by reaction of perfluoroglutaryl chloride with silver cyanate (20,21):

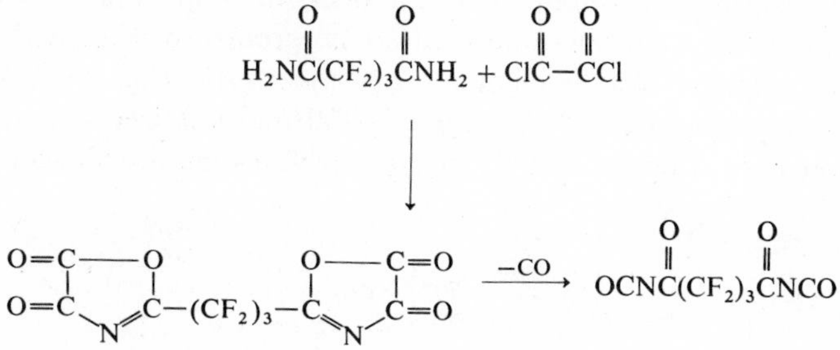

$$\mathrm{ClC(O)(CF_2)_3C(O)Cl + AgOCN \longrightarrow OCNC(O)(CF_2)_3C(O)NCO}$$

Tetrafluoro-*p*-phenylene diisocyanate was prepared by the Curtius reaction on tetrafluoroterephthaloyl chloride, by the phosgenation of tetrafluoro-*p*-phenylenediamine (20,21) and by the reaction of the diamine with chlorocarbonyl pyridinium chloride (21,24):

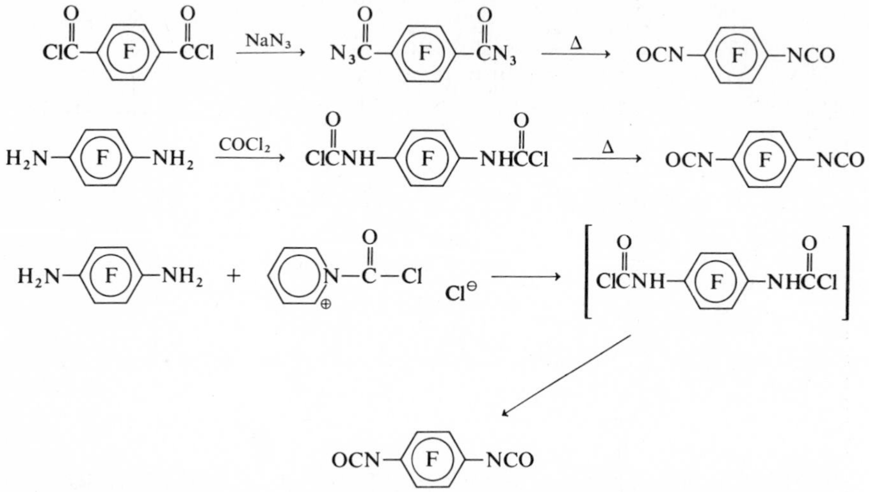

$$ClC(=O)-[F]-CC(=O)Cl \xrightarrow{NaN_3} N_3C(=O)-[F]-C(=O)N_3 \xrightarrow{\Delta} OCN-[F]-NCO$$

$$H_2N-[F]-NH_2 \xrightarrow{COCl_2} ClC(=O)NH-[F]-NHC(=O)Cl \xrightarrow{\Delta} OCN-[F]-NCO$$

$$H_2N-[F]-NH_2 + \overset{\oplus}{[Py]}N-C(=O)-Cl \;\; Cl^{\ominus} \longrightarrow \left[ClC(=O)NH-[F]-NHC(=O)Cl \right]$$

$$OCN-[F]-NCO$$

The preparation using chlorocarbonyl pyridinium chloride gave by far the best yield. The yields of fluorinated diisocyanates by normal type phosgenation of the diamines were quite low due to the serious problem of polyurea formation. This was due to the low reactivity of fluorinated diamines with phosgene, coupled with the spontaneous dehydrochlorination of the intermediate bis-carbamyl chlorides (this is in contrast to nonfluorinated bis-carbamyl chlorides, which require heat to lose hydrogen chloride) and the rapidity of reaction of diisocyanates with unreacted amines. Since the reaction of chlorocarbonyl pyridinium chloride with fluorinated diamines was rapid even below room temperature, there were few unreacted amine groups to react with the diisocyanates, thereby virtually eliminating polymer formation.

Tetrafluoro-*m*-phenylene diisocyanate was also synthesized by reaction of its corresponding diamine with chlorocarbonyl pyridinium chloride [21,33]:

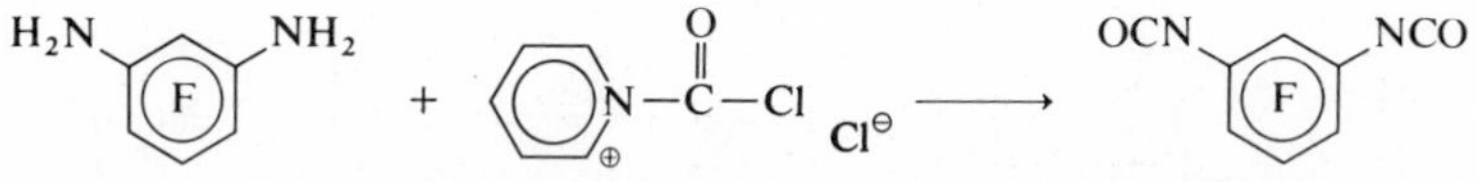

$$H_2N-[F]-NH_2 + \overset{\oplus}{[Py]}N-C(=O)-Cl \;\; Cl^{\ominus} \longrightarrow OCN-[F]-NCO$$

4,4'-Diisocyanatooctafluorobiphenyl was prepared by this chlorocarbonylation method (21,24) and by normal phosgenation (34):

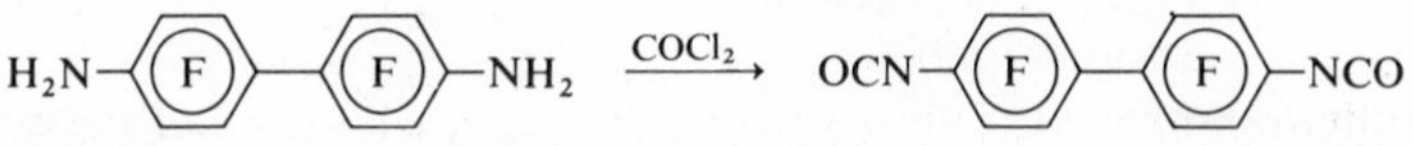

$$H_2N-[F]-[F]-NH_2 \xrightarrow{COCl_2} OCN-[F]-[F]-NCO$$

A chloroperfluoroaryl diisocyanate, 1-chloro-2,4-diisocyanato-3,5-6-trifluorobenzene, was recently prepared using chlorocarbonyl pyridinium chloride (35):

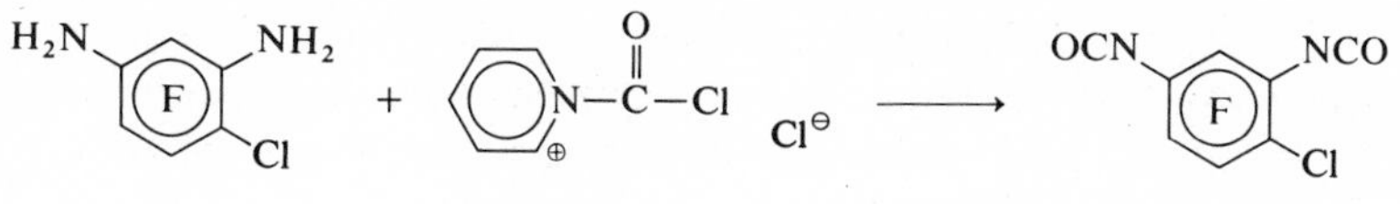

Some properties of these new fluorinated diisocyanates are shown in Table 2.

TABLE 2

Properties of New Fluorinated Diisocyanates

Diisocyanate	Melting point (°C)	Boiling point (°C)	Characteristic NCO infrared absorption (μ)
Perfluorotrimethylene diisocyanate	—	75–76/760 mm	4.5
Hexafluoropentamethylene diisocyanate	—	62/1 mm	4.45 $\overset{\text{O}}{\underset{\|}{}}$ (5.54 and 5.75 $-\text{C}-$)
Perfluoroglutaryl diisocyanate	—	55/760 mm	4.5 $\overset{\text{O}}{\underset{\|}{}}$ (5.6 $-\text{C}-$)
Tetrafluoro-p-phenylene diisocyanate	67–68	70–71/4 mm	4.5
Tetrafluoro-m-phenylene diisocyanate	—	61.5–62/1 mm	4.5
4,4'-Diisocyanato-octafluorobiphenyl	68.5–69	—	4.5
1-Chloro-2,4-diisocyanato-3,5,6-trifluorobenzene	—	74–77/1 mm	4.4

B. Diols

Fluorinated diols of the type $HOCH_2(CF_2)_xCH_2OH$ have been reported in which $x = 1-4$ (36). A general method of synthesis was described involving esterification of perfluorodicarboxylic acids followed by reduction of the esters with lithium aluminum hydride (37,38). An improved process devised for the preparation of hexafluoropentanediol involved vapor-phase oxidation of 1,2-dichlorohexafluorocyclopentene and catalytic reduction of the resulting perfluoroglutaryl chloride to the diol (39,40). A recent review describes the synthesis of other fluorinated diols (41).

The dihydroxy tetrafluorobenzenes have recently come under study. Tetra-fluoro-*p*-benzoquinone, prepared by treatment of octafluorocyclohexa-1,4-diene with oleum, was reduced with hydrogen and Raney nickel to give tetra-fluoro-*p*-hydroquinone (42):

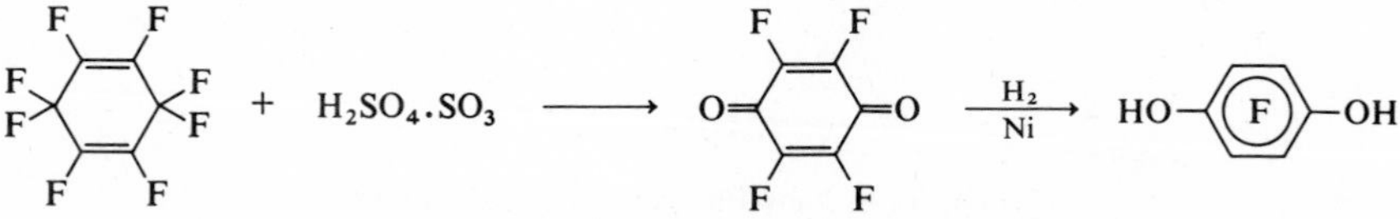

The same diol was prepared by cleavage of the diether formed by reaction of hexafluorobenzene with ethylene glycol and potassium hydroxide (43):

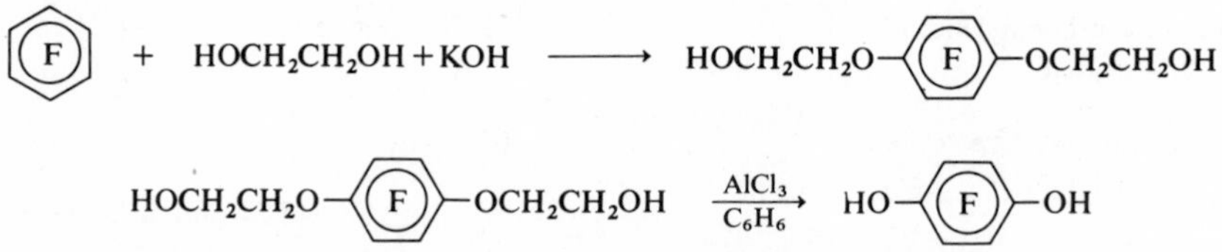

Tetrafluororesorcinol was synthesized by reaction of potassium hydroxide with both hexafluorobenzene (44) and pentafluorophenol (45).

Since hexafluoropentanediol and tetrafluoro-*p*-hydroquinone are the only available fluorinated diols, almost all the reported polyurethanes have been prepared from these two diols.

C. Polyesters

Three methods of synthesis of polyesters from fluorine-containing diols have been reported (46). The methods investigated were (1) direct esterification with dicarboxylic acids using zinc chloride catalyst; (2) transesterification of diethyl esters using various catalysts; and (3) reaction with dicarboxylic acid chlorides. By far the best method was through use of dicarboxylic acid chlorides. All of the fluorinated polyesters synthesized have been prepared using hexafluoropentanediol and octafluorohexanediol. Polyesters have been prepared from these diols and both fluorinated and nonfluorinated dicarboxylic acids (or derivatives). Typical fluorinated carboxylic acids used were perfluoroglutaric acid (20,46,47), perfluoroadipic acid (46,47), and 3-perfluoropropylglutaric acid (46,47). Nonfluorinated diacids used were 3,6-dithiaoctanedioic acid and aliphatic diacids of the type $HOOC(CH_2)_x$-COOH, where x ranged from 1 to 8 (20,24,46,48). Terephthalate and isophthalate polyesters (49) and adipate-isophthalate copolyesters of hexa-fluoropentanediol were also prepared (50).

An interesting fluorinated polycarbonate, poly(hexafluoropentamethylene carbonate), has been prepared by direct phosgenation of hexafluoropentane-diol (51) and by reaction of hexafluoropentanediol with hexafluoropenta-methylene bischloroformate (24):

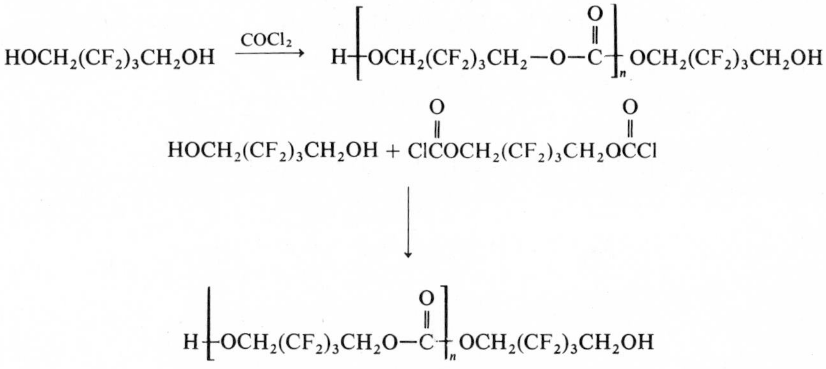

$$HOCH_2(CF_2)_3CH_2OH \xrightarrow{\text{COCl}_2} H\left[OCH_2(CF_2)_3CH_2-O-\overset{\overset{\displaystyle O}{\|}}{C}\right]_n OCH_2(CF_2)_3CH_2OH$$

$$HOCH_2(CF_2)_3CH_2OH + Cl\overset{\overset{\displaystyle O}{\|}}{C}OCH_2(CF_2)_3CH_2O\overset{\overset{\displaystyle O}{\|}}{C}Cl$$

$$\downarrow$$

$$H\left[OCH_2(CF_2)_3CH_2O-\overset{\overset{\displaystyle O}{\|}}{C}\right]_n OCH_2(CF_2)_3CH_2OH$$

D. Polyethers

Polymerization studies of fluorinated 1,2-epoxypropanes of the type

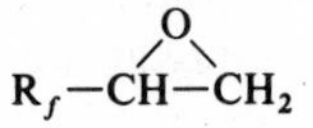

$$R_f-\overset{\overset{\displaystyle O}{\diagup \diagdown}}{CH-CH_2}$$

have been described using aluminum chloride and ferric chloride as catalysts (52). 3,3,3-Trifluoro-1,2-epoxypropane and 2-methyl-3-3,3-trifluoro-1,2-epoxypropane were used in this study. The polymerization of a series of fluori-nated 2,3-epoxybutanes was also studied using aluminum chloride, ferric chloride, boron trifluoride, and potassium hydroxide (53). Recently, the polymerization of 3,3,3-trifluoro-1,2-epoxypropane was reinvestigated using both cationic initiators (boron trifluoride and aluminum chloride) and anionic initiators (potassium hydroxide and the monosodium salt of hexafluoropentanediol) (24,54). A completely hydroxyl-terminated polyether was obtained with the last-mentioned catalyst. A similar hydroxyl-terminated polyether was prepared using ethylene glycol and boron trifluoride-etherate (55,56).

Other fluorinated polyethers based on epoxides have been prepared from hexafluoropentylene diglycidyl ether (57) and copolymers of ethylene oxide with vinylidene fluoride and with hexafluoropropylene (58).

A fluorine-containing polyformal was synthesized by reaction of hexafluoropentanediol with either trioxane or dibutyl formal (24,54):

$$HOCH_2(CF_2)_3CH_2OH + (CH_2O)_3$$

$$\Big\downarrow \text{\scriptsize p-Toluenesulfonic acid}$$

$$HOCH_2(CF_2)_3CH_2O + CH_2-O-CH_2(CF_2)_3CH_2O\frac{}{}_nH$$

$$HOCH_2(CF_2)_3CH_2OH + C_4H_9O-CH_2-OC_4H_9$$

$$\downarrow$$

$$HOCH_2(CF_2)_3CH_2O + CH_2-OCH_2(CF_2)_3CH_2O\frac{}{}_nH$$

Polyethers of hexafluorobenzene and hexafluoropentanediol have been described recently (25,33). Completely hydroxyl-terminated polyethers of hexafluorobenzene and hexafluoropentanediol have been synthesized and have been used to prepare polyurethanes. In a four-step synthesis, the hydroxyl groups of the polyether were protected by reaction with dihydropyran, the remaining monosubstituted aromatic ring capped with hexafluoropentanediol, and the protecting groups removed (59):

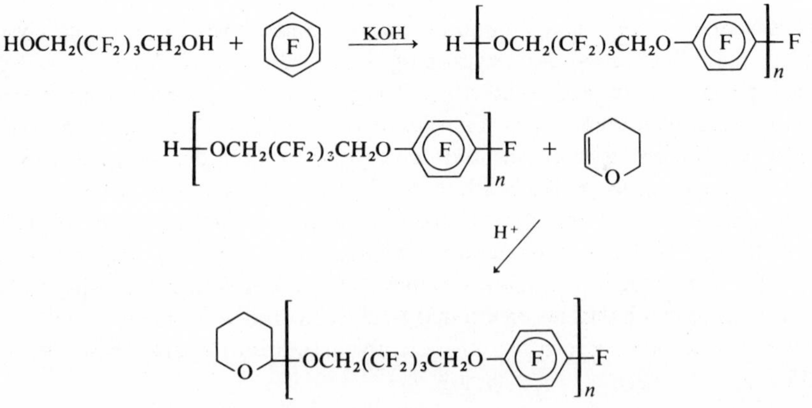

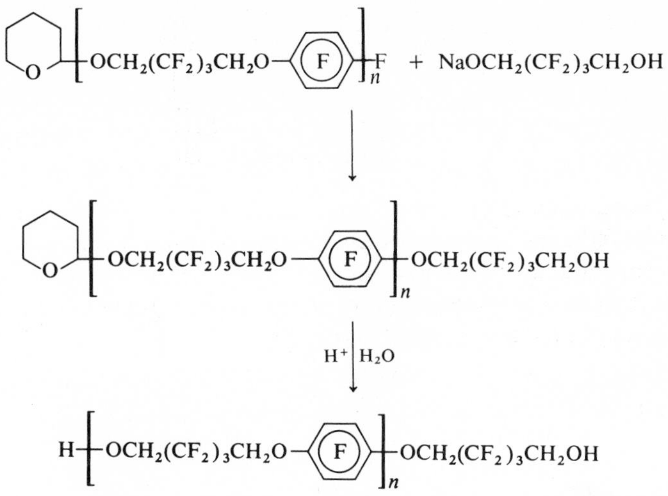

$$\text{(cyclic)} \left[-OCH_2(CF_2)_3CH_2O-\underset{\displaystyle\bigcirc}{(F)}- \right]_n F \;+\; NaOCH_2(CF_2)_3CH_2OH$$

$$\downarrow$$

$$\text{(cyclic)} \left[-OCH_2(CF_2)_3CH_2O-\underset{\displaystyle\bigcirc}{(F)}- \right]_n OCH_2(CF_2)_3CH_2OH$$

$$\downarrow H^+ \mid H_2O$$

$$H\left[-OCH_2(CF_2)_3CH_2O-\underset{\displaystyle\bigcirc}{(F)}- \right]_n OCH_2(CF_2)_3CH_2OH$$

A more convenient, one-step synthesis was described using copper (I) iodide as a catalyst (35):

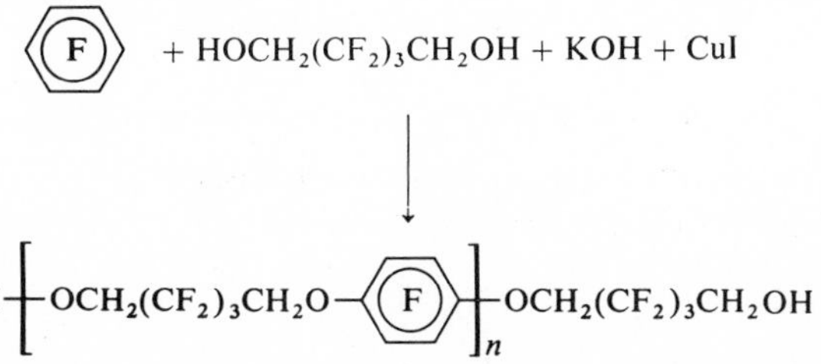

$$\underset{\displaystyle\bigcirc}{(F)} \;+\; HOCH_2(CF_2)_3CH_2OH + KOH + CuI$$

$$\downarrow$$

$$H\left[-OCH_2(CF_2)_3CH_2O-\underset{\displaystyle\bigcirc}{(F)}- \right]_n OCH_2(CF_2)_3CH_2OH$$

A series of recent patents described the synthesis of perfluorinated polyethers based on perfluoropropylene oxide, tetrafluoroethylene oxide, and other perfluorinated epoxides (60–63). Diacid fluoride-terminated polyethers were obtained when perfluoropropylene oxide was polymerized using perfluorinated diacid fluorides and cesium fluorides (64):

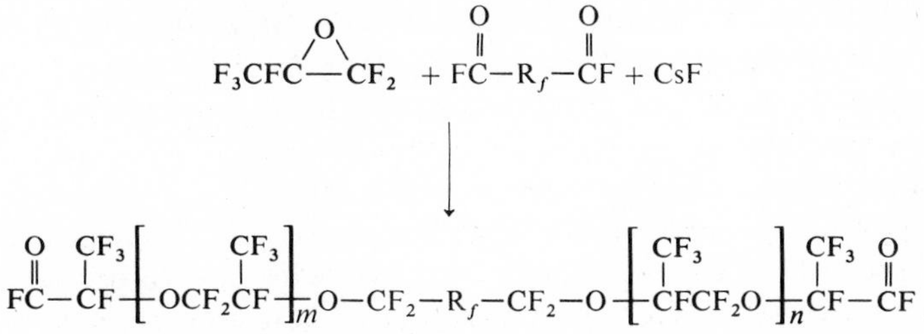

$$F_3CFC\overset{\displaystyle O}{\overbrace{}}CF_2 \;+\; \underset{\displaystyle \overset{\|}{O}}{FC}-R_f-\underset{\displaystyle \overset{\|}{O}}{CF} + CsF$$

$$\downarrow$$

$$\underset{\displaystyle O}{FC}-\underset{\displaystyle CF_3}{CF}\left[-O\underset{\displaystyle CF_3}{CF_2CF} \right]_m O-CF_2-R_f-CF_2-O\left[\underset{\displaystyle CF_3}{CFCF_2O} \right]_n \underset{\displaystyle CF_3\ \ O}{CF}-CF$$

Hydroxyl-terminated polyethers usable as polyurethane prepolymers were prepared by reduction of the diacid fluoride- or the dimethyl ester-terminated polyethers of perfluoropropylene oxide (59).

E. Bischloroformates

Fluorinated bischloroformates have been used in the preparation of fluorinated polyurethanes as intermediates in the preparation of polycarbonate prepolymers and as reactants with diamines. Hexafluoropentamethylene bischloroformate was synthesized by reaction of hexafluoropentanediol with chlorocarbonyl pyridinium chloride (65,66):

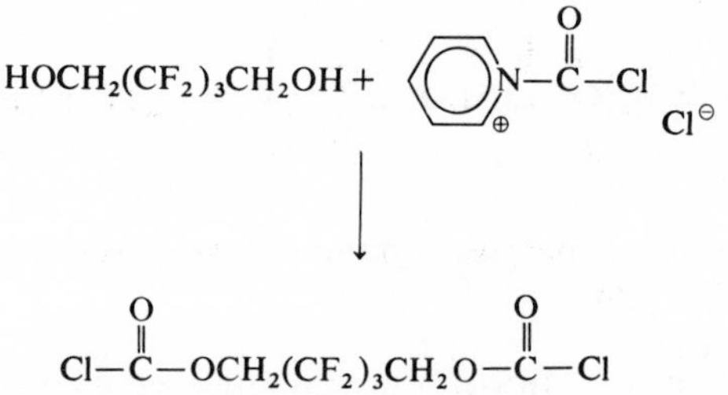

$$
HOCH_2(CF_2)_3CH_2OH + \underset{\oplus}{\bigcirc}N-\overset{\overset{O}{\|}}{C}-Cl \quad Cl^{\ominus}
$$

$$
Cl-\overset{\overset{O}{\|}}{C}-OCH_2(CF_2)_3CH_2O-\overset{\overset{O}{\|}}{C}-Cl
$$

Other fluorinated bischloroformates that have been synthesized in a similar manner are tetrafluoro-p-phenylene bischloroformate (33,67) and tetrafluoro-m-phenylene bischloroformate (68).

F. Diamines

Fluorinated diamines have been used to prepare diisocyanates and to synthesize linear polyurethanes by reaction with bischloroformates. Very few fluorinated diamines are known. The syntheses of hexafluoropentanediamine and octafluorohexanediamine have been reported by reduction of the corresponding dinitriles using platinum oxide catalyst (69,70). Pure hexafluoropentanediamine has been synthesized by reduction of perfluoroglutaramide with lithium aluminum hydride (20,71). The lithium aluminum hydride reaction of N-substituted diamides of perfluoroglutaric acid to secondary diamines has also been reported (72).

The synthesis of fluorinated aromatic diamines by a variety of methods have been reported recently. Tetrafluoro-p-phenylenediamine has been prepared by a series of reactions involving ammonation of hexafluorobenzene, oxidation, further ammonation, and then reduction (73):

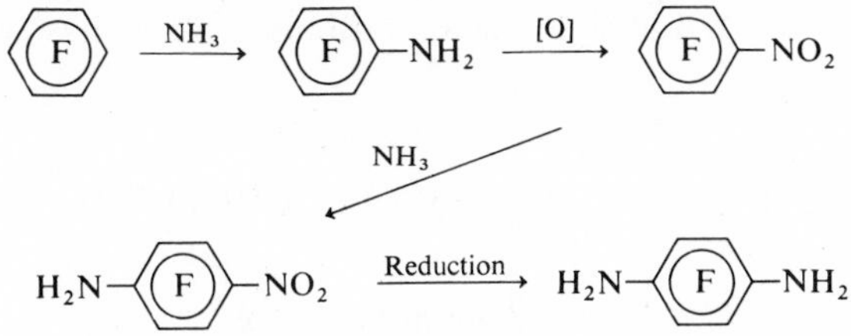

Tetrafluoro-*p*-phenylenediamine has also been synthesized from hexafluorobenzene by preparation and cleavage of the 1,4-dihydrazino derivative (74) and by reaction with potassium phthalimide (Gabriel synthesis) (20,71):

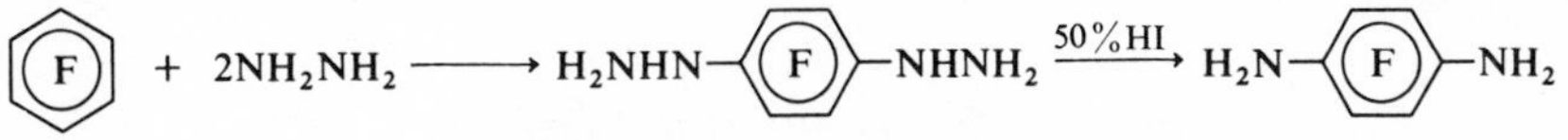

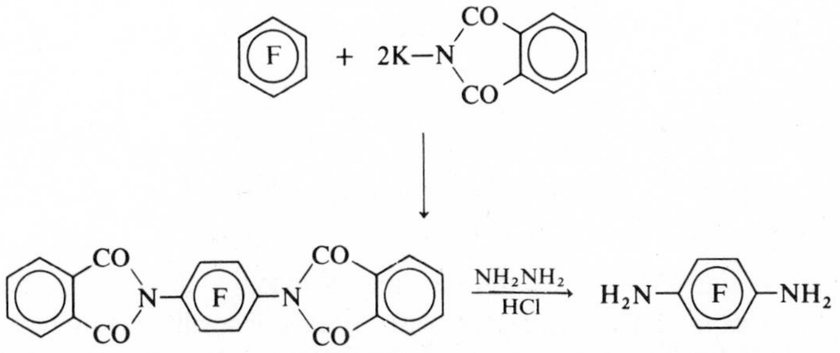

Tetrafluoro-*m*-phenylenediamine has been prepared by reaction of hexafluorobenzene with both alcoholic ammonia (73) and aqueous ammonia (33,71,75). It has also been synthesized by reaction of pentafluoroaniline with hydrazine, followed by cleavage of the *m*-hydrazino-tetrafluoroaniline (74) and with potassium phthalimide (20,71):

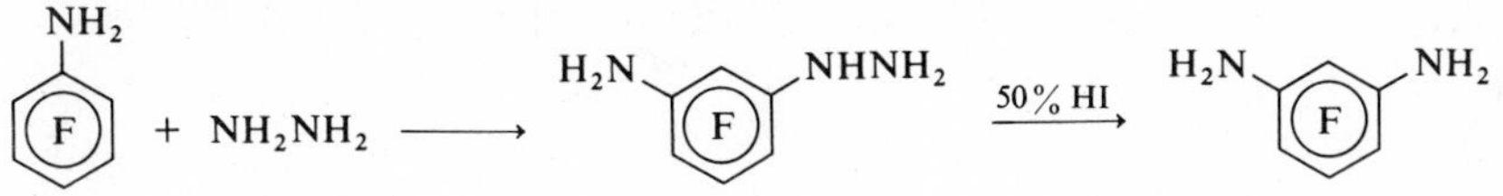

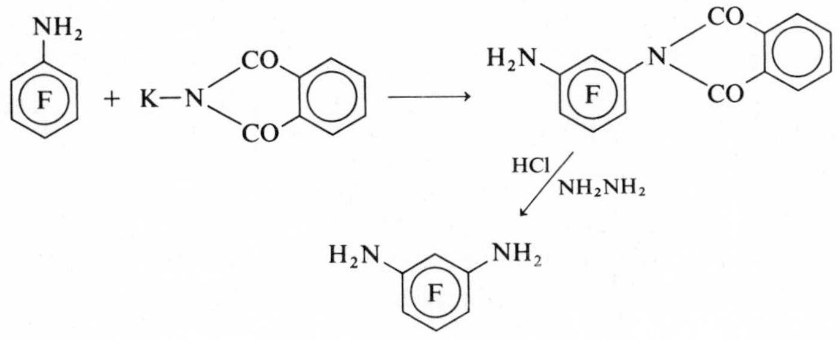

The preparation of 4,4′-diaminooctafluorobiphenyl also has been reported (76–78).

III. POLYURETHANES FROM FLUORINATED DIOLS

A. Hexafluoropentanediol

The first polyurethane reported which contained fluorine was synthesized by reaction of hexafluoropentanediol and hexamethylene diisocyanate (15, 20,33,79):

$$OCN(CH_2)_6NCO + HOCH_2(CF_2)_3CH_2OH$$

$$\left[\begin{array}{c} \overset{O}{\overset{\|}{C}}NH(CH_2)_6NH\overset{O}{\overset{\|}{C}}-OCH_2(CF_2)_3CH_2O \end{array} \right]_n$$

Soon after, polyurethanes based on aromatic diisocyanates were synthesized. Hexafluoropentanediol was reacted with 3,5-tolylene diisocyanate and with a mixture of 2,4-tolylene diisocyanate and 3,3′-bitolylene-4,4′-diisocyanate to give polyurethanes (80):

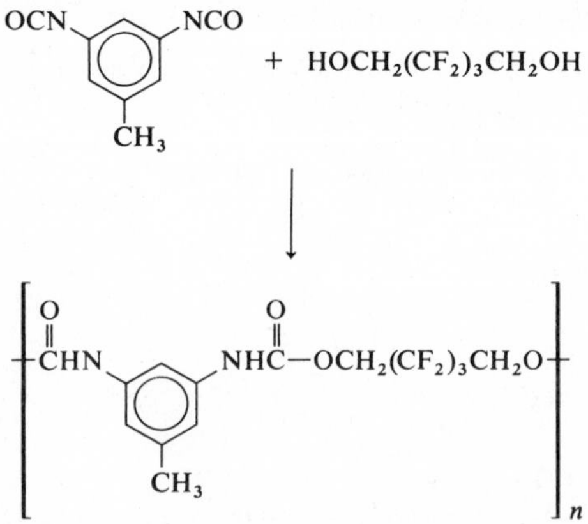

With the advent of the space program, renewed interest in chemically resistant materials led to a study of more highly halogenated polyurethanes. Some of the earliest studied were those prepared from the highly chlorinated diisocyanates, tetrachloro-p-phenylene diisocyanate and tetrachloro-p-xylylene-α,α'-diisocyanate (24,33,81):

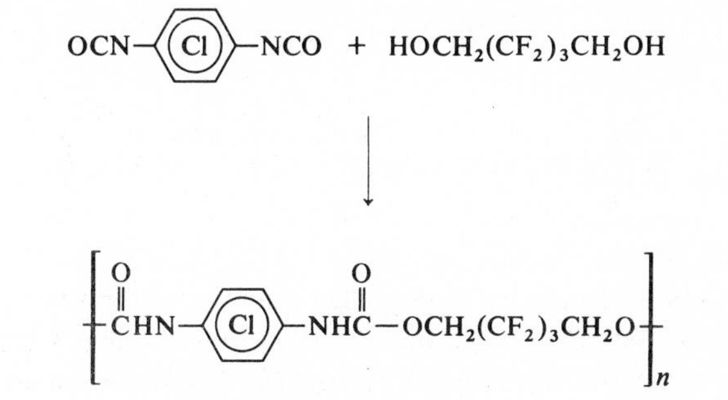

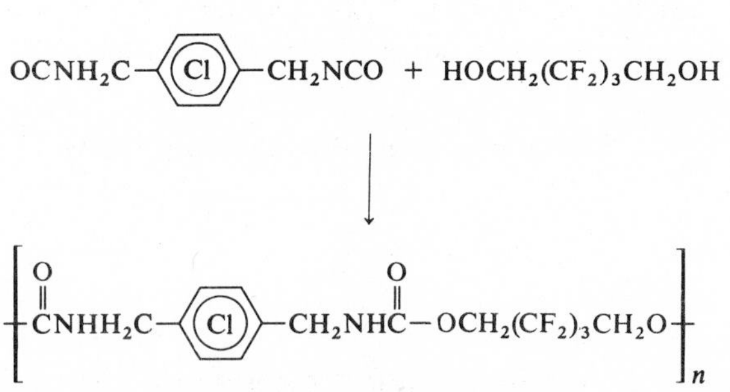

$$\text{OCN}-\underset{\text{(Cl)}}{\bigcirc}-\text{NCO} \;+\; \text{HOCH}_2(\text{CF}_2)_3\text{CH}_2\text{OH}$$

$$\left[\begin{array}{c}\overset{\text{O}}{\overset{\|}{\text{CHN}}}-\underset{\text{(Cl)}}{\bigcirc}-\text{NH}\overset{\text{O}}{\overset{\|}{\text{C}}}-\text{OCH}_2(\text{CF}_2)_3\text{CH}_2\text{O}\end{array}\right]_n$$

$$\text{OCNH}_2\text{C}-\underset{\text{(Cl)}}{\bigcirc}-\text{CH}_2\text{NCO} \;+\; \text{HOCH}_2(\text{CF}_2)_3\text{CH}_2\text{OH}$$

$$\left[\begin{array}{c}\overset{\text{O}}{\overset{\|}{\text{CNHH}_2\text{C}}}-\underset{\text{(Cl)}}{\bigcirc}-\text{CH}_2\text{NH}\overset{\text{O}}{\overset{\|}{\text{C}}}-\text{OCH}_2(\text{CF}_2)_3\text{CH}_2\text{O}\end{array}\right]_n$$

An interesting polyurethane was prepared by reaction of hexafluoropentanediol with butyl phosphonic diisocyanate (25):

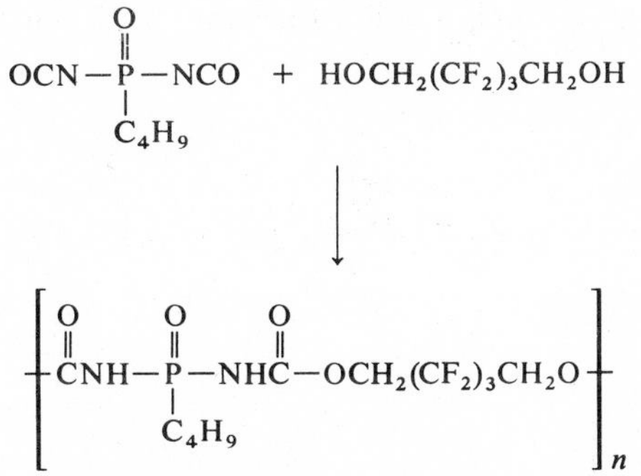

$$\text{OCN}-\overset{\text{O}}{\overset{\|}{\underset{\underset{\text{C}_4\text{H}_9}{|}}{\text{P}}}}-\text{NCO} \;+\; \text{HOCH}_2(\text{CF}_2)_3\text{CH}_2\text{OH}$$

$$\left[\begin{array}{c}\overset{\text{O}}{\overset{\|}{\text{CNH}}}-\overset{\text{O}}{\overset{\|}{\underset{\underset{\text{C}_4\text{H}_9}{|}}{\text{P}}}}-\text{NH}\overset{\text{O}}{\overset{\|}{\text{C}}}-\text{OCH}_2(\text{CF}_2)_3\text{CH}_2\text{O}\end{array}\right]_n$$

A completely substituted fluorinated polyurethane was prepared by reaction of hexafluoropentamethylene bischloroformate with N,N'-diphenyl-p-phenylenediamine (82):

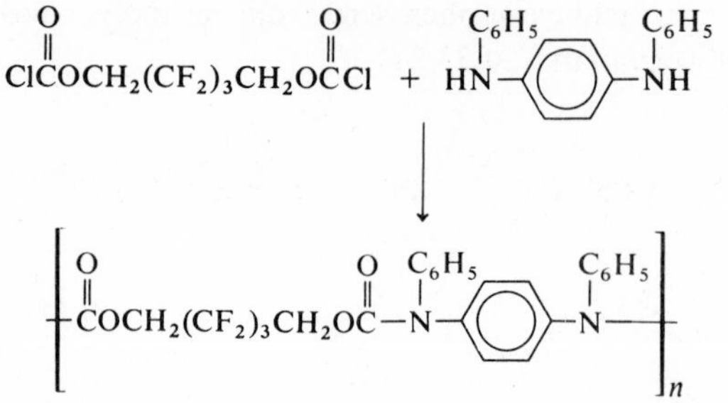

$$
\underset{\text{Cl}}{\overset{\text{O}}{\underset{\|}{\text{C}}}}\text{OCH}_2(\text{CF}_2)_3\text{CH}_2\text{O}\underset{\|}{\overset{\text{O}}{\text{C}}}\text{Cl} \;+\; \text{HN}\overset{\text{C}_6\text{H}_5}{-}\!\!\!\bigcirc\!\!\!\overset{\text{C}_6\text{H}_5}{-}\text{NH}
$$

$$
\left[\!\!\begin{array}{c}\overset{\text{O}}{\underset{\|}{\text{C}}}\text{OCH}_2(\text{CF}_2)_3\text{CH}_2\text{O}\overset{\text{O}}{\underset{\|}{\text{C}}}-\overset{\text{C}_6\text{H}_5}{\underset{|}{\text{N}}}\!\!\!\bigcirc\!\!\!-\overset{\text{C}_6\text{H}_5}{\underset{|}{\text{N}}}-\end{array}\!\!\right]_n
$$

In a systematic study of highly fluorinated polyurethanes, a series of hexafluoropentanediol-based fluorinated polyurethanes were synthesized by reaction with both aliphatic and aromatic fluorinated diisocyanates. The reaction of hexafluoropentanediol with perfluorotrimethylene diisocyanate yielded a polyurethane which was hydrolytically unstable (20,81):

$$
\text{OCN}(\text{CF}_2)_3\text{NCO} + \text{HOCH}_2(\text{CF}_2)_3\text{CH}_2\text{OH}
$$

$$
\left[\!\!\begin{array}{c}\overset{\text{O}}{\underset{\|}{\text{C}}}\text{NH}(\text{CF}_3)_2\text{NH}\overset{\text{O}}{\underset{\|}{\text{C}}}-\text{OCH}_2(\text{CF}_2)_3\text{CH}_2\text{O}\end{array}\!\!\right]_n
$$

An aliphatic polyurethane synthesized based on hexafluoropentamethylene diisocyanate was hydrolytically stable. It was prepared by reaction of hexafluoropentamethylene bischloroformate with hexafluoropentanediamine (24,33,81).

$$
\underset{\text{Cl}}{\overset{\text{O}}{\underset{\|}{\text{C}}}}\text{OCH}_2(\text{CF}_2)_3\text{CH}_2\text{O}\overset{\text{O}}{\underset{\|}{\text{C}}}\text{Cl} + \text{H}_2\text{NCH}_2(\text{CF}_2)_3\text{CH}_2\text{NH}_2
$$

$$
\left[\!\!\begin{array}{c}\overset{\text{O}}{\underset{\|}{\text{C}}}\text{OCH}_2(\text{CF}_2)_3\text{CH}_2\text{O}\overset{\text{O}}{\underset{\|}{\text{C}}}-\text{NHCH}_2(\text{CF}_2)_3\text{CH}_2\text{O}\end{array}\!\!\right]_n
$$

Aromatic-aliphatic polyurethanes were obtained by reaction of hexafluoro-pentanediol with tetrafluoro-*p*-phenylene diisocyanate (24,33,81), tetrafluoro-*m*-phenylene diisocyanate (33,81), and 4,4'-diisocyanato-octafluoro-biphenyl (34):

$$OCN-Ar_f-NCO + HOCH_2(CF_2)_3CH_2OH$$

$$\left[\begin{array}{c} O \\ \parallel \\ CNH-Ar_f-NHC-OCH_2(CF_2)_3CH_2O \end{array} \right]_n$$

$$Ar_f = \quad , \quad , \quad and$$

A mixed polyurethane was prepared by reaction of hexafluoropentane-diamine with excess hexafluoropentamethylene bischloroformate and further reaction of the resulting hydroxyl-terminated polyurethane with tetrafluoro-*m*-phenylene diisocyanate (59):

$$Xs \; ClCOCH_2(CF_2)_3CH_2OCCl + H_2NCH_2(CF_2)_3CH_2NH_2$$

$$\Big\downarrow H_2O$$

$$H \left[OCH_2(CF_2)_3CH_2OCNHCH_2(CF_2)_3CH_2NHC \right]_m OCH_2(CF_2)_3CH_2OH$$

$$\Big\downarrow \; OCN \quad NCO$$

$$\left\{ \begin{array}{c} O \\ \parallel \\ CHN \end{array} NHC \left[OCH_2(CF_2)_3CH_2OCNHCH_2(CF_2)_3CH_2NHC \right]_m OCH_2(CF_2)_3CH_2O \right\}_n$$

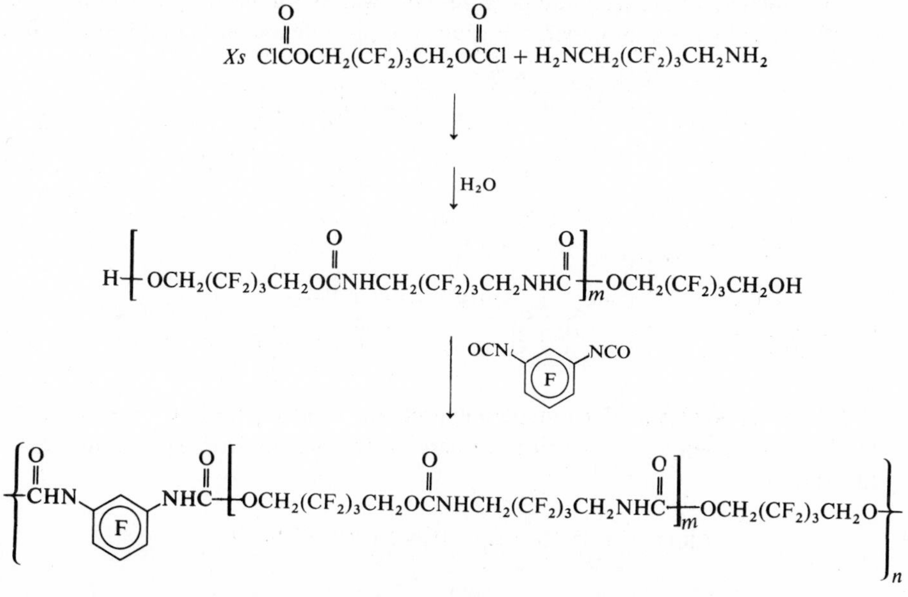

A similar polyurethane-urea was prepared by reaction of the amino-termina-ted polyurethane, formed from hexafluoropentamethylene bischloroformate and excess hexafluoropentanediamine, with tetrafluoro-*m*-phenylene diiso-cyanate (59):

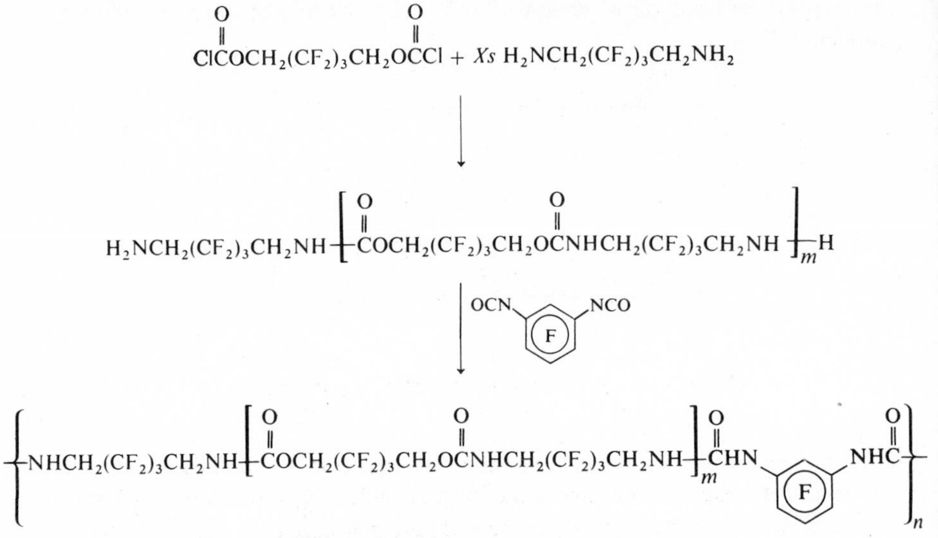

$$\overset{O}{\overset{\|}{Cl}}COCH_2(CF_2)_3CH_2O\overset{O}{\overset{\|}{C}}Cl + Xs\ H_2NCH_2(CF_2)_3CH_2NH_2$$

$$\downarrow$$

$$H_2NCH_2(CF_2)_3CH_2NH\left[\overset{O}{\overset{\|}{C}}OCH_2(CF_2)_3CH_2O\overset{O}{\overset{\|}{C}}NHCH_2(CF_2)_3CH_2NH\right]_m H$$

$$\downarrow\ OCN{-}\langle\!\langle F\rangle\!\rangle{-}NCO$$

$$\left\{{-}NHCH_2(CF_2)_3CH_2NH\left[\overset{O}{\overset{\|}{C}}OCH_2(CF_2)_3CH_2O\overset{O}{\overset{\|}{C}}NHCH_2(CF_2)_3CH_2NH\right]_m\overset{O}{\overset{\|}{C}}HN{-}\langle\!\langle F\rangle\!\rangle{-}NH\overset{O}{\overset{\|}{C}}{-}\right\}_n$$

B. Tetrafluoro-*p*-Hydroquinone

A completely fluoroaromatic polyurethane was synthesized by reaction of tetrafluoro-*p*-hydroquinone with tetrafluoro-*m*-phenylene diisocyanate (33,81):

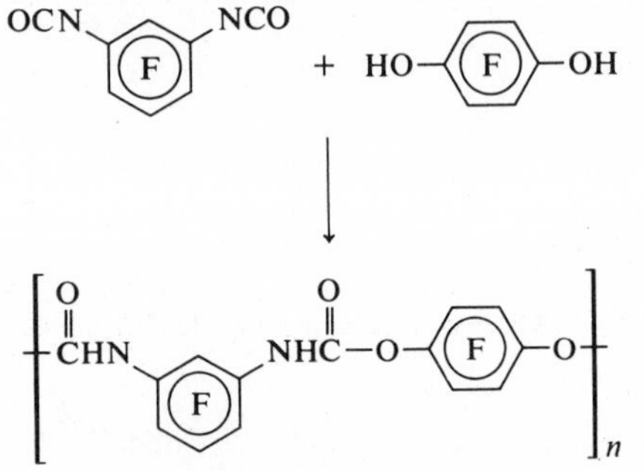

$$OCN{-}\langle\!\langle F\rangle\!\rangle{-}NCO\ +\ HO{-}\langle\!\langle F\rangle\!\rangle{-}OH$$

$$\downarrow$$

$$\left[{-}\overset{O}{\overset{\|}{C}}HN{-}\langle\!\langle F\rangle\!\rangle{-}NH\overset{O}{\overset{\|}{C}}{-}O{-}\langle\!\langle F\rangle\!\rangle{-}O{-}\right]_n$$

A fluorinated aliphatic-aromatic polyurethane was prepared by reaction of tetrafluoro-*p*-phenylene bischloroformate with hexafluoropentanediamine (33,81):

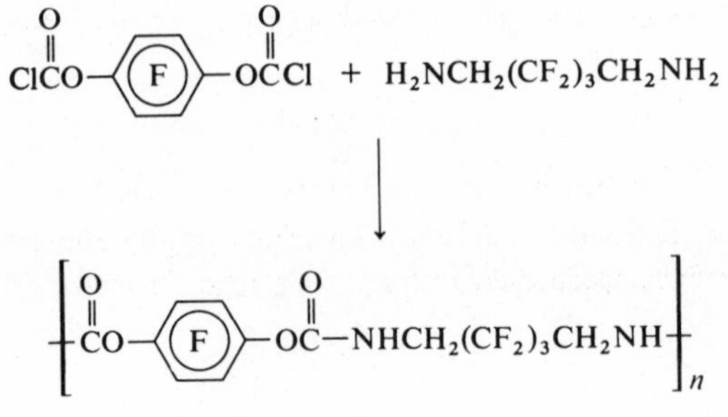

$$Cl\overset{O}{\overset{\|}{C}}O{-}\langle\!\langle F\rangle\!\rangle{-}O\overset{O}{\overset{\|}{C}}Cl\ +\ H_2NCH_2(CF_2)_3CH_2NH_2$$

$$\downarrow$$

$$\left[{-}\overset{O}{\overset{\|}{C}}O{-}\langle\!\langle F\rangle\!\rangle{-}O\overset{O}{\overset{\|}{C}}{-}NHCH_2(CF_2)_3CH_2NH{-}\right]_n$$

IV. POLYURETHANES FROM POLYESTERS OF FLUORINATED DIOLS

All the fluorinated hydroxyl-terminated polyesters that have been used as polyurethane prepolymers were prepared from hexafluoropentanediol or octafluorohexanediol. Polyesters prepared from these two fluorinated diols and nonfluorinated diacid chlorides have been reacted with a variety of non-fluorinated and fluorinated diisocyanates.

Polyurethanes were prepared by reaction of 2,4-tolylene diisocyanate with low-molecular-weight poly(hexafluoropentamethylene adipate) (24) and poly(hexafluoropentamethylene malonate) (24):

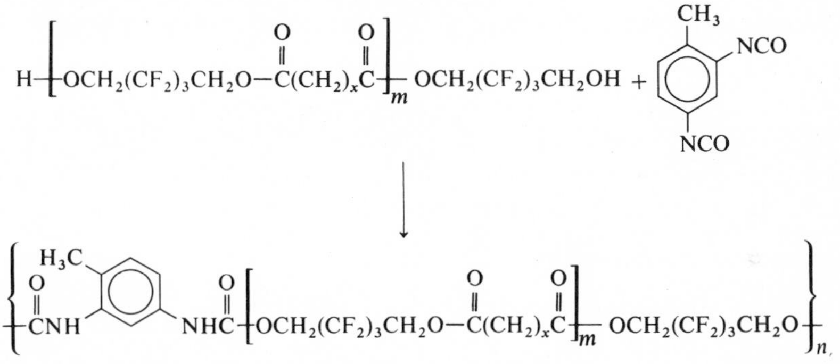

$X = 1$ or 4

Moderately high-molecular-weight polyesters prepared from hexafluoro-pentanediol or octafluorohexanediol with glutaryl, adipyl, suberyl, azelayl, or sebacyl chlorides were chain extended with either 2,4-tolylene diisocyanate, 2,6-tolylene diisocyanate, or diphenylmethane-4,4'-diisocyanate and then cured with hexamethylene diamine (48).

Polyesters prepared from perfluorinated acids have not been used as polyurethane prepolymers because of their hydrolytic instability. Polyesters synthesized from hexafluoropentanediol or octafluorohexanediol and the partially fluorinated diacid chlorides, 3-perfluoropropyl-glutaryl chloride, 3-perfluoroheptyl-glutaryl chloride, 3,5-bis(perfluoropropyl)-4-thiapimelyl chloride, or 5,5,6,6,7,7-hexafluoro-3,9-dioxaundecanedioyl chloride have been chain extended and cured. The diisocyanates used for chain extension were 2,4-tolylene diisocyanate, 2,6-tolylene diisocyanate, or diphenyl-methane-4,4'-diisocyanate (47).

A polyurethane was reported as the reaction product of hexafluoro-pentanediol, isophthaloyl chloride, and diphenylmethane-4,4'-diisocyanate cross-linked with resorcinol (83).

In an attempt to prepare more highly halogenated polyurethanes by reaction of tetrachloro-*p*-phenylene diisocyanate with poly(hexafluoro-pentamethylene malonate) and poly(hexafluoropentamethylene adipate), isocyanate-terminated polyurethanes were obtained which were difficult to cure. This lack of reactivity was explained as being due to steric hindrance by the large chlorine atoms to attack on the isocyanate groups (24).

More highly fluorinated polyester-based polyurethanes were synthesized from poly(hexafluoropentamethylene adipate) and tetrafluoro-*p*-phenylene diisocyanate (24,84), tetrafluoro-*m*-phenylene diisocyanate (33,84), and 4,4′-diisocyanatooctafluorobiphenyl (24,84). An extensive study on curing methods and agents was carried out on the polyurethane from poly(hexa-fluoropentamethylene adipate) and tetrafluoro-*p*-phenylene diisocyanate (24,84):

$$\text{OCN-Ar}_f\text{-NCO} + \text{H}\left[\text{OCH}_2(\text{CF}_2)_3\text{CH}_2\text{O}-\overset{\overset{\text{O}}{\|}}{\text{C}}(\text{CH}_2)_4\overset{\overset{\text{O}}{\|}}{\text{C}}\right]_m\text{OCH}_2(\text{CF}_2)_3\text{CH}_2\text{OH}$$

$$\Big\{\overset{\overset{\text{O}}{\|}}{\text{C}}\text{NH-Ar}_f\text{-NH}\overset{\overset{\text{O}}{\|}}{\text{C}}\left[\text{OCH}_2(\text{CF}_2)_3\text{CH}_2\text{O}-\overset{\overset{\text{O}}{\|}}{\text{C}}(\text{CH}_2)_4\overset{\overset{\text{O}}{\|}}{\text{C}}\right]_m\text{OCH}_2(\text{CF}_2)_3\text{CH}_2\text{O}\Big\}_n\text{H}$$

$$\text{Ar}_f = \quad \text{—}\langle\!\langle F \rangle\!\rangle\text{—} \, , \quad \langle\!\langle F \rangle\!\rangle \, , \text{ or } \text{—}\langle\!\langle F \rangle\!\rangle\text{—}\langle\!\langle F \rangle\!\rangle\text{—}$$

A polyurethane was also synthesized by reaction of poly(hexafluoropenta-methylene malonate) with tetrafluoro-*p*-phenylene diisocyanate (24,84):

$$\text{OCN}\text{—}\langle\!\langle F \rangle\!\rangle\text{—}\text{NCO} + \text{H}\left[\text{OCH}_2(\text{CF}_2)_3\text{CH}_2\text{O}-\overset{\overset{\text{O}}{\|}}{\text{C}}\text{CH}_2\overset{\overset{\text{O}}{\|}}{\text{C}}\right]_m\text{OCH}_2(\text{CF}_2)_3\text{CH}_2\text{OH}$$

$$\Big\{\overset{\overset{\text{O}}{\|}}{\text{C}}\text{NH}\text{—}\langle\!\langle F \rangle\!\rangle\text{—}\text{NH}\overset{\overset{\text{O}}{\|}}{\text{C}}\left[\text{OCH}_2(\text{CF}_2)_3\text{CH}_2\text{O}-\overset{\overset{\text{O}}{\|}}{\text{C}}\text{CH}_2\overset{\overset{\text{O}}{\|}}{\text{C}}\right]_m\text{OCH}_2(\text{CF}_2)_3\text{CH}_2\text{O}\Big\}_n$$

A series of polyurethanes were prepared by reaction of poly(hexafluoro-pentamethylene carbonate) with 2,4-tolylene diisocyanate (24), tetrachloro-*p*-

phenylene diisocyanate (24), tetrafluoro-*p*-phenylene diisocyanate (24,33,84), and tetrafluoro-*m*-phenylene diisocyanate (33,84).

V. POLYURETHANES FROM FLUORINATED POLYETHERS

Most of the fluorinated polyether-based polyurethanes have been synthesized from hydroxyl-terminated polyethers derived from 1,2-epoxides. A polyether of 3,3,3-trifluoro-1,2-epoxypropane, initiated with ethylene glycol and boron trifluoride-etherate, was converted to a polyurethane by reaction with the commercial 80/20 mixture of 2,4- and 2,6-tolylene diisocyanates (55,56). A similar polyurethane was prepared from a polyether of 1,1,2-trihydroperfluoro-1,2-epoxynonane, initiated with antimony pentachloride and water, and the same commercial tolylene diisocyanate mixture (57). Another similar polyether of 3,3,3-trifluoropropylene oxide, initiated with the sodium salt of hexafluoropentanediol, was reacted with 2,4-tolylene diisocyanate to yield a polyurethane (84,88):

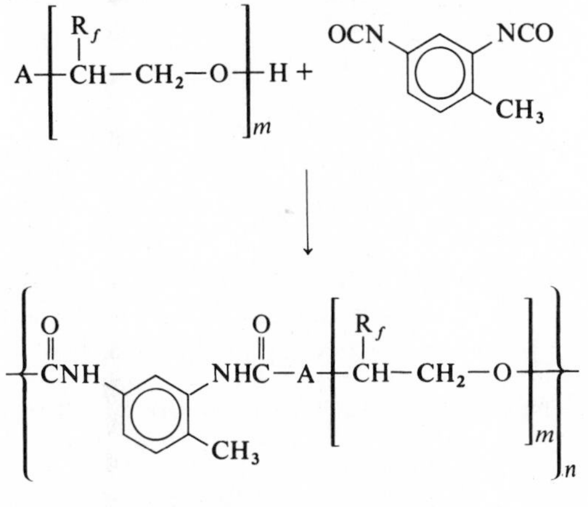

$$R_f = CF_3, \ C_7F_{15}$$
$$A = HOCH_2CH_2O-, \ HO-, \ \text{or} \ HOCH_2(CF_2)_3CH_2O-$$

More highly fluorinated polyurethanes of this type were synthesized by reaction of tetrafluoro-*p*-phenylene diisocyanate with polyethers of 3,3,3-trifluoro-1,2-epoxypropane, which had been prepared using aluminum chloride and the sodium salt of hexafluoropentanediol as initiators (33,84,88).

Polyurethanes have also been prepared from glycidyl ethers of 1,1-dihydroperfluoroalcohols (57). Polyether diols and triols, prepared from the glycidyl ethers of 2,2,2-trifluoroethanol or 1,1-dihydroperfluorooctanol and boron trifluoride-etherate and either water (for diols) or glycerol (for triols), were converted to polyurethanes by reaction with the commercial tolylene diisocyanate mixture and moisture cured:

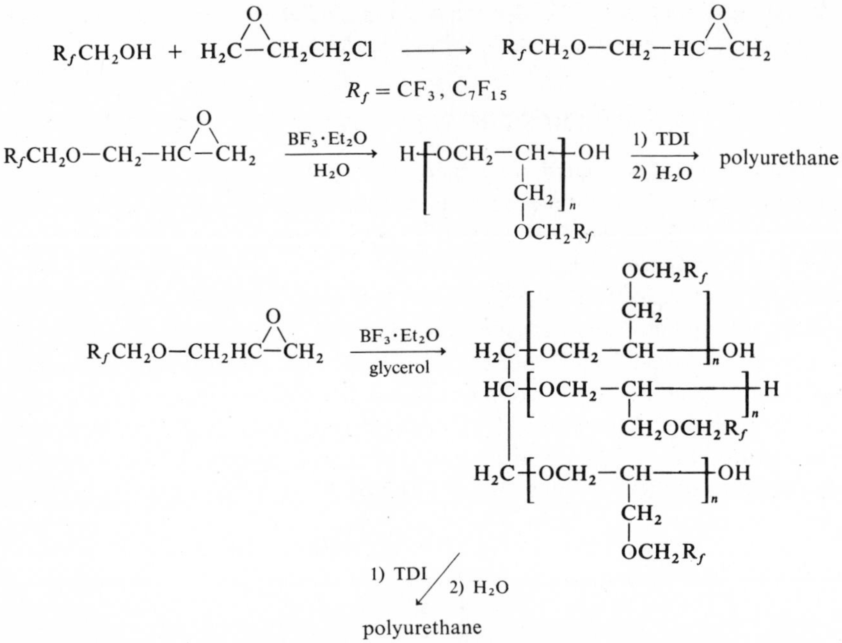

$$R_fCH_2OH + H_2C\overset{O}{\overset{\diagdown\diagup}{C}}{-}CH_2CH_2Cl \longrightarrow R_fCH_2O{-}CH_2{-}HC\overset{O}{\overset{\diagdown\diagup}{C}}H_2$$

$$R_f = CF_3, C_7F_{15}$$

$$R_fCH_2O{-}CH_2{-}HC\overset{O}{\overset{\diagdown\diagup}{C}}H_2 \xrightarrow[H_2O]{BF_3\cdot Et_2O} H\left[OCH_2{-}\underset{\underset{OCH_2R_f}{\overset{|}{CH_2}}}{\overset{|}{CH}}\right]_n OH \xrightarrow[2)\ H_2O]{1)\ TDI} \text{polyurethane}$$

$$R_fCH_2O{-}CH_2HC\overset{O}{\overset{\diagdown\diagup}{C}}H_2 \xrightarrow[\text{glycerol}]{BF_3\cdot Et_2O}$$

$$H_2C\left[OCH_2{-}\underset{\underset{OCH_2R_f}{\overset{|}{CH_2}}}{\overset{\overset{OCH_2R_f}{|}}{CH}}\right]_n OH$$

$$HC\left[OCH_2{-}\underset{CH_2OCH_2R_f}{\overset{|}{CH}}\right]_n H$$

$$H_2C\left[OCH_2{-}\underset{\underset{OCH_2R_f}{\overset{|}{CH_2}}}{CH}\right]_n OH$$

$$\xrightarrow[\ \ 2)\ H_2O]{1)\ TDI}$$

polyurethane

Polyether diols and triols were also prepared from mixtures of the two fluorinated alcohols, and from mixtures of the fluorinated alcohols with propylene oxide.

The synthesis of polyurethanes has been described by reaction of poly-(hexafluoropentamethylene formal) with 2,4-tolylene diisocyanate (84,89) and with tetrafluoro-p-phenylene diisocyanate (24,84,89):

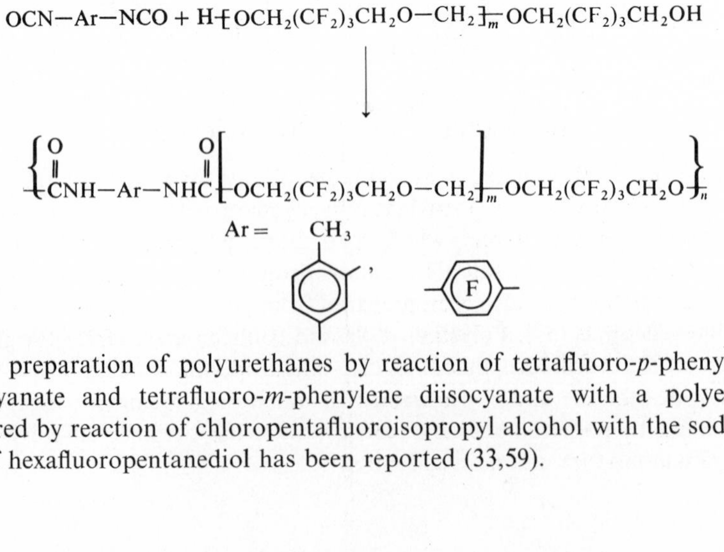

$$OCN{-}Ar{-}NCO + H[OCH_2(CF_2)_3CH_2O{-}CH_2]_m OCH_2(CF_2)_3CH_2OH$$

$$\downarrow$$

$$\left\{\overset{O}{\overset{\|}{C}}NH{-}Ar{-}NH\overset{O}{\overset{\|}{C}}\left[OCH_2(CF_2)_3CH_2O{-}CH_2\right]_m OCH_2(CF_2)_3CH_2O\right\}_n$$

$$Ar =$$

The preparation of polyurethanes by reaction of tetrafluoro-p-phenylene diisocyanate and tetrafluoro-m-phenylene diisocyanate with a polyether prepared by reaction of chloropentafluoroisopropyl alcohol with the sodium salt of hexafluoropentanediol has been reported (33,59).

A series of highly fluorinated polyurethanes have been synthesized from a polyether of perfluoropropylene oxide. The hydroxyl-terminated polyether, prepared by polymerization of perfluoropropylene oxide with perfluoroglutaryl chloride and cesium fluoride, then reduction to the diol, was reacted with tetrafluoro-*p*-phenylene diisocyanate (59,84), tetrafluoro-*m*-phenylene diisocyanate (59,84,85), and 1-chloro-2,4-diisocyanato-3,5,6-trifluorobenzene (85):

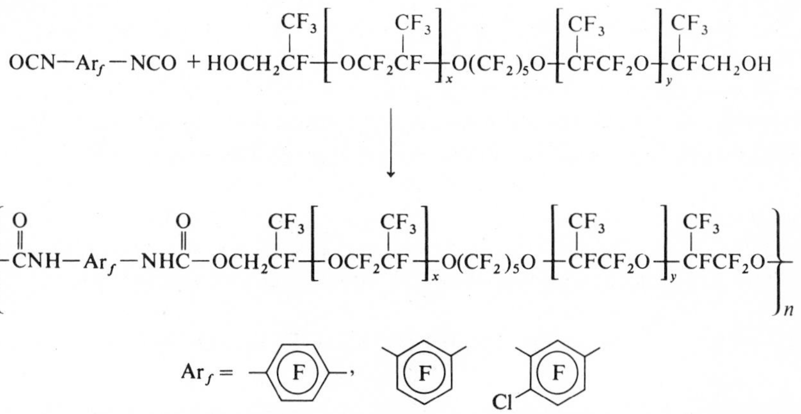

A slightly different perfluoropropylene oxide polyether, prepared by photopolymerization of the diether of perfluoropropylene oxide and perfluoro glutaryl fluoride, then reduction to the diol, was reacted with tetrafluoro-*m*-phenylene diisocyanate to yield a polyurethane (86).

Polyurethanes have been synthesized by reaction of the hydroxyl-terminated polyether of hexafluorobenzene and hexafluoropentanediol with tetrafluoro-*p*-phenylene diisocyanate (59,84), tetrafluoro-*m*-phenylene diisocyanate (59,84), and 1-chloro-2,4-diisocyanato-3,5,6-trifluorobenzene (85):

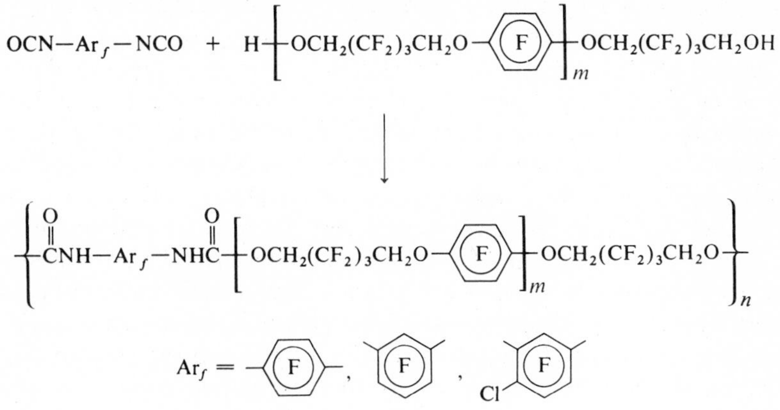

VI. MISCELLANEOUS POLYURETHANES

The only polyurethane reported from a fluorinated diisocyanate and a nonfluorinated diol was synthesized from tetrafluoro-*p*-phenylene diisocyanate and 1,5-pentanediol (24,81):

$$OCN-\underset{}{\bigcirc}(F)-NCO + HO(CH_2)_5OH \longrightarrow \left[\underset{}{\overset{O}{\underset{\|}{C}}}NH-\underset{}{\bigcirc}(F)-NH\overset{O}{\underset{\|}{C}}-O(CH_2)_5O\right]_n$$

An isocyanate-terminated polyurethane of 2,4-tolylene diisocyanate and glycerol-based poly(propylene oxide) was capped with 1,1,11-trihydroperfluoroundecanol. A very small amount of this fluorine-containing polyurethane was used as an additive in the preparation of cellular polyurethanes based on 2,4-tolylene diisocyanate and poly(propylene oxide) (87).

A dihemiketal prepared from hexafluoropentanediol and hexafluoroacetone was converted to polyurethanes by reaction with tetrafluoro-*p*-phenylene diisocyanate followed by curing with tetrafluoro-*p*-phenylenediamine (24) and by reaction with 2,4-Tolylene diisocyanate (90).

VII. STRUCTURE AND PROPERTIES OF FLUORINATED POLYURETHANES

Since fluorinated polyurethanes have only recently been synthesized, very few data are available on the relationship between structure and properties. Many of the relationships that hold for nonfluorinated polyurethanes certainly will hold for the fluorinated ones. Indeed, some of the trends which have been recognized in polyurethanes can be seen with some of the new fluorinated analogs.

In the synthesis of nonfluorinated polyurethanes by the reaction of diisocyanates with diols, the fastest reactions are between aromatic diisocyanates and aliphatic diols and the slowest between aliphatic isocyanates and phenols. In the synthesis of fluorinated polyurethanes, a similar trend can be seen. The fastest reactions are those between fluoroaromatic diisocyanates and nonfluorinated aliphatic diols (e.g., tetrafluoro-*p*-phenylene diisocyanate and 1,5- pentanediol). Very close behind in speed of reaction are those of fluoroaromatic diisocyanates with fluorine-containing diols and diol-terminated prepolymers (e.g., tetrafluoro-*m*-phenylene diisocyanate and hexafluoropentanediol), and those between perfluoroaliphatic diisocyanates and fluorinated diols (e.g., perfluorotrimethylene diisocyanate and hexafluoropentanediol). The reaction of 1-chloro-2,4-diisocyanato-3,5,6-trifluorobenzene with fluorinated diols (e.g., hexafluoropentanediol) is considerably slower. This is believed to be due to steric hindrance by the chlorine atom toward attack of the alcohol on the isocyanate group rather than to the decreased electron-withdrawing ability of the chlorine atom as opposed to the fluorine atom. Slower yet are the reactions of fluoroaromatic diisocyanates with fluorophenols (e.g., tetrafluoro-*m*-phenylene diisocyanate and tetrafluoro-*p*-hydroquinone) and nonfluorinated diisocyanates with fluorinated diols (e.g., tolylene diisocyanate with hexafluoropentanediol and hexamethylene diisocyanate with hexafluoropentanediol). The slowest reaction involving fluorinated materials is that between tetrachloro-*p*-phenylene diisocyanate and hexafluoropentanediol. This diisocyanate has been shown to be quite sterically hindered (24), which accounts for its extremely low reactivity toward fluorinated diols.

An additional indication of the reduced reactivity of fluorinated diols as compared to their nonfluorinated counterparts in the reaction with diisocyanates can be seen from findings during work on fluorinated polyurethanes for use as improved laminating resins (80). Whereas the polyurethane from 2,4-tolylenediisocyanate and 2-butene-1,4-diol had a pot life of less than 1 hour and that from 2,4-tolylene diisocyanate and dibromobutene diol a pot life of between 2 and 3 hours, the polyurethane from 2,4-tolylene diisocyanate and hexafluoropentanediol remained a liquid for several days if the resin was kept in a moisture-free container.

A similar indication of the difference in reactivity of fluorinated diisocyanates can be seen from the pot lives of adhesives prepared from two diisocyanates (85). The polyurethane adhesive prepared from tetrafluoro-*m*-phenylene diisocyanate and the hydroxyl-terminated polyether of perfluoropropylene oxide had a pot life of 30 minutes, whereas the adhesive prepared from the same polyether and 1-chloro-2,4-diisocyanato-3,5,6-trifluorobenzene had a pot life of three to four hours.

In his book (5), Morgan discusses the experimentally found fact that a slight excess of diisocyanate was beneficial in obtaining maximum molecular weight and optimum properties. This effect was ascribed to the loss of isocyanate groups by hydrolysis to amine or to formation of isocyanate dimers. In the preparation of polyurethanes from fluorinated diisocyanates this effect should be enhanced because of the much greater reactivity of fluorinated diisocyanates over their nonfluorinated analogs. A finding during work on improved laminating resins (80) may be an example of this enhanced effect. The flexural values of glass cloth laminates, made with the polyurethane of 2,4-tolylene diisocyanate and hexafluoropentanediol, increased dramatically when the NCO to OH ratio used in preparation of the polymer was increased from 1:1 to 1.1:1.

Very few of the data available on the fluorinated polyurethanes relate melting or softening points to molecular structure. Most of the polyurethanes prepared were cross-linked and insoluble, and very few molecular weights have been determined. The melting or softening points of a few polyurethanes have been reported and these are shown in Table 3.

TABLE 3

Melting or Softening Points of Fluorinated Polyurethanes

Polyurethane	Melting or softening point (°C)
Poly(hexafluoropentamethylene hexamethylene dicarbamate)	90–95 (M.P.)
Polyurethane of hexafluoropentanediol and 2,4-tolylene diisocyanate	241 (S.P.) (NCO/OH ratio = 1.1:1)
	248 (S.P.) (NCO/OH ratio = 1.2:1)
	365 (S.P.) (NCO/OH ratio = 1.3:1)
Poly(hexafluoropentamethylene tetrachloro-p-phenylene dicarbamate)	190–196 (M.P.)
Poly(hexafluoropentamethylene hexafluoropentamethylene dicarbamate)	Room temp. (M.P.)
Poly(hexafluoropentamethylene tetrafluoro-p-phenylene dicarbamate)	280–281 (M.P.)
Poly(tetrafluoro-p-phenylene tetrafluoro-m-phenylene dicarbamate)	Infusible; decomp. at 225
Poly(tetrafluoro-p-phenylene hexafluoropentamethylene dicarbamate)	175–196 (S.P.)
Polyurethane of poly(hexafluoropentamethylene isophthalate) and diphenylmethane-4,4′-diisocyanate	Does not soften below 260
Polyurethane of poly(hexafluoropentamethylene carbonate) and tetrachloro-p-phenylene diisocyanate	125–135 (M.P.)

The infrared spectra of a number of fluorinated polyurethanes indicate that the characteristic absorption bands are almost identical to the bands in nonfluorinated polyurethanes. The N-H absorption of nonfluorinated polyurethanes is located between 3.05 and 3.1μ. All of the fluorinated polyurethanes synthesized from the nonfluorinated diisocyanates, hexamethylene diisocyanate and 2,4-tolylene diisocyanate, have N-H absorptions at 3.0–3.05 μ. All of the fluorinated polyurethanes based on the chloroaromatic diisocyanates, tetrachloro-p-phenylene diisocyanate and tetrachloro-p-xylylene-α,α'-diisocyanate, and the fluorinated diisocyanates, hexafluoropentamethylene diisocyanate, tetrafluoro-p-phenylene diisocyanate, and tetrafluoro-m-phenylene diisocyanate, have N-H absorptions at 3.1 μ. Only one polyurethane, poly(hexafluoropentamethylene perfluorotrimethylene dicarbamate), prepared from perfluorotrimethylene diisocyanate, had an N-H absorption which varied from the nonfluorinated polyurethanes. Its N-H absorption band is at 3.2 μ.

The carbonyl (amide I) bands of all of the fluorinated polyurethanes are between 5.7 and 5.85 μ. For nonfluorinated polyurethanes, this band is between 5.75 and 5.90 μ.

The amide II band of all the fluorinated polyurethanes are between 6.4 and 6.65 μ. This band in nonfluorinated polyurethanes is at 6.5 μ.

Thermogravimetric analysis data on selected fluorinated polyurethanes indicate that their thermal stabilities are of the same order as those of nonfluorinated polyurethanes. The polyurethanes based on tetrafluoro-p-hydroquinone are less thermally stable than those from hexafluoropentanediol. This is analogous to the lower thermal stability of phenol-based polyurethanes compared to aliphatic diol-based polyurethanes. The hexafluoropentanediol or diol-terminated prepolymer-based polyurethanes have about the same polymer decomposition temperatures as the nonfluorinated polyurethane synthesized from 2,4-tolylene diisocyanate and poly(tetrahydrofuran). One fluorinated polyurethane, poly(hexafluoropentamethylene N,N'-diphenyl-p-phenylene dicarbamate) shows exceptional stability. This is most likely because it is completely N-substituted, and its fluorine content plays only a small role, if any, in its stability. The reported polymer decomposition temperatures of the fluorinated polyurethanes, determined by thermogravimetric analysis, are shown in Table 4. Included in the table is the polyurethane from 2,4-tolylene diisocyanate and poly(tetrahydrofuran) and various prepolymers used in the preparation of the polyurethanes.

Other information relating structure to properties has been reported. Polyurethanes having the structure

$$-CF_2-NH-\overset{\overset{\displaystyle O}{\|}}{C}-O-$$

were shown to be unstable hydrolytically and thermally (20,81). Poly(hexafluoropentamethylene perfluorotrimethylene dicarbamate) hydrolyzed on exposure to atmospheric moisture to hexafluoropentamethylene biscarbamate and difluoromalonic acid. This polyurethane was stable if protected from moisture, but it lost hydrogen fluoride on heating to 80°C.

In an extensive study, thermal stability, hydrolytic stability, and crystallinity of a series of fluorinated polyurethanes were investigated (92–101). A few generalizations arose from this study. Within the all aliphatic polyurethane series, those polymers derived from fluorinated diisocyanates and nonfluorinated diols are the most thermally stable. Polyurethanes derived from tetrafluoro-m-phenylenediisocyanate (with either fluorinated or nonfluorinated aliphatic diols) are more thermally stable than the corresponding polymers derived from m-phenylenediisocyanate. Polyurethanes derived from trifluoromethyl or α,α,ω-trihydroperfluoroalkoxy substituted aromatic diisocyanates (with nonfluorinated aliphatic diols) are more thermally stable than the corresponding polymers derived from nonfluorinated aromatic diisocyanates.

TABLE 4

Thermogravimetric Analysis Results

Polymer	Polymer decomposition temp. (°C)[a]	
	In air	In helium
Polyurethane from 2,4-tolylene diisocyanate and poly(tetramethylene oxide)	280	280
Poly(hexafluoropentamethylene hexafluoropentamethylene dicarbamate)	215	200
Poly(tetrafluoro-*p*-phenylene hexafluoropentamethylene dicarbamate)	190	200
Poly(hexafluoropentamethylene tetrafluoro-*p*-phenylene dicarbamate)	300	285
Poly(tetrafluoro-*p*-phenylene tetrafluoro-*m*-phenylene dicarbamate)	225	200
Polyurethane from 2,4-tolylene diisocyanate and poly(hexafluoropentamethylene adipate)	375	300 450
Polyurethane from tetrafluoro-*m*-phenylene diisocyanate and poly(hexafluoropentamethylene adipate)	290 400	300 450
Poly(hexafluoropentamethylene adipate)	385	380
Polyurethane from 2,4-tolylene diisocyanate and poly(hexafluoropentamethylene carbonate)	280 395	285 450
Polyurethane from tetrafluoro-*m*-phenylene diisocyanate and poly(hexafluoropentamethylene carbonate)	285 405	280 450
Polyurethane from tetrafluoro-*p*-phenylene diisocyanate and poly(hexafluoropentamethylene carbonate)	265 400	260 445
Poly(hexafluoropentamethylene carbonate)	450	450
Polyurethane from 2,4-tolylene diisocyanate and poly(hexafluoropentamethylene formal)	250	260
Polyurethane from 2,4-tolylene diisocyanate and poly(3,3,3-trifluoro-1,2-epoxypropane)	290	250 325
Poly(3,3,3-trifluoro-1,2-epoxypropane)	240 340	240 340
Polyurethane from tetrafluoro-*m*-phenylene diisocyanate and poly(perfluoropropylene oxide)	290	290
Polyurethane from tetrafluoro-*p*-phenylene diisocyanate and poly(perfluoropropylene oxide)	245	240
Poly(hexafluoropentamethylene N,N'-diphenyl-*p*-phenylene dicarbamate)	410	480

[a] When two numbers are given, there were two breaks in the thermogravimetric analysis curve.

222

TABLE 5

Properties of Poly(Hexafluoropentamethylene Adipate) Based Polyurethanes

Prepolymer molecular wt.	Diisocyanate	NCO/OH ratio	Curing agent	Percentage of theoretical amount of curing agent	Properties
1400	 OCN–[F]–NCO	1.3	Heat	—	Soft elastomer
		2.0	 H_2N–[F]–NH_2	40	Tough elastomer
		3.0	$HOCH_2(CF_2)_3CH_2OH$	90	Very flexible poor tear resistance
		3.0	 HOH_2C–[Cl]–CH_2OH	75	Tough, flexible elastomer
2000	 OCN–[F]–NCO	2.0	Heat	—	Flexible elastomer
2800		2.0	MOCA	80	Tough, flexible elastomer
3200	 OCN–[F]–NCO	2.0	 H_2N–[F]–NH_2	50	Tough, very flexible elastomer
1800		1.0	Heat	—	Flexible elastomer
1700	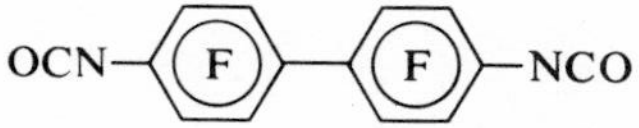 OCN–[F]–[F]–NCO	2.0	 H_2N–[F]–NH_2	40	Very tough elastomer

TABLE 6

Properties of Poly(Hexafluoropentamethylene Carbonate) Based Polyurethanes

Prepolymer molecular wt.	Diisocyanate	NCO/OH ratio	Curing agent	Percentage of theoretical amount of Curing agent	Properties
1000	 OCN—⟨F⟩—NCO	1.5	Heat	—	Tough elastomer
1000		1.8	 H₂N—⟨F⟩—NH₂	14	Tough elastomer
1800	 OCN—⟨F⟩—NCO	1.2	Heat	—	Tough flexible elastomer
1500		1.0	Heat	—	Soft elastomer

TABLE 7

Properties of Poly(Perfluoropropylene Oxide) Based Polyurethanes

Prepolymer molecular wt.	Diisocyanate	NCO/OH ratio	Curing agent	Percentage of theoretical amount of curing agent	Properties
800	 OCN–⬡(F)–NCO	1.1	Heat	—	Brittle solid
1500		1.5	Heat	—	Tough elastomer
800	 OCN–⬡(F)–NCO	1.5	 H_2N–⬡(F)–NH_2	40	Brittle solid
1500		1.25	Heat	—	Soft elastomer
1500		1.2	 H_2N–⬡(F)–NH_2	40	Tough elastomer

Only limited data relating mechanical properties to structure are available. Qualitative data on the flexibility of highly fluorinated polyester and polyether based polyurethanes were reported (84). The effect of polymer structure on flexibility is illustrated in Tables 5, 6, and 7 for polyurethanes of poly-(hexafluoropentamethylene adipate), poly(hexafluoropentamethylene carbonate), and poly(perfluoropropylene oxide).

These data show that, within a given polyurethane system, the flexibility increases with increasing prepolymer molecular weight. In general, the polyurethanes that were cured with the aromatic curing agents, tetrafluoro-p-phenylenediamine, methylene bis(o-chloroaniline) (MOCA), or tetrachloro-p-xylylenediol were tougher than the corresponding ones that were heat cured. In contrast, the poly(hexafluoropentamethylene adipate) based polyurethane cured with an aliphatic diol, hexafluoropentanediol, was very flexible but had considerably poorer tear resistance than the corresponding heat cured elastomers. The toughening effect of aromatic groups is also seen in the observation that the poly(hexafluoropentamethylene adipate) based polyurethane prepared from 4,4′-diisocyanato-octafluorobiphenyl was substantially tougher than the corresponding polyurethane prepared from tetrafluoro-p-phenylene diisocyanate.

Physical properties were reported on a series of moisture cured polyurethanes prepared from tolylene diisocyanate and polyether triols (57). The polyether triols were synthesized by reaction of glycidyl ethers of

TABLE 8

Properties of Polyurethanes Based on Polyether Triols of $R_fCH_2O—CH_2—CH\overset{O}{\overbrace{\qquad}}CH_2$

R_f	Number of ethylene oxide units per polyol chain	Tensile strength[a] (psi)	Elongation (%)
CF_3	3.5	1530	380
CF_3	5.6	380	375
75% CF_3 + 25% C_7F_{15}	4.5	900	365
25% CF_3 + 75% C_7F_{15}	2.8	1270	170
70% CF_3 + 30% $CH_3CH\overset{O}{\overbrace{\qquad}}CH_2$	4.5	1970	375
50% C_7F_{15} + 50% $CH_3CH\overset{O}{\overbrace{\qquad}}CH_2$	4.7	3440	440

[a] Test specimens were 0.02–0.06-in.-thick dumbells, $\frac{1}{8} \times 3$ in.

TABLE 9

Fuel and Oil Resistance of Polyester Based Polyurethanes

Diisocyanate	Polyester	Tensile strength (psi)		Elongation (%)		Hardness (shore A)	
		Original	After immersion	Original	After immersion	Original	After immersion
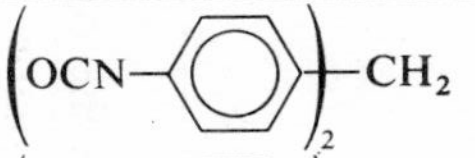 $OCN-\!\!\bigcirc\!\!-CH_2\big)_2$	Poly(hexafluoropentamethylene adipate)	634	350[a]	360	360[a]		
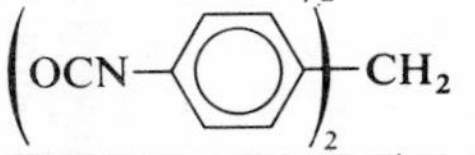 $OCN-\!\!\bigcirc\!\!-CH_2\big)_2$	Poly(hexafluoropentamethylene 3-perfluoropropylglutarate)			420	320[b]	45–50	42–47[b]

[a] Seven days in 70% isooctane-30% toluene mixture at room temperature.
[b] Seventy-two hr in Plexol 201 diester lubricating oil at 350°F.

1,1-di-hydroperfluoro alcohols, glycerol, and boron trifluoride. These properties are shown in Table 8.

These results show several structure-property relationships for this type of polyurethane. An increase in molecular weight of prepolymer, as indicated by the number of ethylene oxide units per polyol chain, results in a decrease in tensile strength with essentially no change in elongation. Incorporation of propylene oxide into the polyether prepolymer results in a marked increase in tensile strength of the polyurethane.

Excellent resistance toward fuels and oils has been reported for two fluorinated polyester based polyurethanes (47,48). This resistance is illustrated in Table 9.

The flexibilizing effect of ether groups in polyurethanes has been discussed quite extensively in Chapter VI of Saunders' and Frisch's book (3) on polyurethanes. This flexibilizing character can be seen as accounting for the excellent low temperature properties of the polyurethanes prepared from polyethers of perfluoropropylene oxide. A polyurethane prepared from photopolymerized perfluoropropylene oxide and tetrafluoro-m-phenylene diisocyanate was reported to have a very low glass transition temperature, $-73°C$ (86).

Excellent low temperature adhesive properties have been obtained with the polyurethane prepared from the polyether of perfluoropropylene oxide and 1-chloro-2,4-diisocyanato-3,5,6-trifluorobenzene (102). A study of the relationship between lap shear strength of aluminum to aluminum bonds and isocyanate to hydroxyl ratio is shown in Table 10.

TABLE 10

Lap Shear Strength versus NCO/OH Ratio of Polyurethane Bonds Made from 1200 Molecular Weight Poly(Perfluoropropylene Oxide) and 1-Chloro-2,4-Diisocyanato-3,5,6-tri fluorobenzene

NCO/OH ratio	Lap shear strength	
	Room temperature	$-320°F$
1.2	456	0
1.4	1365	2000
1.6	1770	4595
1.8	2025	4730
2.0	1940	3650

These data indicate that both the room temperature and low temperature lap shear strength of bonds made with this polyurethane increases with increasing NCO/OH ratio to a maximum value, then decreases. The cross-link density increases with increasing NCO/OH ratio; therefore, the bond strength increases to a certain point after which any increase results in a progressively more brittle material.

The result of a study on the relationship between the lap shear strength of these bonds and molecular weight of the polyether prepolymer are shown in Table 11.

TABLE 11

Lap Shear Strength versus Polyether Molecular Weight of Polyurethane Bonds Made from Poly(Perfluoropropylene Oxide) and 1-Chloro-2,4-Diisocyanato-3,5,6-Trifluorobenzene in Ratio of 1:1.8

Polyether prepolymer molecular weight	Lap shear strength	
	Room temperature	−320°F
1200	2025	4730
1614	1760	4630
2349	573	1865

The results indicate that the room temperature and low temperature lap shear strength of bonds made with this polyurethane decreases with increasing molecular weight in the range from 1160 to 2350.

In another study (103), the adhesion of polyurethanes with long fluorinated side chains was investigated. Copolymer triols were prepared from propylene oxide and glycidyl (α,α,ω-trihydrododecafluoroheptyl) ether or pentadecafluoroheptylepoxyethane; these triols were then converted to polyurethanes by reaction with an 80/20 mixture of 2,4- and 2,6-tolylenediisocyanates. These polyurethanes showed increasing adhesion to anodized aluminum with increasing urea bond content of the polymer. In the range of 0 to 45% fluorine in these polymers, no dependency could be detected between adhesion and fluorine content.

The relationship between structure and compatibility with liquid oxygen was an integral part of the study on fluorinated polyurethanes for use as liquid oxygen compatible adhesives (20,24,33,59). Compatibility means that the material shall not ignite, burn, char, or explode when tested in accordance with Marshall Space Flight Center Specification 106.

Results show that a urethane linkage flanked on the alcohol side by a 1,1-dihydroperfluoroalkyl chain, such as

$$-(CF_2)_xCH_2- \quad \text{or} \quad -\overset{\displaystyle CF_3}{\underset{\displaystyle |}{CF}}-CH_2-,$$

is liquid oxygen compatible if the nitrogen atom is attached to a perfluoro-alkyl, a 1,1-dihydroperfluoroalkyl, a tetrafluorophenylene, or a tetrachloro-phenylene group, but not if it is attached to a nonfluorinated alkyl or aryl group or to a benzyl methylene group. Four polyurethanes based on hexafluoropentanediol were liquid oxygen compatible. They were poly(hexafluoropentamethylene perfluorotrimethylene dicarbamate), poly-(hexafluoromethylene hexafluoropentamethylene dicarbamate), poly(hexa-fluoropentamethylene tetrafluoro-p-phenylene dicarbamate), and poly(hexa-fluoropentamethylene tetrachloro-p-phenylene dicarbamate). Two other polyurethanes, prepared by reaction of poly(perfluoropropylene oxide) with tetrafluoro-p-phenylene diisocyanate and tetrafluoro-m-phenylene diisocyanate, were liquid oxygen compatible.

Fluorinated polyurethanes have found uses in several applications other than as adhesives. Binders for solid propellants were prepared from poly-urethanes based on a poly(tetrafluoroethylene oxide) diol and a variety of fluorinated and nonfluorinated diisocyanates (104).

Polyurethane films were prepared from tolylene diisocyanate and fluori-nated polyol prepolymers (105). These prepolymers were obtained by reaction of diglycidyl ethers of 4,4'-dihydroxyoctafluorobiphenyl, 1,3-bis(2-hydroxy-hexafluoro-2-propyl) benzene, or 1,4-bis(2-hydroxyhexafluoro-2-propyl) ben-zene with either hexafluoropentanediol or 1,3-bis(2-hydroxyhexafluoro-2-propyl) benzene.

Several polyurethanes containing pendent fluorocarbon chains have been claimed as water repellents for textile fabrics. One such polyurethane was synthesized from polymethylene poly(phenylisocyanate) and a fluoroalcohol having the structure $(CF_3)_2CF(CF_2)_4CH_2CH_2OH$ (106). Another was pre-pared from tolylene diisocyanate and a diol of the structure $(CF_3)_2CF(CF_2)_4$-$CH_2CH(OH)CH_2OH$ (107). Two other polyurethanes, also containing pendent fluorocarbon chains in the diol portion, were prepared from a diol of the structure $F(CF_2)_nCH_2CH_2CH(CH_2OH)_2$ (108). This diol could be reacted directly with tolylene diisocyanate to give a polyurethane or could be converted (by reaction with phosgene) to a bischloroformate which could then be reacted with a diamine (such as piperazine) to yield a poly-urethane.

A polyurethane foam, claimed to be water and oil repellent, was prepared by reaction of a nonfluorine containing polyester with a diisocyanate of the structure $(CF_3)_2CF(CF_2)_4C(O)NH(CH_2)_2OC(O)NHCH_2CH(NCO)CH_2$ NCO (109).

Future research on fluorinated polyurethanes is certain to be forthcoming which will provide more information on their uses and especially on the relationships between structure, reactivity, and properties.

References

1. B. A. Dombrow, "Polyurethanes," Reinhold, New York, 1957.
2. W. R. Sorenson and T. W. Campbell, "Preparative Methods of Polymer Chemistry," Interscience, New York, 1961.
3. J. H. Saunders and K. C. Frisch, "Polyurethanes: Chemistry and Technology," Vol. I, Interscience, New York, 1962.
4. *Ibid.*, Vol. II, Interscience, New York, 1964.
5. P. W. Morgan, "Condensation Polymers: By Interfacial and Solution Methods," Interscience, New York, 1965.
6. O. Bayer, *Angew. Chem.*, **A59**:257 (1947).
7. A. Höchtlen, Kunstaffe, **40**:221 (1950); **42**:303 (1952).
8. E. Müller, O. Bayer, S. Petersen, H. F. Piepenbrink, F. Schmidt, and E. Weinbrenner, *Angew. Chem.*, **64**:523 (1952); *Rubber Chem. Technol.*, **26**:493 (1953).
9. O. Bayer and E. Müller, *Angew. Chem.*, **72**:934 (1960).
10. J. H. Saunders, *Rubber Chem. Technol.*, **33**:1259, 1293 (1960).
11. H. Rinke, *Chimia (Aarau)*, **16**:93 (1962); *Angew. Chem. Intern. Ed.*, **1**:419 (1962); *Rubber Chem. Technol.*, **36**:719 (1963).
12. E. Müller, in Houben-Weyl, "Methoden der organischen Chemie, Vol. 14, Part 2, Thieme-Verlag, Stuttgart, 1963, p. 57.
13. E. M. Hicks, Jr., A. J. Ultee, and J. Drougas, *Science*, **147**:373 (1965).
14. D. J. Lyman, *Reviews in Macromolecular Chemistry*, **1**(1):191 (1966).
15. British Patent 797,795 (July 9, 1958).
16. J. H. Saunders and R. J. Slocombe, *Chem. Rev.*, **43**:203 (1948).
17. W. Siefken, *Ann.*, **562**:75 (1949).
18. S. Peterson and H. Piepenbrink, in Houben-Weyl, "Methoden der organischen Chemie," E. Müller (Ed.), Vol. 8, Thieme-Verlag, Stuttgart, 1952, p. 75.
19. R. G. Arnold, J. A. Nelson, and J. J. Verbanc, *Chem. Rev.*, **157**:47 (1957).
20. R. B. Gosnell, E. S. Harrison, J. Hollander, B. Y. Sanders, and F. D. Trischler, "The Development of Structural Adhesive Systems Suitable for Use with Liquid Oxygen," Contract No. NAS 8-11068, Annual Summary Report I, July 1964.
21. J. Hollander, R. B. Gosnell, F. D. Trischler, and E. S. Harrison, "The Synthesis of Fluorinated Diisocyanates," Abstract, 152nd American Chemical Society Meeting, New York, N.Y., Sept. 1966, Paper No. 18K.
22. J. J. Tazuma, U.S. Patent 2,915,545 (Dec. 1, 1959).
23. H. Holtschmidt, O. Bayer, and E. Degener, U.S. Patent 3,277,138 (Oct. 4, 1966).
24. J. Hollander, F. D. Trischler, Edward S. Harrison, Richard T. Rafter, and Luis Acle, "The Development of Structural Adhesive Systems Suitable for Use with Liquid Oxygen," Contract No. NAS 8-11068, Annual Summary Report II, July 1965.
25. E. W. Cook, C. A. Erickson, and J. A. Gannon, "Synthesis of High Strength Chemical-Resistant Elastomers for Extreme Temperature Service," Contract No. DA 19-129-AMC-147(N), Report No. C & OM4, Jan. 1965.
26. A. H. Ahlbrecht and D. R. Husted, U.S. Patent 2,617,817 (1952).
27. D. R. Husted and A. H. Ahlbrecht, *J. Am. Chem. Soc.*, **75**:1605 (1953).
28. A. L. Henne and J. J. Stewart, *J. Am. Chem. Soc.*, **77**:1901 (1955).
29. D. A. Barr and R. N. Hazeldine, *J. Chem. Soc.*, **1956**:3428.
30. N. N. Yarovenko, S. P. Motornyi, L. I. Kirenskaya, and A. S. Vasil'eva, *Zhur. Obschchei Khim.*, **27**:2243 (1957).
31. British Patent 689,425 (1953).

32. T. S. Reid, U.S. Patent 2,706,733 (Apr. 19, 1955).
33. J. Hollander, F. D. Trischler, and E. S. Harrison, "The Development of Structural Adhesive Systems Suitable for Use with Liquid Oxygen," Contract No. NAS 8-11068, Annual Summary Report III, July 1966.
34. S. B. Eglin, R. I. Akawie, and L. J. Miller, "Development of Structural Foams for Cryogenic Applications," Contract No. NAS 8-11406, QR3, May 1965.
35. J. Hollander, F. D. Trischler, E. S. Harrison, R. M. DeBorde, and B. Y. Sanders, "The Development of Structural Adhesive Systems Suitable for Use with Liquid Oxygen," Contract No. NAS 8-11068, Q.R. No. 11, June 1967.
36. A. M. Lovelace, D. A. Rausch, and W. Postelnek, "Aliphatic Fluorine Compounds," Reinhold, New York, 1958, p. 151.
37. E. T. McBee, W. F. Marzluff, and O. R. Pierce, *J. Am. Chem. Soc.*, **74**:444 (1952).
38. E. T. McBee, O. R. Pierce, and D. D. Smith, *J. Am. Chem. Soc.*, **76**:3725 (1954).
39. C. F. Baranackas and S. Gelfand, Canadian Patent 525,287 (August 8, 1961).
40. C. F. Baranackas and R. L. K. Carr, *2nd Int. Sym. on Fluorine Chemistry, Preprints*, July 1962, p. 296.
41. I. L. Knunyants, Li Chih-Yüan, and V. V. Shokina, *Uspekhi Khimii (Advances in Chemistry)*, **32**(9):1502 (1963); Translation RSIC-165 (Redstone Information Center), 22 April 1964.
42. E. Nield and J. C. Tatlow, *Tetrahedron*, **8**:38 (1960).
43. L. Denivelle and R. Chesneau, *Compt. Rend.* **254**:1646 (1962).
44. W. J. Pummer and L. A. Wall, *J. Res. NBS*, **68A**(3):277 (May-June 1964).
45. J. Burdon, W. B. Hollyhead, and J. C. Tatlow, *J. Chem. Soc.*, **1965**:5152.
46. G. C. Schweiker and P. Robitschek, *J. Poly Sci.*, **24**:33 (1957).
47. G. C. Schweiker and P. Robitschek, U.S. Patent 3,016,361 (Jan. 9, 1962).
48. G. C. Schweiker and P. Robitschek, U.S. Patent 3,016,360 (Jan. 9, 1962).
49. British Patent 798,824 (July 30, 1958).
50. E. V. Gouinlock, Jr., C. J. Verbanic, and G. C. Schweiker, *J. App. Poly. Sci.*, **1**(3):361 (1959).
51. D. B. G. Jaquiss, U.S. Patent 3,220,978 (Nov. 30, 1965).
52. D. D. Smith, R. M. Murch, and O. R. Pierce, *Ind. and Eng. Chem.*, **49**:1241 (1957).
53. F. B. Jones, P. B. Stickney, L. E. Coleman, Jr., D. A. Rausch, and A. M. Lovelace, *J. Poly. Sci.*, **26**:81 (1957).
54. F. D. Trischler and J. Hollander, *J. Poly Sci.*, **5** (A-1):2343 (1967).
55. H. A. Bruson, and J. S. Rose, French Patent 1,371,674 (Sept. 9, 1964).
56. H. A. Bruson and J. S. Rose, U.S. Patent 3,269,961 (Aug. 30, 1966).
57. M. M. Boudakian, M. C. Raes, and S. V. Urs, "Development of Solvent-Resistant Sealants," Report No. NOw 66-0323d, Naval Air Systems Command, Feb. 15, 1967.
58. M. Hauptschein and J. M. Lesser, *J. Am. Chem. Soc.*, **78**:676 (1956).
59. J. Hollander, F. D. Trischler, E. S. Harrison, R. M. DeBorde, and B. Y. Sanders, "The Development of Structural Adhesive Systems Suitable for Use With Liquid Oxygen," Contract No. NAS 8-11068, Summary Report IV, January 1967.
60. W. T. Miller, U.S. Patent 3,242,218 (March 22, 1966).
61. J. L. Warnell, U.S. Patent 3,250,806 (May 10, 1966).
62. E. P. Moore, Jr., A. S. Milian, Jr., and H. S. Eleuterio, U.S. Patent 3,250,808 (May 10, 1966).
63. D. P. Carlson, U.S. Patent 3,303,145 (Feb. 7, 1967).
64. C. G. Fritz and E. P. Moore, U.S. Patent 3,250,807 (May 10, 1966).

65. R. B. Gosnell, E. S. Harrison, J. Hollander, R. T. Rafter, and F. D. Trischler, "Development of Structural Adhesive Systems Suitable for Use with Liquid Oxygen," Contract No. NAS 8-11068, Q.R. No. 4, Oct. 1964.
66. S. B. Eglin, R. I. Akawie, and L. J. Miller, "Development of Structural Foams for Cryogenic Applications," Contract No. NAS 8-11406, Q.R. 2, Feb. 1965.
67. E. R. Lynch and S. A. Evans, "Perfluorinated Polymers and Fluids," AFML-TR-65-13, [Contract No. AF 33(615)-1344, Phase II], Part III, April 1967.
68. S. A. Evans and E. R. Lynch, "Perfluorinated Aromatic Compounds," Contract No. AF 33(615)-1344, Phase II, P.R. No. 6, Oct. 1965.
69. E. T. McBee, P. A. Wiseman, and G. B. Bachman, Ind. and Eng. Chem., 39:415 (1947).
70. E. T. McBee and P. A. Wiseman, U.S. Patent 2,515,246 (July 18, 1950).
71. R. B. Gosnell, J. Hollander, F. Trischler, and E. S. Harrison, "The Synthesis of Fluorinated Diamines," Abstract, 152nd American Chemical Society Meeting, New York, Sept. 1966, Paper 17K.
72. B. S. Marks and G. C. Schweiker, J. Am. Chem. Soc., 80:5789 (1958).
73. G. M. Brooke, J. Burdon, and J. C. Tatlow, J. Chem. Soc., 1961:802.
74. D. G. Holland, G. J. Moore, and C. Tamborski, J. Org. Chem. 29:1562 (1964).
75. G. G. Yakobson, V. D. Shteingarts, G. G. Furin, and N. N. Vorozhtsov, Jr., Zhur. Obshchi. Khim., 34(10):B514 (1964).
76. G. G. Yakobson, V. D. Shteingarts, A. I. Miroshnikov, and N. N. Vorozhtsov, Jr., Dokl. Akad. Nauk, SSSR, 159(5):1109 (1964); Chem. Abstr., 62:9040C.
77. G. M. Brooke and W. R. Musgrave, J. Chem. Soc., 1965:1864. (1965).
78. D. H. Holland, G. J. Moore, and C. Tamborski, Chem. Ind. (London), 31:1376 (1965).
79. D. D. Smith, U.S. Patent 2,911,390 (Nov. 3, 1959).
80. A. P. Bonanni, U.S. Naval Air Engineering Center Report No. NAEC AML 1636, March 13, 1963.
81. J. Hollander, F. D. Trischler, and R. B. Gosnell, "The Synthesis of Polyurethanes from Fluorinated Diisocyanates," J. Poly Sci., 5 (A-1):2757 (1967).
82. J. Hollander, Whittaker Corporation, Narmco Research & Development Division, unpublished.
83. H. J. Lee, U.S. Naval Air Engineering Center Report No. NAEC AML 2492, August 2, 1966.
84. J. Hollander, F. D. Trischler, and E. S. Harrison, Am. Chem. Soc. Polymer Preprints, 8(2):1149 (1967).
85. J. Hollander, "The Development of Structural Adhesive Systems Suitable for Use with Liquid Oxygen," Contract No. NAS 8-11068, Q.R. No. 12, Sept. 1967.
86. E. C. Stump and S. E. Rochow, "Preparation of Fluoroelastomers for Use with Cryogenic Storable Propellants," Contract No. NAS 7-476, Q.R. 4, May, 1967.
87. British Patent 994,411 (June 10, 1965).
88. F. Trischler, U.S. Patent 3,463,762, Aug. 26, 1969.
89. F. Trischler, U.S. Patent 3,475,384, Oct. 28, 1969.
90. W. E. Weesner, U.S. Patent 3,413,271, Nov. 26, 1968.
91. I. L. Knunyants, M. P. Krasuskaya, and D. P. Delitsova, Izv. Akad. Nauk. SSSR; Ser. Khim., 6:1110 (1966); Chem. Abstr., 65:10482c (1966).
92. B. F. Malichenko, E. V. Shelvd'Ko, and Yu. Y. Kercha, Vysokomol. Soedin., Ser. A, 9 (11):2482 (1967); Chem. Abstr., 68:30251d (1968).
93. Yu. Y. Kercha, L. I. Ryabokon, and B. F. Malichenko, Sin. Fiz. Khim. Polim., 5:198 (1968); Chem. Abstr., 70:4197y (1969).

94. B. F. Malichenko and V. V. Penchuk, *Vysokomol. Soedin., Ser. B.,* 11 (1):37 (1969); *Chem. Abstr.,* 70:97295f (1969).
95. B. F. Malichenko, O. N. Tsypina, and A. E. Nesterov, *Vysokomol. Soedin., Ser. B,* 11 (1):67 (1969); *Chem. Abstr.* 70:97296g (1969).
96. B. F. Malichenko et al., *Vysokomol. Soedin., Ser. A,* 11 (7):1518 (1969); *Chem. Abstr.,* 71:102319u (1969).
97. B. F. Malichenko et al., *Vysokomol. Soedin., Ser. B,* 11 (8):566 (1969); *Chem. Abstr.* 72:3854d (1970).
98. L. L. Chervyatsova et al., *Vysokomol. Soedin., Ser. A,* 12 (5):981 (1970); *Chem. Abstr.,* 73:56542c (1970).
99. B. F. Malichenko, A. V. Yazlovitskii, and A. E. Nesterov, *Vysokomol. Soedin., Ser. A,* 12 (8):1700 (1970); *Chem. Abstr.* 73:120958x (1970).
100. B. F. Malichenko et al., *Vysokomol. Soedin., Ser. A,* 13 (1):213 (1971): *Chem. Abstr.,* 74:76949q (1971).
101. A. V. Yazlovitskii and B. F. Malichenko, *Vysokomol. Soedin., Ser. A,* 13 (3):734 (1971); *Chem. Abstr.,* 74:142711j (1971).
102. J. Hollander, "The Development of Structural Adhesive Systems Suitable for Use with Liquid Oxygen," Contract No. NAS 8-11068, Monthly Report No. 32, Oct. 1967; NASA Contract Report 1967, NASA-CR-61743.
103. M. C. Raes, M. M. Boudakian, and S. V. Urs, *J. Appl. Poly. Sci.,* 14 (3), 699 (1970).
104. R. A. Mitsch and J. L. Zollinger, *Ger. Offen.* 2:011,744 (September 17, 1970).
105. J. G. O'Rear, J. R. Griffith, and S. A. Reines, *J. Paint Technol.,* 43, (552):113 (1971).
106. A. Katsushima et al., Japan Pat. 67 21,331 (Oct. 21, 1967).
107. A. Katsushima et al., Japan. Pat. 68 26,518 (Nov. 14, 1968).
108. K. C. Smeltz, U.S. Patent 3,547,894 (Dec. 15, 1970).
109. A. Katsushima et al., Japan. Pat. 69 20,639 (Sept. 4, 1969).

7. FLUORINATED POLYETHERS

NORMAN L. MADISON. *Dow Chemical Company, Midland, Michigan*

Contents

1. INTRODUCTION

The need for a thermally stable polymer with a relatively flexible backbone has led to considerable research on fluorinated polyethers. Materials are required which have broad application in the areas of stable and solvent resistant elastomers and high-temperature fluids. Although the investigation of fluorinated polyethers began in 1956 with the work of Hauptschien (1), this area has grown into a major effort only with the very recent discovery of the facile preparation and polymerization of perfluoroepoxides. The technical literature has provided only a meager amount of information, probably due to secrecy policies of the major companies involved in the research and development of these materials. Most of the references therefore, have come from the patent literature. This chapter briefly reviews the significant developments in this area through 1967.

Although polymeric fluorinated vinyl ethers are an important class of materials, this chapter focuses only on those polymers that contain oxygen in the backbone structure. The earlier work in which the backbone of the polymer was partially hydrogenated is emphasized here less than new polymers with perfluorinated backbones, since these are more important because of increased thermal and chemical stability. The subject matter covers foreign and United States patents, but, in many cases where the patents are duplicated in various countries, only the United States issue is cited.

II. COPOLYMERS OF EPOXIDES AND FLUOROOLEFINS

Hauptschein and Lesser (1) have reported the preparation of fluorinated polyethers from the free-radical reaction of fluorinated olefins with aliphatic epoxides.

Polymerization of perfluoropropylene and trifluorochloroethylene with ethylene oxide initiated with ultraviolet light or di-*t*-butyl peroxide gave fluids, jellies, or waxlike products, depending on the composition. Copolymers of perfluoropropene contained as much as 50% of combined fluoroolefin. Ultraviolet initiation, which apparently produced purer polymers, gave a 40–50% conversion to polymer in ~5 days by irradiation through heavy walled pyrex.

A polymer containing 50/50 epoxide and perfluoropropene was unaffected by concentrated H_2SO_4 and 47% hydoiodic acid, indicating that it contained alternating monomers since nonfluorinated polyalkyl ethers are readily degraded by this treatment.

All of the polymers are probably of low molecular weight, since large low-boiling fractions are readily removed by distillation. Although the polymers are believed to be polyethers, the only characterization performed was carbon and hydrogen analysis. The utility of these polymers has never been disclosed.

III. POLYMERS OF PARTIALLY FLUORINATED EPOXIDES

Smith and co-workers (2) have prepared several fluorinated epoxides by well-established procedures and were able to affect polymerization with Lewis acids.

Aluminum chloride and ferric chloride were the most effective for achieving maximum conversions of 3,3,3-trifluoro-1,2-epoxypropane and 2-methyl-3,3,3-trifluoro-1,2-epoxypropane to polymers. When chlorine was present in the monomers, however, the rate of polymerization was considerably slower and conversion was very poor.

Aluminum chloride catalysis gave very rapid polymerization but resulted in low-molecular-weight liquids. Ferric chloride resulted in much slower rates; however, much higher molecular-weight materials were obtained with the propylene oxide monomers. For example, with 2-3 mole % of ferric chloride at 90–100°C for 64 hr, polymers having an average molecular weight of 230,000 were obtained with 3,3,3-trifluoro-1,2-epoxy propane. Very few data were reported on the properties of the polymers and, although the polymers are said to be thermally stable, no quantitative data are presented. By present-day standards, these materials would probably not qualify for high-temperature applications.

Smith was unsuccessful in attempts to extend the polymers by reacting them with hexamethylene diisocyanate. However, recently Trishchler (3), who prepared low molecular-weight polymers of 3,3,3-trifluoro-1,2-epoxy propane.

$$CF_3HC\overset{\displaystyle O}{\overbrace{}}CH_2$$

using the sodium salt of hexafluoropentanediol as catalyst, obtained materials that contained hydroxyl on each end and were readily extended with perfluoro-aromatic diisocyanates to give polyurethanes which were tough elastomers. Other urethane elastomers were prepared from prepolymer ethers (4):

$$\boxed{F} + (CF_2)_3(CH_2OH)_2$$

$$\downarrow KOH$$

$$HOCH_2(CF_2)_3CH_2O\left[\boxed{F}-OCH_2(CF_2)_3CH_2\right]_n OH$$

Jones and co-workers (5) prepared and polymerized several highly fluorinated epoxides. The polymerization effort was concentrated on 1,1,1-trifluoro-2,3-butylene oxide, since this was available in greatest quantity. With ferric chloride as catalyst, only low-molecular-weight liquid products were obtained, and under optimum conditions (91 % yield) polymers with an inherent viscosity of 0.024 were achieved. When $FeCl_3 \cdot 6H_2O$ was used as the catalyst, the yields of polymer were sharply reduced.

Other catalysts, such as BF_3, BF_3. etherate, KOH, $AlCl_3$, $Al(Et)_3 \cdot TiCl_4$, gave inferior polymers to those obtained with the $FeCl_3$. Free-radical sources were totally ineffective for initiating polymerization.

All of the other polyepoxides prepared by Jones (5) were low-molecular-weight liquids. It was concluded that substitution larger than one CF_3 group alpha to the oxide ring prevents occurrence of extensive polymerization.

Pittman, however, has recently shown that heptafluoroisopropylglycidyl ether, prepared from hexafluoroacetone and epibromohydrin.

$$\underset{F_3C}{\overset{F_3C}{>}}C=O + BrCH_2-\overset{\displaystyle O}{\overbrace{CH-}}KH_2 \xrightarrow{KF} \underset{F_3C}{\overset{F_3C}{>}}CF-OCH_2-\overset{\displaystyle O}{\overbrace{CH-}}CH_2$$

can be polymerized to a high-molecular-weight (DP-350) elastomer with diethyl zinc (6). This elastomer was readily soluble in benzotrifluoride and had a glass transition temperature of $-43°C$ and thermal stability to $270°C$.

When the polymerization was carried out with $BF_3 \cdot$ etherate, a low-molecular-weight liquid polymer was obtained, indicating that a coordination catalyst is more effective than strong Lewis acids.

IV. COPOLYMERS OF FORMALDEHYDE AND FLUOROOLEFINS

A recent patent by Miller et al. (7) describes the copolymerization of perfluoroolefins with formaldehyde. Both oily and waxy polymers were obtained with perfluoroisobutylene and a cesium fluoride initiator. These polymers were of relatively low molecular weight (<4000) and always contained a greater percentage of formaldehyde than fluoroolefin. Perfluoropropene was also found to copolymerize with formaldehyde, but a description of the polymer was not given. DMF was sufficiently basic to catalyze the copolymerization with perfluoropropene.

V. PHENYLENE POLYETHERS

Jones (8) found that α,α,α-trifluoro-p-cresol undergoes rapid polymerization to give a polyether in cold dilute alkali, via the elimination of hydrogen fluoride:

These polymers, once formed, were found to be resistant to further loss of HF.

The o-cresol under the same polymerizing conditions is overhydrolyzed to the acid, whereas the m-cresol was unaffected by hot sodium hydroxide solution. Interestingly, the polymerization of the para isomer could also be catalyzed by a small amount of HF.

Further work on this system by McMasters (9) showed that these polymers were of only moderate molecular weight. High-polymer formation was apparently inhibited by hydrolysis of the CF_3 group, since considerable amount of COOH linkage was discernible in the infrared spectrum of the polymer. However, even when the polymerization was carried out with anhydrous bases such as amines or cesium fluoride, the molecular weights were still only moderately high.

Pummer and co-workers (10) polymerized perfluoro-p-cresol in aqueous base and achieved molecular weights of only 3000–5000. The polymets are claimed to be rubbery above $100°C$ when heated initially above $200°C$.

Pummer was also able to obtain polyperfluorophenylene oxide polymers by reaction of pentafluorophenol with aqueous potassium hydroxide.

These materials, with molecular weights up to 12,500, were low-melting glasses (<100°C) and were soluble in all common organic solvents. The 12,500 molecular weight polymers, after heating for 18 hr at 260°C, become rubbery between 90 and 300°C; however, the thermal instability within this range precludes application as an elastomer.

VI. POLYKETALS

Alternating 1/1 copolymers of perhaloketones and epoxides were recently reported (11,12).

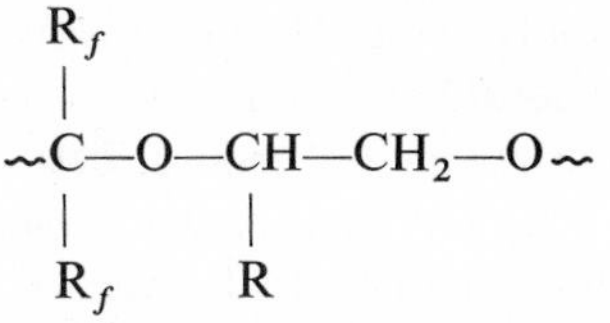

$$
\overset{\displaystyle R_f}{\underset{\displaystyle R_f \qquad R}{\sim\!C\!-\!O\!-\!CH\!-\!CH_2\!-\!O\sim}}
$$

The polymerization is initiated by a variety of nucleophilic reagents. Cesium fluoride is most effective and gives molecular weights inversely proportional to the amount of fluoride used. The epoxide and the fluoroketone have been shown to be joined in a one to one alternating fashion in the chain. The system is a living one since monomer added to the polymer produces more polymer with increased molecular weights. Although the polymerization of hexafluoroacetone can be readily effected in nonpolar solvents, tetrahydrofuran totally inhibits the reaction.

The proposed mechanism, based on the coordination of the epoxide with the ketone, is consistent with the alternating nature of the monomers. There is no reaction when ethylene oxide is treated with independently prepared cesium perfluoroisopropoxide, consistent with the mechanism. If this mechanism is valid, then THF, as observed, would act as a polymerization inhibitor by competing with the ethylene oxide for coordination with the ketone:

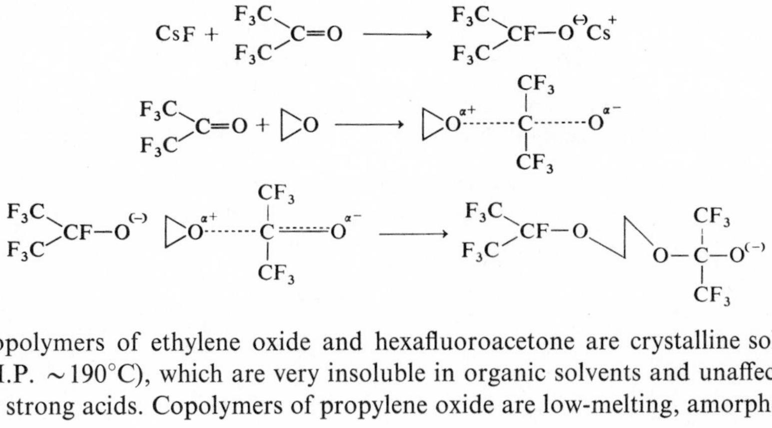

Copolymers of ethylene oxide and hexafluoroacetone are crystalline solids (M.P. ~190°C), which are very insoluble in organic solvents and unaffected by strong acids. Copolymers of propylene oxide are low-melting, amorphous solids which are completely soluble in common organic solvents. Terpolymers using combinations of epoxides are readily prepared with physical properties dependent upon the ratio of epoxides used.

VII. COPOLYMERS OF PERHALOKETONES AND OLEFINS

Polyethers prepared from perhaloketones have also been obtained with various olefinic materials such as styrene, acrylonitrile isoprene, butadiene, and methyl methacrylate, by catalysis with sodium biphenyl or sodium naphthalene in THF (13). The polymer of hexafluoroacetone and styrene was shown to be a true copolymer by its complete insolubility in THF, whereas polystyrene is completely soluble in THF. The infrared spectrum is similar to polystyrene but contains strong absorption between 7 and 11 μ characteristic of C—F bonds. With methyl methacrylate and HFA, a 1:1 copolymer has been prepared which is claimed to be very strong and notably flame resistant. Polymers of 85% butadiene/15% HFA have good resistance to hydrocarbon solvents.

In a more recent patent (14), it has been disclosed that vinyl acetate or ethylene can be copolymerized, although in very poor conversion, with hexafluoroacetone or dichlorotetrafluoro acetone in the presence of free-radical catalysts, such as benzoyl peroxide or di-tertiary butyl peroxide. All of the polymers were solid materials containing a much higher percentage of the olefin than ketone. The copolymers of vinyl acetate resemble the homopolymer, but have lower water vapor transmission and lower combustibility. Vapor-phase osmometry gave a molecular weight of 437 for a copolymer of ethylene and hexafluoroacetone.

However, in light of the work by Howard et al. (15), the structure of the vinyl polymers claimed in the patent may be inaccurate, since it has been shown that the methyl group of peroxides, such as di-tertiary butyl peroxide, enters into a reaction with hexafluoroacetone to give alcohols:

$$\text{HFA} + \text{CH}_3\cdot \longrightarrow (\text{CF}_3)_2\overset{\overset{\displaystyle \text{OH}}{|}}{\text{C}}-\text{CH}_3$$

and the fluoroketone can react with preformed hydrocarbon polymers to give polymers with pendant fluorinated alcohol or ether side chains:

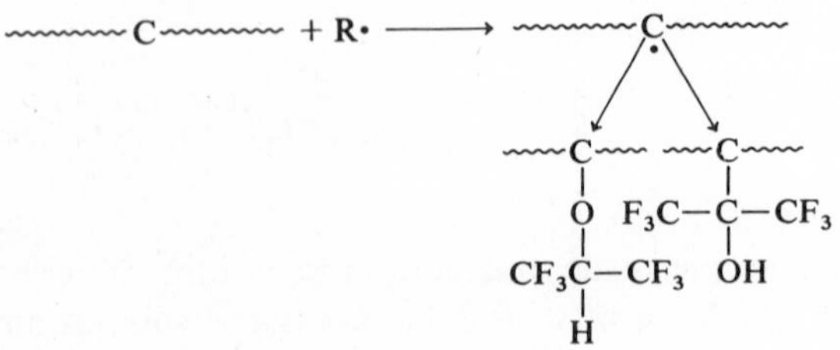

The polymerization of vinyl monomers with perfluoroketones in the presence of peroxides may therefore give materials with oxygen in the backbone, as shown in the Allied patent (14), but also with a considerable amount of pendant fluoroether or fluoroalcohol.

Howard (15) has also found that hexafluoroacetone and tetrafluoroethylene form a copolymer, but little has been reported about this potentially important polymer except that infrared analysis confirmed the presence of $C-O-C$ bonds.

VIII. COPOLYMERS OF HEXAFLUOROACETONE AND FORMALDEHYDE

Copolymers prepared from hexafluoroacetone and formaldehyde (16) are hard, resilient materials which are insoluble in common organic solvents and are unaffected by cold concentrated sulfuric acid. The polymers are prepared by reaction of pure monomeric formaldehyde with hexafluoroacetone and cesium fluoride as a catalyst. With very pure monomers, 1/1 alternating copolymers are obtained; however, even the best polymers readily unzip to monomers at comparatively low temperatures (130°C). All attempts at end-capping have not improved this stability.

IX. HOMOPOLYMERS OF PERFLUOROKETONES

Although several perfluorinated ketones are known, only two, perfluorocyclobutanone and perfluorocyclobutanedione (17,18), have been successfully homopolymerized. The pulled back CF_2 groups probably provide sufficient sterochemical freedom to allow polymerization to a polyether. Although these polymers can be capped with an acetate end group, they also readily decompose. Upon standing at room temperature, polyperfluorocyclobutanone unzips to the monomer. This polymer is prepared at $-78°C$ using sodium acetate as catalyst.

X. POLYMERS OF FLUOROALDEHYDES

The preparation of polymers of fluorocarbon aldehydes has been described in a British patent (19). The polymerization of trifluoroacetaldehyde is initiated by organic peroxides such as acetyl peroxide, benzoyl peroxide, or cumene hydroperoxide, whereas higher perfluoroaldehydes require acid catalysts. The fluorocarbon aldehyde polymers are solids which neither melt nor support combustion.

Trifluoroacetaldehyde has been copolymerized with monomeric formaldehyde with a complex quaternary ammonium acetate initiator (20). The inherent viscosity is 2.85 (in 0.5% dimethyl formamide) at 150°C. Analysis indicated that the product contained 29.5% of the aldehyde. Attempting to end cap the polymer with acetic anhydride reduced the aldehyde content to 14.2%. The acetylated copolymers are about as stable as acetylated polyformaldehyde.

XI. COPOLYMERS OF FLUORINATED DIOLS AND TRIOXANE

Hollander has prepared low-molecular-weight polyformals ($\overline{M}_n 835$) from the copolymerization of trioxane and hexafluoropentadiol with p-toluenesulfonic acid as catalyst (3,4). These polymers are very unstable and liberate CH_2O on standing .

XII. PERFLUORINATED POLYETHERS FROM PERFLUOROEPOXIDES

The patent literature, both American and foreign, on perfluorinated polyethers has grown voluminous since the initial discovery of the preparation of perfluoropropene oxide and subsequent polymerization of this monomer.

The preparation of a perfluoroepoxide in liquid hydrogen fluoride from epichlorohydrin electrolysis was claimed in 1952 in a British patent granted to Simons (21). However, the more recent disclosure (22) convincingly refutes this claim by showing that the material prepared earlier was actually hexafluorotrimethylene oxide (hexafluorooxetane).

Perfluoropropylene oxide (PFPO) is conveniently prepared with alkaline hydrogen peroxide oxidation of perfluoropropylene at reduced temperatures ($-30°C$) in aqueous methanol (22). Although the epoxide is obtained as a mixture with the olefin which is unseparable by distillation, the epoxide can be isolated by titrating the remaining olefin with bromide and subsequently fractionally distilling the product mixture. The epoxide is a gas which undergoes several nucleophilic ring-opening reactions and is also a convenient source of difluorocarbene (23–25).

Tetrafluoroethylene oxide (TFEO) was first reported in a patent to Warnell (26). This epoxide, which is considerably less stable than perfluoropropylene oxide was prepared by direct oxidation of tetrafluoroethylene with UV light in the presence of a small amount of a chain stopper such as bromine.

These two epoxides have formed the basis for many other patents on perfluorinated polyethers. Most of the work has been done by workers at duPont, although the literature indicates that other companies such as Montecatini and 3M have considerable interest in this area (21,27,28).

Sianisi has succeeded in preparing perfluoropolyethers from the direct oxidation of perfluoropropylene with oxygen and a mercury lamp initiator (40). The polymers obtained resemble the perfluorinated polyethers prepared by the duPont workers from the perfluorinated epoxides, although under certain conditions a polyperoxide is formed. The polymerization to the polyether is believed to proceed via a polyperoxide which then decomposes photolytically to the more stable polyether. In one case, when only short wave length light was used (<300 mu), a very stable polymer corresponding to $(CF_2O)_n$ was obtained.

A. Polymers Based on Polyperfluoropropylene Oxide (PFPO)

Polymerization of PFPO was first reported to be initiated by a complex formed from the reaction of an alkali metal fluoride such as cesium fluoride in the presence of an acid fluoride such as CF_3CF in diglyme (28,29):

$$CF_3\overset{\overset{O}{\|}}{C}F + CsF \longrightarrow CF_3CF_2O^{(-)}Cs^+$$

$$CF_3CF_2O^{(-)} + F_2C\overset{\overset{O}{\triangle}}{\underset{\underset{CF_3}{|}}{C}}F \longrightarrow CF_3CF_2-O\left[CF-CF_2-O\atop\underset{CF_3}{|}\right]_n \overset{CF_3}{\underset{\overset{\|}{O}}{CFCF}}$$

Polymers are obtained in which the n values are 1-6 and all materials with $n = 1,2,3$, have been isolated and identified. These liquid polymers are all acid fluoride-terminated, but in general they are thermally stable except at the end group.

A broad duPont patent for preparing the polymers indicates that the conversion of PFPO can be carried out using any of the alkali metal fluorides in small amounts of diluent such as a polar aprotic solvent (30). Other catalysts successfully used are quaternary salts such as quaternary ammonium halides (Br, Cl, I), acetates, and cyanides, and quaternary phosphorus or arsenic salts. With the quaternary salts, polymerization could be effected even if nonpolar diluents such as chlorocarbons were used. The degree of polymerization obtained with these initiators or combinations of them ranged from 0 to ~35. The highest degree of polymerization was obtained using dried cesium fluoride in ultradry tetraglyme and carrying out the reaction at a temperature of -32 to $-38°C$. A large fraction of this material had an average DP of 33.5. Apparently, most of the polymers obtained by this technique are liquids with an acid fluoride end group and an inert fluoroalkyl end group.

The highest molecular weight polymers of PFPO reported result from the treatment of the monomer with a Darco 12 × 20 charcoal catalyst which has been thoroughly devolatilized (31). About half of the polymer, with a DP of greater than 35, is a nonpourable oil, grease, or a waxy solid.

B. Polymers Based on Tetrafluoroethylene Oxide (TFEO)

Polymers of TFEO were prepared first by Warnell by employing the Darco charcoal catalyst and, although a broad molecular weight range of products was obtained, fractions as high as 4700 M.W. (M.P. 39–41°C) were isolated (26). Again these polymers were all acyl fluoride-terminated. In a French patent (32), Warnell has disclosed that much higher molecular weight polymers have been prepared by irradiaton of TFEO at −190°C with 2 MeV electrons from a van de Graaff generator or merely by treating the epoxide with fluorine at −196°C (33). The polymers obtained in this manner appear to be greater than 100,000 molecular weight with some fractions apparently as high as 725,000. These materials melt at 38–42°C, are white solids which have no carbonyl functionality, and are very stable. They can be solvent cast into films which are elastomeric below their melting points.

C. Modification of Perfluoroepoxide Polymers

The acyl fluoride termination of the polyperfluoroepoxide provides a convenient handle for modifying the polymers. Probably the most actively pursued modification is the conversion of the terminus to a vinyl ether. Thermal decarboxylation of the dry salts of the terminal carboxylic acids obtained from the oligomers of perfluoropropylene oxide converts them to vinyl ethers in excellent yield (37):

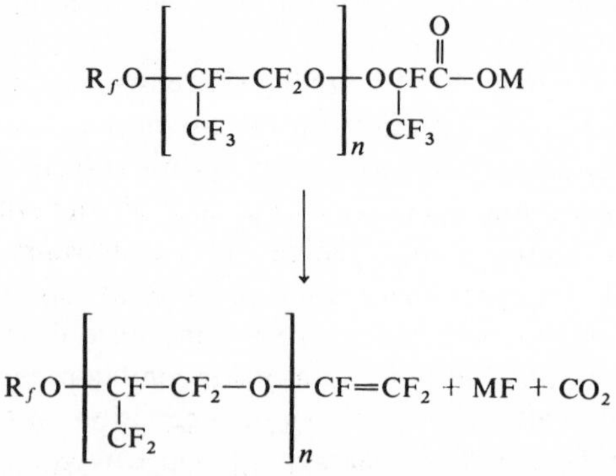

$$R_fO\left[CF-CF_2O\right]_n\overset{O}{\overset{\|}{OCFC}}-OM$$

$$\downarrow$$

$$R_fO\left[CF-CF_2-O\right]_nCF=CF_2 + MF + CO_2$$

However, a far more convenient method of preparing these ethers was disclosed in a patent involving direct decomposition of the terminal acid fluorides (34). Lorenz found that by passing the acyl fluoride oligomers over a bed of a dried metal oxide such as ZnO, CaO, SiO_2 at temperatures of 275–400°C, quantitative conversions and yields of vinyl ethers up to 95% could be obtained. Yields and conversions do not appear to be dependent upon the molecular weight of the oligomers but rather on the conditions employed. Silicon dioxide appears to require higher temperature than zinc oxide (300 versus 390°C) to obtain optimum yields and conversions.

Fritz and Selman found that the vinyl ethers could be prepared by pyrolysis of the acyl fluorides over such salts as sodium sulfate or alkali metal carbonates at elevated temperatures (35). This process is noncatalytic since stoichiometric quantities of the salts are required. The process is believed to be one of decarboxylation of an intermediate double salt:

$$
\underset{\substack{\| \\ R_fCF}}{O} + NaOCONa \longrightarrow \underset{\substack{\| \quad \| \\ R_fOC-OCONa}}{O \quad O} + NaF
$$

$$
\underset{\substack{\| \; \| \\ R_fCOCONa}}{O \; O} \longrightarrow R_fCO^-Na^+ + CO_2
$$

The yields and conversions to vinyl ethers are as good as with the metal oxides discussed earlier. It was also observed that the ethers could be prepared from the acyl fluorides at much lower temperatures in the liquid phase using a polar aprotic solvent which slightly solubilizes the metal salt. For example, with diglyme and sodium carbonate nearly quantitative yields and conversions could be obtained with perfluoro-2-propoxypropionyl fluoride at 60°C compared to 300°C when the dry salt was used. These vinyl ethers have been copolymerized with olefins such as TFE (36) and have been converted to sulfonyl fluorides with sulfuryl fluoride (37).

Several divinyl ethers have been prepared from polymers of diacid fluorides (38):

$$
\underset{\substack{\| \; \| \\ FCR_fCF}}{O \; O} \xrightarrow{CsF} Cs^{(+)(-)}OCF_2R_fCF_2O^{(-)}Cs^{(+)} \xrightarrow{\overset{\triangle}{CF_3}}
$$

$$
\text{divinyl ether} \xleftarrow{-COF_2} \underset{\substack{| \\ CF_3}}{\overset{\substack{O \\ \|}}{FCCF}} \left[O-\underset{\substack{| \\ CF_3}}{CF_2CF} \right]_m \left[OCF_2R_fCF_2O-\underset{\substack{| \\ CF_3}}{CF}-CF_2O \right]_n \underset{\substack{| \\ CF_3}}{\overset{\substack{O \\ \|}}{CFCF}}
$$

The oligomers of PFPO and TFEO have been converted to alcohols by reducing of the acyl fluoride end groups with solid borohydride in dry dioxane. Yields varied from 40 to 86% of isolated alcohol. The alcohols have been converted to carboxylic esters or phosphate diesters by standard techniques, yielding lubricating oils and hydraulic fluids (39).

D. Capping of Fluoropolymers

Since most of the polyners produced are terminated with an unstable $\underset{\substack{\| \\ CF}}{O}$ group, methods were derived for capping the liquid polymers to make them suitable for applications requiring both thermal and chemical stability. Selman has reported conversion of the terminal groups of polyepoxides to hydrogen by hydrolysis and the pyrolysis of the potassium salts at 115–200°C (37,38,40):

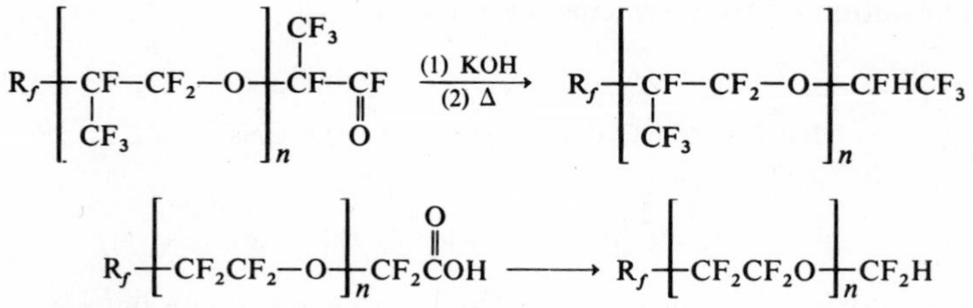

$$R_f\left[CF-CF_2-O\left]\begin{array}{c}CF_3\\|\\CF-CF\\\|\\O\end{array}\right.\right. \xrightarrow[(2)\ \Delta]{(1)\ KOH} R_f\left[CF-CF_2-O\right]CFHCF_3$$

with CF_3 and CF_3 substituents, subscript n

$$R_f\left[CF_2CF_2-O\right]_n\overset{O}{\overset{\|}{CF_2COH}} \longrightarrow R_f\left[CF_2CF_2O\right]_nCF_2H$$

The resulting oils are produced in greater than 90% yield and are reported to be very stable.

Miller was able to prepare an even more stable end group by hydrolyzing the polymers, again, to the pure acids and fluorinating the resulting acids at elevated temperatures (184–187°C) in an inert solvent such as perfluoro-dimethyl cyclobutane (41):

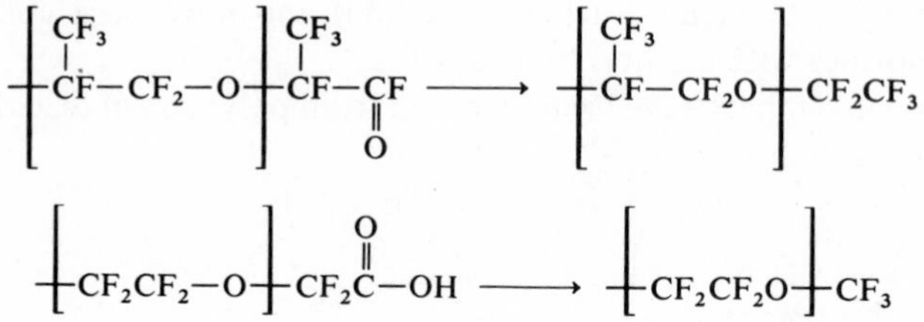

$$\left[\begin{array}{c}CF_3\\|\\CF-CF_2-O\end{array}\right]\left[\begin{array}{c}CF_3\\|\\CF-CF\\\|\\O\end{array}\right] \longrightarrow \left[\begin{array}{c}CF_3\\|\\CF-CF_2O\end{array}\right]CF_2CF_3$$

$$\left[CF_2CF_2-O\right]\overset{O}{\overset{\|}{CF_2C-OH}} \longrightarrow \left[CF_2CF_2O\right]CF_3$$

Yields and conversions were quantitative. The products capped in this manner are useful as high-temperature lubricants or as heat transfer fluids.

An interesting method of capping was reported by Milian, who found that photolytic treatment of the terminal acid fluorides results in a decarbonylation coupling (31):

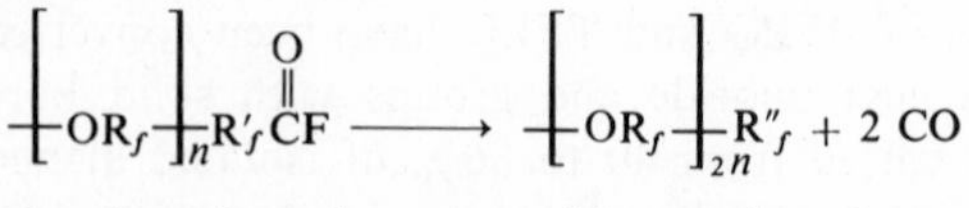

$$\left[OR_f\right]_n\overset{O}{\overset{\|}{R'_fCF}} \longrightarrow \left[OR_f\right]_{2n}R''_f + 2\ CO$$

This procedure also has the advantage of increasing the molecular weights of the polymers.

References

1. M. Hauptschein and J. M. Lesser, *J. Am. Chem. Soc.*, **78**:676 (1956).
2. D. D. Smith, R. M. Murch, and O. R. Pierce, *Ind. and Eng. Chem.*, **49**:124 (1957).
3. F. D. Trischler and J. D. Hollander, *Am. Chem. Soc. Polymer Preprints*, **8**(1):419 (1967).

4. J. D. Hollander, F. D. Trischler, and E. S. Harrison, *Am. Chem. Soc. Polymer Preprints*, **8**(2):1149 (1967).
5. F. B. Jones, P. B. Stickney, L. E. Coleman, Jr., D. A. Rausch, and A. M. Lovelace, *J. Polymer Sci.*, **26**:81 (1957).
6. A. G. Pittman and D. L. Sharp, *Polymer Letters*, **3**:379 (1965).
7. D. L. Miller and N. L. Madison, U.S. Patent 3,330,808 (1967).
8. R. G. Jones, *J. Am. Chem. Soc.*, **69**:2346 (1947).
9. E. McMasters, Technical Documentary Report No. ML TDR 64-112, Part III, 1966, Air Force Materials Laboratory, Wright-Patterson Air Force Base, Ohio.
10. W. J. Pummer and J. M. Antonucci, *Am. Chem. Soc. Polymer Preprints*, **7**(2):1071 (1966).
11. N. L. Madison, *Am. Chem. Soc. Polymer Preprints*, **7**(2):1099 (1966).
12. F. S. Fawcett and E. G. Howard. French Patent 1,433,649 (1966): U.S. Patent 3,311,216 (1967).
13. M. W. B. Baker, Jr., French Patent 1,357,933 (1964).
14. British Patent 1,051,239 (1966).
15. E. G. Howard, P. B. Sargent, and C. G. Krespan, *Am. Chem. Soc. Polymer Preprints*, **7**(2):1091 (1966).
16. N. L. Madison and D. L. Miller, Technical Documentary Report No. ML-TDR-64-140, Part II, (1965), Air Force Materials Laboratory, Wright-Patterson Air Force Base, Ohio.
17. D. C. England, U.S. Patent 3,039,995 (1962).
18. D. C. England, U.S. Patent 3,133,046 (1966).
19. D. R. Husted and A. H. Albrecht, British Patent 717,877 (1954).
20. T. L. Cairns, E. T. Cline, P. J. Graham, U.S. Patent 2,828,287 (1958).
21. British Patent 672,720 (1952).
22. British Patent 904,877 (1962).
23. D. Sianisi, A. Pasetti, and Franco Tardi, *J. Org. Chem.*, **31**:2312 (1966).
24. D. P. Carlson and A. S. Milian, *Abstracts, Fourth International Symposium on Fluorine Chemistry*, Estes Park, Colo., 24–28, 1967, p. 146.
25. F. C. McGrew, U.S. Patent 3,136,744 (1964).
26. J. L. Warnell, U.S. Patent 3,321,532 (1964).
27. D. Sianisi, A. Pasetti, and C. Corti, *Die makromolekulare Chemie*, **86**:308 (1966).
28. D. E. Morin, U.S. Patent 3,213,134 (1965).
29. E. P. Moore, French Patent 1,362,548 (1964).
30. E. P. Moore, U.S. Patent 3,322,826 (1967).
31. A. S. Milian, U.S. Patent 3,214,478 (1965).
32. J. L. Warnell, French Patent 1,342,523 (1963).
33. R. A. Darby, French Patent 1,341,087 (1963).
34. C. E. Lorenz, U.S. Patent 3,321,532 (1967).
35. C. G. Fritz and S. Selman, U.S. Patent 3,291,843 (1966).
36. J. F. Harris, U.S. Patent 3,132,123 (1964).
37. S. Selman, *Abstracts, Fourth International Symposium on Fluorine Chemistry*, Estes Park, Colo., July 24–28, 1967, p. 112.
38. C. G. Fritz and E. P. Moore, Canadian Patent 707,360 (1965).
39. R. E. LeBleu, U.S. Patent 3,293,306 (1966).
40. S. Selman and W. S. Smith, Jr., French Patent 1,373,014 (1964).
41. W. T. Miller, U.S. Patent 3,242,218 (1966).

8. FLUOROTHIOCARBONYL POLYMERS

WILLIAM H. SHARKEY, *E. I. du Pont de Nemours and Company, Experimentat Station, Wilmington, Delaware*

Contents

I. INTRODUCTION

Considerable activity in recent years resulted in the preparation and elucidation of chemistry of a large number of fluorothiocarbonyl compounds. This extension of research into new areas of fluorine chemistry was rewarded by the discovery that many of these compounds polymerize easily and that some of the polymers have most unusual properties. Indeed some of these materials, the fluorothioaldehydes, polymerize so rapidly that the monomers have not been characterized other than by being recognized as colored liquids.

The fluorothiocarbonyl compound that has received most attention is poly(thiocarbonyl fluoride), $+CF_2S+_n$. Polymerization mechanism, polymer structures, and the nature of the C—S bond in a polymer chain have been deduced from the preparation and behavior of this material. It is probable that other fluorothiocarbonyl polymers in the main follow the same pattern.

Differences in degree are already known, and future work will undoubtedly uncover more.

Since routes to fluorothiocarbonyl compounds required the development of special methods, which differ greatly from those used to synthesize their more familiar oxygen counterparts, it seems desirable to review monomer preparation first. This is followed by discussion of polymerization and by a few comments on polymer properties.

II. PREPARATION OF FLUOROTHIOCARBONYL MONOMERS

A. Thiocarbonyl Fluoride

The simplest of the perfluorothiocarbonyl compounds is thiocarbonyl fluoride, $CF_2=S$, a material that has been prepared in a number of ways (1). One involves reaction of thiophosgene with chlorine, fluorination of the product, and dehalogenation of the fluorinated material (2):

$$CCl_2=S \xrightarrow{Cl_2} CCl_3SCl \xrightarrow{SbF_3} \begin{array}{l} CF_2ClSCl \xrightarrow{Sn} CF_2=S \\ + \\ CFCl_2SCl \xrightarrow{Sn} CFCl=S \end{array}$$

Thiocarbonyl fluoride has also been prepared from bis(trifluoromethylthio) mercury and iodosilane (3):

$$(CF_3S)_2Hg + SiH_3I \longrightarrow \begin{array}{l} SiH_3SCF_3 \\ + \\ HgI_2 \end{array} \longrightarrow SiH_3F + CF_2=S$$

A more direct route is reaction of sulfur with tetrafluoroethylene, which gives thiocarbonyl fluoride together with appreciable amounts of trifluoro-thioacetyl fluoride and bis(trifluoromethyl) disulfide (4):

$$CF_2=CF_2 \xrightarrow[500-600°]{S} CF_2=S + CF_3CF=S + CF_3SSCF_3$$

Another direct route that gives very high yields of thiocarbonyl fluoride is reaction of sulfur with chlorodifluoromethane at very high temperatures (5):

$$CF_2HCl \xrightarrow[700-900°]{S} CF_2=S + HCl$$

In the laboratory the best method appears to be one in which thiophosgene is dimerized, the dimer is fluorinated, and the fluorinated dimer is pyrolyzed (4):

$$CCl_2{=}S \;\underset{}{\rightleftharpoons}^{h\nu}\; Cl_2C{\overset{S}{\underset{S}{<}}}CCl_2 \;\xrightarrow{SbF_3}\; F_2C{\overset{S}{\underset{S}{<}}}CF_2 \;\xrightarrow{500^\circ}\; CF_2{=}S$$

Because the first step is reversed by heat, temperature control during thiophosgene dimerization is very important. The temperature should not be allowed to rise much above 20°. The fluorination step is straightforward, and the product is chemically and thermally stable. It usually contains small amounts of monochloro- and dichlorodifluoro-1,3-dithietanes, which can be removed by distillation. Pyrolysis of highly purified 2,2,4,4-tetrafluoro-1,3-dithietane gives thiocarbonyl fluoride, which, after fractionation in a low-temperature column (B.P. −54°), contains no impurity other than about 15–25 ppm of carbon oxysulfide.

Pyrolysis of 2-chloro-2,4,4-trifluoro-1,3-dithietane yields a mixture of thiocarbonyl chlorofluoride (B.P. 6–7°), $CFCl{=}S$, and thiocarbonyl fluoride. Because of the great disparity in boiling points, these compounds are easily separated by distillation.

Thiocarbonyl cyanofluoride, $CF(CN){=}S$, has also been synthesized and polymerized (6). It is prepared by the high-temperature reaction of fluorochloroacetonitrile or dichlorofluoroacetonitrile and sulfur:

$$FClHC{-}CN \;\xrightarrow[600{-}750]{S}\; F{-}\overset{\overset{\displaystyle S}{\|}}{C}{-}CN + HCl$$

B. Fluorothioacid Halides

Fluoroalkylthioacid halides are prepared by reaction of appropriate fluorocarbons with sulfur at high temperatures. As already discussed, reaction of tetrafluoroethylene with sulfur ordinarily gives thiocarbonyl fluoride along with minor amounts of trifluorothioacetyl fluoride. However, when this reaction is carried out by reaction of the fluorocarbon with sulfur vapor over a bed of activated charcoal, perfluorothioacetyl fluoride is the major product of the reaction (7):

$$CF_2{=}CF_2 + S \;\xrightarrow[\text{charcoal}]{\text{act.}}\; CF_3CF{=}S$$

Chloro- and bromofluoroethylenes also react with sulfur vapor, and the products are fluorothioacetyl halides. These transformations take place in the absence of a catalyst. An example is passage of chlorotrifluoroethylene through boiling sulfur, which leads to chlorodifluorothioacetyl fluoride. Similarly, bromotrifluoroethylene and sulfur gives bromodifluorothioacetyl fluoride. The product from the reaction of sulfur and 1,2,2,2-dichlorodifluoro-ethylene is chlorodifluorothioacetyl chloride. This reaction has been cited as evidence that the first product is an episulfide that immediately rearranges to a thioacid halide:

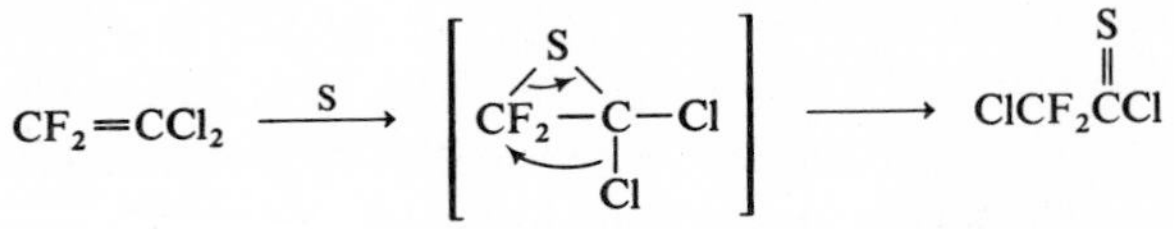

$$CF_2=CCl_2 \xrightarrow{\;S\;} \left[CF_2-C-Cl \atop Cl \right]^{S} \longrightarrow ClCF_2CCl^{S}$$

A very useful route for the preparation of thioacetyl fluorides is reaction of bis(perfluoroalkyl)mercury with sulfur. These materials are obtained by reaction of a fluoroethylene with mercuric fluoride in anhydrous hydrogen fluoride (8):

$$CF_2=CF_2 + HgF_2 \xrightarrow{HF} (CF_3CF_2)_2Hg$$

$$CF_2=CFCl + HgF_2 \xrightarrow{HF} (CF_3CFCl)_2Hg$$

Reaction of either of the mercuric fluoride addition products with sulfur at 450° gives trifluorothioacetyl fluoride in very good yields. Use of bis(1,1-dichloroperfluoroethyl)mercury, obtained from 1,1-dichloro-2, 2-difluoro-ethylene and mercuric fluoride, in this reaction gives trifluorothioacetyl chloride.

Another specialized method is reaction of phosphorus pentasulfide with perfluoroalkyl iodides at 550°. Examples are conversion of perfluoroethyl iodide to perfluorothioacetyl fluoride, and 1-iodoperfluoropropane to per-fluorothiopropionyl fluoride. The usefulness of this method is limited by the general unavailability of perfluoroalkyl iodides. This appears to be a reaction between an iodide and sulfur with the phosphorus compound serving as a high-temperature source of sulfur.

In at least one case, milder conditions are sufficient for reaction of an iodide and sulfur (4). This case is conversion of 4,4-diiodoperfluoro-1-butene to perfluorothio-3-butenoyl fluoride:

$$CF_2=CFCF_2CFI_2 \xrightarrow[450°]{S} CF_2=CFCF_2CF^{S}$$

The ability of the double bond to resist attack by sulfur is remarkable.

A unique preparative procedure for obtaining hydrogen-containing fluoro-thioacid fluorides has been reported by Harris and Stacey (9). High-energy radiation is employed to promote the addition of hydrogen sulfide to a fluoro-olefin to obtain a terminally substituted thiol, which is then dehydrofluorin-ated by reaction with sodium fluoride. The following compounds are among those that have been prepared in this manner:

$$CF_2{=}CF_2 \xrightarrow{\text{H}_2\text{S}} HCF_2CF_2SH \xrightarrow{\text{NaF}} HCF_2C\overset{\displaystyle S}{\overset{\|}{F}}$$

$$CFCl{=}CF_2 \xrightarrow{\text{H}_2\text{S}} HCFClCF_2SH \xrightarrow{\text{NaF}} HCFClC\overset{\displaystyle S}{\overset{\|}{F}}$$

C. Fluorothio Ketones

A method of wide applicability for the preparation of fluorothio ketones is reaction of appropriate fluoroalkyl mercurials with sulfur (4). This is done in the same way as related for fluorothioacid fluorides. An illustration is the preparation of hexafluorothioacetone. The mercurial is prepared by addition of mercuric fluoride to hexafluoropropene, a reaction that is carried out in anhydrous HF:

$$2CF_3CF{=}CF_2 + HgF_2 \xrightarrow{\text{HF}} (CF_3)_2CFHgCF(CF_3)_2$$

Reaction with sulfur is done in a second step:

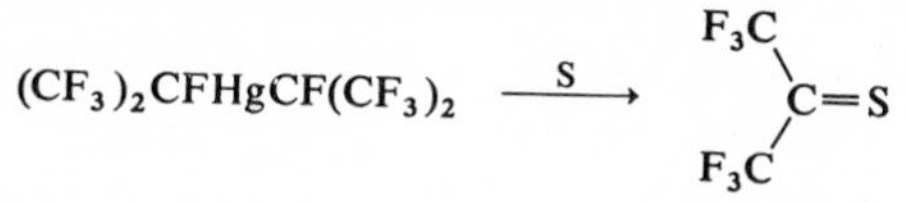

$$(CF_3)_2CFHgCF(CF_3)_2 \xrightarrow{\text{S}} \begin{array}{c} F_3C \\ \diagdown \\ C{=}S \\ \diagup \\ F_3C \end{array}$$

It is important to use high temperatures in this reaction since different products are formed at lower temperatures. For example, at 200° a mixture of perfluoroisopropyl disulfides and polysulfides is the product:

$$(CF_3)_2CFHgCF(CF_3)_2 \xrightarrow[200°]{\text{S}} (CF_3)_2CF{-}{(}S{)}_n{-}CF(CF_3)_2 \quad (n = 2, 3, 4)$$

These polysulfides can be defluorinated by reaction with triphenylphosphine, in which case the product is 2,2,4,4-tetrakis(trifluoromethyl)-1,3-dithietane formed by dimerization of the primary product, hexafluorothioacetone:

$$(CF_3)_2CF{-}{(}S{)}_n{-}CF(CF_3)_2 + (C_6H_5)_3P \longrightarrow \left[\begin{array}{c} CF_3 \\ | \\ C{=}S \\ | \\ CF_3 \end{array}\right] + (C_6H_5)_3PF_2$$

$$\downarrow$$

$$(CF_3)_2C\overset{\textstyle S}{\underset{\textstyle S}{\diagup\diagdown}}C(CF_3)_2$$

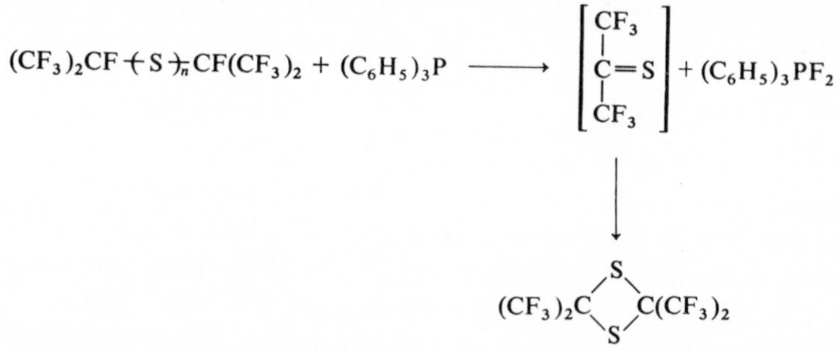

It is surprising that triphenylphosphine removes fluorine rather than sulfur This has been explained as nucleophilic attack of fluorine by phosphorus:

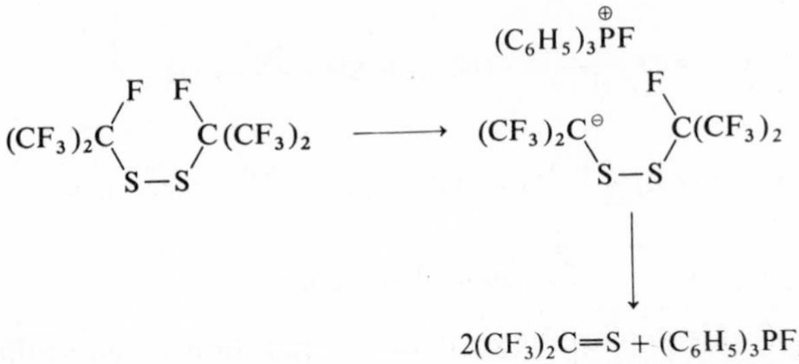

$$(CF_3)_2C\overset{F\quad F}{\underset{S-S}{\diagup \diagdown}}C(CF_3)_2 \xrightarrow{\quad (C_6H_5)_3\overset{\oplus}{P}F \quad} (CF_3)_2\overset{\ominus}{C}\overset{F}{\underset{S-S}{\diagup \diagdown}}C(CF_3)_2$$

$$\downarrow$$

$$2(CF_3)_2C{=}S + (C_6H_5)_3PF_2$$

Sulfurization of secondary perfluoroalkyl iodides with refluxing phosphorus pentasulfide (4) provides another route to perfluorothio ketones. An example is conversion of 2-iodoperfluorobutane to perfluorobutane-2-thione, which occurs in very high yields:

$$C_2F_5CFICF_3 \xrightarrow[550°]{P_2S_5} C_2F_5\overset{S}{\overset{\|}{C}}CF_3$$

The most direct method for preparing perfluorothioacetone is reaction of hexafluoropropene with sulfur over a bed of activated charcoal (7):

$$CF_3CF{=}CF_2 \xrightarrow[\text{act. charcoal}]{S} CF_3\overset{S}{\overset{\|}{C}}CF_3$$

These fluorothio ketones all dimerize with great ease. This reaction is rapid at room temperature but can be greatly slowed by cooling to −80°. However, even at low temperatures dimerization is rapid in the presence of compounds having unshared electrons on such heteroatoms as oxygen, nitrogen, and phosphorus. Included among these compounds are ethers, amides, amines, phosphines, and sulfur dioxide. Hexafluorothioacetone dimerizes rapidly to 2,2,4,4-tetrakis(trifluoromethyl)-1,3-dithietane in the presence of a trace of diethyl ether:

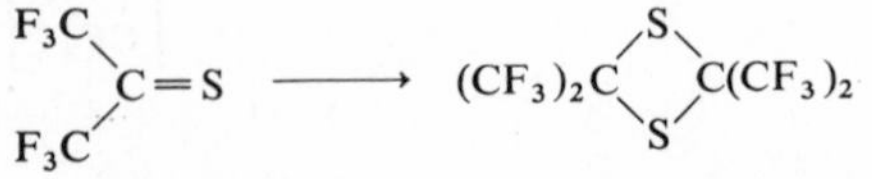

$$\overset{F_3C}{\underset{F_3C}{\diagdown \diagup}}C{=}S \longrightarrow (CF_3)_2C\overset{S}{\underset{S}{\diagup \diagdown}}C(CF_3)_2$$

The reaction can be reversed by pyrolysis at 500°. Ordinarily it is most convenient to keep hexafluorothioacetone in the form of its dimer and then pyrolyze it to monomer just before use in a reaction.

1,1,1-Trifluorothioacetone has also been prepared (10) and polymerized. Hydrogen sulfide is first added to a trifluoromethyl acetone to obtain either an olthiol or dithiol:

$$\begin{array}{c}
F_3C \\
\quad\diagdown \\
\qquad C=O + H_2S \\
\quad\diagup \\
H_3C
\end{array}
\quad
\begin{array}{c}
\xrightarrow[\text{HCl}]{\text{CaCl}_2} \quad
\begin{array}{c} F_3C \quad OH \\ \diagdown\,|\,\diagup \\ \qquad C \\ \diagup\,|\,\diagdown \\ H_3C \quad SH \end{array} \\[2em]
\xrightarrow[\text{HCl}]{\text{P}_2\text{O}_5} \quad
\begin{array}{c} F_3C \quad SH \\ \diagdown\,|\,\diagup \\ \qquad C \\ \diagup\,|\,\diagdown \\ H_3C \quad SH \end{array}
\end{array}$$

These compounds are then decomposed by heat.

$$\begin{array}{c} F_3C \quad OH \\ \diagdown\,|\,\diagup \\ \qquad C \\ \diagup\,|\,\diagdown \\ H_3C \quad SH \end{array}
\xrightarrow{\text{distil}}
\begin{array}{c} F_3C \\ \diagdown \\ \qquad C=S \\ \diagup \\ H_3C \end{array} +
\begin{array}{c} F_3C \\ \diagdown \\ \qquad C=O \\ \diagup \\ H_3C \end{array} + H_2O + H_2S$$

$$\begin{array}{c} F_3C \quad SH \\ \diagdown\,|\,\diagup \\ \qquad C \\ \diagup\,|\,\diagdown \\ H_3C \quad SH \end{array}
\xrightarrow{550^\circ}
\begin{array}{c} F_3C \\ \diagdown \\ \qquad C=S \\ \diagup \\ H_3C \end{array} + H_2S$$

Trifluorothioacetone can be stored for short periods at -80°. It poly- merizes when it is allowed to warm. Addition of hydrogen sulfide followed by heating has been used to synthesize such other thioketones as

$$\begin{array}{c} HCF_2F_2C \\ \diagdown \\ \qquad C=S \\ \diagup \\ C_2H_5 \end{array}
\quad \text{and} \quad
\begin{array}{c} F_3C \\ \diagdown \\ \qquad C=S \\ \diagup \\ C_6H_5 \end{array}$$

D. Fluorothioaldehydes

Fluorothioaldehydes are also derived from oxygen analogs (11,12). Hydrogen sulfide is added to a fluoroaldehyde to obtain an olthiol, which is then heated to remove water. Excess hydrogen sulfide is required in the first step; a 2:1 aldehyde/H_2S adduct results from a 1.7:1 aldehyde/H_2S reaction mixture:

$$CF_3CHO
\quad
\begin{array}{c}
\xrightarrow[\text{H}_2\text{S}]{\text{excess}} \quad
\begin{array}{c} OH \\ | \\ CF_3CH \\ | \\ SH \end{array} \\[2em]
\xrightarrow[\text{H}_2\text{S}]{\text{limited}} \quad
\begin{array}{c} OH \quad OH \\ | \qquad | \\ CF_3CHS-CHCF_3 \end{array}
\end{array}$$

Loss of water from the olthiol occurs spontaneously, or more rapidly if this compound, an evil-smelling liquid, is heated under reflux. The thioaldehyde formed apparently polymerizes spontaneously to a mixture of low polymers:

$$CF_3-\overset{\overset{\displaystyle OH}{|}}{CH}-SH \longrightarrow \left(\overset{\overset{\displaystyle CF_3}{|}}{CHS}\right)_n + H_2O$$

Vacuum pyrolysis of 2,2,2-trifluoroethan-1-ol-1-thiol at 600° gives a mixture of the thioaldehyde and aldehyde, which begin to copolymerize immediately. The example reported consisted of 82% thioaldehyde and 18% aldehyde. HCF_2-$CF_2CH{=}S$ and $H(CF_2)_4CH{=}S$ also have been converted to thioaldehyde polymers by addition of hydrogen sulfide followed by removal of water. $HCF_2CH{=}S$ polymers have been prepared from trifluoroethylene by addition of H_2S followed by removal of HF (9):

$$CF_2{=}CFH \xrightarrow{\ H_2S\ } HCF_2CFHSH \xrightarrow{\ NaF\ } [HCF_2CH{=}S]$$

$$\left(\overset{\overset{\displaystyle HCF_2}{|}}{\underset{\underset{\displaystyle CH-S}{|}}{}}\right)_n$$

The ease with which fluorothioaldehydes polymerize is not greatly affected by steric factors. Perfluorothioisobutyraldehyde (13) exists only transitorily; it polymerizes almost as fast as formed, even at $-80°$:

$$(CF_3)_2CF-\overset{\overset{\displaystyle OH}{|}}{\underset{\underset{\displaystyle SH}{|}}{CH}} \xrightarrow{\ \Delta\ } [(CF_3)_2CFCH{=}S] \longrightarrow \left(\overset{\overset{\displaystyle CF_3CFCF_3}{|}}{\underset{\underset{\displaystyle CH-S}{|}}{}}\right)_n$$

III. ANIONIC POLYMERIZATION

A. Introduction

The ability of fluorothiocarbonyl compounds to undergo polymerization is illustrated by $CF_2{=}S$, which polymerizes rapidly in the presence of anionic initiators (14). Polymerization is favored by low temperatures, proceeding best at $-78°$. The preferred medium is anhydrous diethyl ether, and the preferred initiator is dimethylformamide. High-molecular-weight polymers are readily formed; number average molecular weights of 300,000–400,000 and weight average molecular weights in excess of 1 million are easily obtained.

In addition to dimethylformamide, tetraisopropyl titanate has also been investigated as an anionic initiator for $CF_2{=}S$ polymerization. Among other effective compounds are triisopropyl aluminate, quaternary ammonium salts, amines, and phosphines.

Although anhydrous diethyl ether is the preferred solvent, polymerizations have been run in petroleum ether, methylene chloride, and chloroform.

At polymerization temperatures higher than $-78°$, the molecular weight of the product formed is greatly reduced. For example, tetraisopropyl titanate initiation of $CF_2=S$ in chloroform at $-25°$ gave a low-molecular-weight syrup.

Molecular weights of poly(thiocarbonyl fluoride) prepared by polymerization at $-78°$ are such that their inherent viscosities, measured on 0.1% chloroform solutions, range from about 2 to 5. Somewhat lower-molecular-weight polymers prepared at higher temperature have inherent viscosities around 1, which corresponds to a $\overline{M}_n$ of about 200,000 as determined by osmometry.

A supplemental way of controlling molecular weight is addition of a chain transfer agent. Conditions that ordinarily lead to a polymer of inherent viscosity 2.88 have been modified by addition of 7% isopropyl alcohol to reduce inherent viscosity to 1.04 (14).

B. Polymer Structure

Spectral and chemical studies have shown the main chain of poly(thiocarbonyl fluoride) to be $+CF_2S\rightarrow_n$. The F^{19} NMR spectrum of chloroform solution of the polymer contains a peak for $-S-CF_2-S-$ at 43.5 ppm higher field relative to CCl_3F used as an external standard. Reaction of the polymer with SbF_5 degrades the polymer to completely fluorinated chains low enough in molecular weight for cryoscopic characterization:

$$+CF_2S\rightarrow_n \xrightarrow{\quad SbF_5 \quad} CF_3S+CF_2S\rightarrow_x CF_3$$

$$\mathbf{I}$$

The F^{19} NMR spectra of I shows resonance at 43.5 ppm for $-S-CF_2-S-$ and also at 37.6 ppm for $-SCF_3$, both at higher field relative to CCl_3F as an external standard. Molecular weight calculated from the ratios of the NMR peaks compare favorably with molecular weights determined cryoscopically.

Infrared spectra of I confirms the assigned structure (14). Absorption is obtained at $4.75\,\mu$ for the CF overtone, at $7.65\,\mu$ for CF_3-S-C, at $8.9\,\mu$ for CF, at $12.25\,\mu$ for CS in $-CF_2S-$, and at $13.13\,\mu$ for CS in CF_3S-.

End groups on poly(thiocarbonyl fluoride) obtained by dimethylformamide initiation are $-CF=O$, $-CF=S$, and CF_3S-. Although the infrared spectrum of high-molecular-weight polymer is practically undecipherable, lower-molecular-weight material prepared at $225°$ gives a spectrum with bands at 5.4, 8.2, 10.1, and $13.1\,\mu$. The 5.4 and $8.2\,\mu$ bands are assigned to $-SCF=O$ and $-SCF=S$, respectively. The $10.1\,\mu$ band has not been assigned, but the $13.1\,\mu$ is the same as was seen in the spectrum of I for CF_3S-. Nuclear magnetic resonance of F^{19} gave, in addition to higher field peaks discussed earlier, a resonance peak at 71.5 ppm lower field that is associated with F in $-SCF=S$. The F in $-SCF=O$ has not been identified by NMR, probably because it is present in such low concentration.

The $-SCF=O$ and $-SCF=S$ end groups are removed by reaction with methanol. The bands identifying these groups are not present in the infrared spectra of the methanol-reacted products. By reaction of methanol $-C^{14}$ it has been established that $-SCF=O$ and $-SCF=S$ together account for one end of the polymer. It appears likely that $-SCF=O$ arises by hydrolysis of a small amount of $-SCF=S$.

Poly(thiocarbonyl fluoride) prepared by tetraisopropyl titanate initiation contains isopropoxy end groups. This was shown by infrared examination of low-molecular-weight polymer. The spectrum contains a double peak at 7.24 μ, characteristic of *gem* dimethyl groups, CH absorption at about 3.3 μ, and CH_2 at about 6.8 μ. These absorptions were absent in the spectrum of polymer that had been treated with thionyl chloride, which contained instead a band at 5.4 μ for $-SCF=O$.

On the basis of these studies II is given as the structure for polymer formed by dimethylformamide initiation and III for polymer obtained using tetra-isopropyl titanate.

$$CF_3S + CF_2S \xrightarrow{}_n CF=S$$
$$(CF=O)$$

II

$$(CH_3)_2CHO + CF_2S \xrightarrow{}_n CF=S$$

III

C. Polymerization Mechanism

The anionic polymerization of thiocarbonyl fluoride is believed to involve initiation by a basic species, denoted here by $B^\ominus$, which gives an intermediate that propagates until termination occurs by expulsion of a fluoride ion.

Initiation
$$B^\ominus + CF_2=S \longrightarrow BCF_2S^\ominus$$

Propagation
$$BCF_2S^\ominus + nCF_2=S \longrightarrow BCF_2S(CF_2)_{n-1}CF_2S^\ominus$$

Termination
$$BCF_2S(CF_2S)_{n-1}CF_2S^\ominus \longrightarrow BCF_2S(CF_2S)_{n-1}CF=S + F^\ominus$$

The fluoride ion, $F^\ominus$, also initiates polymerization and leads to CF_3S-$+CF_2S+_nCF=S$. It appears that about ten chains are initiated by $F^\ominus$ for every one started by the primary initiator.

Dimethylformamide does not appear to be a primary initiator. Attempts to identify it in the polymer through use of a C^{14} tag gave polymers that are completely free of radioactivity. It is believed that dimethylformamide reacts with something else in the solution to give the active initiator. This "something" has not been identified. A speculative possibility is a product (III) formed by reaction with $CF_2=S$:

$$(CH_3)_2NCH = O + CF_2 = S \longrightarrow \left[\begin{array}{c} H_3C \quad\; CH=O \\ \diagdown \overset{+}{N} \diagup \\ H_3C \diagup \quad\; \diagdown CF=S \end{array} \right] F^-$$

$$\mathbf{III}$$

Tetraisopropyl titanate initiation appears to be more clear-cut, since polymers obtained with this initiator have isopropoxy groups on one end. Their formation is visualized as follows:

$$CF_2 = S + [(CH_3)_2CHO]_4Ti \longrightarrow (CH_3)_2CHOCF_2S^{\ominus}Ti^{\oplus}[OCH(CH_3)]_3$$
$$\mathbf{IV}$$

$$IV + CF_2=S \longrightarrow (CH_3)_2CHO(CF_2S)_nCF_2S^{\ominus}$$
$$\mathbf{V}$$

$$V \longrightarrow (CH_3)_2CHO(CF_2S)_nCF=S + F^{\ominus}$$

It may be significant that no evidence has been found for the presence of $-SCF=O$ in polymers made by tetraisopropyl titanate initiation.

D. Fluorothioacid Fluorides

Anionic polymerization of fluorothioacid fluorides is generally more sluggish than that of $CF_2=S$. However, such compounds as $CF_3CF=S$, $ClCF_2CF=S$, and $CF_3CF_2CF=S$ form polymers at low temperatures upon initiation with dimethylformamide. The closely related thiocarbonyl cyanofluoride, $NCCF=S$, also polymerizes under these conditions. Pentafluoro-3-thiobutenoyl fluoride, $CF_2=CFCF_2CF=S$, polymerizes adventitiously upon storage at room temperature for long periods (14).

E. Fluorothioketones and Fluorothioaldehydes

The polymerization of perfluorothioketones appears to require exceptionally low temperatures. At $-78°$ anionic initiators catalyze dimerization of these compounds to 1,3-dithietanes. However, at about $-110°$ hexafluorothioacetone polymerizes when initiated by either dimethylformamide or BF_3 etherate:

$$\begin{array}{ccc}
 & & S \\
 & & \diagup \diagdown \\
 & (CF_3)_2C & \;\; C(CF_3)_2 \\
 \overset{DMF}{\underset{-78°}{\nearrow}} & & \diagdown \diagup \\
 & & S
\end{array}$$

$$\begin{array}{c}
CF_3 \\
| \\
C=S \\
| \\
CF_3
\end{array}
\quad
\overset{-110°}{\underset{DMF}{\searrow}}
\quad
\left[\begin{array}{c}
CF_3 \\
| \\
\!\!-\!C-S\!-\!\! \\
| \\
CF_3
\end{array} \right]$$

The polymer is colorless and insoluble in most organic solvents. It can be pressed into films. At room temperature the polymer slowly reverts to monomer, which dimerizes as formed to tetrakis(trifluoromethyl)-1,3-dithietane.

1,1,1-Trifluoroacetone polymerizes much more readily (10). When prepared it is collected in a receiver cooled to at least $-80°$. As it is allowed to warm it polymerizes spontaneously. Such higher-molecular-weight thioketones as 1,1,2,2-tetrafluoro-3-pentanethione and thioacetophenone have been converted to polymer by irradiation with ultraviolet in CF_2Cl_2 solution (10).

As has already been discussed, fluorothioaldehydes polymerize spontaneously to give products quite high in molecular weight. In many instances this polymerization is so rapid it makes monomer purification difficult. The polymers that have been prepared range from poly(trifluorothioacetaldehyde), which is a film- and fiber-forming elastomer (12), to poly(hexafluoroisobutyrthioaldehyde), which is highly crystalline, insoluble, hard polymer (13).

F. Copolymers

Thiocarbonyl fluoride copolymerizes anionically with other fluorothioacid fluorides and with perfluorocyclobutanone. In certain cases monomer reaction rates are such that polymer composition is nearly the same as that of the monomer mixture employed. For example, a 60:40 $CF_2=S/CF_3CF=S$ copolymer has been obtained from a 62:38 mixture and an 88:12 $CF_2=S/CF_3CF=S$ copolymer was prepared from an 85:15 mixture. Some thioacid fluorides copolymerize more slowly than $CF_2=S$. A 96:4 mixture of $CF_2=S$ and $HCFClCF=S$ gave a 98:2 $CF_2=S/HCFClCF=S$ copolymer.

Copolymers, particularly the $CF_2=S/HCFClCF=S$ copolymer, have somewhat better resistance to degradation by heat or amines than poly(thiocarbonyl fluoride). It has been postulated that thermal degradation of these polymers involves hydrolysis of the $-SCF=S$ end by adventitious water absorbed in the polymer followed by loss of COS:

$$CF_3S \text{\small\char`\~\char`\~} CF_2SCF=S \xrightarrow{H_2O} CF_3S \text{\small\char`\~\char`\~} CF_2-S-C\underset{S}{\overset{O}{\diagup}} \ominus H^\oplus + HF$$

$$\downarrow -COS$$

$$\overset{\frown}{F}-CF_2-\overset{\frown}{S}{}^\ominus + CF_2=S \longleftarrow CF_3S \text{\small\char`\~\char`\~} CF_2-\overset{\frown}{S}-CF_2-\overset{\frown}{S}{}^\ominus H^\oplus$$

$$\downarrow$$

$$CF_2=S + F^\ominus$$

Amine degradation, which occurs very rapidly in the case of poly(thiocarbonyl fluoride), is believed to follow a similar course:

$$CF_3S \text{\small\sim\sim} CF_2SCF=S \xrightarrow{R_3N} \left[CF_3S \text{\small\sim\sim} CF_2S\overset{\overset{\displaystyle S}{\|}}{C}-N^{\oplus}R_3 \right] F^{\ominus}$$

$$FCF_2 \enspace \overset{\frown}{S} \text{\small\sim\sim} CF_2 \overset{\frown}{-} S - CF_2 \overset{\frown}{-} S^{\ominus} \qquad\qquad \Big\downarrow H_2O$$

$$\overset{\nwarrow\; -COS}{}$$

$$CF_3S \text{\small\sim\sim} CF_2 \overset{\frown}{-} S - CF_2 \overset{\frown}{-} S - C \overset{\overset{\displaystyle S}{\diagup}}{\underset{\underset{\displaystyle O}{\diagup}}{\raisebox{0pt}{$\cdot$}\ominus}} \qquad R_3\overset{\oplus}{N}H + HF$$

The hypothesis that $-S-CF=S$ (and $-SCF=O$) ends are the weak part of the molecule is supported by the behavior of $CF_3S \displaystyle(CF_2S)_x CF_3$, which is stable at 300° in the absence of air and is not degraded by amines. The improved stability of the $CF_2=S/HCFClCF=S$ copolymer may be related to the presence of blocks in the chain that degrade to give dithiocarbamate ends, which are more stable than thioacid fluorides.

$$CF_3S \text{\small\sim\sim} S\overset{\overset{\displaystyle H-\overset{\displaystyle F}{\overset{|}{C}}-Cl}{|}}{\underset{\underset{\displaystyle F}{|}}{C}}=S^{\ominus} \longrightarrow CF_3S \text{\small\sim\sim} S - \overset{\overset{\displaystyle H-\overset{\displaystyle F}{\overset{|}{C}}-Cl}{|}}{C}=S \;\; + F^{\ominus}$$

IV. FREE-RADICAL POLYMERIZATION

A. Introduction

Probably the most surprising chemical property of the $C=S$ group is its ability to undergo free-radical addition polymerization (15). This has been studied using $CF_2=S$ and $CFCl=S$ as members of the fluorothiocarbonyl compound class.

Also surprising was the discovery that thiocarbonyl fluoride and thiocarbonyl chlorofluoride readily copolymerize with vinyl monomers in free-radical systems. This has led to the preparation of a large number of new materials, many of which are unusual elastomers.

B. Free-Radical Initiators

Almost any source of free radicals is effective in promoting the polymerization of $CFCl=S$ or $CF_2=S$. Benzoyl peroxide and ultraviolet irradiation have been used. However, these free-radical polymerizations proceed best at low temperatures. For this reason, the R_3B/O_2 redox couple, which generates radicals at low temperature, is the most satisfactory initiator that has been uncovered to date. Although this couple generates radicals at temperatures as low as $-100°$, $CF_2=S$ polymerizations are usually carried out at $-80°$. In such polymerizations, $CF_2=S$ and $CFCl=S$ are converted to high-molecular-weight products quickly and smoothly.

The reaction of R_3B and O_2 at $-80°$ takes place in two stages. The first results in a dialkyl(alkylperoxy)borane:

$$R_3B + O_2 \longrightarrow R_2BOOR$$

In the presence of excess R_3B a second reaction occurs:

$$R_2BOOR + 2R'_3B \longrightarrow R_2BOBR'_2 + R'_2BOR + 2R'\cdot$$

Radicals generated by the second reaction are responsible for initiation of polymerization. Owing to the preceding stoichiometry, it is necessary to use more R_3B than O_2, a mole ratio of at least 2:1 usually being preferred.

Rate studies on diethyl(ethylperoxy)borane and dibutyl(butylperoxy)-borane have established that the half-lives of these compounds at $-80°$ is appreciably longer than one week. Accordingly, it is most convenient to prepare a standard solution of the peroxyborane in a hydrocarbon solvent, which is then kept cold in a solid carbon dioxide bath. Aliquots are then taken as needed and combined with R_3B to initiate polymerizations.

C. Copolymers with Unsaturated Compounds

$CF_2=S$ copolymerizes with a very large number of compounds that have carbon-carbon unsaturation. These include olefins, vinyl halides, vinyl esters, acrylates, vinyl ethers, and even allyl derivatives.

Olefins that copolymerize with $CF_2=S$ include ethylene, propylene, and isobutylene. Also included are such olefins as 2-butene, cyclohexene, and tetramethylethylene, which are usually unreactive in polymerizations because they are sterically hindered. As might be expected, nonconjugated dienes lead to cross-linked products. Curiously, conjugated dienes not only do not copolymerize but also inhibit free-radical polymerization of $CF_2=S$.

Propylene copolymerization is a special case. $CF_2=S$/propylene mixtures lead to copolymers containing approximately two molecules of $CF_2=S$ for each propylene. These copolymers have $\overline{M}_n$ as high as 820,000. By use of a large excess of $CF_2=S$ it is possible to increase the $CF_2=S$ content of the copolymer. Use of a large excess of propylene, however, does not increase propylene content beyond the 2:1–2.3:1 $CF_2=S$/propylene mole ratio given by 1:1 mixtures.

This 2:1 copolymerization of $CF_2=S$ and propylene has not been satisfactorily explained and is a feature not shared by other comonomers. For example, such compounds as t-butylethylene and vinyl acetate form copolymers with compositions near that of the monomer charge.

Vinyl acetate appears to copolymerize with $CF_2=S$ in a fashion to give acetate and sulfur on the same carbon. This is thought to be because hydrolysis leads to production of a low molecular weight oil. It has been proposed that chain scission occurs as follows:

$$\sim CH_2-\overset{\overset{\displaystyle OAc}{|}}{CH}-S-CF_2\sim \longrightarrow \sim CH_2-\overset{\overset{\displaystyle OH}{|}}{CH}-S-CF_2\sim$$

$$\downarrow$$

$$\sim CH_2\overset{\overset{\displaystyle O}{\parallel}}{CH} + HSCF_2\sim$$

$$\downarrow$$

$$HF + S=CF\sim$$

This is surprising since the copolymerization appears to require addition of a free-radical to a sulfur atom:

$$S=CF_2 \xrightarrow{R\cdot} RSCF_2\cdot \xrightarrow{CH_2=\overset{\overset{\displaystyle OAc}{|}}{CH}} RSCF_2CH_2-\overset{\overset{\displaystyle OAc}{|}}{CH}\cdot$$

$$CF_2=S\downarrow \qquad \overset{\overset{\displaystyle OAc}{|}}{}$$

$$RCF_2SCH_2-CHSCF_2\sim$$

The expected mode of polymerization based on energetics and analogy with mercaptan addition to double bonds is

$$RCF_2S\cdot \xrightarrow{CH_2=\overset{\overset{\displaystyle OAc}{|}}{CH}} RCF_2SCH_2\overset{\overset{\displaystyle OAc}{|}}{CH}\cdot \xrightarrow{CF_2=S} RCF_2SCH_2\overset{\overset{\displaystyle OAc}{|}}{CH}CF_2S\sim$$

Perhaps this is a case of abnormal addition or perhaps high molecular weight is built up by combinations of relatively low-molecular-weight chains.

$$R(CF_2SCH_2\overset{\overset{\displaystyle OAc}{|}}{CH})_xCF_2S\cdot + \cdot\overset{\overset{\displaystyle OAc}{|}}{CH}CH_2(SCF_2\overset{\overset{\displaystyle OAc}{|}}{CH}CH_2)_ySCF_2R$$

$$\downarrow$$

$$R(CF_2SCH_2\overset{\overset{\displaystyle OAc}{|}}{CH})_xCF_2S\overset{\overset{\displaystyle OAc}{|}}{CH}CH_2(SCF_2\overset{\overset{\displaystyle OAc}{|}}{CH}CH_2)_ySCF_2R$$

Insulation of the acetate group from the chain by interposing methylene groups gives copolymers that are not degraded by hydrolysis. $CF_2=S/3$-butenyl acetate copolymers of $\overline{M}_n$ 800,000 have been hydrolyzed without degradation.

Free-radical copolymerization of $CFCl=S$ with vinyl compounds is complicated by the rapid polymerization rate of this thiocarbonyl. Copolymers are readily obtained, but usually the products contain 90% or more $CFCl=S$. By use of an excess of methyl acrylate, $CFCl=S$/methyl acrylate copolymers containing 70% of the thiocarbonyl have been prepared.

By matching CFCl=S with a monomer that has a very rapid polymerization rate it has been possible to prepare products with compositions approximating those of the monomer mixture. For example, an 80:20 CFCl=S/2,-3-dichloro-1,3-butadiene gave a product containing 75% CFCl=S and a 20:80 mixture gave a copolymer containing 17% CFCl=S.

V. POLYMER PROPERTIES

A brief discussion of the properties of poly(thiocarbonyl fluoride) may be of some interest. This polymer can be pressed into films or molded into shapes. It is of such high molecular weight that it is viscous and flows slowly. As a consequence it is kept in a mold under pressure at 100–150° for several hours to obtain shapes with good surfaces. As prepared, these films and moldings are elastomeric because poly(thiocarbonyl fluoride) is a highly resilient rubber in the amorphous state. However, at room temperature poly(thiocarbonyl fluoride) gradually crystallizes. Its crystalline melting point is 35°. In the crystalline state the polymer is opaque and nonelastic.

Rubbery films at room temperature can be oriented by cold drawing, which is accompanied by crystallization. Oriented crystalline films have a tensile strength of about 11,000 psi and an elongation of about 90%. An interesting property of poly(thiocarbonyl fluoride) is an exceptionally low (−118°) glass transition temperature.

The chemical resistance of poly(thiocarbonyl fluoride) is good. It suffers no apparent damage upon brief exposure to boiling fuming nitric acid. It is degraded upon long exposure as illustrated by an inherent viscosity decrease from 3.45 to 0.54 over a 21-hr period. The polymer also resists boiling aqueous sodium hydroxide. No change in inherent viscosity occurs upon boiling 21 hr in 10% NaOH, although the polymer loses about 9% of its weight. As discussed earlier, the polymer is rapidly degraded by amines.

Inasmuch as poly(thiocarbonyl fluoride) is an elastomer, some attention has been given to the possibility of curing it. By mixing in divinylbenzene and benzoyl peroxide a composition has been obtained that can be cured in a positive-pressure mold at 100°. Cured polymer has very good resistance to abrasion and compression set and has good strength and elongation. Its resilience is reduced somewhat over that of virgin polymer and it retains the disadvantages of uncured polymer, which are slow crystallization below 35°, thermal decomposition above 175°, and sensitivity to amines.

The loss of elastomeric properties at room temperature can be overcome by use of copolymers. For example, the CF_2=S/allyl chloroformate copolymers have received attention because they are easily cured. Such copolymers having as little as 2–3 mole % of allyl chloroformate melt well below 0° and have high resilience. They can be cured by incorporation of 2–5% zinc oxide followed by heating to 100° under pressure in a mold for an hour. These compositions do not harden upon storage and retain most of the good properties of poly(thiocarbonyl fluoride).

References

1. W. Sundermeyer and W. Meise, *Z. anorg. allgem. Chem.*, **317**:334 (1962).
2. N. N. Yarovenko and A. S. Vasil'eva, *J. Gen. Chem.* (*USSR*), **29**:3754 (1959). Eng. Transl.
3. A. J. Downs and E. A. V. Ebsworth, *J. Chem. Soc.*, **1960**:3516.
4. W. J. Middleton, E. G. Howard, and W. H. Sharkey, *J. Org. Chem.*, **30**:1375 (1965).
5. D. M. Marquis, U.S. Patent 2,962,529 (September 29, 1960).
6. S. Proskow, U.S. Patent 3,026,304 (March 20, 1962).
7. K. V. Martin, U.S. Patent 3,048,629 (1962).
8. P. E. Aldrich, E. G. Howard, W. J. Linn, W. J. Middleton, and W. H. Sharkey, *J. Org. Chem.*, **28**:184 (1963).
9. J. F. Harris, Jr., and F. W. Stacey, *J. Am. Chem. Soc.*, **85**:749 (1963).
10. T. J. Kealy, U.S. Patent 3,069,397 (December 18, 1962).
11. J. F. Harris, Jr., *J. Org. Chem.*, **30**:2190 (1965).
12. J. F. Harris, Jr., U.S. Patent 3,047,545 (July 31, 1962).
13. B. C. Anderson, *Polymer Letters*, **4**:283 (1966).
14. W. J. Middleton, H. W. Jacobson, R. E. Putnam, H. C. Walter, D. G. Pye, and W. H. Sharkey, *J. Polymer Sci.*, **A3**:4115 (1965).
15. A. L. Barney, J. M. Bruce, Jr., J. N. Coker, H. W. Jacobson, and W. H. Sharkey, *J. Polymer Sci.*, **A1 4**:2617 (1966).

9. PERFLUOROALKYLENETRIAZINE ELASTOMERS AND RELATED HETEROCYCLIC POLYMERS

JOHN A. YOUNG, *Denver Research Institute, University of Denver, Colorado*

I. INTRODUCTION

Perfluoroalkylenetriazine elastomers are structurally composed of *s*-triazine rings linked by two perfluoroalkylene bridges with a perfluoroalkyl group as the third substituent on the ring, as follows:

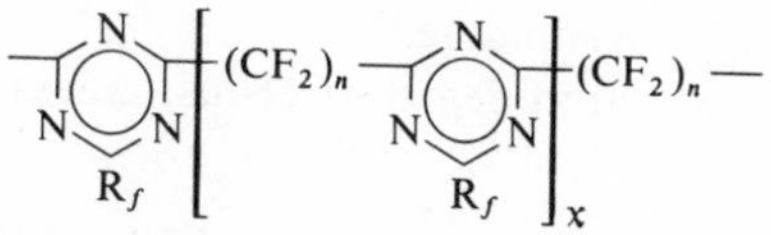

Organic polymers containing a repeating triazine structure, expecially the thermosetting melamine-formaldehyde type, have been commercially valuable for many years. Linear triazine polymers have never received comparable attention, however, and the perfluoroalkylene triazine elastomers represent the first real structural innovation fluorine chemistry has contributed to the field of polymers. In this respect they antedate perfluoro nitroso polymers by several years, although these polymers have already made the transition from research laboratory to pilot plant.

From the very outset, perfluoroalkylenetriazines have promised clear superiority over other known elastomers in thermal and oxidative stability. Their reduction to practice, however, has been frustratingly slow both because of the complex nature of the chemical system originally used for polymerization and because of the nonavailability of desirable synthetic intermediates. These difficulties must be overcome if the potential value of triazine elastomers is to be realized and they are to become commercial materials.

This chapter reviews the chronological development of the perfluoroalkylenetriazine polymers, discusses the various methods used for their synthesis, and examines some of the difficulties involved. Where data are available, polymer properties are presented and compared with those of other fluoropolymers. Finally, a section describing preliminary results on other perfluoro heterocyclic polymers is included.

A few words should be said about terminology. If the most recently proposed system of polymer nomenclature (1) is followed, the correct name for the polymer illustrated above is "poly(2-perfluoroalkylene-4-perfluoroalkyl-s-triazin-6-yl)." In the past the structure has been referred to variously as "poly(perfluoroalkylene triazine)," "poly(perfluoroalkyl triazine)," or simply "polytriazine" with the presence of perfluoroalkylene groups tacitly understood. It is called here "poly(perfluoroalkylenetriazine)," omitting the locants and the third substituent on the ring except under special circumstances.

The polymers to be described contain no hydrogen and no halogen other than fluorine, except in end groups, and can unequivocally be called "perfluoro." Many intermediates or related compounds, however, contain both hydrogen and halogen and the rather noncommittal term "fluoroalkyl" is used here to include such types along with perfluoro compounds, as long as relatively few atoms of either hydrogen or halogen are present.

Purely for the sake of convenience, the name "N'-(perfluoroacylimidoyl)-perfluoroalkylamidine" for the structure $R_fC(NH_2)=N—C(=NH)R_f$ is here abbreviated to "imidoylamidine."

All temperatures are given in degrees Celsius, unless otherwise noted.

II. SYNTHESIS OF FLUOROALKYL-s-TRIAZINES

A great deal of information is to be found in the literature concerning the synthesis of s-triazine derivatives with alkyl or functional substituent groups on the ring, the best single source being the book on s-triazines by Smolin and Rappaport (2). The following discussion will be limited to synthetic methods useful in the preparation of fluoroalkyltriazines, or their precursors, which can be converted to polymers by presently known techniques.

The simplest synthesis of a tris(fluoroalkylkyl)triazine is the trimerization of a fluoroalkyl nitrile, which produces a symmetrical molecule containing three identical R_n groups:

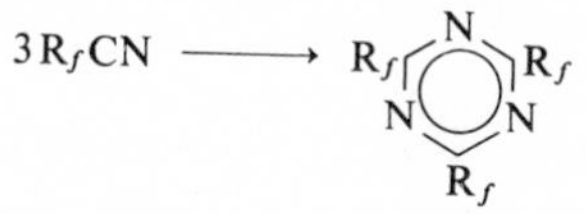

$$3 R_f CN \longrightarrow R_f \underset{N \diagdown N}{\overset{N}{\bigcirc}} R_f$$
$$R_f$$

This reaction requires rather rigorous conditions, giving conversions of only 8–30 % after 30–120 hr at 300–350° (3), but it is fortunately amenable to catalysis by both acids and bases. Among the additives which have been found most effective are HCl, HCl/AlCl₃. AgF, AgF₂, Ag₂O, (C₆H₅)₄Sn, and amines (4–6). A 54 % conversion to triazine was obtained with C₇F₁₅CN after 17 hr at 100° in the presence of 5 wt % of silver oxide (7). Table 1 lists triazines which have been made by nitrile trimerization.

TABLE 1

s-Triazines Made by Nitrile Trimerization

Nitrile	B.P.	*s*-Triazine conversion (%)	Reference
CCl_3CN	m. 90–92	92	4
CF_3CN	96	25	3
		72	8
C_2F_5CN	121	58	3
$n-C_3F_7CN$	165	10	3
$n-C_7F_{15}CN$		54	7

Symmetrically substituted triazines can also be made directly from fluoro-olefins by reaction with ammonia; the α-haloamines first formed lose hydrogen halide to give nitriles, which then trimerize to triazines (9). If this method is used, however, the final fluoroalkyl group retains an α-hydrogen atom. Triazines made by this method include $(CHFBrCN)_3$, $(CHFClCN)_3$, $(CHF_2CN)_3$, and $(CH_2FCN)_3$ (9).

A milder and more generally useful preparation of fluoroalkyl-substituted triazines is afforded by the condensation of amidines (3). The amidines are easily prepared by the addition of ammonia to fluoroalkyl nitriles and the condensation is carried out merely by heating them above their melting points for several hours, ammonia being liberated as cyclization occurs:

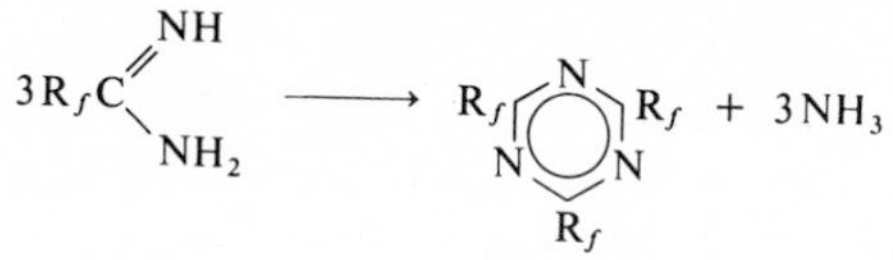

$$3 R_f C \underset{NH_2}{\overset{NH}{\diagup}} \longrightarrow R_f \underset{N \diagdown N}{\overset{N}{\bigcirc}} R_f + 3 NH_3$$
$$R_f$$

Tris(perfluoroethyl) and tris(perfluoropropyl) triazines have been made in 35 and 64% yields, respectively, by this method (3). When the amidine of O_2NCF_2CN was heated, however, HCF_2NO_2 was the only volatile product (10).

Thermal cyclization of amidines takes place in steps (11), the first being elimination of 1 mole of ammonia from 2 moles of amidine to form an intermediate imidoylamidine. The imidoylamidine can also be obtained by reaction of an amidine with a nitrile, or by reaction of a nitrile (2 moles) with ammonia (1 mole).

$$
\begin{array}{c}
2\ R_fC\!\!\begin{array}{l}\nwarrow NH\\[2pt] \searrow NH_2\end{array} \quad -NH_3 \\[10pt]
R_fC\!\!\begin{array}{l}\nwarrow NH\\[2pt] \searrow NH_2\end{array} \ +\ R_fCN \\[10pt]
2\ R_fCN + NH_3
\end{array}
\longrightarrow
R_fC\!\!\begin{array}{l} NH_2 \\ N\!-\!C\!\!\begin{array}{l} NH\\ R_f\end{array}\end{array}
$$

If heating of the amidine is continued, the imidoylamidine formed condenses with itself or with the original amidine to eliminate ammonia and close the triazine ring:

$$
R_fC\!\!\begin{array}{l} NH_2 \\ N\!-\!C\!\!\begin{array}{l} NH\\ R_f\end{array}\end{array}
\ +\ R_fC\!\!\begin{array}{l}\nwarrow NH\\[2pt] \searrow NH_2\end{array}
\longrightarrow
\underset{R_f}{\overset{R_f\ \ \ R_f}{\bigcirc}}
$$

Conversion of an imidoylamidine to a triazine can also be effected by reacting it with a nitrile (12).

All of these reactions involving ammonia and a carbon-nitrogen multiple bond are at least partially reversible, as has been shown by scrambling reactions (see below). It can therefore be seen that a mixture which initially contains only a fluoroalkyl nitrile and ammonia can give rise to a very complex system involving ammonia, nitrile, amidine, imidoylamidine, and triazine, whose composition at a given moment is a function of elapsed time, relative concentrations, temperature, and pressure.

It is apparent from the steps shown that it should be possible to vary the nature of the R_f substituent groups in a triazine merely by choosing different nitriles in three sequential reactions—with ammonia, with the resulting amidine, and with the resulting imidoylamidine—and so to synthesize not only triazines with all groups alike but also those with two like groups and one unlike or those with three unlike groups. However, difficulties arise when one attempts to synthesize an asymmetrically substituted triazine by this procedure. The addition of ammonia or an amidine to a nitrile proceeds readily at room temperature or below, but quantitative elimination of ammonia during cyclization to a triazine requires heating. At the higher temperatures thus encountered, the liberated ammonia catalyzes reorganization reactions and a random mixture of all possible triazines, rather than a single species, is obtained (12):

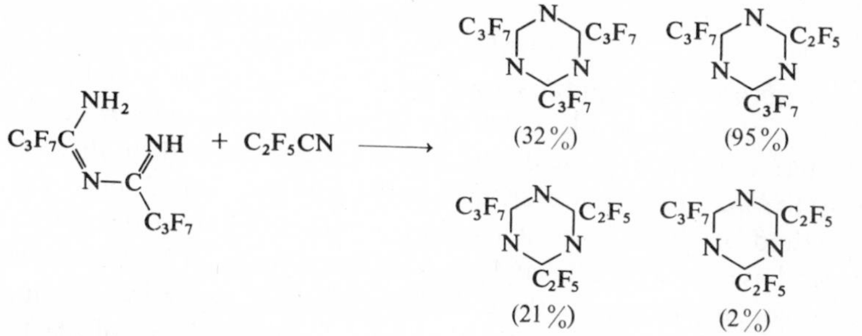

Since the electronegativity of $CF_3(CF_2)_n$ does not vary greatly with the magnitude of n, the resulting product distribution is controlled by relative abundances of the several R_f moieties present. The same type of scrambling can be observed when a mixture of two different symmetrically substituted triazines, $(R_fCN)_3$ and $(R_f'CN)_3$, is heated with ammonia or some other base of comparable strength; the product is a random mixture of all possible triazines.

Fortunately, this complication can be avoided. Reaction of the silver salt of an imidoylamidine with a perfluoroacyl halide was found to give a single triazine in high purity, and it was later found that preliminary conversion to a metal derivative was unnecessary as the acyl halide could be used directly with the imidoylamidine (12).

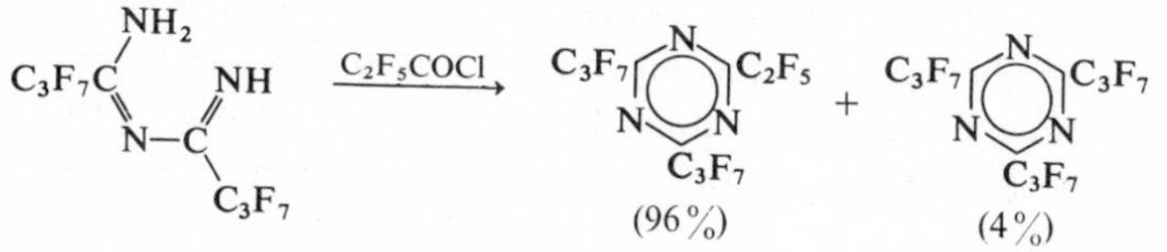

Acyl halides are somewhat undesirable as cyclization agents, however, since amine salts of one type or another are unavoidably formed, resulting in low yields and difficult separation of products; in the last reaction shown, for instance, conversion of imidoylamidine to triazine was only 50% of theory (12). The further discovery that an acid anhydride could be substituted for the acyl halide (12) constituted a great improvement, since when an ether solution of an imidoylamidine is slowly added to a large excess of well-stirred anhydride at a temperature near 0°, cyclization to a single triazine is virtually quantitative. The cyclization proceeds by initial acylation of the midoylamidine and subsequent dehydration of this intermediate by a second molecule of anhydride, resulting in ring closure.

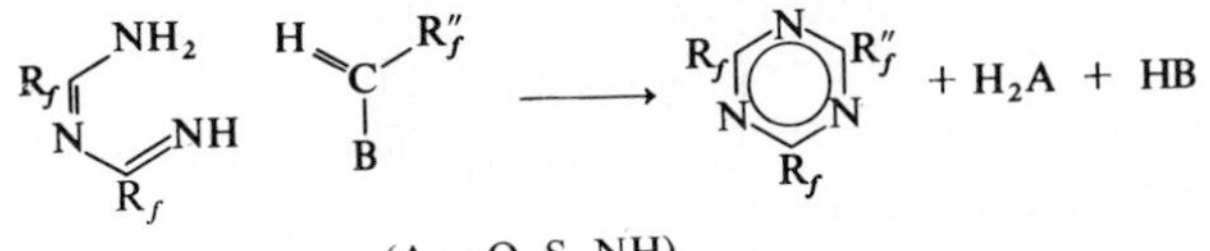

$$(A = O, S, NH)$$
$$(B = Cl, NH_2, R_fCOO)$$

As the preceding diagram shows, reaction of an imidoylamidine with an amidine or with a nitrile eliminates ammonia, that with an acid halide eliminates water and an acid, etc., triazine being formed in all cases. For preparative purposes, however, cyclization using an anhydride is definitely the method of choice since it requires no gas handling, causes no loss of nitrogen bases as acid salts, and gives excellent yields. It is prudent, when a triazine containing two or three different fluoroalkyl groups is being synthesized, to introduce the cheapest or most readily available of these in the anhydride form, since efficient cyclization requires a large excess of anhydride, whereas amidine and imidoylamidine formation require only stoichimetric amounts of reactants. For the preparation of perfluoro-[2-methyl-4-propyl-6-heptyl]-s-triazine, for instance, one of the two longer-chain nitriles should be converted to an amidine, this added to the second longer-chain nitrile, and the resulting imidoylamidine cyclodehydrated with trifluoroacetic anhydride. When a rare or expensive anhydride must be used for cyclization, it is possible to use just enough of it to insure acylation of the imidoylamidine and then to cyclize this intermediate (without isolation) by an excess of cheaper trifluoroacetic anhydride. The main product will be the desired triazine, although some of the trifluoromethyl triazine will invariably be present and rectification of the mixture is difficult even by gas-phase chromatography (13).

This route to fluoroalkyl s-triazines affords excellent overall yields from available or easily accessible starting materials, with few restrictions on structure. The following figures illustrate the four-step synthesis of a rather bulky triazine molecule with an overall yield, based on the orginal ester, of 88% (14):

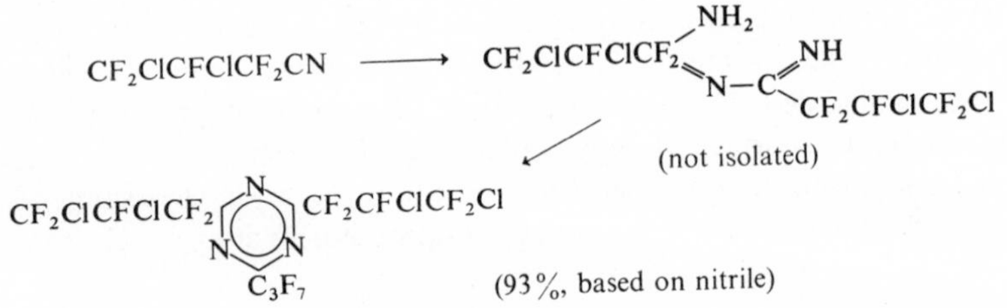

Table 2 lists some triazines which have been made by acylation-cyclo-dehydration of imidoylamidines.

TABLE 2

s-Triazines Made by Acylation-Cyclodehydration of Imidoylamidines

Triazine			Cyclization agent	Yield (%)	Reference
2	4	6			
C_3F_7	C_3F_7	C_2F_5	C_2F_5COCl	51	12
C_3F_7	C_3F_7	C_2F_5	$(C_2F_5CO)_2O$	91	12
C_7F_{15}	C_3F_7	C_3F_7	$(C_3F_7CO)_2O$	80	12
C_3F_7	C_3F_7	CF_3	$(CF_3CO)_2O$	80	12
C_3F_7	C_3F_7	CF_2Cl	$(C_3F_7CO)_2O$	70	12
C_3F_7	C_3F_7	CH_3	$(CH_3CO)_2O$		12
C_2F_5	C_2F_5	CH_3	$(CH_3CO)_2O$		12
CF_2Br	CF_3	CF_3	$(CF_3CO)_2O$		8
CF_2Br	CF_2Br	CF_3	$(CF_3CO)_2O$	71	8
CF_2Br	CF_2Br	CF_2Br	$(CF_2BrCO)_2O$		8
CF_2Br	CF_2Br	C_7F_{15}	$(C_7F_{15}CO)_2O$	low	8
$CF_2ClCFClCF_2$	$CF_2ClCFClCF_2$	C_3F_7	$(C_3F_7CO)_2O$	93	14
$CF_2ClCFClCF_2$	$CF_2ClCFClCF_2$	CF_3	$(CF_3CO)_2O$	80	14

One other method of synthesis of fluoroalkyl triazines offers some degree of flexibility, though it is greatly inferior in this respect to imidoylamidine acylation-cyclodehydration. The remarkable new field of fluoroanion chemistry (for a review of recent developments in this area, see reference 15) includes many reactions in which a perfluoroalkyl carbanion displaces a halogen atom via an S_N2 mechanism. Halogens attached directly to carbon in a triazine ring are quite labile and readily undergo such reactions; fluoroolefins which are capable of forming stable fluorocarbanions will therefore alkylate a halo-triazine according to the following reaction (16):

$$R_fCF{=}CF_2 + F^- \longrightarrow R_f\underset{\underset{CF_3}{|}}{C}F^-$$

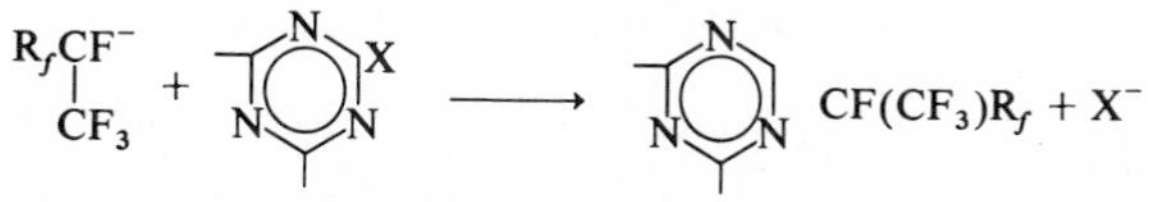

$$R_f\underset{\underset{CF_3}{|}}{C}F^- + \text{(triazine)}X \longrightarrow \text{(triazine)}CF(CF_3)R_f + X^-$$

Cyanuric fluoride, 2,4,6-trifluoro-*s*-triazine, undergoes stepwise replacement of nuclear fluorine to form the mono-, di-, and trialkylated products, the extent of replacement depending on reaction conditions and reactant ratios. Perfluoroazomethines react similarly, giving perfluoro(dialkylamino)triazines (16):

$$F \underset{N \underset{F}{\bigcirc} N}{\overset{N}{\bigcirc}} F \xrightarrow[CsF]{CF_3N=CF_2} F \underset{N \underset{N(CF_3)_2}{\bigcirc} N}{\overset{N}{\bigcirc}} F + \underset{N \underset{N(CF_3)_2}{\bigcirc} N}{\overset{N}{\bigcirc}} N(CF_3)_2 + (CF_3)_2N \underset{N \underset{N(CF_3)_2}{\bigcirc} N}{\overset{N}{\bigcirc}} N(CF_3)_2$$

Analogously, perfluoroalkoxides would yield completely fluorinated cyanurates (alkoxytriazines). It has been reported (17) that this reaction succeeds with cyanuric chloride but not with cyanuric fluoride, as fluoride ion reverses the reaction and rapidly decomposes the cyanurate if added to a pure sample.

The alkylation reaction gives good conversions and is convenient to carry out, but it is subject to all the structural restrictions common to fluorocarbanion reactions. A chain branch at the α-carbon atom of the R_f substituent is apparently unavoidable except with two-carbon olefins, since carbanion formation proceeds by attack of fluoride ion at the terminal CF_2 group of an olefin and linear perfluorocarbanions are consequently not observed when the olefin chain exceeds two carbons. Furthermore, the formation of fluorocarbanions from perfluoro α-olefins suffers from the competing reaction of double-bond migration, which usually predominates, and a much less reactive internal olefin results. Reaction of perfluoro-1,3-butadiene with cyanuric fluoride, for instance, yielded only the migration product, $CF_3C=CCF_3$, and no alkylation was observed. With perfluoroallyl cyanide, however, the desired alkylation product was obtained in 62% yield (16).

III. REACTIONS LEADING TO POLYMERIC FLUOROALKYLENETRIAZINES

Polymerization methods for the production of poly(perfluoroalkylenetriazine)s can be divided into three classes, in which triazine ring formation (1) precedes chain extension, (2) occurs simultaneously, or (3) follows chain extension. At present none of the three is entirely satisfactory; consequently the material properties exhibited by current polymer specimens remain rather distant from the optimum values to be expected from this polymer system.

A. Chain Extension by Cyclization

This method has a history almost as long as that of fluoroalkyl triazines themselves, The attainment of a truly linear polymer of high molecular weight is dependent upon the efficiency of the cyclization reaction, and polymer improvement has gone hand in hand with successive improvements in ring closure techniques.

The earliest triazine synthesis, nitrile trimerization, revealed the excellent thermal and oxidative stability of fluoroalkyl triazines and aroused the interest of Henry C. Brown in developing triazine polymers exploiting these properties, but the reaction itself was not attractive for polymer formation since conversions were low and experimental conditions very severe. Brown's first perfluoroalkylene triazine polymers consequently were made by thermal cyclization of bisamidines (11).

As described previously, amidines when heated lose ammonia and condense to form triazines. Bisamidines made from perfluoroalkylene dinitriles, and also the cyclic imidine resulting from the addition of ammonia to perfluoroglutaronitrile, were found to undergo similar condensation to give polymers in which triazine rings were linked by perfluoroalkylene bridges. These materials were inert to oxidizing agents and displayed excellent thermal stability, being stable apparently indefinitely at 359° (650°F), but they were insoluble and infusible because of their highly cross-linked structure. Since the functionality of a triazine derivative formed by condensation of three difunctional molecules would be 3, an idealized structure would show a cross-link at every ring:

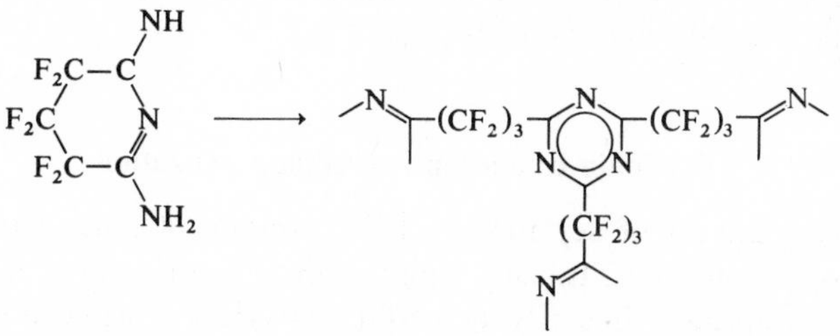

It is apparent that inclusion of a monofunctional amidine in the reaction mixture would decrease the number of cross-links, and polymers obtained in this way by Brown (11) did exhibit elastomeric behavior at no cost in thermal stability; however, random interaction of a mixture of mono- and difunctional amidines does not afford an effective pathway to high molecular weight and is not capable of generating a highly ordered, purely linear polymer structure. The four possible modes of condensation involving three amidine molecules in such a mixture are (1) three bisamidine molecules, (2) two bisamidines and one monoamidine, (3) one bisamidine and two monoamidines, and (4) three monoamidines. Of these only the second would lead to linearity; the first would result in a network; (3) would act as a chain stopper; and (4) would give only a discrete triazine molecule with a single ring.

Elucidation of the formation and structure of imidoylamidines (18) provided what seemed to be a much more elegant way of attaining polymer linearity. Hypothetically, a dinitrile could be reacted with a monoamidine to form a cyano-terminated imidoylamidine, and this difunctional monomer would then self-condense to give a linear polymer:

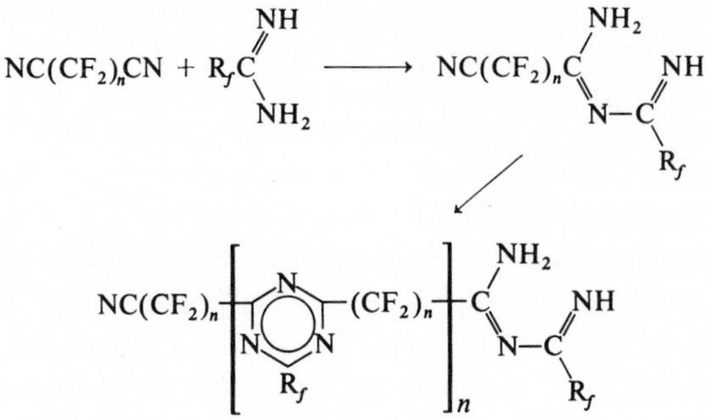

$$NC(CF_2)_nCN + R_fC\begin{matrix} NH \\ \\ NH_2 \end{matrix} \longrightarrow NC(CF_2)_nC\begin{matrix} NH_2 \\ \\ N-C \end{matrix}\begin{matrix} \\ NH \\ \\ R_f \end{matrix}$$

$$\left[NC(CF_2)_n \underset{N \underset{R_f}{\nearrow} N}{\overset{N}{\bigcirc}} (CF_2)_n \right]_n C \begin{matrix} NH_2 \\ \\ N-C \end{matrix} \begin{matrix} \\ NH \\ \\ R_f \end{matrix}$$

This scheme fails because the ammonia liberated during cyclization causes reorganization reactions of the type previously described, randomizing the assimilation of mono- and difunctional compounds into triazine rings, altering the functionality of the reaction system, and leading to undesired cross-linking and shortened chains (11).

B. Chain Extension Preceding Cyclization

The recent advent of acylation-cyclodehydration of imidoylamidines for the purpose of forming triazine rings caused another surge of optimism concerning improved linearity in perfluoroalkylene triazine polymers. As noted previously, ring closure of monofunctional imidoylamidines with perfluoro acid anhydrides gives excellent yields of specific products. The application of this reaction to polymeric systems is presently under intensive investigation by the Hooker Chemical Corporation (6), and materials made by this route represent the closest approaches to practicable elastomers.

The imidoylamidine acylation-dehydration reaction cannot be used effectively for simultaneous cyclization and chain extension since the cyclic or polymeric anhydride necessary would be very apt to form a carboxy-terminated molecule, reducing chain length as shown below:

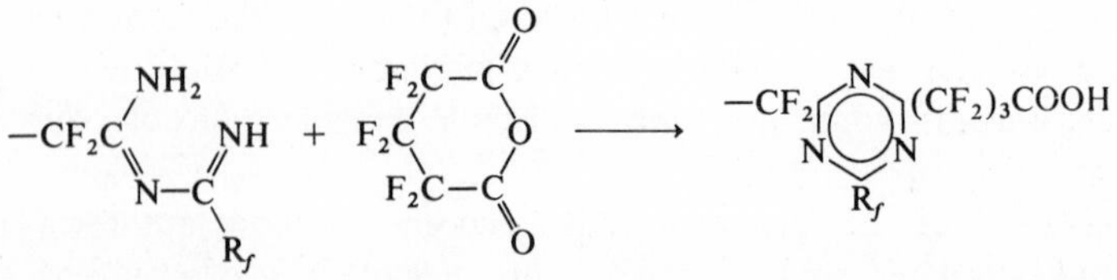

$$-CF_2C\begin{matrix} NH_2 \\ \\ N-C \\ \\ R_f \end{matrix}\begin{matrix} \\ NH \\ \\ \end{matrix} + \begin{matrix} F_2C-C \\ \\ F_2C \\ \\ F_2C-C \end{matrix}\begin{matrix} O \\ \\ O \\ \\ O \end{matrix} \longrightarrow -CF_2\underset{N\underset{R_f}{\searrow}N}{\overset{N}{\bigcirc}}(CF_2)_3COOH$$

A better way, and the one whereby present polymers are made, is to synthesize a poly(imidoylamidine) and then close the rings along the chain with the anhydride of a monofunctional acid:

$$\left[\!\!-(CF_2)_h\,C\overset{N}{\underset{NH_2}{\diagup}}\,C\overset{}{\underset{NH}{\diagdown}}\!\!-\right]_n \quad \xrightarrow{(R_fCO)_2O} \quad \left[\!\!-(CF_2)_h\,\overset{N}{\underset{N\diagdown\diagup N}{\diagup}}\,\underset{R_f}{}\!\!-\right]_n$$

The requisite poly(imidoylamidine)s can be made by the methods previously described, using difunctional reactants. The reaction of ammonia with a dinitrile, forming amidine groups *in situ* which then react with the nitrile group of another molecule, has an advantage over the reaction of a dinitrile with a bisamidine in that it avoids the necessity of preparing and purifying the bisamidine (6). However, the first method has the limitation that all of the perfluoroalkylene bridges between rings will be of the same length, since they originate from only one species, whereas the second permits alternate bridges of two different lengths.

As might be expected, the length of the inter-ring bridges has a great effect on polymer properties, although it has little effect on thermal stability. The greater stiffness of perfluoro polymer chains, when compared with hydrocarbon chains, and the bulkiness of the triazine rings, make mandatory a chain length of at least five carbon atoms between rings for good elastomeric properties. Very few dibasic fluorocarbon acids are available, and this circumstance has always seriously inhibited research on perfluoroalkylene triazine polymers. The most common acid, perfluoroglutaric, is unsatisfactory for the system just described because of the great tendency of perfluoroglutaronitrile, when treated with ammonia, to cyclize intermolecularly and form an imidine:

$$\begin{array}{c} F_2C-C\equiv N \\ F_2C \\ F_2C-C\equiv N \end{array} \quad \xrightarrow{NH_3} \quad \begin{array}{c} F_2C-C\overset{NH}{\diagup} \\ F_2C \qquad N \\ F_2C-C \\ \qquad NH_2 \end{array}$$

The fact that a desirable dibasic fluorocarbon acid, such as perfluorosebacic, may cost as much as $1000 per pound is apt to dampen enthusiasm for this particular area of polymer chemistry.

The incorporation of sites for cross-linking in a linear triazine polymer can be accomplished by including in the anhydride used for cyclization a small amount of an acid derivative which contains, or gives rise to, two different functionalities. The use of two such types has been investigated (6); a cyclic anhydride (a polymeric anhydride might also be used) gives a carboxy-pendant polymer, and an ω-cyanoacyl halide gives a cyano-pendant polymer:

$$\left[(CF_2)_n C \overset{N}{\underset{NH_2}{\diagdown}} \overset{N}{\underset{NH}{C}}\right]_n + \overset{O}{\underset{F_2C}{\diagup}} \overset{F_2C-C}{\underset{F_2C-C}{\diagdown}} \overset{O}{\diagup} \longrightarrow \left[(CF_2)_n \overset{N}{\underset{N}{\diagdown}} \overset{N}{\underset{N}{\diagup}}\right]_n$$
$$(CF_3)_3COOH$$

$$\left[(CF_2)_n C \overset{N}{\underset{NH_2}{\diagdown}} \overset{N}{\underset{NH}{C}}\right]_n + NC(CF_2)_3COCl \longrightarrow \left[(CF_2)_n \overset{N}{\underset{N}{\diagdown}} \overset{N}{\underset{N}{\diagup}}\right]_n$$
$$(CF_2)_3CN$$

The first of these can be cured by conversion to the silver salt of the acid and subsequent thermal decarboxylation, according to the equation $2R_fCF_2COOAg \rightarrow R_fCF_2-CF_2R_f$, and the second can be cured by catalytic trimerization of the pendant nitrile groups. Both of these methods form cross-links which are identical with the main polymer structure and ideally should have equal strength and thermal stability; however, the first requires rather high temperatures (250–300°) and is not a very clean reaction, and the second, to be fully effective, requires interaction of three polymer chains rather than the two usually involved in cross-linking. Of the two methods, trimerization of pendant nitrile groups has given the higher tensile strengths (6).

C. Chain Extension Following Cyclization

Perfluoroalkylenetriazine elastomers made by the foregoing process show outstanding thermal and oxidative stability, and materials recently obtained have fairly good physical properties, but it cannot be denied that the nitrile-ammonia system is a very complex and sensitive one with which to work. It should be possible to develop methods whereby functionally active perfluoroalkyl triazines could be first synthesized and then polymerized, thus obtaining the benefits of the perfluoroalkylenetriazine structure but circumventing the difficulties associated with ring closure in the nitrile-ammonia system. Considerable work has been done along this line in the last few years.

Free-radical coupling of fluorocarbon derivatives is often nearly quantitative and uncomplicated by side reactions, since perfluoroalkyl radicals under normal circumstances do not abstract fluorine from other perfluoroalkyl chains in the way that alkyl radicals abstract hydrogen from hydrogen-containing chains. Such coupling reactions are consequently attractive for the purpose of chain extension in a polymer-forming system.

The Dow Corning Corporation is investigating the preparation of polymers from perfluoro-2,4-bis(bromomethyl)-6-alkyl-s-triazines (8,13), proceeding on the initial assumption that the bromine in a CF_2Br group, normally unreactive, would be sufficiently activated by its benzylic relationship with the triazine ring to undergo coupling. The triazine monomers are made by acylation-cyclodehydration of the requisite imidoylamidines, as described in detail previously.

$$BrCF_2CN \longrightarrow BrCF_2\overset{NH_2}{\underset{N-C}{\overset{|}{C}}}\overset{NH}{\underset{CF_2Br}{}} \xrightarrow{(R_fCO)_2O} BrCF_2\underset{N\underset{R_f}{\diagdown}N}{\diagup N} CF_2Br$$

It was found that such compounds undergo metal-induced coupling, although the extreme severity of conditions necessary for polymer formation well illustrates both the reluctant reactivity of the CF_2Br group and the remarkable stability of the perfluoroalkyl triazine structure. 2,4-Bis(trifluoro methyl)-6-bromodifluoromethyl-s-triazine was first examined, as a model compound. This gave 50% conversion to a dimer when heated with excess mercury for 6 hr at 250°. Extension of the reaction to a difunctional compound capable of polymer formation, 2,4-bis(bromodifluoromethyl)-6-trifluoro-methyl-s-triazine, gave a spectrum of products ranging from dimer through low polymers to high polymer, according to the following equation (8):

$$BrCF_2\underset{N\underset{R_f}{\diagdown}N}{\diagup N}CF_2Br \longrightarrow \left[CF_2\underset{N\underset{R_f}{\diagdown}N}{\diagup N}CF_2 \right]_n$$

It was subsequently found that a narrower molecular-weight range could be obtained in a two-step polymerization, first converting the bis(bromo-difluoromethyl(triazine at 160° to a mixture of dimers and oligomers (DP 2-8) and further polymerizing this material at 200–330°. When assisted by ultraviolet illumination, coupling occurred somewhat more easily and the dimer of the mono(bromodifluoromethyl)triazine was obtained in almost 50% conversion after 68 hr exposure at 50°, even in the absence of mercury (8).

2,4-Bis(chlorodifluoromethyl)-6-trifluoromethyl-s-triazine was much less reactive than the corresponding bromo compound, giving only a 4% yield of dimer after treatment with mercury for 16 hr at 250°. Iododifluoromethyl compounds have not been investigated because of difficulties in their syn-thesis, but the facile coupling of aliphatic perfluoroalkyl iodides, as well as some of the work to be described later, indicates that they would be far more susceptible to coupling than the bromo compounds.

Iodoperfluoroalkyltriazines have been made and coupled in a program at the Denver Research Institute (14). In this work, perfluoroallyl-substituted triazines were made by dehalogenation of the vicinal dichloroperfluoro-alkyltriazines whose synthesis was described earlier. Conventional methods of incorporating an iodine atom, such as addition of iodine monofluoride, reaction with fluoride ion and iodine, and S_N2' reaction with iodide ion, were all unsuccessful, but mercuric fluoride could be added and the resulting mercurials treated with iodine to give the desired compounds:

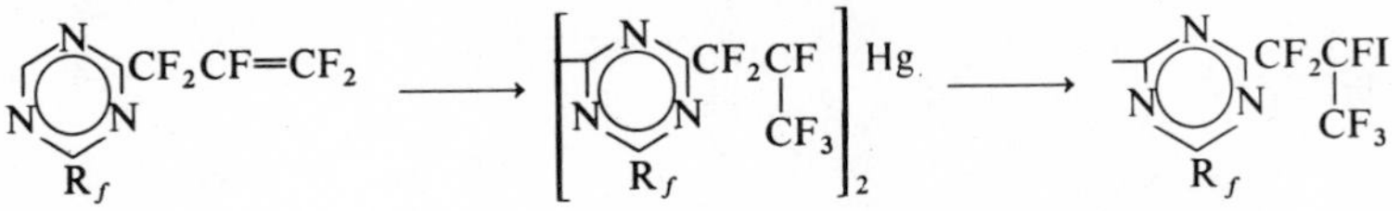

Coupling occcurred readily with mercury and ultraviolet light, and an 83% yield of dimer was obtained from a mono(2-iodoperfluoropropyl)-triazine (14). Polymerization was observed when a bis(iodoperfluoroalkyl) compound was similarly treated, and also when the latter was merely irradiated *in vacuo* and the iodine was continuously pumped away (14).

The most serious problem at present in polymer formation by coupling is the severe structure restriction placed on the perfluoroalkylene bridges between rings, either because of synthetic difficulties or because of variation in reactivity with structure. As has been noted, a bridge of at least five carbon atoms is necessary to produce a good elastomer. This bridge should be an unbranched chain for optimum thermal stability, since structural studies on other fluoropolymers have shown that branches constitute weak points in a chain. Coupling of two triazine rings by means of pseudo-benzylic halogen atoms limits a normal chain between the rings to two carbon atoms, with resultant lack of elastomeric properties in a polymer; materials prepared in this way have been leathery but not elastomeric at room temperature. If the halogen atom is removed farther from the ring, its reactivity falls off very sharply, so that β- and γ-bromoperfluoroalkyl triazines show much less tendency to couple (13). The reactivities of α-, β-, and γ-bromo compounds are compared in Table 3.

TABLE 3

Coupling of 2-R_f-4,6-bis(trifluoromethyl)-s-triazines[a] (13)

R_f	Temp. (°C)	Time (hr)	Conversion (%)
—CF_2Br	60	24	100
—$(CF_2)_2Br$	60	24	50
—$(CF_2)_4Br$	60	24	trace
	175	71	25

[a] Irradiated with ultraviolet light in the presence of mercury.

Iodoperfluoroalkyltriaznes couple much more readily than the corresponding bromo compounds, but only the secondary iodoperfluoroalkyl derivatives, which lead to branched chain coupled products, have been made, since the preparation of perfluoroalkyltriazines with iodine in a terminal position is very difficult by presently known methods.

One possible improvement in this situation lies in the use of "dumbbell" molecules with an adequately long chain between two bromodifluoro-methyl-substituted triazine rings:

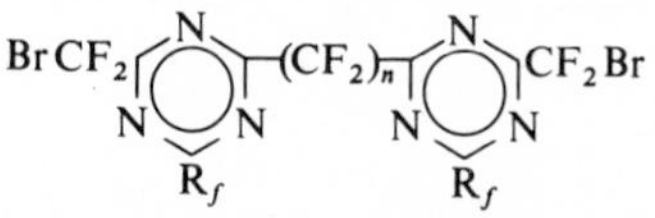

$$BrCF_2 \underset{\underset{R_f}{N \diagdown N}}{\overset{N}{\diagup}} (CF_2)_n \underset{\underset{R_f}{N \diagdown N}}{\overset{N}{\diagup}} CF_2 Br$$

Such monomers can yield elastomeric high polymers, but only one half of the bridges in the resultant polymers are of the longer type; furthermore, the synthesis of these monomers necessitates recourse to acylation-cyclo-dehydration of a bis(imidoylamidine), with its inherent problems. This variation would seem to be a compromise combining the least attractive aspects of both methods.

Perfluoroalkyl triazines other than halogen derivatives have also been converted to polymers. It has been reported (14) that both monofunctional and difunctional cyanoperfluoroalkyl triazines on treatment with silver difluoride give azo compounds which on irradiation or heating eliminate nitrogen and presumably form poly(perfluoroalkylenetriazine)s:

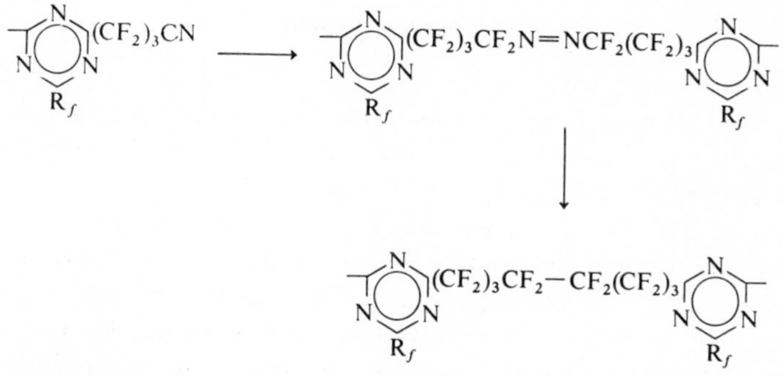

$$\underset{\underset{R_f}{N \diagdown N}}{\overset{N}{\diagup}}(CF_2)_3CN \longrightarrow \underset{\underset{R_f}{N \diagdown N}}{\overset{N}{\diagup}}(CF_2)_3CF_2N{=}NCF_2(CF_2)_3\underset{\underset{R_f}{N \diagdown N}}{\overset{N}{\diagup}}$$

$$\downarrow$$

$$\underset{\underset{R_f}{N \diagdown N}}{\overset{N}{\diagup}}(CF_2)_3CF_2{-}CF_2(CF_2)_3\underset{\underset{R_f}{N \diagdown N}}{\overset{N}{\diagup}}$$

The perfluorodiallyl triazines previously alluded to reacted when irradiated in the presence of perfluoroglutaryl fluoride to give solids and high-boiling liquids which, in the light of known reactions of this type, were probably perfluorooxetanes (14):

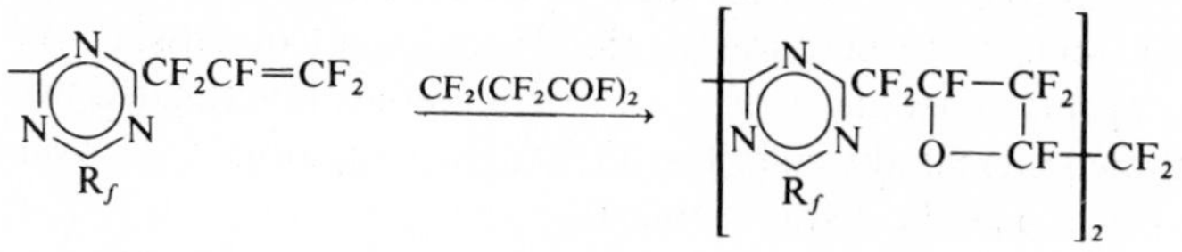

$$\underset{\underset{R_f}{N \diagdown N}}{\overset{N}{\diagup}}CF_2CF{=}CF_2 \xrightarrow{CF_2(CF_2COF)_2} \left[\underset{\underset{R_f}{N \diagdown N}}{\overset{N}{\diagup}}\underset{\underset{O - CF}{\overset{|}{}}}{\overset{CF_2CF - CF_2}{\overset{|}{}}}CF_2 \right]_2$$

Neither of these last two polymer-forming reactions has been well-documented, nor have such polymers been prepared in large enough quantity for evaluation, but the results, taken in conjunction with those on the coupling of halogen derivatives, seem to indicate a distinct possibility of preparing elastomers from functionally substituted perfluoroalkyltriazines by established reactions of fluorine chemistry. A great deal of work is still needed, however, to transform the poly(perfluoroalkylenetriazine)s from an interesting laboratory study to a practicable polymer system, regardless of the method used for chain extension.

IV. POLYMER PROPERTIES

Since the poly(perfluoroalkylenetriazine)s are very definitely an exploratory area, relatively few data are available for comparison with other materials. Those properties that depend mainly on chemical constitution, such as thermal or oxidative stability, are fairly well-documented, and the results will presumably hold true for polymer samples obtained from this system in the future as well as for present samples. Those properties that depend mainly on chain length and aggregate structure, however, such as tensile strength or percent elongation, are certain to improve with continued development of the area; consequently, figures that appear here reflect only the present state of the art and are not to be construed as optimum values.

A. Thermal Stability

As now prepared, triazine elastomer gums have a very wide molecular-weight distribution and contain appreciable amounts of light ends. A sample prepared by the imidoylamidine method, which on thermogravimetric analysis showed initial weight loss at 307°, after preliminary vacuum treatment for 30 hr at 200° lost no weight below 362° (6). Brown's early elastomers made by amidine condensation were also stable at 350° after an initial weight loss (19). It would seem that perfluoroalkylenetriazine elastomers, when practical materials become available, will have service temperatures of at least 300–350°, compared with 200° for poly(1,1-dihydroperfluorobutyl acrylate) (3M Fluororubber 1F4), 227° for $CF_2=CH_2/CF_2=CFCl$ copolymers (3M Kel-F elastomer), 250° for poly(3,3,3-trifluoropropyl methyl siloxane) (Dow-Corning Silastic LS-53), and 300° for $CF_3CF=CF_2/CF_2=CF_2$ copolymers (3M Fluorel, duPont Viton) (20).

The temperatures given here for the triazine elastomers refer to crude gums. Much work remains to be done on the development of satisfactory curing and filling agents, since most of the cured materials investigated by Hooker (6) have shown increased weight loss at 300° by a factor of 5–10 over that of the crude stock, as indicated in Table 4.

TABLE 4

Weight Loss Rate of Perfluoroalkylenetriazine Elastomers (6)

Material	Weight loss (%/hr.)		
	250°	300°	350°
Gum	.00067	.009	.09
Cured gum	.003	.04	.15
Gum + carbon		.10	
Gum + pumice		.06	

B. Chemical Stability

The resistance of the perfluoroalkylenetriazine elastomers to strong acids and oxidizing agents is excellent, approaching that of fluorocarbons themselves. The polymers obtained by coupling bis(bromodifluoromethyl) triazines with mercury at 200–300° can be freed of unreacted mercury by digesting them with aqua regia (8), and other triazine elastomers have been refluxed with fuming nitric acid for 24 hr with no effect except color removal (19). No charring was observed when samples were heated in air as high as 500° (19).

Resistance to bases is not likely to be as good, although test data are not available. The triazine ring is very easily hydrolyzed by alkali, and the resistance to alkaline hydrolysis of polymers containing this structure is probably due only to lack of access to the ring by the hydrolytic agent, because of steric protection by the fluoroalkyl chains, rather than to any fundamental increase in resistance of the ring itself. A similar situation can be found in perfluoroalkyl derivatives of SF_4; sulfur tetrafluoride itself and $(CF_3)_2CFSF_3$ are very reactive toward water, but the disubstituted compound $(CF_3)_2$-$CFSF_2CF(CF_3)_2$ can be steam-distilled with negligible loss (7). It is likely that the perfluoroalkylenetriazine elastomers will share with most other fluoropolymers a sensitivity toward compounds which are both strongly basic and strongly reducing, such as hydrazine, but will be adequately resistant toward dilute aqueous alkali at moderate temperatures.

C. Physical Properties

Since it is only recently that true linearity has been achieved in poly(perfluoroalkylenetriazine)s, the art of compounding or curing them is still in its infancy. Values for measurements of physical properties of gum stock, and even more so for those of cured samples, should be taken as minimal, with considerable improvement to be expected in the near future.

TABLE 5

Physical Properties of Fluoroelastomers

	Triazine[a]	Kel-F 3700[b]	Viton, Fluorel[b]	Silastic LS-53[b]	Fluororubber 1F4[b]
Tensile strength (psi)[c]	1,440	3,500	2,300	800	1,250
Elongation (% at break)[c]	70–300	500	320	250	300
Hardness, Shore A[c]	70	58	40	50	55
T_g (°C)[d]	−17 to −5	−15			
Gehman $T_1 0$ (°C)[d]		−14	−17	−60	−9

[a] Ref. 6.
[b] Ref. 20.
[c] Cured gum.
[d] Raw gum.

V. OTHER PERFLUOROALKYLENE HETEROCYCLIC POLYMERS

Although the development of thermally stable linear alkylenetriazine polymers has been limited to fluorinated compounds, the success obtained with ladder polymers over the past few years has shown that excellent thermal stability can be achieved with nonfluorinated nitrogen-heterocyclic polymers as well, if they are restricted to nonoxidizing environments. In an effort to avoid the difficulties associated with polytriazine synthesis and at the same time to retain the superior oxidative stability of fluorine compounds, polymers containing heterocyclic repeating groups other than s-triazines but similarly connected by perfluoroalkylene bridges are also being investigated. Several synthetic routes exist which may prove applicable to polymer formation; the rings in question have been shown to be adequately stable, in general, and a few polymers have been made. However, the exploration of these systems is of much more recent origin than that of triazines and fewer data are available in the open literature.

The general synthetic chemistry of these ring systems is beyond the scope of this chapter. The account here is intended to be only a brief review of work that has an immediate bearing on fluoropolymers.

A. 1,2,4-Triazoles

If hydrazine is substituted for ammonia in reaction with perfluoroalkyl nitriles, addition occurs in somewhat similar manner to give hydrazidines (21) and imidoylyhdrazidines (19), which can be converted to 1,2,4-triazoles by cyclodeammonation just as imidoylamidines are converted to s-triazines (19):

$$2R_fCN \ + \ N_2H_4 \ \longrightarrow \ R_fC \overset{HN \quad NH}{\underset{HN-NH}{\diagup \diagdown}} CR_f \ \overset{-NH_3}{\longrightarrow} \ R_f \overset{H}{\underset{N-N}{\diagup N \diagdown}} R_f$$

Use of a dinitrile in the reaction with hydrazine results in a linear, acyclic polymer which can be cyclodeammonated by heating, preferably in the presence of anhydrous hydrogen chloride. Alternatively, a dihydrazidine can be acylated and cyclodehydrated by reaction with a diacyl halide (19):

$$-(CF_2)_n C \overset{NH}{\underset{NHNH_2}{\diagup \diagdown}} \ + \ Cl\overset{O}{\overset{\|}{C}}(CF_2)_n- \ \longrightarrow \ -(CF_2)_n \overset{H}{\underset{N-N}{\diagup N \diagdown}} (CF_2)_n-$$

Although the triazole system is apparently free of the reorganization reactions that plague polytriazine chemistry, it has certain disadvantages of its own. For one, the deammonation step is not as free of competing reactions as one would desire; second, it would seem that the imino hydrogen of the NH group in the triazole ring, although leading to desirably strong interchain hydrogen-bonding, would constitute a point of inherent thermal weakness in the structure because of the possibility of elimination of hydrogen fluoride. The imino hydrogen atom might also cause oxidative instability at high temperatures, since it seems to do so in nonfluorinated polymers.

B. Oxadiazoles

In 1,3,4-oxadiazoles in ether oxygen replaces the imino NH group of the 1,2,4-triazoles. This change has two definite advantages: it eliminates from the structure all hydrogen and any associated thermal or oxidative instability, and it prevents any complications due to the prototropy which is possible in a triazole system. The synthesis of a 1,3,4-oxadiazole is a little more elaborate than that of a 1,2,4-triazole, and the polymer-forming reaction is rather more exotic, but the system offers good possibilities and has already produced some encouraging results.

One method of oxadiazole synthesis that has been in use for about three-quarters of a century, the dehydration of N,N'-diacylhydrazines (22), is applicable to fluorinated compounds and several 2,5-perfluoroalkyl-substituted compounds have been made in this way (23):

$$R_fCOOR \ \overset{N_2H_4}{\longrightarrow} \ R_fCONHNH_2$$

$$\Big\downarrow (R_f'CO)_2O$$

$$R_f \overset{O}{\underset{N-N}{\diagup \diagdown}} R'_f \ \overset{P_2O_5}{\longleftarrow} \ R_fCONHNHOCR_f'$$

Although no spectral evidence of aromaticity in the ring was observed, the oxadiazoles so made exhibited excellent thermal stability, no decomposition occurring when samples were heated in sealed tubes at 400° for 1 hr. However, resistance to nucleophilic attack was reported to be poor, as is the case for most perfluoroalkyl-substituted heterocycles; treatment with ammonia or amines at room temperature converted the oxadiazoles to triazoles almost quantitatively (24):

$$R_f \underset{N-N}{\overset{O}{\Big\langle}} R_f \quad \xrightarrow{RNH_2} \quad R_f \underset{N-N}{\overset{\overset{R}{N}}{\Big\langle}} R_f$$

The oxadiazole synthesis shown above, if applied to difunctional molecules for the purpose of polymer formation, would necessitate acylation-cyclo-dehydration of an intermediate acyclic polymer analogous to a poly(imidoyl-amidine). Methods of synthesis free from this objection are currently under study, in particular an interesting route involving 1,2,3,4-tetrazoles.

Perfluoroalkyl nitriles react smoothly with azide ion and on subsequent acidification give tetrazoles (25):

$$R_f CN \quad \xrightarrow{N_3^-} \quad \underset{N\sim N\sim N}{R_f C \overset{N^-}{\underset{|}{\Big\langle O \Big\rangle}}} \quad \xrightarrow{H^+} \quad \underset{\underset{H}{N\sim N\sim N}}{R_f \Big[\overset{N}{\underset{}{}} \Big\rangle}$$

The perfluoroalkyltetrazoles are strongly acidic substances of only moderate thermal stability and do not themselves represent attractive polymer possibilities. However, on thermal decomposition a transitory and very reactive intermediate is formed, postulated as a nitrilimine by Huisgen (26) and Brown (27). When a perfluoroalkyltetrazole is acylated, the ring is so weakened that this decomposition occurs at room temperature, cyclization involving the carbonyl oxygen follows, and a 1,3,4-oxadiazole ring appears. It is not even necessary to isolate the acylation product; treatment of a perfluoroalkyl-substituted tetrazole with a perfluoroacyl halide initiates the whole sequence of reactions (19).

The use of difunctional reactants should result in polymerization, and this seems to be a very promising route to perfluoroalkyleneoxadiazole polymers. Synthesis of the required tetrazoles is easy, efficient, and not obviously susceptible to structural restrictions; the polymerization reaction sequence occurs under mild conditions, is gratifyingly irreversible, and does not require prepolymer preparation.

$$-(CF_2)_n \underset{\underset{H}{N\sim N}}{\Big[\overset{N}{\underset{}{}} \overset{N}{\Big\rangle}} \; + -(CF_2)_m COCl \quad \longrightarrow \quad -(CF_2)_n \underset{N-N}{\overset{O}{\Big\langle}} (CF_2)_m -$$

The 1,2,4-oxadiazole system is accessible via cyclodehydration of acylated amidoximes, as shown (28):

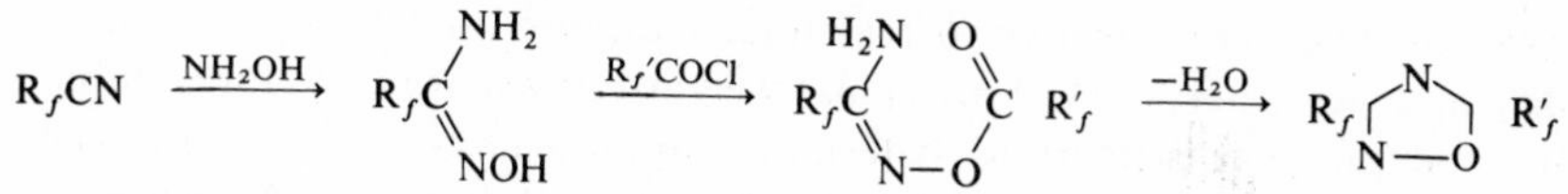

$$R_f CN \xrightarrow{NH_2OH} R_f C \overset{NH_2}{\underset{NOH}{\Big|}} \xrightarrow{R_f'COCl} R_f C \overset{H_2N}{\underset{N-O}{\Big|}} \overset{O}{\underset{}{C}} R_f' \xrightarrow{-H_2O} R_f \overset{N}{\underset{N-O}{\Big[}} R_f'$$

1,2,4-Oxadiazoles are reported to be more resistant to basic hydrolysis than the 1,3,-4oxadiazoles (27). Model compounds have shown good thermal stability but it is doubtful that the 1,2,4-system, containing an N—O bond, is inherently as stable as the 1,3,4 system.

Polymers in which two heterocyclic rings are directly joined, rather than separated by a perfluoroalkylene bridge, can be made either by reaction of bitetrazole with a difunctonal perfluroacyl halide or by reaction of a perfluoro-alkylene-α,ω-bis(tetrazole) with oxalyl chloride, the first method being preferable (19).

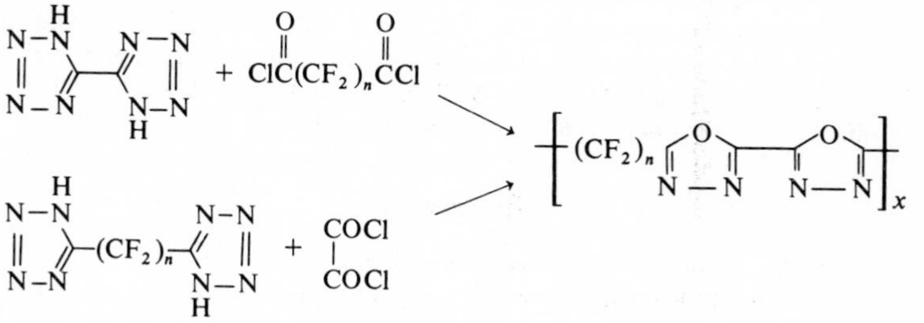

The direct union of the two rings should improve thermal stability because of extension of the conjugation, and increased resistance to basic hydrolysis of polymers containing this system has also been observed (19).

VI. SUMMARY

Perfluoroalkylenetriazine elastomers are, chemically speaking, an accomplished fact. Two separate approaches, the acylation-cyclodehydration of imidoylamidines and the free-radical coupling of preformed triazine derivatives, have been used successfully to produce polymers of good linearity, and cross-linking techniques for the two systems have been developed. Samples already made or imminent should show acceptable glass transition temperature, tensile strength, hardness, and ultimate elongation, although these values will probably not be as good as those of some conventional elastomers. The resistance of perfluoroalkylenetriazine elastomers to heat, oxidation, and radiation is significantly greater than that of other elastomers, including fluoroelastomers, presently known. They are essentially unaffected by strong acids and oxidizing environments and should have service temperatures above 300°C.

However, such elastomers are still some distance from commercialization. The imidoylamidine approach, which is by far the more intensively investigated, is very sensitive to reaction conditions, affords poor reproducibility, and yields polymers of molecular weights only marginally high enough for good physical properties. Alternative approaches, notably those using free radical coupling for chain extension, are severely limited by elaborate syntheses or by excessive restrictions on structure. Present curing treatments and filling agents have serious deleterious effects on elastomer thermal stability, and desirable long-chain perfluorodicarboxylic acids are either unavailable or prohibitively expensive. Any one of these problems, remaining unsolved, could represent the difference between success and failure in commercialization of the perfluoroalkylenetriazines.

Acknowledgments

We who have worked in the field of perfluoro heterocyclic polymers are indebted to the Air Force Materials Laboratory at Wright-Patterson Air Force Base for its commendable patience and persistence in continuing to support research in this area, in the face of serious technical problems, over a number of years. Without this support, perfluoroalkylenetriazines would probably have remained no more than an attractive and elusive possibility.

References

1. P. B. Fox, *Am. Chem. Soc. Polymer Preprints*, **8**:e (April, 1967).
2. E. M. Smolin and L. Rappaport, "*s*-Triazines and Derivativs," Interscience, New York, 1959, p. 215.
3. W. L. Reilly and H. C. Brown, *J. Org. Chem.*, **22**:698 (1957).
4. T. R. Norton, *J. Am. Chem. Soc.*, **72**:3527 (1950).
5. W. E. Emerson and E. Dorfman, Abstr. 152 Meeting, American Chemical Society, New York, September 1966, paper no. 22K.
6. E. Dorfman, W. E. Emerson, R. L. K. Carr, and C. T. Bean, *Rubber Chem. and Tech.*, **39**(4), Pt. 2:1175 (1966).
7. Unpublished observations of the author.
8. G. A. Grindahl, W. X. Bajzer, and O. E. Pierce, *J. Org. Chem.*, **32**:603 (1967).
9. G. W. Rigby, U.S. Patent 2,484,528 (1949).
10. E. R. Bissell, *J. Org. Chem.*, **28**:1717 (1963).
11. H. C. Brown, *J. Polymer Sci.*, **44**:9 (1960).
12. H. C. Brown, P. D. Schuman, and J. Turnbull, *J. Org. Chem.*, **32**:231 (1967).
13. L. H. Toporcer, G. A. Grindahl, B. L. Blanck, and O. R. Pierce, Abstr. 4th International Fluorine Symposium, Estes Park, Colo., July, 1967.
14. J. A. Young and R. L. Dressler, *J. Org. Chem.*, **32**:2237 (1967).
15. J. A. Young, *Fluorine Chem. Rev.*, **1**:359 (1967).
16. R. L. Dressler and J. A. Young, *J. Org. Chem.*, **32**:2004 (1967).
17. R. W. Anderson, N. L. Madison, and C. I. Merrill, Abstr. 4th International Fluorine Symposium, Estes Park, Colo., July 1967.
18. H. C. Brown and P. D. Schuman, *J. Org. Chem.*, **28**:1122 (1963).
19. H. C. Brown, *Am. Chem. Soc. Polymer Preprints*, **5**:243 (1964).
20. H. G. Bryce, in J. H. Symons, "Fluorine Chemistry," Vol. V, Academic Press, New York, 1964, p. 66.
21. H. C. Brown and D. Pilipovich, *J. Am. Chem. Soc.*, **82**:4700 (1960).

22. R. Stolle, *Ber.*, **32**:797 (1899).
23. H. C. Brown, M. T. Cheng, L. J. Parcell, and D. Pilipovich, *J. Org. Chem.*, **26**:4407 (1961); W. J. Chambers and D. D. Coffman, *J. Org. Chem.*, **26**:4410 (1961).
24. H. C. Brown and M. T. Cheng, *J. Org. Chem.*, **27**:3240 (1962).
25. W. G. Finnegan, R. A. Henry, and R. Lofquist, *J. Am. Chem. Soc.*, **80**:3908 (1958).
26. R. D. Huisgen, *Angew. Chem.*, **72**:366 (1960); R. D. Huisgen, J. Sauer, and M. Seidel, *Chem. Ber.*, **93**:2106 (1960); J. Sauer, R. D. Huisgen, and H. J. Sturm, *Tetrahedron*, **11**:241 (1960).
27. H. C. Brown and R. J. Kassall, Abstr. 148th Meeting, American Chemical Society, Chicago, Ill., September 1960, p. 20K.
28. H. C. Brown and C. R. Wetzel, Abstr. American Chemical Society Meeting, New York, 163, p 19M.

10. CHEMICAL CROSS-LINKING OF FLUOROELASTOMERS

KAZIMIERA J. L. PACIOREK, *Dynamic Science, Irvine, California*

Contents

I. INTRODUCTION

Synthetic elastomers, as well as natural rubber, are formed from long, flexible, chainlike molecules which are cross-linked to a three-dimensional network. Some materials exhibit significant elastomeric properties even in the absence of cross-linking, as evident in polymers where the "entanglements" between the chains act as pseudo cross-links; this effect disappears at sufficiently high temperatures. Consequently, the useful temperature range of such materials is necessarily very limited. Cross-linking can be achieved during the polymerization phase using a certain percentage of reagents of functionality greater than two. However, the materials thus produced are usually insoluble and do not lend themselves readily to fabrication of useful end items. The most frequently employed procedure in the preparation of synthetic elastomers is to synthesize a linear chain material, which then is mixed with a suitable cross-linking agent (filler, etc.), molded into the desired shape, and subjected to a cure.

The number of cross-links required to procure the desired strength and resistance to swelling with adequate elasticity can be theoretically derived. Yet the number of cross-links necessary to produce a certain set of properties varies from material to material. It also varies depending on molecular-weight distribution. Accordingly, for polymers of equal number average molecular weights, a broader molecular-weight distribution favors gelation at lower degrees of reaction. Too dense a population of cross-links results in a rigid structure with almost complete resistance to swelling, yet devoid of elastic characteristics.

The theoretical aspects of cross-linking have been developed largely by Flory (1) and Stockmayer (2,3) and have been amply reviewed in the literature (4,5). Consequently, no attempt will be made here to delve into this area. This chapter is entirely limited to fluorine-containing elastomers and to the discussion of cross-linking by chemical means. Included is cross-linking by combination of polymer radicals when initiated by peroxides; radiation-induced curing is covered in Chapters 11 and 12.

Ideally, a cross-link should be some subtle extension of the overall chain, thus avoiding an introduction of weak links or of potential sites for degradative attack. Unfortunately, in practice, this is rarely feasible and usually in the case of fluoroelastomers the creation of cross-linking sites is invariably associated with a decreased stability as compared to the raw gum. This will be made clear later in discussion of individual systems. It should be pointed out at this juncture that materials such as natural rubber have ready cross-linking sites due to the inherent presence of unsaturation. In the fluorinated systems, such sites have to be created either by incorporation into the original polymerization system of comonomers containing active groups or by chemical reaction such as dehydrohalogenation of the preformed chain. Either one of the approaches will supply the sites which can be subsequently utilized in curing.

The interest in fluorinated elastomers stems from their exceptional thermal, chemical, and oxidative stability. Several different types of fluoroelastomers have been synthesized. Some of these, as exemplified by fluorinated polyesters, fluorinated acrylates, and to a certain degree vinylidene fluoride-chlorotrifluoroethylene copolymers, have been superseded by more advanced materials (6,7). However, at some stage of development they served an important function before the advent of the newer compositions. It is to be expected that today's materials will be replaced in the future by systems of improved range and more specific applications.

To facilitate the discussion of the cross-linking processes, the fluoroelastomers are here divided into (1) elastomers formed by copolymerization of vinylidene fluoride with perfluorinated or chlorofluorinated moieties (some terpolymers are also included in this group), (2) nitroso rubbers, the simplest being the copolymer of tetrafluoroethylene and trifluoronitrosomethane, (3) perfluorinated triazine elastomers, (4) fluorinated siloxanes, (5) fluorinated polyesters, (6) fluorinated acrylates, and (7) new elastomeric materials obtained by copolymerization of thiocarbonyl fluorides or fluorinated vinyl ethers with suitable monomers.

II. VINYLIDENE FLUORIDE COPOLYMERS

The vinylidene fluoride-hexafluoropropene copolymers represented by the Viton (E. I. du Pont de Nemours & Company) and Fluorel (Minnesota Mining and Manufacturing Company) series comprise the most widely investigated fluoroelastomer systems. Presently these elastomers are successfully employed in both military applications and commercial uses where thermal and chemical stability combined with good elastomeric properties are required. Early investigations (8,9) have shown that these materials can be cross-linked with a variety of agents: radiation, diamines, and peroxides. In all the compounding formulations an acid acceptor such as magnesium oxide, zinc oxide, basic lead phosphite, or a similar ingredient was employed. Aliphatic diamines, which were the first curing agents used, tended to react prematurely. Accordingly, they were replaced by their Schiff bases as exemplified by cinnamylidene derivatives, carbamates, and others (10). Complexes such as Co-bissalicylalethylenediimine and Ti-salicylalimine have also been employed (11). In addition to diamines, primary monoamines were found to effect cross-linking (12). Certain aromatic amines, particularly in the presence of carbon black, appeared to impart good high-temperature properties (13). The importance of fillers on mechanical properties of Viton vulcanizates was shown by the studies of Novikov et al. (14). The thermodynamic behavior of such a vulcanized Viton sample is qualitatively similar to that observed with a natural rubber network, except that the internal energy makes here a negative contribution (15).

The vinylidene fluoride-chlorotrifluoroethylene copolymer (Kel-F elastomer, Minnesota Mining and Manufacturing Company) exhibits a general behavior, insofar as curing is concerned, closely comparable to that of the vinylidene fluoride-hexafluoropropene copolymer (16–18). In addition, cross-linking was achieved with a chromium complex of $Cl(CF_2CFCl)_2$-CF_2COOH (19) and with diisocyanates in the presence of zinc oxide (20). Both Viton and Kel-F elastomers undergo cure with alkali metal salts of polyphenols, polythiols, and mercaptophenols (21). It should be pointed out that Kel-F elastomers and their vulcanizates, in view of the presence of the relatively labile chlorine atom, do not demonstrate some of the superior properties associated with the chlorine-free fluoroelastomers.

Terpolymers such as the products obtained by copolymerization of tetra-fluoroethylene, 1H-pentafluoropropylene and vinylidene fluoride, undergo curing operations analogous to the Viton and Kel-F elastomers (22).

The elastomeric materials are not necessarily 1:1 copolymers. Actually a whole range of combinations is involved. It has been found that 80–30% of vinylidene fluoride copolymerized with chlorotrifluoroethylene gives elastomers. In the hexafluoropropene series the range is 85–50%. In the commercially available products vinylidene fluoride usually accounts for more than 50%, giving segments of polyvinylidene fluoride. Consequently, any mechanistic considerations of cross-linking processes must make allowance for the various possible arrangements present.

A. Amine Cross-Linking

The vinylidene fluoride-trifluorochloroethylene copolymer was found to be cross-linked in solution by primary mono- and diamines at room temperatures; by secondary mono- and diamines at ca. 50–60°C, by tertiary diamines at ca. 90–100°C, and by tertiary monoamines at ca. 180–190°C (23).

The isolation of amine hydrochlorides and hydrofluorides in conjunction with the facile cross-linking by primary diamines can be readily explained by dehydrohalogenation followed by amine addition and concomitant hydrogen fluoride elimination (23,24):

$$-CF_2-CH_2-CFCl-CF_2- \xrightarrow{H_2N(CH_2)_nNH_2} -CF_2-CH=CF-CF_2-$$
$$\textbf{I} \qquad\qquad\qquad\qquad\qquad \textbf{II}$$

$$2\text{-}CF_2-CH=CF-CF_2- + H_2N(CH_2)_nNH_2 \longrightarrow \begin{array}{c} -CF_2-CH_2-CF-CF_2- \\ | \\ NH \\ | \\ (CH_2)_n \\ | \\ NH \\ | \\ -CF_2-CH_2-CF-CF_2- \end{array}$$

$$\begin{array}{c} -CF_2-CH_2-C-CF_2- \\ \| \\ N \\ | \\ (CH_2)_n \\ | \\ N \\ \| \\ -CF_2-CH_2-C-CF_2- \end{array} \xleftarrow{-HF}$$

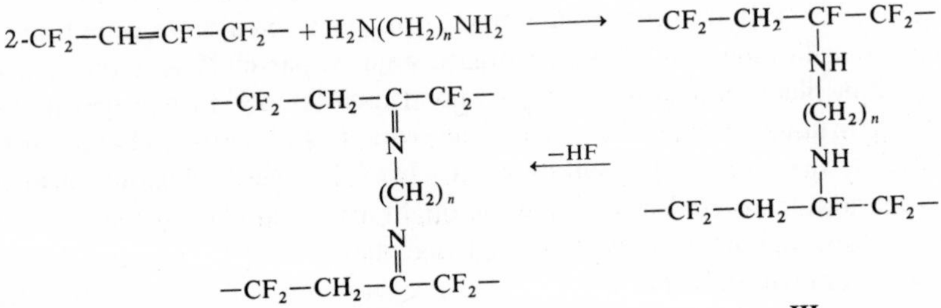

$$\textbf{IV} \qquad\qquad\qquad\qquad\qquad \textbf{III}$$

The dehydrofluorination of III was further supported by Smith and Perkins (25) and by Paciorek et al. (26). The latter found that treatment of $C_2F_5CF=CHC_3F_7$ with butylamine gave two isomers:

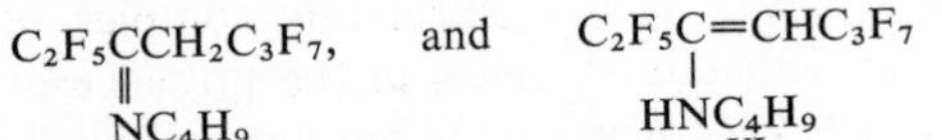

$$\begin{array}{ccc} C_2F_5CCH_2C_3F_7, & \text{and} & C_2F_5C=CHC_3F_7 \\ \| & & | \\ NC_4H_9 & & HNC_4H_9 \\ \textbf{V} & & \textbf{VI} \end{array}$$

together with a diamino-product. The isolation of VI shows a way in which a primary monoamine can indeed constitute a cross-link. Based on the work of Pruett et al. (27), it is obvious that a system where an NH moiety is adjacent to the CF grouping is unstable in the presence of amines. Thus the addition of the derivative

$$\begin{array}{c} -CF-CH_2- \\ | \\ HNR \end{array}$$

to another chain before elimination of hydrogen fluoride seems unlikely. However, the isolation of VI would permit the following process to occur:

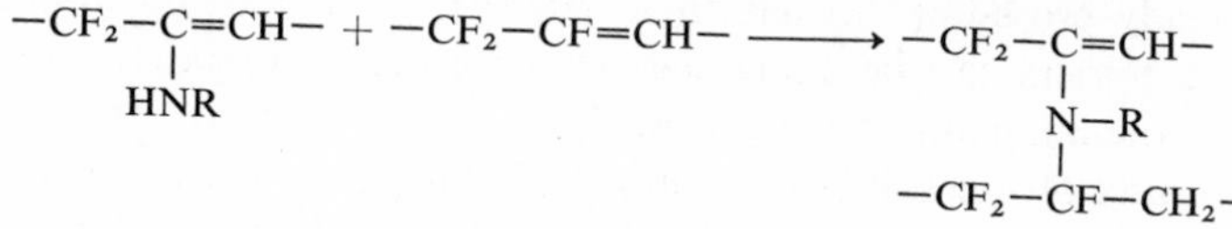

$$-CF_2-\underset{\underset{HNR}{|}}{C}=CH- + -CF_2-CF=CH- \longrightarrow -CF_2-\underset{\underset{N-R}{\overset{}{|}}}{C}=CH-$$

$$-CF_2-\underset{}{C}F-CH_2-$$

The mechanisms of cross-linking by secondary diamines similarly can be visualized simply by dehydrohalogenation followed by addition. The difficulty arises in postulating a mechanism whereby a secondary monoamine, a ditertiary diamine, and in particular a tertiary monoamine effects crosslinking at relatively low temperatures, where amine rearrangements and double-bond interactions are not very likely. The gelation observed with tertiary diamines could be ascribed to quaternary salts:

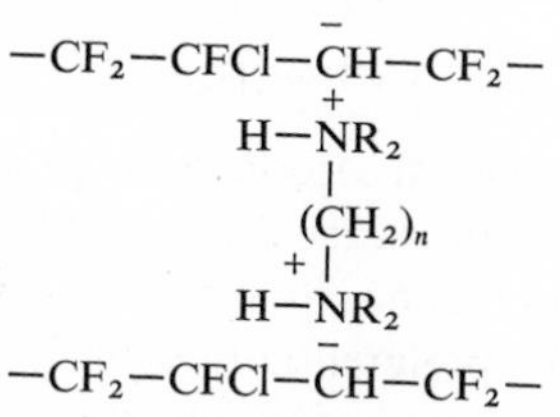

$$-CF_2-CFCl-\overset{-}{C}H-CF_2-$$
$$\overset{+}{H}-NR_2$$
$$|$$
$$(CH_2)_n$$
$$\overset{+}{H}-NR_2$$
$$|$$
$$-CF_2-CFCl-\overset{-}{C}H-CF_2-$$

It has been recognized that the strong electronegative effect of the neighboring perfluoroalkyl group imparts acidic character to the hydrogen atoms. Inasmuch as amines are known to produce salt linkages with carboxyl groups (28), similar behavior would not be surprising in the system represented by the vinylidene fluoride copolymers.

Nimoy proposed (29) a somewhat related arrangement for the crosslinking of polyvinylidene halides (among others vinylidene fluoride-hexafluoropropene copolymer) by imidazolines:

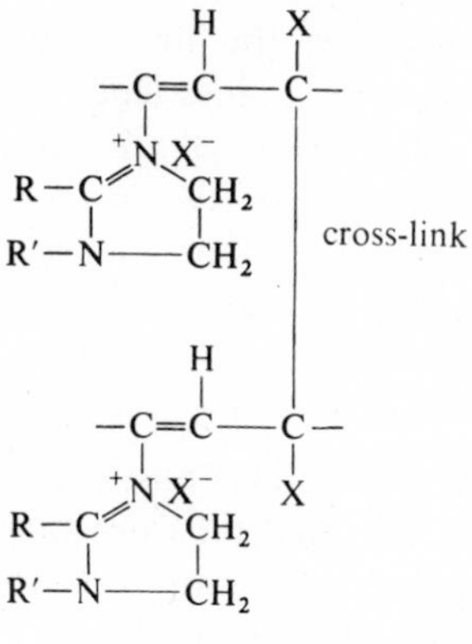

Here the effective cross-link stable at 200°C is believed to be formed between two carbons after the loss of allylic halogens.

The ready cross-link formation at 50–60°C by secondary amines in the Kel-F elastomers can be explained by some degree of interchain bonding involving the amine-modified polymer.

The results of the solution studies performed on the vinylidene fluoride-perfluoropropene copolymer differed from those obtained with the chlorine-containing elastomer. Only primary amines afforded gelation below 190°C, although dehydrohalogenation leading to isolated and conjugated double bonds was observed with all the amines tested and in some instances amounted to 52% of the available hydrogen (30).

In the copolymers of vinylidene fluoride-hexafluoropropene, regardless of the absolute ratio of the two monomers, the arrangement

$$\overset{\displaystyle CF_3}{\underset{\displaystyle |}{-CF_2-CF-CH_2-CF_2-}}$$

accounts for 93% of the hexafluoropropene present (31). The vinylidene fluoride hexafluoropropene ratio is usually much higher than 1 : 1, consequently giving segments of polyvinylidene fluoride. However, dehydrohalogenation would be expected to occur preferentially with fluorine on a tertiary carbon atom rather than with the difluoromethylene groups. This, in turn, should lead to a certain proportion of the amine adduct depicted below:

$$\overset{\displaystyle CF_3}{\underset{\displaystyle |}{\underset{\displaystyle RNH}{\overset{|}{-CF_2-C-CH_2-CF_2-}}}}$$

These compounds will not be expected to eliminate hydrogen fluoride spontaneously. Formation of such intermediates would then explain the ready gelation by primary monoamines in the Viton series. Based on a number of investigations (23–25,32,33), it is obvious that in the amine cross-linking dehydrohalogenation constitutes the first step. The lack of reaction of polychlorotrifluoroethylene with amines tends further to support this postulate (23). The next step does not necessarily need to include the amine reagent. The simplest interaction of two double bonds would give four-membered ring structure:

$$2-CF_2-CF=CH-CF_2- \longrightarrow \begin{array}{c} -CF_2-CF-CH-CF_2- \\ | \quad\;\; | \\ -CF_2-CF-CH-CF_2- \end{array}$$

or another, similar arrangement. This postulate has to be rejected in view of the findings of Park et al. (34) that the cyclobutane structure does not form where one of the unsaturated carbon atoms bears a hydrogen substituent.

Mechanisms have been proposed whereby conjugated unsaturation introduced by amine dehydrohalogenation results in the establishment of an aromatic ring arrangement via Diels-Alder reaction (33):

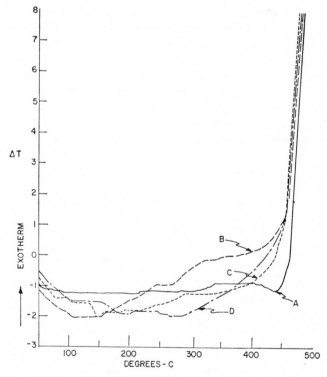

Yet amine-free materials which were definitely conjugated failed to cross-link in solution, refuting the preceding postulations (23). On the other hand, differential thermal analyses (35) of variously treated vinylidene fluoride-hexafluoropropene copolymer-magnesium oxide mixtures indicate that a certain degree of cross-linking is reached with magnesium oxide alone. In Fig. 1 the exotherm observed in curve B near 160 to *ca.* 250°C can be

Fig. 1. Differential thermal analysis of Viton A–HV-Maglite D mixture after various heat treatments: (A) Viton A–HV; (B) Viton A–HV + 15 phr Maglite D (mill-mixed); (C) Viton A–HV + 15 phr Maglite D (press-cured); (D) Viton A–HV + 15 phr Maglite D (oven-cured) (35).

attributed to thermal cross-linking of the copolymer with magnesium oxide. Novikov (36) proved that heating of Viton at 150°C results in unsaturation; metal oxides promote this reaction even at lower temperatures. Above 220°C thermal cross-linking appears to occur as indicated by the loss in solubility. Work conducted with model compounds shows definitely that magnesium oxide can act as dehydrofluorinating agent (37). Consequently, it would appear that interaction of unsaturated centers is in some way responsible for cross-link formation. These are somewhat conflicting results, which unfortunately have not been satisfactorily resolved to date.

Once unsaturation is introduced, agents which are known to react with double bonds will form a cross-link. This was demonstrated by Smith (38) using dithiol-tertiary amine-magnesium oxide combinations. From the data given it is obvious that no practical cure can be reached in the absence of magnesium oxide, as evidenced by Fig. 2.

For better understanding of the attendant processes, the conditions and typical recipes used in conventional curing are briefly outlined. The generally recommended procedure is to formulate the elastomer with a curing agent, an acid acceptor, and a filler on conventional processing equipment, then to mold the required articles in a press using a curing cycle of 5–60 min at 100–150°C. The purpose of this press cure is to develop sufficient cross-links in the sample to prevent sponging through release of trapped air during the early stages of the subsequent oven cure. Optimum properties are then developed by heating the partially cured samples in an air oven at a temperature of 200°C for a period of 15–24 hr.

The action of acid acceptor and the importance of postcure or oven-cure

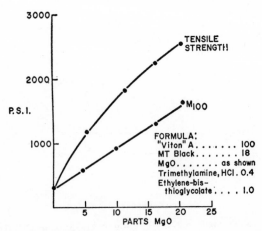

Fig. 2. Effect of magnesium oxide on Viton A cures with dithiol (38). (Reprinted by permission of the copyright owner, Bill Brothers Publ. Corp., Rubber World.)

in the vulcanization of Viton was subject of a detailed study (25). The cross-link density was determined following the method of Cluff et al. (39) from the measurement of compression modulus of swollen elastomer pellets. The effect of hexamethylene-diamine carbamate concentration on cross-linking density of Viton A elastomer is represented by Fig. 3. Fair agreement is seen between the experimental curve and the theoretical line representing direct equivalence between cross-linking density and curing agent concentrations, especially at lower levels of the curing agent. Furthermore, it was found that the level of the acid acceptor seemed also to influence, up to a certain point, the density of cross-links, as illustrated by Fig. 4. This is to be expected since the amine is employed in the form of its Schiff base. Consequently, to effect any cure the free amine has to be produced. This is accomplished by water liberated through a reaction of magnesium oxide with hydrogen fluoride:

$$MgO + 2HF \longrightarrow MgF_2 + H_2O$$

Whether the initially needed catalytic amounts of water are produced from the magnesium oxide-copolymer interaction is immaterial. In the absence of magnesium oxide, providing that hydrolysis of the Schiff base took place, only a certain amount of amine can be utilized for cross-link formation; the remainder will be tied up in the form of amine hydrofluoride. Interaction of amine hydrofluoride with magnesium oxide will regenerate the free amine:

$$MgO + 2RNH_3F \longrightarrow MgF_2 + H_2O + 2RNH_2$$

Thus magnesium oxide serves a double purpose, freeing the amine from the derivative Schiff base and regenerating the amine from the hydrofluoride.

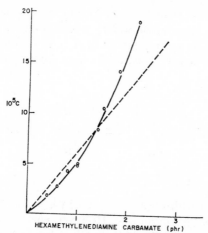

Fig. 3. Effect of HMDA–C concentration on the cross-linking density of Viton A: (——) cross-linking density corrected for loose chain ends; (— —) curve calculated assuming one mole of cross-links per mole of diamine curing agent (25).

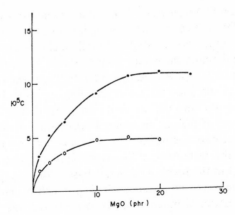

Fig. 4. Effect of magnesium oxide on the cross-linking density of Viton A with (○) 1.0 phr and (●) 1.5 phr hexamethylenediamine carbamate (25).

However, other workers (40) view the action of Schiff bases not merely as moderators of the amine action, but as the actual participants in the cross-link formation. The mechanism proposed involves only a partial hydrolysis of the Schiff base into the free amine, which then dehydrohalogenates the polymer. But it is the Schiff base itself which is believed to add to the double bonds. The higher thermal stability of materials vulcanized by Schiff bases as compared to those treated by free amines is attributed to the connecting C—N bonds. Contrary evidence from stress-relaxation studies (41) indicates a deteriorating action of the cinnamaldehyde residues in the polymer.

The importance of postcure, as seen from the findings of Smith (25), lies mainly in the slow removal of water, whose presence in the polymer is detrimental to attaining an optimum state of vulcanization. It was postulated that the presence of water actually promotes hydrolysis of the cross-links initially formed:

$$
\begin{array}{ccc}
\mid & \mid & \mid \\
CH_2 & CH_2 & CH_2 \\
\mid & \mid & \mid \\
C=N-(CH_2)_n-N=C & \xrightarrow{H_2O} & 2C=O + H_2N(CH_2)_nNH_2 \\
\mid & \mid & \mid \\
CH_2 & CH_2 & CH_2 \\
\mid & \mid & \mid \\
\textbf{VII} & & \textbf{VIII}
\end{array}
$$

This is in agreement with the recovery on hydrolysis of 70% of the amine originally present in the vulcanizate. However, since Viton A is believed to consist of a 4.5:1 ratio of $CH_2=CF_2$ to $CF_3 CF=CF_2$, not all of the amine present will be in the form of arrangement VII, which is derived from the polyvinylidene segment. The structure

$$CF_3$$
$$|$$
$$-CF_2-CF-CH_2-CF_2-$$

is expected to participate also. Consequently, some of the amine addition compound will be in a form

$$
\begin{array}{ccc}
 | & & | \\
H_2C \quad H & & H \quad CH_2 \\
 | \quad | & & | \quad | \\
CF_3-C-N-(CH_2)_n-N-C-CF_3 \\
 | & & | \\
CF_2 & & CF_2 \\
 | & & |
\end{array}
$$

which should be resistant to hydrolysis. The recovery of only 70% of the amine present supports this postulate. It should not be overlooked that any fluoroelastomer vulcanizate cured with amines probably contains a certain percentage of cross-links derived from the interaction of unsaturated centers.

The postcure behavior of the Vitons treated with secondary diamines was not extensively investigated; the limited work conducted by Spain (42) points to the formation of good vulcanizates using piperazines. The susceptibility of these systems to hydrolysis was not evaluated.

B. Peroxide-Induced Cross-Linking

Peroxides in the presence of acid acceptors appear to afford good vulcanizates, both with the Viton and Kel-F fluoroelastomers. This is mainly true of aromatic peroxides since the results with the aliphatic materials do not seem to be very encouraging (42). Investigations were essentially centered on benzoyl peroxide formulations, although a limited effort was expended using 2,4-dichlorobenzoyl peroxide. A much better cure was attained with the first agent (37).

Attempts were made to determine the extent of peroxide cross-linking in the absence of an acid acceptor, but the samples thus produced did not lend themselves to swelling measurements (due to the spongy nature), nor could the tensile strengths and elongations be determined. From the studies performed on the polymer-magnesium oxide-benzoyl peroxide mixture, it is evident that the maximum cure takes place between 140 and 150°C; no cross-linking seems to occur below 85°C (30). However, claims have been made that a certain amount of cross-linking and chain scission takes place below 80°C during milling of Viton elastomers (43). In analogy to the amine series, only after oven postcure are the optimum properties developed.

Several mechanisms can be postulated for the peroxide cross-linking. The most obvious way of attack would be to abstract one of the hydrogens from the methylene group giving rise to a free radical, which could then form a cross-link by interaction with a similar polymer radical:

$$(PhCOO)_2 \longrightarrow Ph\cdot + CO_2 + PhCOO\cdot$$

$$R_fCXYCH_2R_f \xrightarrow{Ph\cdot \text{ or } PhCOO\cdot} R_fCXY\dot{C}HR_f + C_6H_6 \text{ or } PhCOOH \qquad \begin{array}{l}(X = F \text{ or } Cl) \\ (Y = CF_3 \text{ or } F)\end{array}$$

$$2R_fCXY\dot{C}HR_f \longrightarrow \begin{array}{c} R_fCXYCHR_f \\ | \\ R_fCXYCHR_f \end{array}$$

Another mode of attack would be the addition of a benzoyl peroxide fragment to the double bond, produced by magnesium oxide dehydrohalogenation:

$$2R_fCXYCH_2R_f \xrightarrow{MgO} 2R_fCY{=}CHR_f + MgF_2 + H_2O$$

$(X = F \text{ or } Cl)$
$(Y = CF_3 \text{ or } F)$

$$\downarrow (PhCOO)_2$$

$$\begin{array}{c} Ph \\ | \\ R_f\dot{C}YCHR_f \end{array} \text{ and/or } \begin{array}{c} \overset{O}{\overset{\|}{OCPh}} \\ | \\ R_f\dot{C}YCHR_f \end{array}$$

$$\begin{array}{c} Ph \\ | \\ R_f\dot{C}YCHR_f \end{array} \xrightarrow{R_fCY=CHR_f} \begin{array}{c} Ph \\ | \\ R_fCYCHR_f \\ | \\ R_fCH\dot{C}YR_f \end{array} \xrightarrow{-H\cdot} \begin{array}{c} Ph \\ | \\ R_fCYCHR_f \\ | \\ R_fC{=}CYR_f \end{array}$$

Model compound studies conducted with $C_2F_5CF{=}CHC_3F_7$ and benzoyl peroxide failed to give reaction and investigations performed using the saturated analog, $C_2F_5CF_2CH_2C_3F_7$, did not yield any of the expected dimer. To prove that the cross-linking was due to free radicals, studies were carried out with substances known to readily liberate phenyl radicals, triphenyl antimony and triphenyl arsenic (37). The results with model compounds again were negative; in the fluoroelastomer series some cure was attained with these reagents, but the degree of cross-linking achieved was definitely lower than with the corresponding amounts of benzoyl peroxide. The results of parallel curing operations conducted on Viton with magnesium oxide alone (Maglite D, Merck and Company), magnesium oxide-benzoyl peroxide mixture, and magnesium oxide bis-cinnamylidenehexamethylenediamine (LD-214, E. I. du Pont de Nemours & Company) are summarized in Tables 1 and 2.

Examining the physical characteristics of the vulcanizates presented in Table 2, it is obvious that benzoyl peroxide produces a vulcanizate definitely superior to that obtainable with magnesium oxide alone. On this basis, it is clear that cross-linking must be due to interaction of the polymer chain either with benzoyl peroxide or its radicals or via some secondary radicals produced. Unfortunately, none of the studies performed to date was successful in elucidating either the type of cross-link formed or the operative mechanism.

TABLE 1

Formulations of Viton A-HV Vulcanizates (35)

Component parts	Vulcanizate		
	I	II	III
Viton A-HV	100	100	100
Maglite D	15	15	15
LD-214			4
Benzoyl peroxide		4	

Press cure: 30 min at 150°C.
Oven cure: 25°C/min from 120 to 200°C; then 24 hr at 200°C.

TABLE 2

Physical Test Data for Viton A-HV Vulcanizates (35)

Property	Vulcanizate I		Vulcanizate II		Vulcanizate III	
	Press cure	Oven cure	Press cure	Oven cure	Press cure	Oven cure
Tensile strength (psi)	270	1,080	3,550	3,650	2,230	3,250
Elongation (%)	1,480	820	460	420	460	310
Hardness, Shore A	66	70	60	63	64	68

C. Conclusions and Stability Considerations

Amine cross-linking of fluorinated polymers, wherein a methylene moiety is adjacent to a chlorine or fluorine-bearing carbon atom, occurs via dehydrohalogenation followed in the case of primary and secondary amines by amine addition to the unsaturation thus formed. This does not necessarily exclude interaction of the unsaturated centers. Most likely even with primary diamines this type of action does occur. However, no definite proof of the latter mechanism is available at present. Yet only this type of reaction could explain the cross-linking action of magnesium oxide and would apply also to vulcanization by tertiary amines, although in the latter case rearrangements cannot be entirely rejected. Once a double bond has been created any molecule containing active hydrogen would be expected to add. This has been found to be true of thiols and should occur with primary and secondary phosphines.

Acid acceptors have been found to play a multirole, since these materials, particularly magnesium oxide, can effect dehydrohalogenation, assist in hydrolysis of the Schiff base, and regenerate the free amine from its hydrofluoride.

The cross-linking by peroxides appears to be limited largely to benzoyl peroxide; the operative mechanisms have not been clarified to date. The presence of magnesium oxide is necessary to attain a practical degree of cure. The cross-link density as determined by stress relaxation is lower in the peroxide-cured formulations than in the amine vulcanizates (41).

Cross-linking of an arrangement such as that present in saturated fluoroelastomers, where ready cross-linking sites are not available but have to be created, will result necessarily in a somewhat degraded system. Obviously, susceptibility to chemical and thermal bond scissions will be vastly increased. The differential thermograms of various Viton elastomer formulations illustrate this point amply (see Fig. 5). The amine cross-linked material exhibits

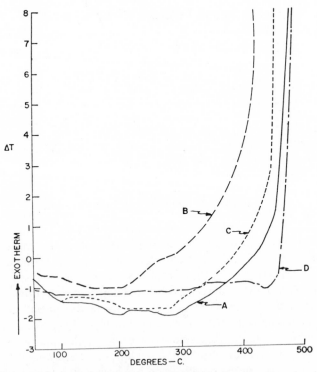

Fig. 5. Differential thermal analysis of various Viton A–HV formulations after oven cure: (A) Viton A–HV + 15 phr Maglite D; (B) Viton A–HV + 15 phr Maglite D + 4 phr LD–214; (C) Viton A–HV + 15 phr Maglite D + 4 phr benzoyl peroxide; (D) Viton A–HV (35).

the lowest thermal stability, which can be readily ascribed to the presence of residual unsaturation in the form of

$$-CF=CH- \qquad \text{or} \qquad \begin{array}{c} CF_3 \\ | \\ -C=CH- \end{array}$$

moieties as well as conjugated systems including the amine-containing structures of the type

$$-(C-CH=CF)- \qquad \text{and} \qquad -(C-CH=C)- \\ \quad \| \qquad\qquad\qquad\qquad\qquad \| \quad\; | \\ \quad NR \qquad\qquad\qquad\qquad\qquad NR \; NHR$$

The degradative effect of benzoyl peroxide and magnesium oxide is of a much lesser magnitude than the amine action. However, even these vulcanizates are thermally less stable than the raw gum.

III. NITROSO-FLUOROPOLYMERS

A. General Considerations

The most important of this series is the copolymer of trifluoronitrosomethane and tetrafluoroethylene (44,45); other analogs utilizing variously substituted nitroso monomers and different olefins were also prepared (46,47). Initial vulcanization studies (48) showed that the nitroso fluoropolymer did not respond to conventional sulfur vulcanization, which is not surprising in view of the absence of residual unsaturation. Divalent metallic oxides also failed to produce cross-linking, indicating that the fluorine substituents were not susceptible to this type of attack. Irradiation, by both beta and gamma radiation, resulted in degraded products (49). The only effective cross-linking agents were found to be diamines. Early compounding mixtures comprised a combination of triethylenetetramine and hexamethylenediamine-carbamate in conjunction with a fine silica filler (48–50). The tensile strength obtained was very low, 700 psi, with ultimate elongation of 600; however, the vulcanizate exhibited very good low-temperature properties and remarkable resistance to swelling by hydrocarbon solvents. Of particular significance was the inertness toward carbonyls and oxidizing agents.

No mechanism was postulated for the amine cure and no studies were performed to determine whether the cross-link is established at the tetrafluoroethylene or the nitroso site. The most obvious point of attack would be the fluorine atoms in the chain

$$\left[\begin{array}{c} CF_3 \\ | \\ -N-O-CF_2-CF_2- \end{array} \right]_n$$

Yet, in view of the inertness of the completely fluorinated polymers (32) and polytrifluorochloroethylene (23) toward primary amines, this does not appear to be the case. It should be pointed out that amines were found not only to cross-link but also to degrade the nitroso rubber (51).

Another related system was synthesized by copolymerization of $ONCF_2$-CF_2NO_2 with a variety of unsaturated materials. The copolymers with either tetrafluoroethylene or trifluoroethylene showed definite elastomeric properties and could be cured at room temperature with hexamethylenediamine (52).

B. Cross-Linking of Modified Polymers

The low tensile strength exhibited by the tetrafluoroethylene-trifluoronitrosomethane-amine systems was attributed to inferior cross-linking. To provide polymeric systems more amenable to curing, terpolymers with pendant functional groups were synthesized. One approach involved the polymerization of 4-nitrosotetrafluorobenzoic acid or 4-nitrosotetrafluorobromobenzene with tetrafluoroethylene and trifluoronitrosomethane (53). A terpolymer containing 0.5–2% of nitrosoperfluorobutyric acid or nitrosoperfluoropropionic acid was prepared in an analogous manner (54). The latter materials were readily cross-linked with chromium triperfluoroacetate. A tensile strength up to 2170 psi was attained with this system (using fine silica as the filler). Swelling in hydrocarbon solvents was negligible; however, the swelling in nitrogen tetroxide was somewhat higher than that of the tetrafluoroethylene-trifluoronitrosomethane copolymer. This effect was quite significant in carbonyl solvents.

As reviewed by Brown (55), a free carboxyl group provides a functional site amenable to a large variety of cross-linking processes. The nitroso rubber terpolymer $CF_3NO/HOOC(CF_2)_3NO/C_2F_4$ is no exception. Sheehan et al. (56) have obtained vulcanizates stable to the oxidizing action of nitrogen tetroxide by cross-linking this copolymer either with chromium trifluoroacetate or with dicyclopentadiene dioxide. Metallic oxides as represented by magnesium oxide, zinc oxide, and sodium aluminate also afforded a cross-linked product. Using magnesium oxide together with silica filler (Silstone 110) resulted in tensile strength of 3200 psi. The magnesium oxide-cured vulcanizates were thermoplastic at 300°F, showing the temperature dependence of an ionic bond.

Polyamines were found to cure the terpolymer, although only piperazine gave a satisfactory vulcanizate. As expected, aliphatic and aromatic hydroxy compounds functioned as cross-linking agents; however, the end-product did not demonstrate very good characteristics.

To preserve the optimum properties exhibited by the original tetrafluoroethylene-trifluoronitrosomethane copolymer, and yet use the carboxyl group in the establishment of nonfunctional cross-links, the Hunsdiecker reaction followed by zinc dehalogenation was applied:

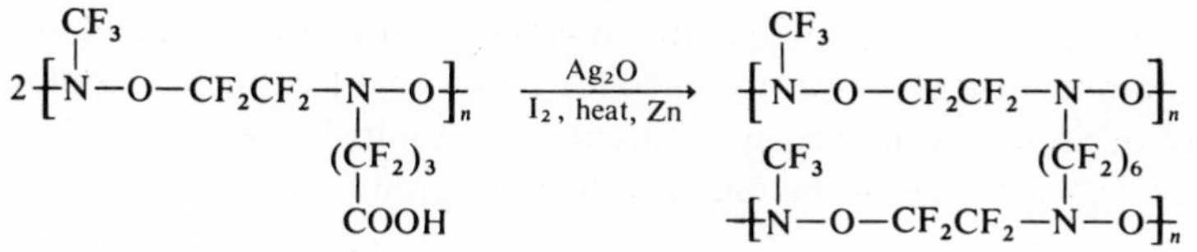

$$2 \left[\begin{array}{c} CF_3 \\ | \\ N-O-CF_2CF_2-N-O \\ | \\ (CF_2)_3 \\ | \\ COOH \end{array} \right]_n \xrightarrow[I_2,\ heat,\ Zn]{Ag_2O} \begin{array}{c} CF_3 \\ | \\ \left[N-O-CF_2CF_2-N-O \right]_n \\ | \\ CF_3 \\ | \\ \left[N-O-CF_2CF_2-N-O \right]_n \\ | \\ (CF_2)_6 \end{array}$$

Unfortunately, the product was weak and porous.

In another modification a terpolymer was prepared in which the free carboxylic acid was substituted by its methyl ester, e.g., the formulation $CF_3NO/CH_3OC(O)(CF_2)_2NO/CF_2CF_2$. This polymer could be cross-linked with peroxides (57). A number of mechanisms could be postulated, yet in view of the absence of more detailed experimental results this would amount to pure speculation.

A variety of other terpolymers containing cross-linking sites were prepared by copolymerizing tetrafluoroethylene and trifluoronitrosomethane with dienes or with fluorinated nitroso compounds containing hydrogen or one of the heavier halogens. None of the vulcanizates obtained with these polymers measured up to those prepared from the polymers containing the free or esterified carboxyl group.

C. Conclusions and Stability Considerations

The cross-linking of the copolymer of trifluoronitrosomethane and tetrafluoroethylene was accomplished with diamines but some degradation took place during the cure and the final product had relatively low tensile strength. A terpolymer in which pendant carboxyl groups were incorporated afforded a system susceptible to vulcanization by a large selection of agents. However, the presence of carboxyl groups resulted in somewhat decreased oxidative stability as compared to the original copolymer system.

IV. PERFLUORINATED TRIAZINE ELASTOMERS

Polyperfluoroalkyl s-triazines exhibit elastomeric characteristics in addition to a remarkable thermal and oxidative stability. The early synthetic procedures were developed by Brown (58–60). The elastomeric character of these materials is due to a three-dimensional network represented by IX.

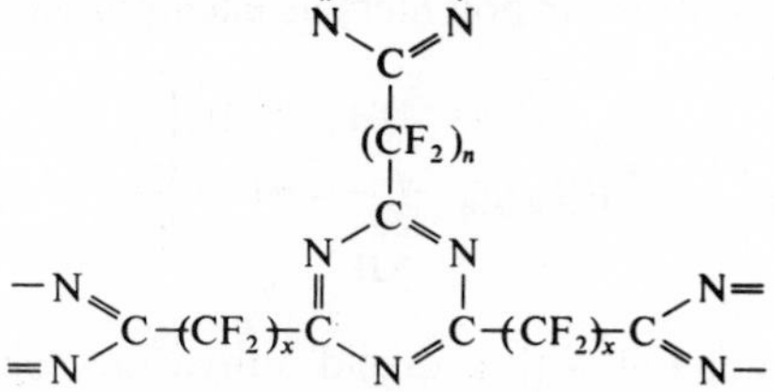

IX

Basically the synthesis involves the interaction of a perfluorinated dinitrile with ammonia at low temperatures followed by heat treatment. Employing a mixture of amidines derived from mono- and dinitriles, a certain degree of linearity was achieved in the final product, as shown by X.

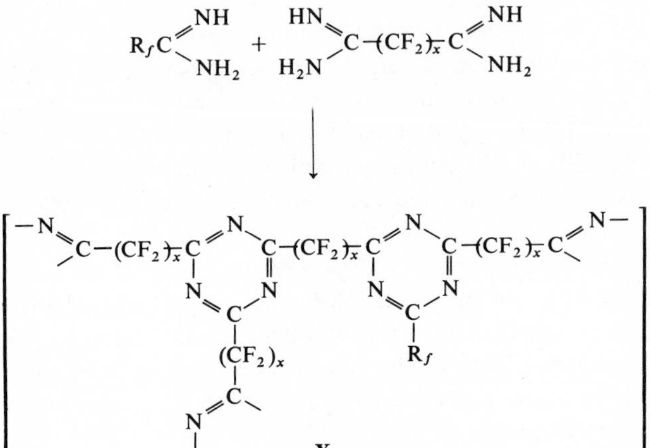

Investigations of these polymers disclosed in addition to cross-links the presence of pendant amidine groups, which provided sites adaptable to a variety of curing processes (61). Cross-linking was achieved with diamines, epoxides, diisocyanates, dinitriles, and diamidines. Actually any reagent susceptible to reaction with an active hydrogen was found to be effective. Brown and Schuman (62), in their triazine polymerization studies, found that a stable intermediate of form XI can be prepared. This material on treatment

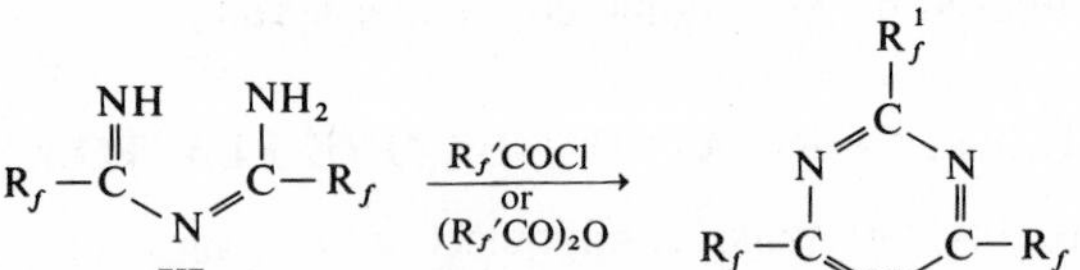

with acyl chlorides or preferably acid anhydrides gave essentially only one type of triazine. This concept was successfully expanded by Dorfman et al. (63) into the synthesis of linear polymers as exemplified by XII.

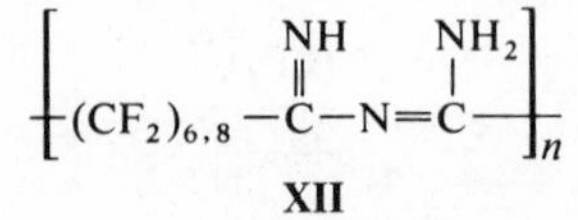

Interaction of this material with a diacid anhydride, e.g., perfluoroglutaric anhydride, produced a system wherein pendant carboxyl groups furnished cross-linking sites (64). By admixing a certain proportion of monofunctional perfluoroacid anhydride the number of eventual cross-links could be controlled. The subsequent cross-linking reaction via decarboxylation of the silver salt with formation of perfluoroalkane chains required temperatures above 265°C:

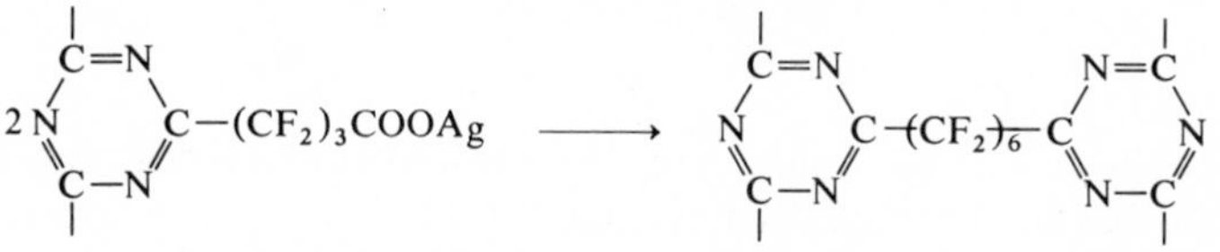

$$2\ \underset{\substack{\\ C-N}}{\overset{\substack{C=N \\ }}{N}}\ \ C-(CF_2)_3COOAg \longrightarrow \underset{\substack{C-N}}{\overset{\substack{C=N}}{N}}\ C-(CF_2)_6-C\ \underset{\substack{N-C}}{\overset{\substack{N=C}}{N}}$$

but the elastomers derived by this procedure did not exhibit the desired properties.

Another approach to introduce crosslinking sites into the polymer XII utilized cyanoperfluorobutyryl chloride instead of the difunctional acid anhydride:

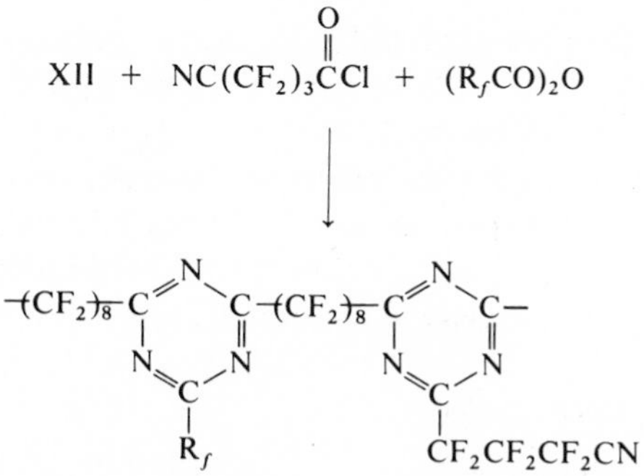

$$XII\ +\ NC(CF_2)_3\overset{\overset{\displaystyle O}{\|}}{C}Cl\ +\ (R_fCO)_2O$$

$$\downarrow$$

$$-(CF_2)_8-C\underset{\substack{N\\ \|\\ C}}{\overset{\substack{N}}{\diagdown}}C-(CF_2)_8-C\underset{\substack{N\\ \|\\ C}}{\overset{\substack{N}}{\diagdown}}C-$$

$$\quad\quad\quad R_f \quad\quad\quad\quad\quad\quad CF_2CF_2CF_2CN$$

The free cyano-groups could be then cross-linked catalytically into triazine structures (65), affording elastomers of varying tensile strengths, generally from 300 to 1500 psi. A variety of metal oxides, i.e., silver oxide, cadmium oxide, and lead oxide, were found to effectively catalyze this process. Materials such as zirconium acetylacetonate, zinc acetylacetonate, and tetraphenyl tin also promoted the cyclization (64).

A system where a cross-link is a logical extension of the polymer backbone, thus avoiding any structure disruption, should insure maximum thermal stability. Cross-linking through the triazine ring formation as discussed previously gives just such a system. However, even here the residues of the catalysts or cyclization promoters definitely enhance thermal degradation as compared to the untreated triazine gum (64). This effect was particularly evident in the case of silver oxide, yet almost negligible with tetraphenyl tin. From this finding it can be deduced that at the temperatures where the onset of decomposition occurs, the relatively volatile catalysts are no longer present and thus can have no influence on the thermal degradation.

V. FLUOROSILICONE RUBBER

The general form

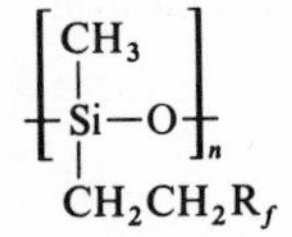

$$\left[\begin{array}{c} CH_3 \\ | \\ Si-O \\ | \\ CH_2CH_2R_f \end{array}\right]_n$$

represents a family of fluorinated polysiloxanes. These materials exhibit resistance to fuels and chemical solvents in addition to the thermal stability and low-temperature characteristics associated generally with silicone rubbers. Methyl 3,3,3-trifluoropropyl siloxane is available commercially as Silastic LS-53 (Dow Corning Corporation). Both aromatic and aliphatic peroxides —2,4-dichlorobenzoyl peroxide, benzoyl peroxide, ditertiarybutyl peroxide, and 2,5-dimethyl-2,5-di(t-butyl-peroxy) hexane (66)—were found to produce an effective cure.

The peroxide-induced cross-linking in linear polysiloxanes occurs via hydrogen abstraction from suitable carbon atoms by the initiator radicals. The subsequent interaction of the polymer radicals establishes the cross-links. In the case of methyl 3,3,3-trifluoropropyl siloxane, several potential sites exist for hydrogen abstraction. However, studies (67) have shown that the methyl group is attacked in preference to the 3,3,3-trifluoropropyl chain. The following schematic gives a general picture of the cross-linked chains:

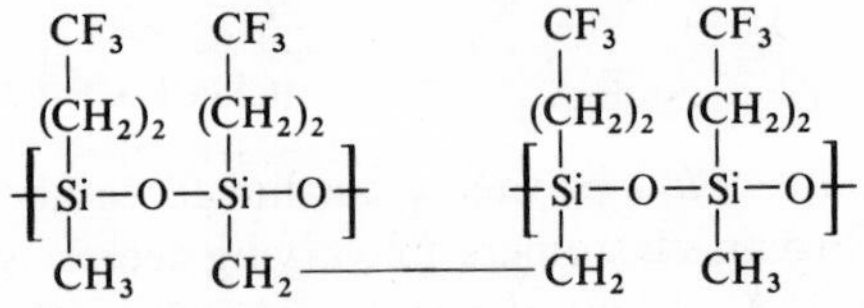

$$\begin{array}{cccc} CF_3 & CF_3 & CF_3 & CF_3 \\ | & | & | & | \\ (CH_2)_2 & (CH_2)_2 & (CH_2)_2 & (CH_2)_2 \\ | & | & | & | \\ \left[Si-O-Si-O\right] & & \left[Si-O-Si-O\right] \\ | & | & | & | \\ CH_3 & CH_2 ——— & CH_2 & CH_3 \end{array}$$

VI. FLUORINATED POLYESTERS

A whole spectrum of partially fluorinated polyesters was synthesized by Schweiker and Robitschek (68) by interaction of aliphatic dicarboxylic acids, their dithia derivatives, e.g. $(CH_2SCH_2COOH)_2$, and their fluorinated analogues, e.g., $(CF_2)_3(COOH)_2$, with fluorinated diols. Of the polymers obtained, the most promising were the hexafluoropentylene adipate elastomers. However, these materials do not exhibit as good a thermal stability and solvent resistance as the fluoroelastomers discussed in the preceding sections.

Polyesters generally can be effectively cured by peroxides. The reaction proceeds by the usual path: (1) decomposition of the initiator to yield free-radical fragments; (2) abstraction of hydrogen to form a polymer radical; and (3) combination of two polymer radicals to give a carbon-carbon cross-link. It has been established that in polymers in which the interunit functional groups connect linear methylene chains, the favored site for hydrogen abstraction is usually in alpha position to the activating functional group.

Polyhexafluoropentylene adipate was successfully cured with dicumyl peroxide (69). In view of the mechanism operative in the polyester series a similar process is postulated for the fluorinated materials affording a system represented in a simplified form by XIII.

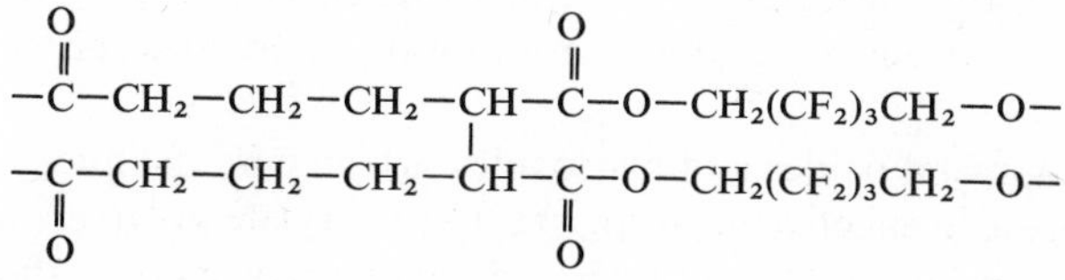

$$-\underset{\underset{O}{\|}}{C}-CH_2-CH_2-CH_2-\underset{|}{CH}-\underset{\underset{O}{\|}}{C}-O-CH_2(CF_2)_3CH_2-O-$$
$$-\underset{\underset{O}{\|}}{C}-CH_2-CH_2-CH_2-\underset{|}{CH}-\underset{\underset{O}{\|}}{C}-O-CH_2(CF_2)_3CH_2-O-$$

XIII

In a modified approach hydroxyl-terminated polyesters derived from adipic acid and fluorinated diols were extended with diisocyanate and then cured with peroxides (70). Using this procedure carboxyl and hydroxyl end groups of potential chemical reactivity are eliminated. The actual cross-linking occurs via CH radicals generated adjacent to the carbonyl group in a manner analogous to the urethane and unextended polyester polymers. Chain extension of hydroxyl-terminated polymers was also accomplished by reaction with epoxides.

Introduction of polyfunctional alcohols or acids into the original polymerization mixture creates pendant functional groups which can be then cross-linked into a three-dimensional network without the need of peroxides or other free-radical initiators. Both cross-linking of the pendant groups and chain extension via the terminal entities have been successfully accomplished by the use of alkylenimine derivatives (71).

VII. FLUORINATED ACRYLIC ELASTOMERS

A series of fluorinated polyacrylates was investigated as candidates for fuel-resistant, thermally stable elastomers with good low-temperature properties (72–76). The basic structure is represented by XIV.

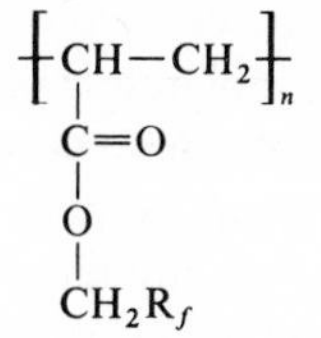

$$\begin{array}{c} \left[\underset{|}{CH}-CH_2\right]_n \\ \underset{|}{C}=O \\ \underset{|}{O} \\ CH_2R_f \end{array}$$

XIV

The R_f grouping can be either a fully fluorinated or partially fluorinated entity, or a perfluorinated ether chain. Of the materials studied, poly (1,1-dihydroperfluorobutyl) and poly (1,1-dihydroperfluoropropylperfluoromethoxy acrylates) exhibited the best properties. The low-temperature behavior of the latter compound was a definite improvement over the 1,1-dihydroperfluorobutyl acrylate elastomer. To improve the low-temperature flexibility even further, copolymerization with butadiene was investigated; the derived products showed markedly increased swelling in solvents. In addition, the butadiene copolymers, in view of residual unsaturation, were susceptible to oxygen attack (77).

In general, metal oxides and hydroxides, which cure the fluorine-free polyacrylates, are also effective in curing the fluoroacrylate polymers. A combination of calcium hydroxide and hydrated sodium metasilicate affords vulcanizates with good tensile strengths. Whether cross-linking occurs via hydrolysis followed by salt formation with divalent metal, or whether Claisen condensation involving the alpha hydrogen atom occurs, is unknown.

Polyfunctional amines were found to be the most promising vulcanizing agents for the development of practical elastomers. A whole spectrum of aliphatic diamines and polyamines, both primary and secondary, was investigated. During the amine cure the fluorinated alcohol is liberated and amide linkages are formed in the polymer. Diimides or imidazoline-rings apparently are not produced (77). The copolymers of butadiene and fluorinated acrylates, due to the presence of unsaturation, are susceptible to sulfur vulcanization in addition to the other cross-linking agents employed usually in the polyacrylate polymers.

In comparison with the other cross-linking systems described, amines in the presence of carbon black reinforcement seem to give the best end-product. However, as noted previously, the fluorinated acrylates were superseded by other, more advanced elastomers.

VIII. NEW FLUORINATED ELASTOMERS

In this series are included the copolymers of trifluorovinyl ether, CF_3-$OCF=CF_2$, trifluorovinyl trifluoroethyl ether, $CF_3CH_2OCF=CF_2$, and materials derived from fluorothiocarbonyl compounds.

The vinyl ethers gave elastomers on copolymerization with vinylidene fluoride; in some instances the polymerization conditions also led to crosslinking (78). It is believed that these copolymers, having essentially the reactive centers of Viton elastomers, should endergo curing processes found effective in the hexafluoropropene-vinylidene fluoride copolymers.

A new and unusual series of polymers has been obtained through homo- and copolymerization of thiocarbonyl fluoride, CF_2S, and its close relatives (79–81).

A number of these exhibited elastomeric properties. Among others the high-molecular-weight polymer of CF_2S was found to be a tough, highly resilient elastomer even when uncured. Copolymerization of thiocarbonyl fluoride with 3-butenyl acetate followed by methanolysis affords an elastomer with pendant hydroxyl groups which provide sites for cross-linking with diisocyanates.

Another copolymer of thiocarbonyl fluoride and allyl chloroformate presents an elastomeric system with functional acid chloride groups which are readily cross-linked with zinc oxide. Incorporation of a given mole percent of the comonomer results in a predetermined number of potential cross-linking sites. Consequently, at the termination of the cure, no active groups are left for a possible degradative action by other chemical agents in the actual application.

References

1. P. J. Flory, "Principles of Polymer Chemistry," Cornell University Press, Ithaca, N.Y., 1953.
2. W. H. Stockmayer, *J. Chem. Phys.*, **11**:45 (1943).
3. W. H. Stockmayer, *J. Chem. Phys.*, **12**:125 (1944).
4. F. W. Billmeyer, Jr., "Textbook of Polymer Science," Interscience, New York, 1971.
5. S. C. Temin, *Encycl. Polymer Sci. Technol.*, **4**:331 (1966).
6. W. Postelnek, *Ind. Eng. Chem.*, **50**:1602 (1958).
7. J. C. Tatlow, *Rubber and Plastics Age*, **39**:33 (1958).
8. S. Dixon, D. R. Rexford, and J. S. Rugg, *Ind. Eng. Chem.*, **49**:1687 (1957).
9. J. S. Rugg and A. C. Stevenson, *Rubber Age (New York)*, **82**:102 (1957).
10. G. A. Gallacher and T. D. Eubank, *Rubber World*, **141**:827 (1960).
11. A. P. Terent'ev *et al.*, USSR Patent 171,567. May 26, 1965.
12. R. G. Spain, "Behavior of Some Elastomers in Petroleum Based Fluids at Elevated Temperatures," Division of Rubber Chemistry, A.C.S. Meeting, Cincinnati, Ohio, May 15, 1958.
13. E. W. Bergstrom, A.D 609,003. Technical Report AD 609,003, Available from NTIS. Compare: CA 63, 1972b (1965).
14. A. S. Novikov, F. S. Tolstukhina, and G. V. Chernov, *Kauchuk i Rezina*, **20**(12)30 (1961).
15. Ryong-Joon Roe and W. R. Krigbaum, *J. Polymer Sci. A*, **1**:2049 (1963).
16. M. E. Conroy, F. J. Honn, L. E. Robb, and D. R. Wolf, *Rubber Age (New York)*, **76**:542 (1955).
17. L. E. Robb, F. J. Honn, and D. R. Wolf, *Rubber Age (New York)*, **82**:286 (1957).
18. W. R. Griffin, *Rubber World*, **136**:687 (1957).
19. C. D. Dipner and N. J. Cranford, U.S. Patent 3,245,968. April 12, 1966.
20. H. R. Davis, Jr., F. J. Honn, C. B. Griffis, and J. C. Montermoso, U.S. Patent 3,071,565. Jan. 1, 1963.
21. P. O. Tawney and R. P. Conger, Belgian Patent 625,285. Nov. 14, 1963.
22. Netherlands Appl. 6,509,095 (Feb. 1, 1966).
23. K. L. Paciorek, L. C. Mitchell, and C. T. Lenk, *J. Polymer Sci.*, **45**:405 (1960).
24. A. S. Novikov, F. A. Galil, N. S. Gilinskaya, and Z. N. Nudel'man, *Kauchuk i Rezina*, **21**(3):4 (1962).

25. J. F. Smith and G. T. Perkins, *J. Appl. Polymer Sci.*, **5**:460 (1961).
26. K. L. Paciorek, B. A. Merkl, and C. T. Lenk, *J. Org. Chem.*, **27**:266 (1962).
27. R. L. Pruett, J. T. Barr, K. E. Rapp, C. T. Bahner, J. D. Gibson, and R. H. Lafferty, Jr., *J. Am. Chem. Soc.*, **72**:3646 (1950).
28. S. C. Temin, *Encycl. Polymer Sci. Technol.*, **4**:385 (1966).
29. M. Nimoy, U.S. Patent 3,183,207. May 11, 1965.
30. K. L. Paciorek, unpublished results.
31. R. C. Ferguson, *J. Am. Chem. Soc.*, **82**:2416 (1960).
32. M. I. Bro, *J. Appl. Polymer Sci.*, **1**:310 (1959).
33. J. F. Smith, *Rubber World*, **142**:102 (1960).
34. J. D. Park, H. V. Holler, and J. R. Lacher, *J. Org. Chem.*, **25**:990 (1960).
35. K. L. Paciorek, W. G. Lajiness, and C. T. Lenk, *J. Polymer Sci.* **60**:141 (1962).
36. A. S. Novikov, F. A. Galil, N. A. Slovokhotova, and T. N. Dyumaeva, *Vysokomolekul. Soedin*, **4**:423 (1962).
37. B. A. Merkl and P. Davis, Technical Report, ASD-TDR 62-30, Part II. Wright-Patterson Air Force Base, Ohio.
38. J. F. Smith, *Rubber World*, **140**:263 (1959).
39. E. F. Cluff, E. K. Gladding, and R. Pariser, *J. Polymer Sci.*, **45**:341 (1960).
40. A. S. Novikov, F. A. Galil, N. A. Slovokhotova, and T. N. Dyumaeva, *Vysokomolekul. Soedin., Khim. Svoistva i Modifikatsiya Polimerov. Sb. Statei*, **1964**:160.
41. N. Luijendijk, *Rev. Gen. Caoutchouc Plastiques*, **43**(7–8):981 (1966).
42. R. G. Spain, K. L. Paciorek, W. G. Lajiness, and B. A. Merkl, WADC Technical Report 55–492, Part VI.
43. A. S. Novikov, F. A. Galil, and N. S. Gilinskaya, *Kauchuk i Rezina*, **21**(2):4 (1962).
44. G. A. Morneau, P. I. Roth, and A. R. Shultz, *J. Polymer Sci.*, **55**:609 (1961).
45. G. H. Crawford, D. E. Rice, and B. F. Landrum, *J. Polymer Sci.* A, **1**:565 (1963).
46. G. H. Crawford, D. E. Rice, and J. C. Montermoso, 6th JANAF Conference on Elastomers, Boston, 1960, p. 644.
47. G. H. Crawford, U.S. Patent 3,213,009. Oct. 19, 1965.
48. J. C. Montermoso, C. B. Griffis, and A. Wilson, 6th JANAF Conference on Elastomers, Boston, 1960, p. 672.
49. A. R. Shultz, N. Knoll, G. A. Morneau, *J. Polymer Sci.*, **62**:211 (1962).
50. A. M. Borders, U.S. Patent 3,072,625. Jan. 8, 1963.
51. J. C. Montermoso, *Chem. Eng. Prog.*, **57**:98 (1961).
52. G. H. Crawford, *J. Polymer Sci.*, **45**:259 (1960).
53. J. Green, N. Mayes, and E. Cottrill, Abstrs. Papers, 152nd Meeting Am. Chem. Soc., New York, September 1960, K-3.
54. C. B. Griffis and M. C. Henry, 7th SAMPE Nat'l. Symp. Los Angeles, 1964.
55. H. P. Brown, *Rubber Chem. Technol.*, **36**:931 (1963).
56. W. R. Sheehan, N. B. Levine, and J. Green, Tech. Report ML-TDR-64-107, Part IV, April 1967.
57. E. C. Stump, W. H. Oliver, C. D. Padgett, and C. B. Griffis, Abstrs. Papers, 154th Meeting Am. Chem. Soc., Div. of Fluorine Chem., September 1967, Chicago, K-11.
58. W. L. Reilly and H. C. Brown, *J. Org. Chem.*, **22**:698 (1957).
59. W. L. Reilly and H. C. Brown, *J. Am. Chem. Soc.*, **78**:6032 (1956).
60. H. C. Brown, *J. Polymer Sci.*, **44**:9 (1960).
61. A. D. Delman and A. E. Ruff, U.S. Patent 3,218,270. Nov. 16, 1965.
62. H. C. Brown and P. D. Schuman, *J. Org. Chem.*, **28**:112 (1963).
63. E. Dorfman, W. E. Emerson, R. L. K. Carr, and C. T. Bean, *Rubber Chem. Technol. Part 2*, **39**(4):1175 (1966).

64. E. Dorfman, W. E. Emerson, and R. L. K. Carr, Fourth Intern. Symposium, Fluorine Chem., Estes Park, Colo., 1967.
65. W. E. Emerson and E. Dorfman, Abstrs. Papers 152nd Meeting Am. Chem. Soc., New York, September 1966, K-22.
66. Dow Corning Bulletin 09–193, December 1966.
67. S. C. Temin, *Encycl. Polymer Sci. Technol.*, **4**: 343 (1966).
68. G. C. Schweiker and P. Robitschek, *J. Polymer Sci.*, **24**: 33 (1957.)
69. J. C. Montermoso, *Rubber Chem. Technol.*, **34**:1530 (1961).
70. G. C. Schweiker and P. Robitschek, U.S. Patent 3,016,360. Jan. 9, 1962.
71. N. L. Watkins, Jr., U.S. Patent 3,198,770. Aug. 3, 1965.
72. D. W. Codding, T. S. Reid, A. H. Ahlbrecht, G. H. Smith, Jr., and D. R. Husted, *J. Polymer Sci.*, **15**: 515 (1955).
73. F. A. Bovey, J. F. Abere, G. B. Rathmann, and C. L. Sandberg, *J. Polymer Sci.*, **15**: 520 (1955).
74. F. A. Bovey and J. F. Abere, *J. Polymer Sci.*, **15**: 537 (1955).
75. A. H. Ahlbrecht, U.S. Patent 2,642,416. June 16, 1953.
76. A. H. Ahlbrecht, U.S. Patent 2,839,513. June 17, 1958.
77. P. J. Stedry, J. F. Abere, and A. M. Borders, *J. Polymer Sci.*, **15**: 558 (1955).
78. W. S. Durell, E. C. Stump, Jr., G. Wextmoreland, and C. D. Padgett, *J. Polymer Sci. Part A*, **3**: 4065 (1965).
79. W. J. Middleton, H. W. Jacobson, R. E. Putman, H. C. Walter, D. G. Pye, and W. H. Sharkey, *J. Polymer Sci. Part A*, **3**: 4115 (1965).
80. A. L. Barney, J. M. Bruce, Jr., J. N. Coker, H. W. Jacobson, and W. H. Sharkey, *J. Polymer Sci. Part A-1*, **4**: 2617 (1966).
81. W. J. Middleton, U.S. Patent 3,240,765. March 15, 1966.

11. RADIATION CHEMISTRY OF FLUOROCARBON POLYMERS

ROLAND E. FLORIN, *Polymer Chemistry Section, National Bureau of Standards, Washington, D.C. 20234*

Contents

I. INTRODUCTION

Fluorocarbon polymers are affected by radiation in one of two distinct ways, depending on their composition. Those that are completely halocarbon, like polytetrafluoroethylene and polychlorotrifluoroethylene, suffer rapid molecular-weight degradation, with very moderate cross-linking or none. Those that contain some hydrogen, like the vinylidene fluoride copolymer elastomers, are rapidly cross-linked into a network. Intermediate cases are rare; the most notable is the mild cross-linking of TFE-HFP copolymer. Even more than with other polymers, the details depend strongly upon conditions of irradiation, such as atmosphere, impurities, temperature history, and crystalline, rubbery, or glassy state of the polymer. Polytetrafluoroethylene is in a class of its own because of the great volume of studies upon it, and the incomplete knowledge gained from them for lack of convenient solvents and reliable molecular-weight data.

In general, radiation damage to organic compounds is an effect not so much of the original particles, such as alpha particles, slow neutrons or X-ray photons, as of the electrons liberated by them in their passage through matter. The absorbed radiation dose is commonly measured in energy units—electron volts per gram, ergs per gram, or rads (1 rad = 100 ergs/g = 6.24 $\times 10^{13}$ eV/g). The *exposure dose* is measured in roentgens; a radiation field involving an exposure dose of 1 roentgen will given an absorbed dose of 0.871 rad in carbon but somewhat different values in other substances. If chain reactions are not present, the amount of chemical change will often be principally a function of the absorbed dose rather than of the dose rate, time, and quality of radiation considered separately. The "radiation yield" of a product is frequently expressed as the G-value, which is the number of molecules per 100 eV absorbed. Observed G-values usually fall in the range from 0.001 to about 5, barring chain reactions or complex energy transfer. As a general mechanism of action, the energetic electrons rapidly cause ionization and excitation to higher electronic states. Decomposition of excited molecules, and ion recombination, yield molecules and free radicals. Overall reactions resemble appreciably those brought about by known free-radical reagents, and free radicals are easily identified in irradiated products, but undetermined amounts of direct molecular and ionic reaction products must exist as well.

Amounts of reaction that may be small from the point of view of preparative organic chemistry can be critical in polymers because of the drastic effect on properties when even a few chain cuts are made in a molecule thousands of units long.

These matters are discussed systematically by Bovey (1), Charlesby (2), and Chapiro (3), for example; the last two contain abundant detail on

PTFE as well. Voluminous data from reports have also been collected and summarized by the REIC (4–6). For reference, some general numerical values are shown in Tables 1 to 3.

TABLE 1

Characteristic Radiation Doses for Various Effects

Dose (rad)	Effect
4	Natural background, average 30 years, sea level
400–700	50% lethal for man
10^5	Gas evolution from water; least stable plastics begin to lose tensile strength
10^8	Stable organic compounds evolve gas
10^9	Metals increase yield stress, most polymers and oils unusable
10^{11}	Carbon steel loses ductility

TABLE 2

Typical Dose Rates of Equipment

Dose rate (rad/hr.)	Apparatus
10^{10}	Electron accelerators
10^5 to 10^9	Reactors
10^8	Soft X-rays
10^5 to 10^6	Gamma-ray sources (Co^{60}, 10,000 curies)

TABLE 3

Estimated Times for Radiation-Chemical Processes

Time (sec)	Process
10^{-18}	Primary electron traverses molecule
10^{-16}	Secondary electron (5 eV) traverses molecule
10^{-14}	Molecular vibration; fast dissociation; electron capture
10^{-12}	Radical makes one diffusive jump; dielectric liquid relaxation
10^{-8}	Lifetime for allowed transitions
10^{-4}	Reaction, radical-solute, 1 M, 5 kcal diffusion between spurs
10^{-2}	Triplet-state radiation

II. POLYTETRAFLUOROETHYLENE

A. Molecular Weight

Profound deterioration of PTFE was noted in the earliest systematic studies (8). This was surprising because of its well-known chemical and thermal resistance. A temporary improvement of tensile impact strength in the range of a few hundred rads is followed by decline of tensile strength and of elongation at break. At very great doses, about 1000 Mrad, chunks of polymer break down to powder, and gas evolution and weight loss become prominent (9). There are also complex changes in the infrared, which may be due to double bonds and other structures. Free radicals are abundant and long-lived. The density increases up to rather large doses, 100 or perhaps 300 Mrad. The increase is associated with an increase in crystallinity, observed independently by either X-ray scattering or diagnostic IR bands. Ultimately, the melting transition temperature decreases.

The indications of molecular-weight decrease necessarily come from specialized, indirect molecular-weight methods. Those most closely evaluated numerically are related to rate of crystallization from the melt under a prescribed annealing schedule, i.e., the standard specific gravity (where crystallinity is observed via density measurement) and the inherent specific gravity (where the observation is made by infrared); in the second, no special precautions are needed against voids. Methods based on viscoelastic properties of the melt have been used as well, including melt viscosity (10,14), zero strength time (10,14), and instantaneous and delayed elastic modulus. The latter changes gradually with degradation from a positive to a negative temperature coefficient (10–13), which is reasonable behavior for rubber elasticity of entanglements (of positive temperature coefficient) that become less numerous as chains become shorter. Some of the viscoelastic methods used should be nearly equivalent to Tobolsky's "maximum relaxation time," which has been correlated with molecular weight. Caution in using the correlation is suggested by the unique molecular-weight exponent, 0.78, versus the more usual 3.4, and by the incidence of permanent thermal degradation at temperatures a little above the best temperature of measurement, 380°C (15,16). On the whole, the standard specific gravity and inherent specific gravity methods are most extensively calibrated, although they too, like most conceivable methods, involve a severe heat treatment before measurement, during which less thermally stable structures such as branches might decompose.

The most explicit molecular-weight data are given by Bro (17) and by Ferse, Koch, and Wuckel (18). Unfortunately, agreement is poor, at least by implication, and the two sets of authors may have recorded essentially different events. When the irradiated sample is subsequently sintered in air to prepare it for measurement, the increase in inherent specific gravity is appreciably greater than when the sample is sintered in vacuum (17). The reality of this difference is confirmed by the enormous differences of melt viscosity, as illustrated in Table 4. The results shown in Table 5 and Fig. 1, where samples were consistently sintered in air, lead to the divergent scission G-values of 1.3 (vacuum, initial, sheet), 0.08 (vacuum, late, sheet), and 4.5 (air, film) for the given conditions, if molecular weights are computed according to Sperati. From the same data the authors using Osten's molecular-weight relation get the higher values 2, 0.15, and 5.6.

TABLE 4

Molecular Weights and other Properties of Vacuum-Irradiated PTFE[a]

Dose (Mrad)[b]	ISG	$\bar{M}_n$	$\eta(380°C)$ (poise)	Remarks
0	2.163	$>10 \times 10^6$	3.2×10^{11}	Initial (Bro)
15	2.240	2.5×10^6	2.8×10^9	Vacuum sinter
75	2.244	2.1×10^6	1.4×10^8	Vacuum sinter
75	2.266	0.9×10^6	8.0×10^6	Air sinter
0	2.237	2.8×10^6	4.9×10^{11}	Reference sample
0	2.250	1.7×10^6	6.7×10^9	Reference sample
0	2.264	1.0×10^6	5.8×10^8	Reference sample
0	2.1795	$[15.1 \times 10^6]$	n.d.	Initial (Ferse)
0.2	2.221	4.8×10^6	n.d.	Air sinter
0.84	2.227	4.0×10^6	n.d.	Air sinter

[a] All samples vacuum irradiated; subsequent sintering as noted. References 17 and 18
[b] Dose calibration from Bro (19).

TABLE 5

Molecular Weights of One Type of Irradiated Polytetrafluoroethylene[a]

Dose (R)	s.s.g.	$10^{-6} M$ (Osten)	$10^{-6} M$ (Sperati)
0	2.1530	13.1	46.8
1.10^3	2.1530	13.1	46.8
5.10^3	2.1555	12.3	43.6
1.10^4	2.1627	10.3	34.3
5.10^4	2.1940	4.7	12.2
1.10^5	2.2090	3.3	7.3
5.10^5	2.2420	1.4	2.5
1.10^6	2.2560	1.0	1.6

[a] Reference 18, their "type C"; low crystallinity, high molecular weight.

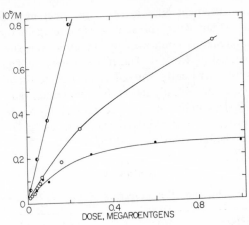

Fig. 1.   Reciprocal of molecular weight of irradiated polytetrafluoroethylene (18). (◐) foil in air; (○) sheet in air; (●) sheet irradiated in vacuum then sintered in air. Radiation, ^{60}Co gamma at 0.6×10^6 r/hr.

These results, together with others not quoted in the table, show a dependence upon sample quality (which we can rationalize as initial molecular weight and crystallinity) and a strong dependence upon the presence of oxygen. The high initial G-value in vacuum is attributed to incompletely removed oxygen, which is used up later. The low steady-state $G_{sc} = 0.08$ or 0.15 is associated with a balance of scission and either recombination or crosslinking after the oxygen is gone. The scission G-values that can be deduced from the sparse high-dose data of Bro (17) are still lower, and by a large factor—0.02 at 0–15 Mrad, 0.0012 at 15–75 Mrad for vacuum-sintered samples, and 0.012 at 75 Mrad for samples sintered in air, as shown in Fig. 2.

Variations with initial character of sample (18,20), and with atmosphere were observed in more detail in the study of changes of density as such, and of tensile strength and elongation at break. An adjustment in which initial quality is represented by a simulated initial dose appears able to bring some elongation-at-break curves on different samples into fairly good correspondence above 10^5 R dose. A similar device, sometimes successful, sometimes not, had been applied by Charlesby to other polymers. Sample quality correlations involving tensile are more complex. Nearly linear plots were given by the reciprocals of tensile and elongation versus dose (18). The extreme influence of atmosphere had been known earlier in studies of tensile strength (22), elongation at break, and density (24), isolated instances being indeed reported but not much noted in the earliest work (8). The halogens have an accelerating effect like oxygen, perhaps a little weaker (20), whereas there are reports that immersion in oil more nearly resembles vacuum (24). A few curves are shown in Figs. 3 and 4.

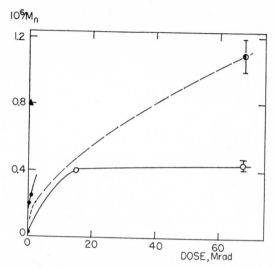

Fig. 2. Reciprocal of molecular weight of irradiated polytetrafluoroethylene (17,18). (▲) irradiated in air (17); (◐) (17), (●) (18), irradiated in vacuum, sintered in air. (○) irradiated and sintered in vacuum (17).

In work on very thick pieces, atmosphere ought to be unimportant (24), since initial oxygen is very unlikely to diffuse out completely enough during preliminary evacuation and new oxygen is likewise unable to penetrate rapidly to the interior during irradiation. Incidental observations on rods seem to confirm this, at least for rapid pile irradiations (26). The ultimate break up of large samples is sometimes attributed to internal stresses and specifically to buildup of enormous pressures by dissolved gaseous products unable to diffuse out rapidly enough. Another factor, at least in air at high doses, is surface embrittlement and crack growth, by which the whole piece is laid open to attack (25). The low G (scissions) and the observed course of mechanical properties displayed above indicate that under irradiation in a good vacuum, PTFE, although nowhere near the highly resistant polymers such as polystyrene and definitely inferior even to polyethylene, is not unsuitable for moderate-dose applications, e.g., out to tens of megarads with partial retention of properties. However, at the much higher but somewhat indefinite exposures used in a 20-year simulated space environment, all useful properties were lost (27).

B. Mechanical Properties

Other tensile and elongation data are given in the references (5,6,21–23, 28,29). In comparing such data, the elongations are usually considerably greater in thin films than in thick sheets. Flex life decreases as well (17). Creep is temporarily accelerated during irradiation (30). The extent is fairly

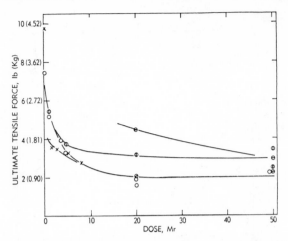

Fig. 3.   Ultimate tensile loads of polytetrafluoroethylene strips irradiated in chlorine (21). (⊖) Cl_2 1 torr, 10 Mr/hr; (○) Cl_2 1 torr, 0.3 Mr/hr; (○) Cl_2 10, 100 and 760 torr, 0.3 Mr/hr; (○) Cl_2 760 torr, 10 Mr/hr; (×) second lot of polymer at 760 torr and 0.3 Mr/hr. Temperature 25°C. (Reprinted from SPE Transactions Vol. 3 No. 4 Oct. 1963.)

small, less than twofold at 4 Mrad hr^{-1}, but greater than for most other polymers, e.g., polyisobutylene at 21 Mrad hr^{-1}. Reversible bond breaks are suggested as the cause.

Viscoelastic properties measured at high temperature, on PTFE which was irradiated in air at room temperature (11), are plotted in Figs. 5 and 6. A somewhat high preliminary $G_{sc} = 10$ was deduced from the data. High-temperature thermomechanical studies, with a penetrometer (31), should reflect the same basic properties and remain feasible on very badly degraded specimens.

High irradiation temperatures speed up loss of tensile and elongation (21,23) and, less markedly, the decrease of the 380°C melt viscosity. A dose of 5 Mrad at 350°C produced the same change as 10 Mrad at room temperature, a lowering to values near 5×10^6 poise (19). One contrary effect is reported—consistometer curves indicated less rapid stiffening than usual when irradiation was conducted above the crystal melting transition (31). According to a few sources (18,29), dose rate has a mild effect on elongation at break (Table 6).

C. Density, Crystallinity, and Permeability

Density changes reflect changes in crystallinity to a large extent (10,23,26, 32,33). They were early correlated with crystallinity increases measured in other ways. Specific volumes have been recorded over a temperature range −80 to 150°C and a dose range to 10^9 rad (26,32,33) (Figs. 7 and 8).

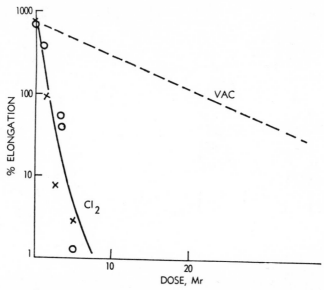

Fig. 4. Elongation at break of polytetrafluoroethylene strips irradiated in chlorine and in vacuum (21). Temperature 25°C, dose rate 0.3 Mr/hr Co⁶⁰ γ. (---) vacuum; (——), ($\times$), ($\bigcirc$), two lots of polymer in Cl_2 at 10,100 and 760 torr (pressure effect nil). (Reprinted from SPE Transactions Vol. 3 No. 4 Oct. 1963.)

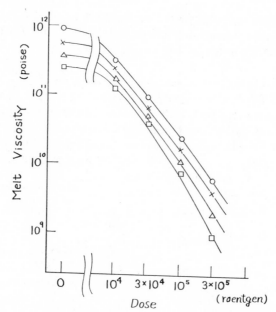

Fig. 5. Melt viscosity of polytetrafluoroethylene (13). Irradiated in air at room temperature; measurement in air at ($\bigcirc$) 335°C; ($\times$) 353°C; ($\triangle$) 371°C; ($\square$) 391°C.

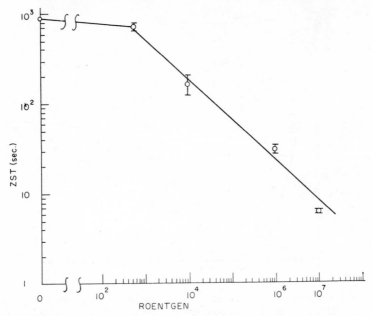

Fig. 6.   Zero strength time of polytetrafluoroethylene irradiated in air at room temperature (14).

Radiation quality has a minor but unequivocal effect, since the functional form of dose-rate dependence differs slightly for γ and reactor radiation (Fig. 9). A maximum density occurs around 100 Mrad, after which density decreases (32).

Crystallinity changes have been measured most directly by X-ray diffraction (10,34), which is also the calibration for infrared (35) and density (36) methods of estimation. Figure 10 shows the course of crystallinity to large doses from X-ray data; some infrared data are also available (32,33). If G_{sc} = 0.3, then at 300 Mrad chains are short enough to broaden the X-ray line. In principle, NMR spectroscopy can also be used to estimate crystallinity,

TABLE 6

Influence of Dose Rate on Elongation of PTFE
Irradiated to 10^5 R (18)

Dose rate, R hr^{-1}	Elongation at break %
(initial value)	344 ± 6
2.5 × 10³	262 ± 8
3.0 × 10⁴	258 ± 10
6.0 × 10⁵	192 ± 8

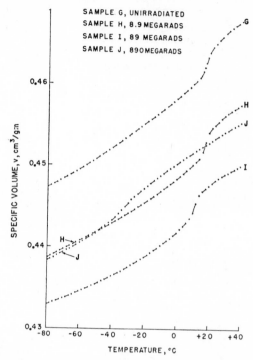

Fig. 7. Specific volume of irradiated polytetrafluoroethylene, lower temperature range (26).

by separating a very wide crystalline peak (*ca.* 10 *G*) from a narrower, amorphous one. This is difficult to apply in PTFE (34).

Furthermore, the area of the melting-transition endotherm, observed in differential scanning calorimetry or differential thermal analysis (18,37), is proportional to the percentage crystallinity, although usually less precisely known than other measures.

An incidental observation in calorimetry is that the melting transition is lowered and broadened (Fig. 11) (18,31,37,38). The data can lead to a further estimate of G_{sc} which, probably because of its poor reliability, seems not to have been done by any investigator. The entropy of fusion is 1.14 cal deg^{-1} per CF_2 group and the accepted T_m is 600°K (39). By applying the earliest notions of thermodynamics of the melting transition (40), the MP lowering should be 1.05×10^3 deg per unit mole fraction of foreign groups. The observed lowering of 26°K after 126 Mrad (18) indicates a mole fraction of .026 foreign groups per CF_2 have been introduced; counting all these as chain ends and two chain ends per molecule, the indicated molecular weight is about 3800. The G_{sc} is about 2.0, which is in the same range as the previous G_{sc} in air. Similar deductions would follow from the other data on melting transition (31,37). A crystal transition near 18°C is also lowered.

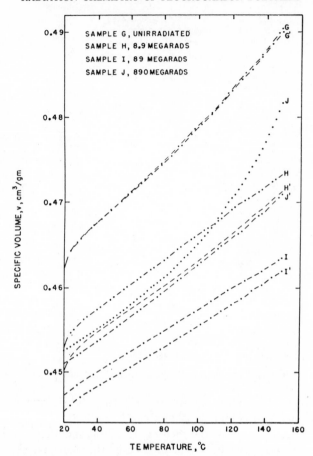

Fig. 8.   Specific volume of irradiated polytetrafluoroethylene, higher temperature range.
G′ to J′ are the same as G to J, but after prolonged heating (33).

It has been speculated several times that although the percentage crystallinity increases, the perfection of crystallites is impaired, which could
ultimately decrease density. Heat treatment after irradiation, 300°C for 4 hr,
can raise crystallinity to very high levels (38). According to Fig. 10 and other
data, the crystallinity itself at least decreases appreciably, but Kline (33)
reports a region, 100–300 Mrad, where density is decreasing while crystallinity,
measured by infrared, continues to increase. The occurrence of low-angle
X-ray scattering, in addition to the major X-ray peak connected with crystallinity, has been attributed to "microvoids" (41). Some such increase of
microvoids may also lie back of the otherwise incredible observation of
increased permeability to Ar and He while density and crystallinity are
increasing with dose (42). For Ar, approximate values are as follows:

Dose (Mrad)	0	1	4
Diffusivity, D (cm^2 sec^{-1})	1×10^{-8}	9×10^{-8}	11×10^{-8}
Solubility, σ [cm^3 cm^{-3} (cm Hg)$^{-1}$]	$<1 \times 10^{-3}$	45×10^{-3}	57×10^{-3}

The permeability to helium exhibited a post effect, increasing from 0.4 $\times 10^{-10}$, immediately after irradiation, to 8×10^{-10} cm^3 cm^{-2} sec^{-1} cm (cm Hg)$^{-1}$, after 3 to 5 days of storage at room temperature. A temporary increase during irradiation was observed for helium at 15°C, from 8×10^{-10} cm^3 cm^{-2} sec^{-1} cm atm^{-1} before and after, to 14×10^{-10} during irradiation at a rate of 560 rad sec $^{-1}$ (43,44). By contrast, unirradiated 90% crystalline PTFE is stated to be 40 times less permeable to CO_2 than the 40% crystalline form (45). More normally behaving transport properties are also implicit in free-radical disappearance rates, presumably diffusion controlled, which are slower in crystalline regions of irradiated PTFE.

TABLE 7

Miscellaneous Properties of Irradiated PTFE (52)

Dose (Mrad)	Shear strength (psi)	Breaking energy (ft-lb in^{-1} of notch)	H$_2$O absorbed (%)	Haze (%)
0	3010	2.65	0.0053	5.5
100	483	0.37	0.014	14.6
1000	94	0.30	0.345	—

D. Molecular Motion

Nuclear magnetic resonance, dielectric relaxation, and dynamic mechanical properties all involve molecular motion. Fragmentary early studies of radiation-induced changes (46,47) in NMR were supplemented by study to 940 Mrad (34). The amorphous line broadens from an initial 0.5 G to 2 G at 20–50 Mrad, then more slowly to about 3 G at 600 Mrad, which indicates an increasing restriction of motion, even in the amorphous region. The change is attributed to a rearrangement with closer packing of chains. The broad, 10-G wide crystalline NMR line was not observed in these instances because of power-saturation, but other studies (48) show that it and the overall second moment in the range -50 to $+50°C$ increase along with crystallinity.

Some dynamic mechanical property curves are shown in Figs. 12 and 13. To secure these data, a bar was excited in its normal transverse vibration, at its resonant frequency. The modulus shows inflections and the internal friction shows peaks in four major regions:

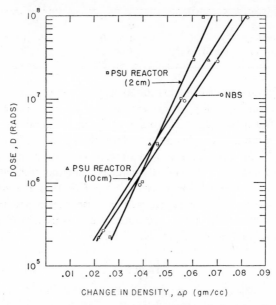

Fig. 9.   Density versus dose behavior for polytetrafluoroethylene irradiated in a reactor
and in a ^{60}Co source (32).

γ	200°K	Glasslike, few units; amorphous
β	300°K	Crystalline phase transition conformational change
α	400°K	Glasslike, or perhaps crystalline; longer chain sequences
(4th)	near 600°K	onset of melting

The modulus and loss peak are generally lowered by radiation and more so by a subsequent heat treatment (Fig. 13). Changes in the 200°K γ peak are most conspicuous. Temperature shifts of the crystal-transition peaks parallel those of the crystal transitions, especially near 300°K. Supplemental data, at very low doses, show that the changes vary with initial crystallinity (20). Thermal conductivity as a function of temperature, although not greatly changed by radiation also correlates with the changes in transition temperatures (49).

The low-speed sliding friction coefficient is raised by radiation (50). The coefficent at 1 cm sec^{-1} rubbing speed rises from an initial 0.08 to 0.16 at 5 Mrad. At intermediate speeds (10 cm sec^{-1}, 90 cm sec^{-1}) the coefficients are initially higher and go through maxima followed by a fall to a plateau. These changes in friction are deducible from the general changes in mechanical

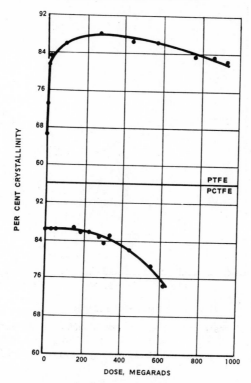

Fig. 10.   Percentage crystallinity (by X-ray) as a function of ^{60}Co γ-ray dose for polytetrafluoroethylene and polychlorofluoroethylene (34).

properties, the increasing hardness, and the transition in behavior from drawing-out under low shear stress to transfer of matter by breaking of chunks.

The frequency spectrum of electrical properties supplies practically no information on relaxation processes, because of the lack of permanent dipoles. The dielectric constant and dielectric loss over the frequency range 1 to 10^9 Hz remain low after radiation doses up to 300 Mrad or 4×10^{17} fast neutrons cm^{-2}. Exceptions that may occur at very low frequency and during irradiation are discussed in the section on electrical properties.

E. Volatile Products

The volatile products are expected to have some relation to the mechanism of radiolysis. This is also true of new functional groups as detected by infrared or titrations, for example. Products often reported are ionic fluoride, SiF_4, COF_2, CO_2, CO, CF_4, other saturated fluorocarbons, and occasionally

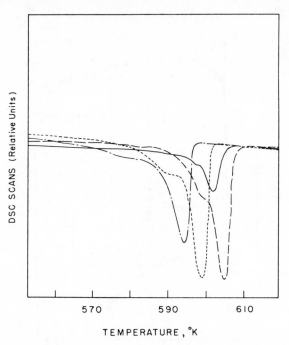

Fig. 11. Differential scanning calorimetry crystal-melting endotherms for irradiated polytetrafluoroethylene (37). (——) initial; (— —) 11.7 Mrad; (---) 78 Mrad; (— · —) 196 Mrad. (Reprinted by permission of the copyright owner, Marcel Dekker, Inc.)

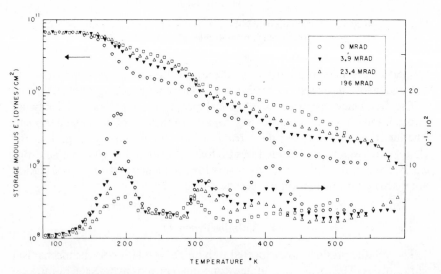

Fig. 12. Storage modulus (E') and internal friction (Q^{-1}) of irradiated polytetrafluoroethylene as a function of temperature (37). (Reprinted by permission of the copyright owner, Marcel Dekker, Inc.)

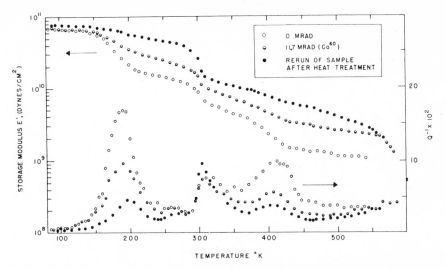

Fig. 13. Effect of heat treatment on the dynamic mechanical behavior of irradiated polytetrafluoroethylene (37). (Reprinted by permission of the copyright owner, Marcel Dekker, Inc.)

unsaturations and carboxylic acid groups. Fluorocarbon and peroxidic radicals are formed in large amount. As with all other radiation changes, the behavior with air and water present is totally different from that in vacuum, provided samples are not too thick. Not all investigators state what atmosphere was present. The agreement among investigators is generally poor, and perhaps worst for the inorganic fluoride yields of Fig. 14. Some of the differences are associated with differences of dimensions and some may be associated with unstated presence or absence of oxygen. Much fluoride continues to escape after stopping irradiation. Carbon tetrafluoride was recognized early and, judging from total gas products and composition, appeared to be proportional to square of absorbed dose (9,38). Two modes of formation were proposed:

1. Random scission sometimes cuts two adjacent C—C bonds. The statistical analysis of this process implied G_{sc} of about 10 (9).

2. Attack by F on the C—C bond forms highly energetic products which split off CF_4 (53–55):

$$F + {\sim}CF_2CF_2{-}CF_2CF_2{\sim} \longrightarrow {\sim}CF_2CF_3^* + \cdot CF_2{\sim}$$
$${\sim}CF_2CF_3^* \longrightarrow CF_4 + {\sim}CF{=}CF_2$$

Later, in vacuum irradiations monitored by refined mass spectrometry, infrared or gas chromatography, it appeared that CF_4 was only one among many fluorocarbon products ranging to C_6 or higher (56). In this work a dose squared law did not seem to hold, up to the smaller doses studied. These

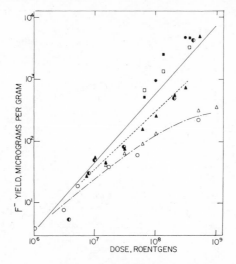

Fig. 14. Ionic fluoride from radiolysis of polytetrafluoroethylene. Solid line, total from solid samples irradiated under aqueous alkali (54); (▲) same, released at once; (△) same, additional after 30 days storage; (○,◐) disks 3 mm thick irradiated under aqueous alkali (52); (○) examined at once, (◐) additional after 30 days; (●) molding powder irradiated under aqueous alkali, no storage (52); (□) powder irradiated in vacuum, and stored 1–3 months in vacuum at room temperature, computed as $\frac{1}{4}$ SiF_4 yield (56); (■) same treatment plus postheating.

TABLE 8

Products from Radiolysis of PTFE

Product	G	Condition	Ref.
F^-	[2.0][a]	Water, air	54
	[0.6, 1.7][a]		52
SiF_4	0.12 to 0.16	γ, vac, 20°C	56
SiF_4	Less than CF_4	Pile	9
CF_4	[.0031 × r(Mrad)][a]	Pile	9
CF_4	.004 to .009	γ	56
$-C=C-$	[about 0.2][a]	Water, air	54
Carbon radicals	0.05[a]	Deuteron	59
Carbon radicals	0.16 to 0.19	γ, vac	56
Peroxide radicals	0.02	γ, vac, then air	60
Peroxide radicals	0.03	γ irradiation in air	60
Total gas	0.30	γ	56
Total gas	0.02 to 0.19	Electrons	61

[a] Computed from data in reference.

results are compared in Table 8. The monomer C_2F_4 is not formed at room temperature. When oxygen is present, the principal products are COF_2, SiF_4 (in glass containers), and CO_2, in high yield (58).

F. Infrared Changes

Infrared spectra furnish somewhat ambiguous evidence on structural changes after irradiation. Spectra of the normal polymer are well known and systematically explained in considerable measure (35,62). The method is less sensitive in PTFE than in some other materials because of scattering from crystallites, and it is also somewhat treacherous because of absorptions specific for amorphous regions. The difficulty with specific amorphous absorptions has been put to advantage in the I Sp G method already mentioned, where the ratio of bands at 778 cm^{-1} (amorphous) and 2367 cm^{-1} (non-specific) is used as the measure of crystallinity. The 778 and weaker amorphous bands at 745 and 715 cm^{-1} were once taken to be crystalline since they disappeared on melting (63).

The best available discussion of structural identification of small molecule spectra (64) is rich in examples but poor in specific structural correlations, which are scattered in the literature. Table 9 shows the very wide disagreement reported for irradiated PTFE, especially as regards double bonds. As usual, irradiation in air is very different from that in vacuum. Figure 15 illustrates spectra of both kinds. Originally, Ryan (54) reported absorptions at 1538 and 1754 cm^{-1} attributed to terminal and internal double bonds, respectively. At least in vacuum irradiations, others fail to find these absorptions, although in air Golden's 1757 band is perhaps identical (58). There is some agreement in the appearance of 980 and 1350 absorptions; the 1350 absorption is attributed to the vinylene group,

$$\begin{matrix} X & X \\ | & | \\ -C & = C - F, \end{matrix}$$

by Slovokhotova (63) but by no others. The absorptions at shorter wave length (2600–3600 in one case, or 3096 and 3472) are attributed to carboxylic acid groups. The formation of such hydrophilic groups should gradually increase the wettability of the surface, as actually observed, After 16 Mrad, the contact angle with aqueous alkali decreased from 90 to 82° in polytetrafluoroethylene and from 74 to 50° in polychlorotrifluoroethylene (165).

The early data on double bonds were supported by permanganate consumption (54,63). If they are invalid, other possible fates for the permanganate might have been reduction by radicals (66) or reduction with associated oxygen liberation by hydroperoxides. The double bonds seem thus doubtful

TABLE 9

Infrared Absorptions Found with Irradiated Polytetrafluoroethylene

Frequency (cm^{-1})	Structure	Irradiation condition
982	Unknown	Air (63,58); vac (58)
1,350	$\overset{\displaystyle X \quad X}{\underset{\displaystyle \vert \quad \vert}{}}$ −C=CF	Air (63); vac (63)
1,450	C−H	Air (63)
1,538	−CF=CF$_2$	Air (54)
1,550	C−H	Air (63)
1,670–1,720	$\overset{\displaystyle X \;\; X \;\; X \;\; X}{\underset{\displaystyle \vert \;\; \vert \;\; \vert \;\; \vert}{}}$ −C=C−C=C−	Vac (63)
1,733	−CF=CF−	Vac (63)
1,754	−CF=CF−	Air (54)
1,757	$\overset{\displaystyle O}{\underset{\displaystyle \Vert}{}}$ −C−OH	Air (58)
1,780	$\overset{\displaystyle O \qquad\quad O}{\underset{\displaystyle \Vert \qquad\quad \Vert}{}}$ −C−OH, −C−H	Air (63)
1,790	−CF=CF$_2$	Vac (weak) (63)
1,870–1,890	$\overset{\displaystyle O}{\underset{\displaystyle \Vert}{}}$ −C−F	Air (63)
2,100–3,600	$\overset{\displaystyle O}{\underset{\displaystyle \Vert}{}}$ −OH, C−OH	Air (63)
3,096	$\overset{\displaystyle O}{\underset{\displaystyle \Vert}{}}$ −C−OH	Air (58)
3,472	$\overset{\displaystyle O}{\underset{\displaystyle \Vert}{}}$ −C−OH	Air (58)

though not disproven. The surest infrared observations are Golden's, since Bro et al. report agreement. Infrared absorption was more specifically successful in identifying gaseous products, notable CF_4, C_2F_6, C_3F_8, SiF_4 (58) and COF_2.

G. Volatile Products at High Temperatures

At high enough temperatures, the radiolysis becomes essentially a pyrolysis with radiolytic initiation (57). Studies in the range above 400°C have thrown light on the mechanism of the related thermal decomposition.

In the interval 300–400°, the radiolysis products are very complicated. The trends of composition with temperature are suggested by the mass spectrometer data of Table 10. With increasing temperature, the rise of C_2F_3 relative

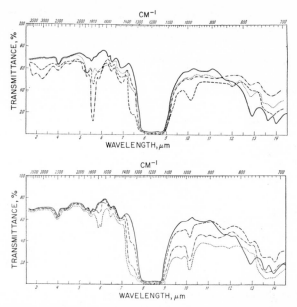

Fig. 15. Infrared absorption of irradiated polytetrafluoroethylene (63). Sample thickness 70 μm. Irradiation as follows. Upper spectra: (——) unirradiated; (— —) fast electrons, air, 40 min; (. . . .) ^{60}Co γ, vacuum, 550 hr; (— · —) ^{60}Co γ, air, 220 hr. Lower spectra: (——) unirradiated; (— —) fast electrons, vacuum, 10 min; (— · —) same, 20 min; (. . . .) same, 30 min.

TABLE 10

Mass Spectra of Volatiles from PTFE Irradiated at 7.66×10^6 Roentgens per Hour[a]

Mass No.	Ion	Relative peak intensities		
		<360°C	407°C	462°C
31	CF^+	1,550	300	390
50	CF_2^+	570	70	177
69	CF_3^+	2,310	234	198
81	$C_2F_3^+$	880	105	250
93	$C_3F_3^+$	132	45	36
100	$C_2F_4^+$	775	84	189
119	$C_2F_5^+$	280	29	22
131	$C_3F_5^+$	3,130	258	197
150	$C_3F_6^+$	184	13	14
181	$C_4F_7^+$	520	558	66
200	$C_4F_8^+$	126	7	7

[a] From ref. 57. Much CO_2, no SiF_4. Below 360°C, also C_5F_9, C_5F_{10}, C_6F_{11}, and C_6F_{12}.

TABLE 11

Weight Loss of Polytetrafluoroethylene Irradiated at 30°C and
Heated Subsequently (37)

Dose (Mrad)	Maximum temp. (°C)	Weight loss (%)
3.9	317	0.094
23.4	297	0.132
78	297	0.34
156	262	0.35
196	237	0.23

Cylinders 6 mm diameter, heated in air. Times various, not stated. Weight loss less than 0.06% after 100 hr at 150°C at all doses up to 196 Mrad.

to CF_3 and high fragments indicates that C_2F_4 is emerging as a significant product, in contrast to room-temperature results, and paraffins and higher olefins are declining.

Small weight losses occur after heat treatment following room-temperature irradiation (Tables 11 and 12). Beside this, the weight-loss rate as a function of irradiation temperature suffers a sudden increase near the crystal melting transition (Table 12). Despite the disparity in observed weight losses in Tables 12 and 13, the rate for unit absorbed dose in both cases lies in the range of hundredths of a percent per megarad. Further comparison is impractical in view of the unknown mechanism and unknown dependence on dose rate and total dose in this intermediate region.

TABLE 12

Weight Losses of Irradiated Polytetrafluoroethylene (31)[a]

Temperature during irradiation (°C)	140	220	260	325	330	345	360
Weight loss (%)	0.02	0.06	0.45	5.5	14.6	45.5	56

Post heating, 7 min (°C)	345[b]	360[b]	360[c]	415[c]
Weight loss (%)	2.2	5.5	0.25	0.95

[a] Dose 21.7×10^{21} eV cm^{-3} (about 750 Mrad) applied in 5 min with electron accelerator, in all cases except blanks marked (c).

[b] Irradiation temperature 30°C, dose as in (a).

[c] Thermal blanks, no radiation.

TABLE 13

High-Temperature Decomposition Rates of Polytetrafluoroethylene during Gamma Irradiation (57)[a]

°C	329	370	400	452
Gas rate [(moles)(base mole)$^{-1}$ sec^{-1} $\times$ 10^{-7}]	15	22	20	37
Weight loss rate (sec^{-1} $\times$ 10^7)	<150	72	110	180

[a] Dose rate, 7.66×10^6 r hr^{-1}.

III. SMALLER FLUOROCARBON MOLECULES

A. Saturated Fluorocarbons and Derivatives

Results of many fluorocarbon irradiations are now available but do not clarify all points regarding PTFE. Early work on pure fluorocarbons showed that perfluoroparaffins, cycloparaffins, and the saturated derivatives $(C_nF_{2n+1})_2O$ and $(C_nF_{2n+1})_3N$ behaved much alike (68–70). All formed both smaller and larger molecules, at G-values often a little lower than in the hydrocarbon series (Table 14). In contrast to hydrocarbon behavior, olefins were usually not formed and F_2, the analog of H_2, was never found, which is not entirely surprising in view of its high reactivity. There is some conflict about attack on metal walls (aluminum, nickel); the weight of evidence is that they are inert or nearly so. When glass was present, SiF_4 was formed in moderate amounts. Later studies put the quantitative observations on firmer ground and demonstrated the multitude of products; however, at extremely high doses, $> 10^3$ Mrad, in reactors, there is in addition a poorly understood solid product (75). A simple but well-studied compound is perfluoroethane (Table 15).

The products were those to be expected from C—F and C—C bond splitting and recombination of the radicals and atoms so formed. The following radical reactions are possible:

$$CF_3 + CF_3 \longrightarrow C_2F_6 \tag{1}$$
$$CF_3 + C_2F_5 \longrightarrow C_3F_8 \tag{2}$$
$$C_2F_5 + C_2F_5 \longrightarrow C_4F_{10} \tag{3}$$
$$CF_3 + F \longrightarrow CF_4 \tag{4}$$
$$C_2F_5 + F \longrightarrow C_2F_6 \tag{5}$$
$$F + F \longrightarrow F_2 \tag{6}$$
$$CF_3 + F_2 \longrightarrow CF_4 + F \tag{7}$$
$$C_2F_5 + F_2 \longrightarrow C_2F_6 + F \tag{8}$$

Abstractions and disproportionations were absent.

$$F + C_2F_6 \longrightarrow F_2 + C_2F_5$$
$$C_2F_5 + C_2F_5 \longrightarrow C_2F_4 + C_2F_6$$

TABLE 14

Irradiation of Pure Fluorocarbons

Compound	Radiation	$G(-M)$	Principal products	Ref.
Tetrafluoromethane	Xenon photosens. 1470 Å	$\Phi < 0.002$	F^-, polymer	71
	UV 1470 Å	—	—	71
	F^{18} hot atoms	—	C_2F_6, CF_3F^{18}	73
	$Co^{60}\ \gamma$	1.5	—	74
	$Co^{60}\ \gamma$	0.02	C_2F_6	72
Hexafluoroethane	$Co^{60}\ \gamma$	2.4–3.0	CF_4, C_3F_8, C_4F_{10}	79
	$Co^{60}\ \gamma$	3.25	CF_4, C_3F_8, C_4F_{10}	74
Octafluoropropane	$Co^{60}\ \gamma$ and pile	4.5	Perfluoroparaffins, C_1 to C_{10}, normal and branched	74
Decafluoro-*n*-butane	$Co^{60}\ \gamma$ and pile	5.4	Perfluoroparaffins, C_1 to C_{10}, normal and branched	74
Perfluoro-*n*-pentane	$Co^{60}\ \gamma$ and pile	4.9	Perfluoroparaffins, C_1 to C_{10}, normal and branched	74
Perfluoro-*n*-hexane	$Co^{60}\ \gamma$ and pile	6.0	Perfluoroparaffins, C_1 to C_{14}, normal and branched	74
Perfluoro-2-methylpentane	$Co^{60}\ \gamma$ and pile	6.0	Perfluoroparaffins, C_1 to C_{14}, normal and branched	74
Perfluoro-2,3-dimethylbutane	$C^{60}\ \gamma$ and pile	9.6	Perfluoroparaffins, C_1 to C_{14}, normal and branched	74
Perfluoro-*n*-heptane	Pile	2–3	Perfluoroparaffins, C_1 to C_{14}; no F^-	68
Perfluoro-*n*-heptane	$Co^{60}\ \gamma$	>2.6	Perfluoroparaffins, C_2 to C_{14}; SiF_4	70
Perfluoro-*n*-octane	γ, spent fuel	4.8	Perfluoroparaffins, C_1 to C_4 and C_5 to C_{20}	69
Perfluorocyclobutane	$Co^{60}\ \gamma$	—	C_5, C_6, and C_8 perfluoroalkylcyclobutanes, C_2F_4, c-C_3F_6	78

TABLE 14 (continued)

Compound	Radiation	$G(-M)$	Principal products	Ref.
Perfluorocyclopentane	Co^{60} γ and pile	3.0	Perfluoroparaffins C_9 to C_{15}	74
Perfluorocyclohexane	1.5 MeV electrons	2.4	Higher and lower boiling fluorocarbons, $C_{12}F_{20}$	76
Perfluorocyclohexane	$Co^{60}\gamma$	3.4	Perfluoroparaffins C_1 to C_9; perfluorobicyclohexyl, perfluoro- cyclohexylhexane, perfluorocyclohexylhexene	78
Perfluorobicyclohexyl	1.5 MeV electrons	1.6	Residue and lower-boiling fluorocarbons	76
Perfluorobicyclohexyl	Unstated; at 350–450°C	50 to 10^5	—	77
Perfluorodecalin	1.5 meV electrons	1.7	Perfluorocarbons C_1 to C_4 and high- boiling residue	76
Cyclic ether $c\text{-}C_8F_{16}O$	γ, spent fuel	6.5	Lower- and higher-boiling liquids, some acidic	69
Perfluorotributylamine	Pile	>1.0	Lower- and higher-boiling liquids, neutral	68
Perfluorotributylamine	γ, spent fuel	11.0	Lower- and higher-boiling liquids	69

TABLE 15

Product G-Values in Hexafluoroethane Radiolysis (79)

Condition	CF_4	C_3F_8	$n\text{-}C_4F_{10}$	$(-C_2F_6)$
Liquid, $-78°C$	1.72	0.87	0.45	3.0
Gas, 2 atm, $+40°C$	2.5	0.45	0.20	2.4
Liquid, $-78°C$; 10% O_2	0.97	0	0	~3

Oxygen tended to be a scavenger of thermal free radicals when present; the total consumption of C_2F_6 remained about the same as in its absence but the usual products were replaced, apparently, by COF_2, $(CF_3)_2O$, and $C_2F_5OCF_3$.

In studies of larger molecules, e.g., $cy\text{-}C_6F_{12}$, as many as 10 or 20 products were isolated by gas chromatography, qualitatively in accord with the preceding discussion of C_2F_6. Some G-values are listed in Table 14.

In the radiolysis of perfluoroparaffins, the G-value for disappearance and the product distribution seem to be roughly predictable from the total mass spectrometer sensitivity and assumed statistical radical recombination (81). Radiolysis products include many branched-chain isomers, although apparently none of *neo* structure (80). Exceptions to the rule of saturated product formation are perfluorocyclobutane, which furnishes appreciable yields of C_2F_4 (78), and the bicyclic $C_{10}F_{18}$, which furnishes traces of unsaturated fluorocarbons (76).

TABLE 16

Irradiation Products from Aromatic Fluorocarbons and Hydrocarbons[a]

Compound	Temp. (°C)	G_{gas}	$G_{polymer}$	Polymer M.W.
C_6F_6	35 ± 5	0.001	2.15	1,140–1,225
C_6F_6	450		5.9	
$C_{12}F_{10}$	105	0.0001	1.4	1,230
$C_{12}F_{10}$	450		2.2	
$C_{12}F_{10}$	500		3.2	
$C_{10}F_8$	108	0.0001	1.28	770
C_6H_6[b]			0.9	
$C_{12}H_{10}$[c]	30(?)		≈ 0.26	

[a] Principally from Mackenzie (76,82).
[b] Gordon (83).
[c] Bolt (84, p. 85).

B. Aromatic Fluorocarbons and Their Mixtures

Aromatic fluorocarbons, like the aromatic hydrocarbons, undergo condensation almost exclusively. Table 16 gives some comparative yields. Compared to ordinary aromatic compounds, the perfluoro aromatics have double the "polymer" yield but an extremely low gas yield.

Gamma radiolysis of C_6F_6, vapor or liquid, also produces octafluoro-cyclohexadiene (85); the Dewar benzene, hexafluorobicyclo[2·2·0]hexa-2,5-diene (I) (85); and another C_6F_6 isomer, perhaps the prismane (V) (86).

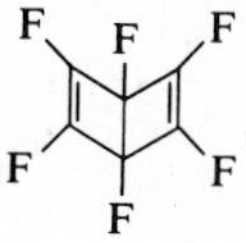

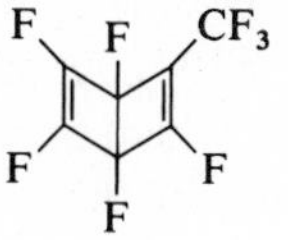

Compound I and its C_7F_8 homologs II, III, are also formed by photolysis of C_6F_6 and $C_6F_5CF_3$ (85,87,88). The compound reverts to hexafluorobenzene at 80°C, sometimes explosively. In mixtures with cyclooctene, a related tetracyclic adduct IV is formed (89). The exact wavelength requirement for isomerization is in dispute; Cammagi et al. (88) found 2537 Å effective in the vapor phase, whereas Haller (87) required unfiltered light of lower wavelength and found several solvent-sensitizer combinations ineffective for isomerization. The latter results suggest that the triplet state cannot be involved, the former that the excited states involved are of low energy, less than 115 kcal, and could therefore be produced either by ion-electron recombination or by F atom-radical recombination. In mixtures, energy transfer involving excited states can either increase or decrease the usual decomposition, "sensitization" or "protection," respectively.

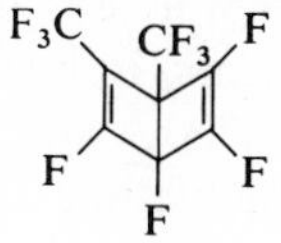

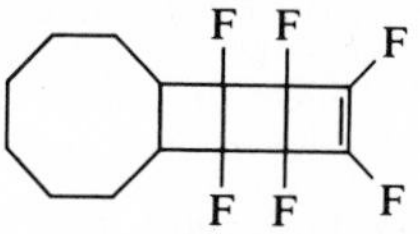

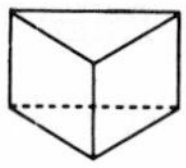

Radiolysis of C_6F_6 mixtures indicated protection of cy-C_6F_{12}, sensitization of n-C_9F_{20} (90), and little effect in several other areas (70), in contrast to the large protective effect of C_6H_6. In photolysis also, both protective (91) and sensitizing (92) effects were discovered. In the photolysis of the ketone $(CF_2Cl)_2CO$, C_6F_6, like C_6H_6, retards the chain reaction component,

$$CF_2Cl + CF_2ClCOCF_2Cl \longrightarrow CF_2Cl_2 + \dot{C}F_2COCF_2Cl$$

Most of this retardation is not a case of energy transfer, but of action as a radical trap.

$$\dot{C}F_2Cl + C_6F_6 \longrightarrow C_6F_6\text{-}CF_2Cl \longrightarrow products$$

The primary decomposition is also retarded somewhat, but less than by C_6H_6, $cy\text{-}C_5H_{10}$ or $cy\text{-}C_5F_{10}$ (93).

In the photolysis of dichloromethane with 2537 Å light, C_6F_6 and C_6H_6 promoted the decomposition, the C_6F_6 perhaps acting by the simple process

$$C_6F_6{}^* + CH_2Cl_2 \longrightarrow C_6F_6 + \dot{C}HCl_2 + H$$

Considerable information relative to the excited states has been accumulated by studying the processes of fluoroescence quenching, and C_6F_6-sensitized phosphorescence and isomerization (94). Apparently two excited singlet states are produced at 2800 Å and below, which go at different rates to a fluorescent state. The triplet lifetime is short, in the range of 10^{-7} sec. Hexafluorobenzene can serve as a solvent for liquid-state scintillators (95).

In addition to excited-state intermediates, hexafluorobenzene irradiated at 77°K also forms free radicals, and it is able to form positive and negative ions without excessive fragmentations.

C. Hydrogen-Containing Molecules and Mixtures

Radiation studies upon molecules containing both fluorine and hydrogen are interesting, because of the partially fluorinated character of several elastomers. Few such studies have been made, except upon the partially fluorinated olefins, and these are not very informative because of the occurrence of radical chain polymerization.

Trifluoroacetaldehyde hydrate (96) and trifluoroethanol (97) have been radiolyzed, the latter pure and in 10% aqueous solution. In trifluoroethanol the G-value for HF was a maximum of 7.9 in 10% aqueous solution and nearer 3 in 99·6% solution. These values, although higher than some occurring in decomposition of fluorocarbons, are much less than for trichloroethanol and tribromoethanol and indicate little or no chain reaction. Appreciable yields of typical alcohol radiolysis products were also found, e.g., $H_2(G = 2.4\text{-}3.2)$, $CF_3CHOHCHOHCF_3$ (0.10–0.27), and carbonyl compounds (0.4–0.9).

In mixed hydrocarbon-fluorocarbon systems, the production of HF is thermodynamically favored. The following paths are among those found for its formation:

Abstraction

$$F + RH \longrightarrow HF$$

Disproportionation

$$CF_3 + CH_3 \longrightarrow CF_2{=}CH_2 + HF$$

Decomposition of excited molecules

$$CFH_2\cdot + CFH_2\cdot \longrightarrow C_2H_4F_2*$$
$$C_2H_4F_2* + M \longrightarrow C_2H_4F_2$$
$$C_2H_4F_2* \longrightarrow CFH=CH_2 + HF$$

Mixtures in which the hydrocarbon component is C_6H_6, cy-C_6H_{12}, or C_6H_{14} have been irradiated by several groups. Unfortunately, modern gas-chromotographic techniques were used to isolate products in only a few cases. Tables 17 and 18 summarize the rather imprecise results.

The relatively high yield of HF and/or HCl is evident in most mixtures. The yield of poorly defined, nonvolatile product is appreciable, the so-called "polymer" fraction, more properly called oligomer or condensation product. In the case of $C_6F_6 + C_6H_6$, the yield of oligomer is higher in most mixtures than in either pure component. In all such mixtures the $(H + F)/C$ ratio in the oligomer is low relative to any mixture of the starting components, suggesting loss of HF from oligomer segments. Nominal protective effects are shown in cyclohexane-hexafluorobenzene and hexane-octafluorocyclobutane; $G(H_2)$ is reduced substantially below the value for the hydrocarbon. A strong case has been made (105) that cy-C_4F_8, cy-C_6F_{12}, and C_6F_6 act as efficient

TABLE 17

Irradiation of Benzene Fluorocarbon Mixtures

Halocarbon	Product (G)	Identification methods	Ref.
CF_4 0.30:1	$C_6H_5CF_3$ (1.0–1.5) C_6H_5F (1.0–1.5) CHF_3 (present), HF (present)	IR	98,99
CHF_3, CF_2Cl_2 $CFCl_3$, $C_2F_4Cl_2$	No aromatic F compound, carbonyl, C_6H_5Cl	IR	99
$C_2F_3Cl_3$ 1:1 static	Principally resins; volatiles form resin and evolve HCl when heated; assume $0.06F^-$, 0.26 Cl^-	Fractionation	100
$C_2F_3Cl_3$ 10:1 circulating K_3PO_4	$C_2F_2Cl_3C_6F_5$; $C_2F_3Cl_2C_6H_5$ $C_2HF_2Cl_2C_6H_5$, as mixtures	Analysis, B.P., properties of fractions	100
C_3F_6 20:1	Products with unsaturated fluorocarbon side chain	Analysis, IR of fractions	100
C_6F_6 20°C	Resin, 0.9 to 2.8; HF, 0.6 to 1.3 (observed as SiF_4)	Analysis, mass spectra	70
C_7F_{16} 0.56:0.44	Oligomer 6.3, HF 2.6, CF_3H 0.16 to 2.0, CF_4 0.19	Analysis, mass spectra	70

TABLE 18

Irradiation of Fluorocarbon-Hydrocarbon Mixtures

Mixtures		Conditions	Products; G-Values					Ref.
Mol. fract. fluorocarbon	Hydrocarbon entity		Oligomer atomic composition		Oligomer G	HF	Other	
			F/C	H/C				
None	cy—C₆H₁₂	—	—	—	1.7	—	H₂ 5.5	70
C₆F₆	cy—C₆H₁₂	Co⁶⁰, 174 Mrad				large, n.d.		70
0.03		Co⁶⁰				1.3	H₂ 2.2; C₆F₅ 1.0	106
0.23		Co⁶⁰	0.37	1.15	6.1	n.d.	H₂ 1.9	
0.65		Co⁶⁰	0.48	0.92	3.1		n.d.	
C₇F₁₆	cy—C₆H₁₂	Co⁶⁰, 339 Mrad				Large, n.d.	n.d.	70
0.19, 2 phase		Co⁶⁰	0.50	1.33	3.0	Large, n.d.	n.d.	
0.68, 2 phase		Co⁶⁰	0.95	0.87	2.9	n.d.		
C₂F₃Cl₃ Unknown	cy—C₆H₁₂	Co⁶⁰				n.d.	Cl + F 7.0	100
None	n—C₆H₁₄	?				—	H₂ 5.24	101
cy—C₄F₈	n—C₆H₁₄	?						
0.037						—	H₂ 2.58	101
n—C₄F₁₀	n—C₆H₁₄					HF present	H₂ lowered	102
Perfluoroolefins	n—C₆H₁₄					HF present	H₂ lowered	102
cy—C₄F₈	CH₄	Gas, 30°C X-rays						
0.75						2.2	H₂ 0.8; C₂F₄ 0.7	103
0.5						2.5	H₂ 1.3; C₂F₄ 0.6	

TABLE 18 (continued)

| Mixtures | | Conditions | Oligomer atomic composition | | Oligomer G | Products; G-Values | | Ref. |
Mol. fract. fluorocarbon	Hydrocarbon entity		F/C	H/C		HF	Other	
0.25			—	—	—	3.2	H_2 2.0 C_2F_4 0.4	103
0 cy—C_4F_8	cy—C_6H_{12}	Co^{60}, 25°C, liquid	—	—	—	0	H_2 5.2	105
cy—C_4F_8	cy—C_6H_{12}	Co^{60}, 77°K					All products very low	105
cy—C_6F_{12}	cy—C_6H_{12}						cy—$C_6F_{11}H$ 3.5 H_2 lowered Fluorocarbons lowered	104
cy—C_6F_{12}	cy—C_6H_{12}	Co^{60}, 25°C liquid					H_2 5.6 ——→2.6	105
(0 to 0.3 M) cy—C_6F_{12}	cy—C_6H_{12}	Co^{60}, 77°K					$C_6F_{11}H$ 0 ——→3.4 All products very low	105

electron scavengers. In the cyclohexane-C_6F_6 mixture the decrease in $G(H_2)$ relative to separate pure components is nearly compensated by the large increases in $G(HF)$.

Many of the mixtures have been examined only at excessively large radiation doses, and therefore the initial Gs may be quite different from those reported.

D. Fluorocarbons and Oxygen

Irradiation of tetrafluoroethylene with oxygen at lower temperatures forms explosive peroxide copolymers (107–109).* At room temperature and a little above, and not too high pressure, the products are the epoxide,

$$\overset{\displaystyle O}{\underset{\displaystyle F_2C-CF_2}{\diagup\diagdown}}$$

the cyclopropane

$$\overset{\displaystyle CF_2}{\underset{\displaystyle F_2C-CF_2}{\diagup\diagdown}}$$

and the acid fluorides, CF_3CFO and COF_2.

The photochemical reactions are closely related. Ultraviolet (108,111,112) acts on C_2F_4 much like ionizing radiation and causes similar reactions in C_3F_6. The epoxides are made into high polymer by the further action of radiation (113,114), or into lower polymer (111), molecular weight about 3000, by charcoal catalysts. At higher temperatures, $CF_3CF{=}CF_2$ with O_2 yields a polyether

$$\underset{\displaystyle CF_3}{\overset{\displaystyle +CF-CF_2-O+_x}{|}}$$

as ultimate product (115,116). Exhaustive studies of the photochemistry of the C_2F_4–O_2 and C_3F_6–O_2 systems have been done (117–124).

When the saturated fluorocarbon C_2F_6 is irradiated at $-80°C$, O_2 scavenges the radicals efficiently, as described earlier. Oxygenated radicals can be observed in such systems, e.g., $FOO\cdot$. Photochemical experiments on perfluoroalkyl iodides are a little like irradiations of perfluoroparaffins. With

oxygen, acid fluorides $RC{-}F$ are formed in some cases (water present, Cl_2 or Br_2 sensitization, restricted O_2) (125), whereas in others the carbon skeleton

---

* The treacherous detonating character of the copolymer has been experienced by the author and fully described by others (110) but is not emphasized in the present references.

is almost all degraded to COF_2 (126–128). It is not evident that the presence of oxygen ever induces chain reactions during these irradiations.

E. Implications for Polymers

Overall, the results on small molecules indicate that radiation yields from fluorocarbons fall in the normal range for organic compounds and tend to be, if anything, somewhat low. Because of the lack of anything analogous to the H_2 production of hydrocarbons, material balance tends to require that condensation be accompanied by a corresponding amount of chain scission in pure, saturated halocarbon systems. The restriction does not apply in mixed systems containing unsaturated aromatic or hydrocarbon units; in the latter, condensation can occur with H_2 or HF evolution. Oxygen can bring about extensive breakdown with formation of COF_2 and acid fluorides.

Applied to PTFE, these general results suggest that breaking of chemical bonds is not very rapid during radiolysis. Deterioration of properties is more probably due to the high molecular weights at which optimal properties are lost.

Chemistry of the small molecules cannot distinguish between several reasons for the nearly stationary value of tensile strength during vacuum irradiations of PTFE in the interval of 5 to 50 Mrad. Both cage recombination and a compensation between scission and cross-linking or end-linking are in agreement with the related chemistry. In either event, oxygen competes and causes extensive degradation. Golden's explanation (58), amplified by Bro (17), seems realistic:

$$\sim CF_2 \cdot + O_2 \longrightarrow \sim CF_2-O-O \cdot \longrightarrow \sim CF_2O \cdot$$
$$\sim CF_2CF_2O \cdot \longrightarrow \sim CF_2 \cdot + CF_2O$$

$$\sim CF_2CFCF_2 \sim \longrightarrow \sim CF_2\overset{\overset{\displaystyle O}{\|}}{C}-F + \sim CF_2 \cdot$$

Recent reviews of free radical evidence suggest only one modification, that the alkoxy radical $\sim CF_2O \cdot$ is short-lived and has probably not been observed as such. Saturated halocarbon polymers in general, e.g., PCTFE, could be expected to behave like PTFE, but in unsaturated and aromatic fluorocarbon polymers, or saturated polymers containing hydrogen, there is no bar to extensive network formation.

IV. POLYCHLOROTRIFLUOROETHYLENE

As far as is known, PCTFE undergoes only scission when irradiated. The G-value for scissions is 0.67 from M_n measured by zero strength time (56) (see Fig. 16). On pieces 0.062 in. thick, the results were unaffected by air, but

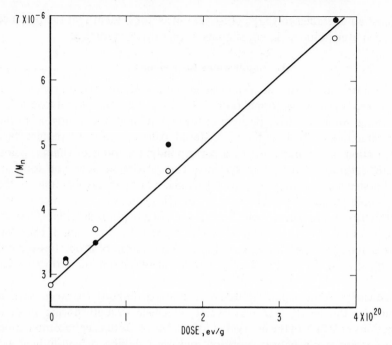

Fig. 16.   Molecular-weight loss of polychlorotrifluoroethylene during gamma radiolysis.
($\bigcirc$) irradiated in vacuum; ($\bullet$) irradiated in air (56).

effects ought to be expected on thin foils, as with PTFE. Melt viscosity (270°C) in irradiations at 250°C went as follows (19):

Dose (Mrad)	0	1	3
Melt viscosity (poises) $\times 10^{-4}$	114	72	7

Flow temperature was reported lowered by 30°C at 120 Mrad (67).

As with PTFE, ordinary or peroxy radicals are formed according to conditions. In vacuum, the volatile products are numerous but unidentified, including molecules containing up to five carbon and two chlorine atoms but not CF_4 or SiF_4, whereas in air, CO_2 is a major product. Some G-values are assembled in Table 19. Since crystallinity is variable, effects of crystallinity ought to be expected but have not been studied. X-ray and NMR were observed to doses near 1000 Mrad (34). Electrical conductivity rises during irradiation, as with PTFE. Deterioration of mechanical properties is sometimes rated as a little less rapid than in PTFE (131). A few examples are quoted in Tables 20 to 22. Bending modulus declines at a rate intermediate between polystyrene and PMMA, and not far different from poly-2,5-dichlorostyrene (Table 22). Like some other irradiated glassy polymers, PCTFE

TABLE 19

Products from Irradiated Polychlorotrifluoroethylene

Product	G	Conditions	Ref.
Total gas	0.11, 0.13	γ, vac., powder	56
Total gas	0.1	γ, vac., film	129
Total gas	0.15	electrons	61
Total gas	[0.9–1.1]	γ, air	129
F$^-$	[3.5]	pile, aq. alkali	130
Cl$^-$	[3.5]	pile, aq. alkali	130
SiF$_4$	0–0.06	γ, vac.	129
SiF$_4$	0.09–0.25	γ, air	129
CO$_2$	0.03–0.09	γ, vac.	129
CO$_2$	Major	electrons, air	61
CO$_2$	0.7–0.9	γ, air	129
Halocarbons	0.004	γ, vac	129
Halocarbons	0.007	γ, air	129

TABLE 20

Mechanical Properties of Irradiated Polychlorotrifluoroethylene (52)

Dose (Mrad)	Tensile strength (psi)	Shear strength (psi)	Elongation (%)
0	2,550	3,410	264
1	2,400	3,650	230
10	1,670	1,850	73
100	Failed	Failed	Failed

TABLE 21

Properties of Irradiated Polychlorotrifluoroethylene[a]

	Control	At 24 Mrad
Tensile strength (psi)	5622	5622
Elongation (%)	45.6	67.2
Impact (in. lb/in. width)	21.3	17.9
Water absorption	0.00	0.048
Specific gravity 25°C	2.12	2.11
Surface resistance (ohm)	17.8×10^6	17.8×10^6
Volume resistance (ohm-in.)	2.5×10^6	2.5×10^6

[a] According to Johnson, quoted in (9).

TABLE 22

Bending Resistance of Irradiated Polychlorotrifluoroethylene (67)

Dose (Mrad)	0	70	120	230
Bending resistance (kg/cm^2)	450	400	220	150

exhibits a curious phenomenon of dendritic crack growth, the subject of several studies, which proceeds at about 5.6 cm min^{-1} with fast electrons at 0.02 Mrad sec^{-1} (67). The mechanism is perhaps supersaturation with gas. The rate of breakdown to powder is said to be inverse to square root of dose rate (Fig. 17) (132). The thermal decomposition of PCTFE can be modified by UV initiation. A preliminary study has been made, along the lines of the gamma-ray initiated decomposition of PTFE (133). The principal observation was the postirradiation decay of depolymerization rate when the light was interrupted. Figure 18 shows the experimental plots and Figure 19 is the Arrhenius plot. Apparent activation energy is negative. The dotted line is attributed to complications due to impeded monomer diffusion at low temperature. The limiting full line, occurring mostly above the softening point, has an activation energy $E_{post} = -3$ kcal, approximately. The steady photo-rate activation energy is $E_{ph} = 13$ kcal.

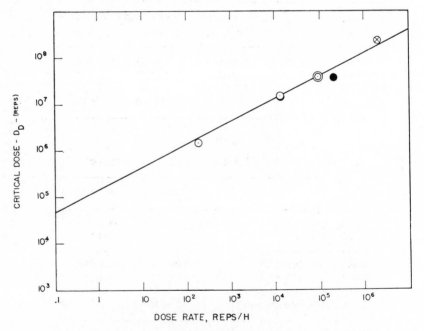

Fig. 17.   Critical dose for failure versus radiation rate for Kel-F (132).

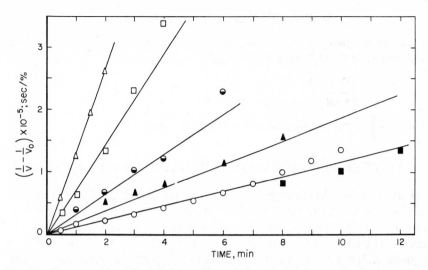

Fig. 18. Decomposition of polytrifluorochloroethylene during the post-irradiation period (133). V = rate of monomer production. (■) 294°C; (○) 277°C; (▲) 247°C and 261°C; (◑) 242°C; (□) 225°C; (△) 217°C.

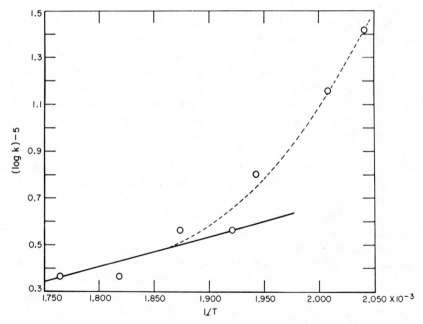

Fig. 19. Arrhenius plot of postirradiation decay of depolymerization rates of polytri-fluoroethylene (133).

V. POLYHEXAFLUOROPROPYLENE

The pure polymer degrades as follows according to measurements of melt viscosity (280°C) (19) of polymer irradiated in nitrogen at 200°C:

Dose (Mrad)	0	1	3
Melt viscosity (poises) $\times$ 10^{-4}	0.40	0.24	0.16

The copolymers are quite different, as will be seen.

VI. PURE FLUOROCARBON COPOLYMERS (TEFLON®100)

Unexpectedly, the copolymer of TFE and HFP (Teflon®100, FEP) and a few related polymers (19) cross-link, although apparently to a moderate extent only (56). The yield of gaseous products is not reliably known. A G_{tg} (i.e., total gas) of 0.10 was observed on beads (56) and should be therefore a gross underestimate, because of diffusion difficulties, yet it agrees with $G = 0.08$ computed from rapid electron irradiation of finely divided material (19). The composition showed CF_4 as the major product, about 60% (56) or 100% as against 10% on PTFE treated in the same way (19). Minor or trace products, noted at once or on postheating, were C_2F_6, C_3F_8, C_3F_6, C_4F_8, i-C_4F_8, C_5F_{12}, COF_2, CO_2, and SiF_4.

If thick strips (0.040 and 0.060 in.) are irradiated *in vacuo* at room temperature and the ZST measured at 280°C, the value first rises and then falls, as in Fig. 20. The rise is associated with cross-linking. The reaction probably occurs only above the so-called glass I temperature, about 80°C for Teflon® 100, which is associated with movements of 10 to 20 segment-lengths in the amorphous regions; thus the real reaction probably occurred during the ZST determination. Figure 21 illustrates this point, showing the effect of temperature. Dramatic improvements were noted in cut-through time of wire insulation (10 mil on No. 22 wire, hung over a 1/4-in. mandrel in a 250°C oven with 3/32-to 3/8-lb tension) which was raised from less than 40 to over 500 hr by irradiation to 7.8 Mrads at 250°C. At low doses, 0.7 Mrad, flow rate was lessened at low shear stresses and unchanged at higher shear stresses. Many other properties are not much changed at doses up to 6 or 8 Mrad. Prolonged UV irradiation produces the same kind of changes, with perhaps less degradation. Improvement of flex life is remarkable, 4415 initial improving to 75,500 flexes to failure (134). Other copolymers that improve in the same way are TFE with perfluoroisobutene and TFE with perfluoroheptene. The copolymer of hexafluoropropene with perfluoroheptene degrades. These results are rationalized (19) by considering reactions of the following radical types in these polymers:

No.	Formula	No.	Formula
1	$\sim CF_2{-}CF_2 \cdot$	5	$\sim CF_2{-}\underset{\cdot}{CF}{-}CF\sim$ with CF_3 above CF
2	$\sim CF_2{-}CF{-}CF_2\sim$ with $CF_2 \cdot$ below	6	$\sim CF_2{-}CF{-}CF_2\sim$ with $CF_2 \cdot$ above
3	$\sim CF_2{-}CF{-}CF_3$ with $CF_2 \cdot$ above	7	$\sim CF_2\underset{\cdot}{C}{-}CF_2\sim$ with CF_3 above
4	$\sim CF_2{-}\underset{\cdot}{CF}$ with CF_3 above	8	$\sim \underset{\cdot}{CF}{-}CF{-}CF\sim$ with CF_3 above each

Of the combinations between these types, some are end-couplings, which keep the molecule linear (1-1, 1-3, 1-4, 3-3, 3-4). A few are branching combinations, which produce branches and ultimately cross-links (1-2, 1-5, 1-6, 1-7, 1-8, 3-6, 6-6), and the others are deemed "forbidden" in view of steric hindrance and reaction rates. From the variety of branched species already

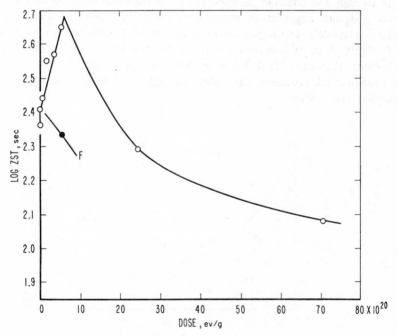

Fig. 20. Zero-strength-time of irradiated copolymer of tetrafluoroethylene and hexafluoropropylene (56). (○) irradiated in vacuum; (●) irradiated in air; (F) too weak to handle.

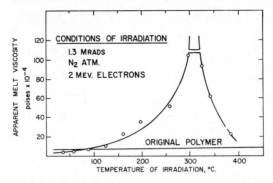

Fig. 21.   Effect of temperature of irradiation on apparent melt viscosity of tetrafluoro-ethylene-hexafluoropropylene copolymer (19). Viscosity measured at 380°C, 1.4×10^5 dynes/cm^2 shear stress.

mentioned as products of liquid fluorocarbon radiolysis, the "forbidden character" must be only relative.

In oxygen and chlorine, as expected from the chemistry, the radicals are scavenged and degradation alone occurs (21). Elongations and relative values of tensile strength go as in Figs. 22 and 23. The decline is slower than with PTFE. A possible reason is that for the same tensile, molecular weights are lower; thus each break has a smaller effect on properties. No G-values of fundamental processes have been derived, since molecular-weight correlations are lacking.

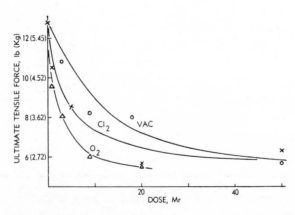

Fig. 22.   Ultimate tensile load of tetrafluoroethylene-hexafluoropropylene copolymer strips irradiated (○) in vacuum, (×) in chlorine at 10 and 760 torr, and (△) in oxygen at 760 torr. Temperature 25°C, dose rate 0.3 Mr/hr (21). (Reprinted from SPE Transactions Vol. 3 No. 4 Oct. 1963.)

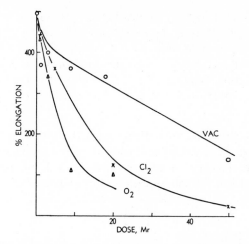

Fig. 23. Elongation at break of gamma-irradiated tetrafluoroethylene-hexafluoropropylene copolymer strips (21). Conditions as in Fig. 22. (Reprinted from SPE Transactions Vol. 3 No. 4 Oct. 1963.)

VII. THE FLUORINATED ETHYLENES

A. Scissions and Cross-Links

When the polymer chain contains hydrogen, the following new effects appear related to the elimination of HF:

1. Extensive cross-linking becomes possible.
2. Further elimination of HF produces unsaturation.
3. Subsequent heating greatly enhances the decomposition.

In the first two effects, elimination of stable HF resembles that of H_2 in hydrocarbon polymers.

The three ethylenes—polytrifluoroethylene, polyvinylidene fluoride, and polyvinyl fluoride—will be considered as a series. Available G-values for cross-links and scissions are collected in Table 23. Two values are given for polyvinylidene fluoride, determined in different laboratories at different temperatures. The values for polyvinyl fluoride involved oriented polymer and are ambiguous within perhaps a factor of three; however, the ratio G_d/G_c should be more reliable. The ratio G_d/G_c was evaluated from the course of sol fraction with dose, assuming that the molecular weight distribution ultimately approaches the "most probable." The data are treated by one of two graphical methods:

1. A plot of log sol fraction against log dose is compared with a series of standard curves for chosen G_d/G_c (141) (shown in Fig. 24 as β/α). The curves are slid to the left or right until a good match to the points is found. It is preferable but not necessary to know the dose for initial gel formation also.

2. The sol-fraction function, $S + \sqrt{S}$, is plotted against reciprocal dose, and the limiting value of $S + \sqrt{S}$, i.e., the intercept in Fig. 25, gives the value of G_d/G_c (142).

As used here, G_c refers to cross-linked chains or cross-linked units, which is twice the often used G-value for cross-link points. A value for G_c can be derived from the increase of rubber elasticity with radiation dose, as the elasticity is related to the concentration of cross-links:

$$\frac{F}{A} = RT\left(\frac{v_e}{V_0}\right)v_2{}^{1/3}\left[\alpha - \frac{1}{\alpha^2}\right]$$

where F/A = force per unit cross section of swollen unstretched polymer (dynes cm^{-2}), R = gas constant (ergs mole^{-1} deg^{-1}), v_e/V_0 = effective cross-linked chain density in unswollen polymer (moles cm^{-3}), v_2 = volume fraction polymer, and α = extension ratio.

For the polymers in Table 23, the rubbery state was brought about by swelling with solvent, but warming above the transition point could have been used. Gel-swelling measurements are also convenient, but their interpretation requires either a calibration with elasticity measurements or else a knowledge of the Flory-Huggins interaction parameter between solvent and polymer segments. Either of these measurements counts only "effective" cross-linked chains of the network, and not dangling chain ends. For use in Tables 23 and 26, the original allowances for these ends were replaced by a more systematic treatment (139,140). As in the measurement of G_d/G_c, it is assumed that the ratio of scission to cross-link formation remains constant, and that the

TABLE 23

Scissions and Cross-Links in Fluorinated Ethylenes[a]

Polymer	G_d/G_c	G_c	G_d	Temp. (°C)	Ref.
CF$_2$CFH	0.14	1.93	0.27	25	135[b]
CF$_2$CH$_2$	0.18	2.2	0.40	25	136, 137[b]
CF$_2$CH$_2$	0.47	—	—	47	138
CH$_2$CHF	0.28	11.4	3.2	25	136[b,c]

[a] G_c refers to cross-linked chains or units per 100 eV, i.e., twice the number of cross-links, cf. Charlesby (2).

[b] Data recomputed according to Mullins (139).

[c] Tentative best values; basis in doubt due to orientation effects.

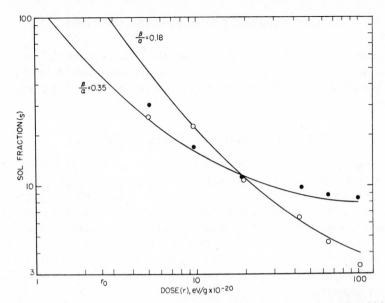

Fig. 24. Sol fraction as function of γ-ray dose for ($\bigcirc$) poly(vinylidene fluoride); ($\bullet$) poly(vinyl fluoride) (136). Dose rate 5×10^{20} eV/g-hr; curves theoretical for indicated β/α (scissions per 100 eV)/(cross-linked units per 100 eV).

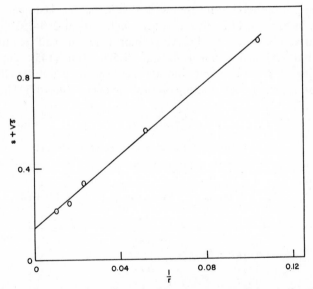

Fig. 25. Determination of ratio of chain fractures to cross-links produced in polytri-fluoroethylene by gamma radiation. Plot of function of sol fraction S, versus reciprocal of dose r. Intercept 0.14, ratio of fractures to cross-linked units (135). (Reprinted from SPE Transactions Vol. 4 No. 1 Jan. 1964.)

359

molecular-weight distribution approaches the "most probable" at least toward the end of the experiment. Other distributions can be handled in principle (2,143) but the methods are difficult and have not been seriously applied to fluorocarbons.

B. Volatile Products from Fluorinated Ethylenes

The volatile yields from polytrifluoroethylene (Table 24) illustrate the predominance of HF which appears in glass systems as SiF_4; thus $G(HF)$ $= 4G(SiF_4)$ if SiF_4 all comes from this source. A part of the H_2 reported could have come from HF acting on the aluminum or silver foil envelopes, but this is unlikely for these metals with anhydrous HF, and no corrosion was evident. The evolution of HF from polyvinyl fluoride, and presumably also from polyvinylidene fluoride, is much greater. Abundant, ill-defined free radicals are also observed in irradiated polytrifluoroethylene and other non-rubbery copolymers.

C. Mechanical Property Changes

The observed mechanical properties agree with the picture of predominant cross-linking, although results are less dramatic than in elastomers. Some are illustrated in Fig. 26. In addition, cut-through tests at 200–255°C improved up to final values of 735 hr, Polyvinyl fluoride retains half its initial tensile strength, but very little elongation, out to 500 Mrad (145). The radiation cross-linking of polyvinylidene fluoride has been used industrially to place an abrasion-resistant outer coating on insulated wire (146,147). The complete

TABLE 24

Gas G-Values from Fluoro-Hydro Polymers (56,144)

Polymer, Gas ———— Product	CF_2CFH, powder	Viton, shreds	Kel-F 3700	CH_2CHF
SiF_4	0.86	0.29		(HF) 4.5
CHF_3	0.016			
CO_2	0.113	0.09		
CO	0.137			
H_2	0.03, 0.12	0.27		
CF_4		0.045		
C_3F_6		0.01		
Total			>2	

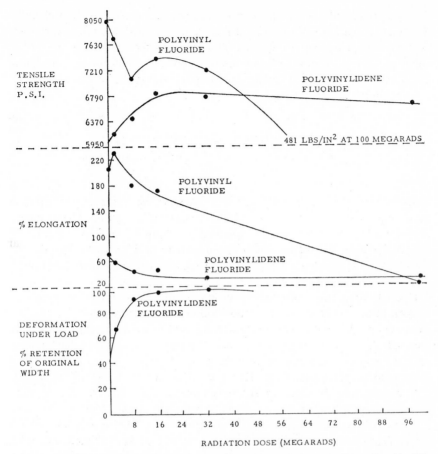

Fig. 26. Physical changes in irradiated polyvinyl fluoride and polyvinylidene fluoride (23).

process consists of extruding a layer of polyethylene insulation over the wire, cross-linking this by irradiation, applying a thin coat, e.g., 0.005 in., of polyvinylidene fluoride containing a little triallyl cyanurate, and then cross-linking the second coat with 6 Mrad of high-energy electrons.

D. Postheating Effects—Hydrogen Fluoride Stripping

Upon heating, these polymers may lose additional HF. Polyvinyl fluoride, when irradiated a second time, sometimes shows about twice the original $G(HF) = 4.5$ (144). If the irradiated sample is heated near 100°C, it may show no change for a while, especially if oxygen has free entry, and then in 5 or 10 days it may evolve 10–14 meq HF per gram, which is about half the total fluorine present, almost independently of the radiation dose received. Beyond this stage, evolution is slow. Similar results, less marked, are found with HCl evolution from polyvinyl chloride. The observed features fit the idea of a chain process of "stripping," ultimately stopped by the occurrence of chain-stopping groups. In the following structure, the atoms F*, near cross-links or unsaturations, are initially labile:

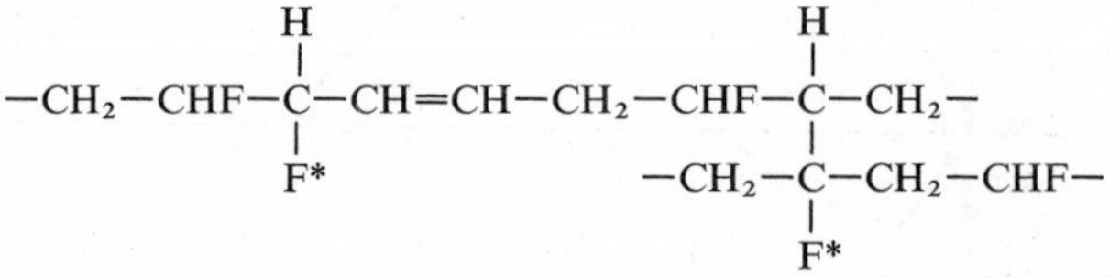

$$-CH_2-CHF-\underset{\underset{F^*}{|}}{\overset{\overset{H}{|}}{C}}-CH=CH-CH_2-CHF-\underset{\underset{-CH_2-\underset{\underset{F^*}{|}}{C}-CH_2-CHF-}{|}}{\overset{\overset{H}{|}}{C}}-CH_2-$$

Stripping off HF creates new labile F sites, which continue the process until some barrier is reached, e.g., another $-CH=CH-$ group. The strangely constant 50% fractional yield is explained if radiation produces one efficient barrier for each two active sites. The proclivity of PVF toward chain decomposition at moderate temperatures suggests that the earlier-mentioned high values for G_c and G_d may be real.

Upon heating the irradiated polymers to the usual range of pyrolysis temperatures, the thermal decomposition becomes complicated. Unlike unirradiated polyvinyl fluoride, which decomposed to yield volatile products in the gaseous to waxy range, the irradiated polyvinyl fluoride tended to strip off HF and leave a carbonaceous residue.

Polyvinylidene fluoride strips HF in all cases, but more if irradiated (136). The temperature for decomposition is lowered also, but because of char formation, volatilization can no longer be complete. Alkali treatments impose further changes on the mechanism. Extraction with hot hexamethylphosphoramide nearly restored the behavior characteristic of unirradiated polyvinyl fluoride, perhaps by removing unsaturated sites at which stripping could be initiated. The changes of pyrolysis rate of polytrifluoroethylene are more moderate, and not greater than can be explained by the introduction of a cross-linked network; thus stripping is apparently unimportant in the post-irradiation heating of this polymer (135).

VIII. ELASTOMERS

A. Vinylidene Fluoride Copolymer Elastomers

Both Viton A ($C_3F_6 + CF_2CH_2$) and Kel-F elastomer ($CF_2CFCl + CF_2CH_2$) can be cross-linked by radiation (29,56,148–151). With Viton A, vulcanization by any process including radiation is incomplete without a "postcure" in which the polymer is heated at about 200°C in an open system (153,154). In addition, high-temperature aging experiments suggest a "two-network" system, in which old cross-links are broken and new ones of a different kind made. Creep is considerable, but total cross-links increase. The initial cross-links formed by amine cures are apparently unstable at high temperatures and the new cross-links formed are more permanent (156). Radiation and peroxide-recipe cross-links are also more stable. The radiation network is thought to resemble the new stable network and to involve C—C bonds. Even for these more stable networks, the activation energy for thermal scission in air is surprisingly low, 28 or 30 kcal (156). Perhaps this is because radiation forms unsaturated sites in the chains, which are open to oxidative attack (156). In vacuum, thermal scission (measured by stress relaxation) is much slower.

Thermal creep rates as a function of radiation dose are listed in Table 25 for a Viton-type elastomer. Absorbers of HF or HCl, such as CaO and MgO, are beneficial in agreement with acid release studies (155), but even then heating in an open system (continuous evacuation, N_2, or air) is needed (163). It is suggested that the water formed must be removed in this case:

$$2HF + MgO \longrightarrow MgF_2 + H_2O$$

The degree of radiation cross-linking will evidently depend upon a number of extraneous factors such as postcure. Mechanical properties will also be heavily dependent upon the carbon black or other loading.

Viton is cross-linked to a useful extent at 10 or 20 Mrads (148) and serves without extreme deterioration to perhaps 100 or 200 Mrad, whereas Kel-F Elastomer deteriorates earlier (29,56).

TABLE 25

Thermal Aging of Irradiated Vinylidene Fluoride Hexafluoropropene Elastomers (158)[a]

Radiation dose (Mrad)	Initial swelling ratio (Q_{in})	Creep rate[b] (sec$^{-1} \times 10^5$)	Permanent set[b] (%)
Filled; 235°C			
10.7	5.5	0.67	81
22.4	3.98	0.41	51
35.0	3.42	0.36	36
Unfilled; 205°C			
10.7	7.90	0.27	24
27.4	6.15	0.18	12
29.0	5.41	0.15	12

[a] These changes measured in air; rates in Ar are much less.
[b] Load 5 kg cm^{-2}, time 3000 min.

TABLE 26

Radiation Yields in Vinylidene Fluoride Copolymer Elastomers[a]

Polymer	G_c	G_d
82 CF_2CH_2/18 C_3F_6 (Viton A-HV)	3.41	1.36
70 CF_2CH_2/30 CF_2CFCl (Kel-F Elastomer 3700)	2.06	1.56

[a] Data from Yoshida (137); revised treatment from Mullins (139).

Radiation yields found under conditions of no postcure and no additives are shown in Table 26. Sol values on the Russian copolymer SKF-32, similar to Kel-F elastomer, lead to a G_d/G_c in fair agreement, as replotted in Fig. 27.

This would also be approximately true of the equilibrium modulus values in Table 27, if adequate correction for scissions were applied. The small reversed air effect, implying more effective cross-links for irradiation in air, is perhaps another exhibition of the need for postcure with escape of volatiles. Scission estimates by stress relaxation were very high (137) and probably vitiated by the inclusion of a dense network of entanglements or crystallites as virtual cross-links. The cross-linking G-values in SKF-32 compositions depended heavily upon the monomer ratio (Table 28) in qualitative agreement with ZST observations on Kel-F elastomer 3700 versus 5500 (56). It is not clear what account if any was taken of concurrent scissions; because of the high scission ratio, they would be important. Infrared has been heavily studied in the effort to demonstrate double-bonded structures and distinguish irradiation from postcure (153,158,159). Principal bands and proposed assignments are listed in Table 29.

Changes in selected mechanical properties, mostly of full compounded

TABLE 27

Equilibrium Modulus of Irradiated Vinylidene Fluoride-Chlorotrifluoro-ethylene Copolymers[a]

Dose (Mrad)	3	41
Equilibrium modulus, vacuum irradiation		
($kg\ cm^{-2}$ at 150°C)	1.3	3.0
Same, air irradiation	3.0	5.0
G_c, uncorrected; vacuum[b]		0.85
Same, air irradiation[b]		1.00

[a] Scaled from Galil-Ogly (153, Fig. 3).

[b] Computed from $E_\infty = 3N_e kT$, assuming 1 R = 0.874 rad, ρ = 1.83 g cm^{-3}, G_d = 0.

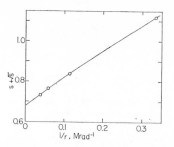

Fig. 27. Cross-links and scissions in irradiated vinylidene fluoride-chlorotrifluoroethylene elastomer, Russian SKF-32. Data (159) replotted as sol-fraction function, $s + \sqrt{s}$, versus reciprocal dose. Intercept, 0.68, ratio of fractures to cross-linked units.

recipes, are shown in Table 30. It is difficult to select representative data because of the considerable variation with recipe. For many other data of this sort, see reference 4,5,29,84,145,160.

The 3M product Fluorel is also described as moderately resistant. A particular recipe—20 MgO, 15 thermal carbon black, 1 hexamethylene diamine carbamate per 100 parts elastomer—gave the following properties:

Dose (Mrad)	10	100
Stress at 100% elongation (psi)	400	—
Tensile strength (psi)	1,675	1,380
Elongation (%)	240	80
Hardness, Shore A	69	76
Set at break (%)	6	0

TABLE 28

Radiation Cross-linking of Vinylidene Fluoride—
Chlorotrifluoroethylene Elastomer, Composition
Dependence (160)[a]

Weight % CF_2CH_2	65	7?
G_c	2.7	4.4

[a] Adjustment for scissions not stated.

B. Other Hydro Fluoro Elastomers

In accord with general principles on hydrogen-bearing polymers, the following elastomers also cross-link with more or less degradation:

FBA(1F4), dihydroperfluorobutyl acrylate:

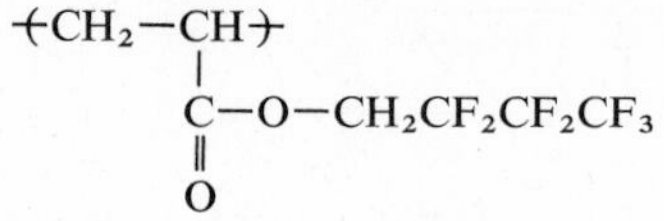

$$-\!\!\left(CH_2-CH\right)\!\!-$$
$$\begin{array}{c} | \\ C-O-CH_2CF_2CF_2CF_3 \\ \| \\ O \end{array}$$

Hooker No. 1, hexafluoropentamethylene adipate:

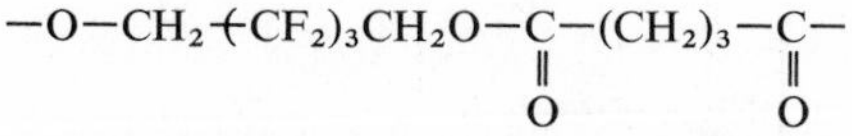

$$-O-CH_2-\!\!\left(CF_2\right)_3CH_2O-\!\!\underset{O}{\underset{\|}{C}}-(CH_2)_3-\!\!\underset{O}{\underset{\|}{C}}-$$

Hooker No. 2, like the above but a mixed adipate phthalate.

Mechanical property changes are listed in table 30. Radiation chemical studies have not been published.

IX. FLUORINATED POLYSTYRENES

Both poly-α,β,β-trifluorostyrene and poly-2,3,4,5,6-pentafluorostyrene yield free radicals when irradiated. The polytrifluorostyrene is insoluble at

TABLE 29

Infrared Spectral Changes of Irradiated Vinylidene Fluoride Elastomers

Wave number (cm^{-1})	Identification	Observed change
	Russian SKF-32; in air (160)	
1,740	Terminal $-CH\!=\!CF_2$	Decreases
1,800	$-CF\!=\!CF_2$, or CFO compounds	Appears low dose, increases at high doses
1,840		Appears then disappears
2,500–2,670	$-OH$ compounds	Appears at high doses, 68 MR
2,700–3,400	$-OH$ compounds	Broadens
	Russian SKF-32; in vacuum (160)	
1,640	$-CH\!=\!CF-$	Appears at 10–20 MR
1,740	$-CH\!=\!CF_2$ and $R\!-\!CF\!=\!CF\!-\!R$	Appears at 10–20 MR, then decreases
1,840	$-CF\!=\!CF_2$, with reservations	Appears at 10–20 MR, then decreases
1,800 wide	$-CF\!=\!CF_2$ (?)	Appears at higher dose
	Viton, γ-irradiated, deoxygenated, in N$_2$ (154)	
1,725		Appears
	Viton irradiated, postcured in N$_2$ 24 hr at 200°C (154) or 250°C (156)	
1,745 (157)	$-CH\!=\!CF_2$	Sharpens, increases
1,668–1,653 (154)	Conjugated double bonds	Appears
1,650 (157)	Isolated $-C\!=\!CF-$	Appears
1,640–1,610 (154)	Aromatic bonds	Appears
1,580 (157)	Conjugated	Appears

20 Mrad and perhaps before, therefore presumably it cross-links. It is probably safe to predict the same for the pentafluorostyrene polymer. The interesting photochemical and radiation behavior of perfluoroaromatic small molecules (isomerization, polymerization) has been mentioned.

X. POLYTETRAFLUOROETHYLENE OXIDE

Well-identified free radicals are formed in this oxide. Viscosity measurements indicate $G_{sc} = 1.8 \pm 0.4$ at room temperature, in rough accord with observed radical yields (166). The empirical relation is

$$[\eta] = (0.7 \pm 0.1)(r + 1.6)^{-0.65 \pm 0.02}$$

An anomolous increase in the Huggins K' value for solution viscosity raises a suspicion of some small component of cross-linking. K' was usually 0.40 but rose to 1.2, 2.0, and 4 at 22.7, 35.5, and 57.7 Mrad. The ratio of cross-links to scissions is judged to be less than 0.1, however. An important volatile product is COF_2.

XI. NITROSO RUBBER

The elastomer from $C_2F_4 + CF_3NO$, repeat unit

$$+CF_2-CF_2-N-O+,$$
$$|$$
$$CF_3$$

is degraded by a rapid chain process with either gamma-ray or 2537 Å ultraviolet initiation (162). At a dose rate of 0.1 Mrad hr^{-1} and a temperature of 25°C, the G-value for chain scission was about 3.7. Each scission is also associated with a decomposition of 5.3 chain units $-CF_2-N(CF_3)-O$ $-CF_2-$ to COF_2 and $CF_3-N=CF_2$. If moisture is present, traces of other products occur, e.g., $(CF_3)_2NH$. It is believed that primary scission occurs at the N—O bond in both radiation and thermal degradation:

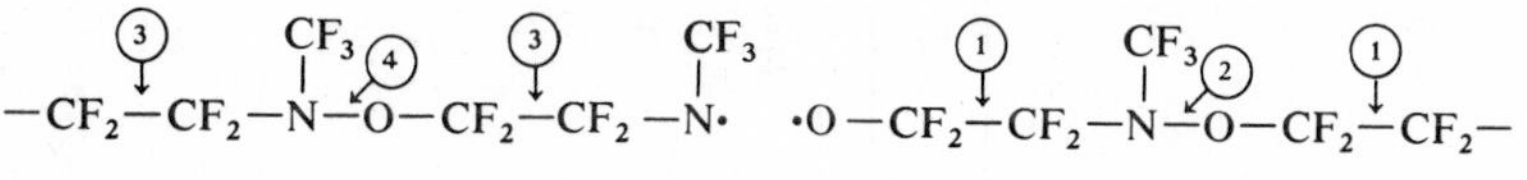

$$-CF_2-CF_2-N-O-CF_2-CF_2-N\cdot \quad \cdot O-CF_2-CF_2-N-O-CF_2-CF_2-$$

The initial break into radicals (indicated at the dots) is followed by splitting sequences of four bond types shown as 1—2—1—2, etc., or 3—4—3—4, etc., to produce the COF_2 and $CF_3N=CF_2$ observed. Decomposition is exothermic. Although the heat of reaction is not established, -8.7 and -24.7 kcal per mole of repeat units are suggested values. Related free radicals have been described.

TABLE 30

Radiation-Induced Changes in Mechanical Properties of Fluorocarbon Elastomers[a]

Polymer[b]:	FBA(1F4)			Hooker 1 Ester			Hooker 2 Ester		
Property[c]:	H	E	T	H	E	T	H	E	T
Dose									
0	71	150	1200	68	225	1830	80	110	1460
1									
3									
5	+5.6	+6.7	+9.7	+2.9	−15.6	−26.6	+6.3	+14.5	−11.2
6									
10	+15.5	−33.3	+2.1						
22				+4.4	−51.0	−55.8	+7.5	−18.2	−17.0
50									
55				+14.7	−68.9	−63.5	+12.5	−54.3	−3.50
60									
100	+23.9	−76.7	−44.9	+17.6	−82.2	−64.7	+13.8	−77.3	−39.9
500	+28.2	−86.7	−33.8	+39.7	−95.6	−36.2	+23.8	−95.5	−64.1

Polymer[b]:	Kel-F Elastomer			Kel-F Elastomer 5500			Viton A-6			Viton, PR 1700-X7		
Property[c]:	H	E	T	H	E	T	H	E	T	H	E	T
Dose												
0	75	640	1100	62	550	1810	80	130	2125	75	250	1865
1				0	+9.1	+44.2						
3	−12.0	+4.4	+13.6									
5							+13.8	−28.5	−2.2	+4.0	−33.2	−4.6
6	−6.7	−10.4	+27.4	0	−8.2	+19.6						
10												
22				+3.2	−41.8	−28.8						
50							+22.5	−80.8	−5.5	+18.7	−79.8	−14.5
55				+16.1	−73.6	−24.9						
60	−10.8	−14.8	−36.4									
100							+25.0	−84.6	+49.2	+25.3	−85.8	+12.5
500				+25.8	−80.0	−13.9						

[a] Selected from Harrington (29,145,160).
[b] FBA: dihydroperfluorobutyl acrylate; Kel-F Elastomer: vinylidene fluoridechlorotrifluoroethylene copolymer; Kel-F 5500: same but higher in chlorotrifluoroethylene; Viton: vinylidene fluoridehexafluoropropene copolymer; Kel-F 5500: same but higher adipate; Hooker 2: like Hooker 1 but containing some phthalate groups.
[c] For initial values, H = Shore hardness, E = elongation at break (%), T = tensile strength (psi), Subsequent values, increase (+) or decrease (−) as percentage of initial value. Doses, megaroentgens.

369

Although chain processes of this sort are rare, other examples seem to be the elimination of HF from irradiated polyvinyl fluoride and the postheating of irradiated polyvinylidene fluoride described earlier.

XII. PHOTOCHEMICAL DEGRADATION

In addition to the rapid photo degradation of nitroso rubber described earlier, a few other instances of ultraviolet action are reported. One example is the UV-initiated thermal decomposition of polychlorotrifluoroethylene already discussed, and another is the slow UV cross-linking of the HFP-TFE copolymer.

A very slow UV-induced degradation of polytetrafluoroethylene is also reported from two sources (19,163,164). The longer study (163,164) indicates that, in curious contrast with other polymers, deterioration is greatest in vacuum, intermediate in oxygen, and least in nitrogen. The considerable influence of wavelength is illustrated by a few results on tensile strength and elongation of 1.0-mil films (163):

Wavelength (Å)		—	3,690	3,140	2,440
Exposure (joules/cm²)	0		6,000	1,300	300
Tensile (10³ psi)	4		4	3.6	3.5
Elongation (%)	1,100		900	700	700

The deduced threshold photon energy is about 3.2 eV, the same as for polyethylene and Mylar polyethylene glycol terephthalate. Since there is no very evident absorption in the wavelength region used, the question arises as to how the photochemical effect could occur. A highly forbidden absorption process, followed by bond breaking with high efficiency, can hardly be excluded, but a more likely route is absorption and activation by trace impurities, such as $C-H$, $C-Cl$ or $C=O$ bonds. Whether characteristic of pure polymer or not, the process is evidently absent under the conditions of long-time outdoor aging tests.

The photochemistry of perfluoroaromatic rings, described in the section on pure compound radiolysis, may have future applications in polymers.

Thermoluminescence and phosphorescence have been observed in irradiated polytetrafluoroethylene, but they are not very prominent. Spectral features, resembling those of trifluoroacetone, suggest that all the observed effect could be due to carbonyl groups occurring as an impurity (167). A specimen irradiated to 1 Mrad has two maxima of thermoluminescence near -125 and $+22°C$, which correlate loosely with other criteria of sudden increase in molecular motion such as viscoelastic loss peaks (168).

XIII. ELECTRICAL PROPERTIES UNDER IRRADIATION

Like many other insulators, polytetrafluoroethylene develops a small electrical conductivity when irradiated (169–180). After irradiation ceases, much of the conductivity dies away in hours or days, but a permanent increase remains if the dose was greater than 10^5 rad. Measurement and description of the time-dependent phenomena are complicated, involving slow polarization of the dielectric, currents induced at the electrodes by charges which have not yet arrived there, and ordinary ohmic conduction modified by disappearance of the charge carriers (178,179). In limited regions of time and temperature, first-order (low-temperature, short-time) and second-order processes seem to control the number of charge carriers (175,176). The steady-state current during irradiation, although conceptually simpler, is nevertheless complicated by thickness and directional effects (171,172), although it obeys Ohm's law approximately. The dose-rate dependence is usually of the form where

$$\sigma - \sigma_0 = bI^\Delta$$

where σ and σ_0 are the steady-state and " dark " conductances, b is a constant, Δ an exponent, and I the dose rate (169,171,172). If the temperature is not too low, over $-20°C$, an Arrhenius law may apply to the steady-state current and to the disappearance (169). Typical values of conductance are around 2×10^{-18} ohm^{-1} cm^{-1} at room temperature and 100r hr^{-1} (182), with exponent Δ in the range of 0.7 to 1.0 and activation energy about 0.5 eV (169).

The classical explanation for the steady-state phenomena has been that radiation promotes electrons from valence states into sparsely occupied conduction bands, difficult to visualize, as in semiconductor theory (169). The temperature dependence involves also metastable electron traps from which thermal liberation occurs.

During storage after irradiation to high doses, the conductance can decay to a minimum and then increase (174,179). The ultimate increase may come from the slow conversion of radicals to acidic or polar groups.

Radiation-induced conduction in polychlorotrifluoroethylene is less by about tenfold than in polytetrafluoroethylene (179,182). In polyvinyl fluoride an ultraviolet induced conductivity has been described (183).

The dielectric strength of polytetrafluoroethylene is unchanged after 5.7×10^7 rad (174). Dielectric constant changes very little, e.g., less than 5% at 800 Hz after 4×10^7 rad (179). Increases in dielectric loss are low absolutely but limits of error do not exclude a severalfold increase after 2×10^8 rad (51,178). The relaxation-time spectrum measured during irradiation was narrowed slightly in a 50% crystalline sample but broadened in one of 19% crystallinity (177).

XIV. MISCELLANEOUS OBSERVATIONS RELATED TO RADIATION

A few other processes have some incidental relation to ionizing radiation. A microwave discharge in a hollow PTFE cylinder forms a number of products (184): CO, CO_2, CF_4, C_2F_6, C_2F_4, C_3F_8, and C_3F_6. The process is a composite of thermal decomposition at the surface,

$$(C_2F_4)_x \longrightarrow C_2F_4$$

and various reactions in the plasma, such as

$$e^- + C_2F_4 \longrightarrow C_2F_4{}^+ + 2e^-$$
$$e^- + CF_2 \longrightarrow CF^+ + F + 2e^-$$

Discharge reactions have been studied earlier in fluorocarbon gases (185, 186). Oxygen, even in excess, does not suppress all fluorocarbon products, as it does in pyrolysis. This suggests that species less reactive with oxygen are present here, such as $C_2F_4{}^+$.

Erosion of PTFE surfaces by a beam of ions is very much greater than on metal or glass surfaces (187). The amount rises parabolically with ion energy, and at 30 keV it reaches values of 15, 7, and 3×10^4 amu per ion for ions of Ar^+, He^+ and H^+, respectively. For PMMA under Ar^+ at 30 keV, the value is only 2.5×10^4. The amount falls off at high angles of incidence.

Occasionally positrons have been absorbed and annihilated in PTFE as well as other media (188). PTFE is also used to study cosmic ray events, as the atomic numbers are close to those of air (189,190). At such low doses, nothing can be said of chemical factors.

Low-energy, short-path particles from the thermal neutron reaction $^{10}B(n,\alpha)^7Li$, have been used in boron sandwich constructions to unite dissimilar polymer surfaces including PTFE (191). The object was to secure localized adhesion without degradation elsewhere. Strength of the joint goes through a maximum, about 90–100 kg cm^{-2} for PTFE on polyethylene, and then declines on further neutron irradiation.

XV. GRAFT POLYMERIZATION

The fluorocarbon polymers PTFE, PCTFE, and VF have been used as substrates for graft polymers, a new monomer being polymerized to form a side chain on the substrate polymer.* With PTFE substrate, the reaction can be written

---

* For a short, comprehensive review see reference 192.

$$\sim CF_2CF_2CF_2\sim \xrightarrow{\quad\gamma\quad} \sim CF_2\overset{\cdot}{C}FCF_2\sim \; + \; F$$

$$\sim CF_2\overset{\cdot}{C}FCF_2\sim \; + \; CH_2{=}CHC_6H_5 \longrightarrow \; \sim CF_2\underset{\underset{\underset{\overset{|}{CH}-C_6H_5}{|}}{\overset{|}{CH_2}}}{C}FCF_2\sim$$

The side chain can modify many properties of the original polymer: solubility, mechanical properties, and perhaps more important, surface, adhesive, and wetting properties.

Sometimes a more complicated system is used. For example, polyvinyl fluoride can be irradiated in the presence of a free-radical initiator such as hydrogen peroxide, and the polymer then warmed with new monomer to initiate the graft polymerization. The fluorocarbon polymers are commonly difficult for monomer to penetrate; therefore only surface grafts are produced unless irradiation is conducted very slowly. The most interesting experiments were done with PTFE irradiated at 48–1860 rad hr^{-1} in styrene (193). In this case, for films 0.040 in. thick, grafting occurred in depth. The product was tear resistant. With a 20–70% styrene content, it behaved like a plasticized PTFE. At higher dose rates, surface grafting became dominant (193,194); however, this also had utility, as dyeability and adhesion to other surfaces improved greatly (195,200). Moderate loss of elongation may be accepted (201). Under surface grafting conditions, 14,000–428,000 rad hr^{-1}, the rate of grafting of vinyl acetate to PTFE was temperature dependent, independent of thickness, and proportional to 3/2 power of monomer concentration and 1/2 power of dose rate (196). Addition of high-molecular-weight polyvinyl acetate did not change the reaction, whereas CCl$_4$ decreased the rate, evidently by chain transfer. Electron spin resonance observations sometimes show the change from substrate to new monomer radicals and have also indicated grafting of TFE on PTFE to produce branched PTFE (17).

Styrene has also been grafted upon PTFE preirradiated in either air (199) or vacuum (198). In the latter work, the ultimate amount of polymer was least at high temperature, 5% at 82°C against 13% at 25°C, but the final amount was reached more rapidly. Empirically, both the dose and the polymerization-time variation fitted first-order curves of the form

$$Q_t = Q_\infty(1 - e^{-2kt})$$

where t is the time and Q_t and Q_∞ are the amounts of graft polymer at time t and at completion. The apparent activation energy was 9 kcal. A few further examples, presumably surface grafts, are listed in Table 31.

TABLE 31

Radiation Graft Polymerization on Fluorocarbon Polymers

Substrate	Monomer	Dose rate (rad hr^{-1} $\times$ 10^{-5})	Dose (rad $\times$ 10^{-5})	Monomer grafted (%)	Ref.
PTFE	Acrylic acid	3.0	3.3	16.0	195
PTFE	Acrylonitrile	—	0.63	1.3	194
PTFE	N-vinylpyrrolidone	1.6	5.2	6.1	194
PTFE	Vinyl acetate	0.14–4.28	—	—	196
PTFE	Styrene	1.6	—	—	194
PTFE	Styrene	(Preirradiate, vacuum)	—	2–13	198
PTFE	Styrene	(Preirradiate, air)	—	—	199
PTFE	Methyl methacrylate	Pretreat, silent electrical discharge	—	3–8	200
PVF	Various	—	—	—	197
PCTFE	Acrylic acid	3.0	0.75	6.0	194
PCTFE	Acrylonitrile	3.0	0.75	1.0	194
PCTFE	Styrene	3.0	72.0	3.4	194
PCTFE	Vinyl acetate	3.0	3.0	1.0	194

References

1. F. A. Bovey, "The Effects of Ionizing Radiation on Natural and Synthetic High Polymers," Interscience, New York, 1958.
2. A. Charlesby, "Atomic Radiation and Polymers," Pergamon Press, New York, 1960.
3. A. Chapiro, "Radiation Chemistry of Polymeric Systems," Interscience, New York, 1962.
4. R. W. King, N. J. Broadway, and S. Palinchak, "The Effect of Nuclear Radiation on Elastomeric and Plastic Components and Materials," REIC Report No. 21, Radiation Effects Information Center, Battelle Memorial Inst., Columbus, Ohio, Sept. 1, 1961.
5. N. J. Broadway and S. Palinchak, REIC Report No. 21 Addendum, Aug. 31, 1964.
6. N. J. Broadway and S. Palinchak, "The Effect of Nuclear Radiation on Fluoropolymers," REIC Memorandum No. 17, June 30, 1959.
7. H. A. Dewhurst, A. H. Samuel, and J. L. Magee, *Radiation Research.* **1**:62 (1964).
8. C. D. Bopp and O. Sisman, "Radiation Stability of Plastics," Oak Ridge National Lab. Reports Nos. ORNL-928 (1951) and ORNL-1373 (July 1953).
9. A. Charlesby, "The Decomposition of Polytetrafluoroethylene by Pile Radiation," Great Britain Atomic Energy Research Establishment Report AERE M/R 978 (1952).
10. A. Nishioka, K. Matsumae, M. Watanabe, M. Tajima, and M. Owaki, *J. Appl. Polymer Sci.*, **2**(4):114 (1959).
11. K. Matsumae, private communication.
12. A. Nishioka and M. Watanabe, *J. Polymer Sci.*, **24**:298 (1957).

13. K. Matsumae, M. Watanabe, A. Nishioka, and T. Ichimiya, *J. Polymer Sci.*, **28**:653 (1958).
14. A. Nishioka, M. Tajima, and M. Owaki, *J. Polymer Sci.*, **28**:617 (1958).
15. A. V. Tobolsky, D. Katz, and A. Eisenberg, *J. Appl. Polymer Sci.*, **7**:469 (1963).
16. M. Takahashi and A. V. Tobolsky, *Polymer Letters*, **2**:129 (1964).
17. M. I. Bro, E. R. Lovejoy, and G. R. McKay, *J. Appl. Polymer Sci.*, **7**:2121 (1963).
18. A. Ferse, W. Koch, and L. Wuckel, *Kolloid Z.*, **219**:20 (1967).
19. E. R. Lovejoy, M. I. Bro, and G. H. Bowers, *J. Appl. Polymer Sci.*, **9**:401 (1965).
20. K. G. McLaren, *Brit. J. Appl. Phys.*, **16**:185 (1965).
21. R. E. Florin and L. A. Wall, *Soc. Plastics Engrs. Trans.*, **3**(4):290 (1963).
22. L. A. Wall and R. E. Florin, *J. Appl. Polymer Sci.*, **2**:251 (1959).
23. R. Timmerman and W. Greyson, *J. Appl. Polymer Sci.*, **6**:456 (1962).
24. C. G. Collins (1957), quoted in R. W. King, N. J. Broadway, R. A. Mayer, and S. Palinchak, "Effects of Radiation on Materials and Components," Ed. J. F. Kircher and R. E. Bowman, Reinhold, New York, 1964.
25. K. Wündrich, *Materialprüfung*, **10**:217 (1968).
26. W. R. Licht and D. E. Kline, *J. Polymer Sci. A*, **2**:4673 (1964).
27. J. V. Pascale, D. B. Herrmann, and R. J. Miner, *Modern Plastics*, **41**(2):239 (Oct. 1963).
28. R. Harrington and R. Giberson, *Modern Plastics*, **36**(3):199 (1958).
29. R. Harrington, *Rubber Age*, **82**(3):461 (Dec. 1957).
30. M. A. Mokulskii, Yu. S. Lazurkin, M. B. Fiveĭskiĭ, and V. I. Kozin, *Doklady Akad. Nauk SSSR*, **125**:1007 (1959).
31. L. P. Yanova and A. B. Taubman, "Deistvie ioniziruĭushchikh izlucheniĭ na neorganicheskie i organicheskie sistemy," Akad. Nauk SSSR, Moscow, 1958, pp. 314–324.
32. A. E. Scherer and D. E. Kline, *J. Appl. Polymer Sci.*, **11**:341 (1967).
33. W. R. Licht and D. E. Kline, *J. Polymer Sci. A2*, **4**:313 (1966).
34. W. M. Peffley, V. R. Honnold, and D. Binder, *J. Polymer Sci. A1*, **4**:977 (1966).
35. R. E. Moynihan, *J. Am. Chem. Soc.*, **81**:1045 (1959).
36. N. G. McCrum, *ASTM Bull. No. 242*, **1959**:80.
37. G. A. Bernier, D. E. Kline, and J. A. Sauer, *J. Macromol. Sci. (Phys.)*, B1(**2**):335 (1967).
38. A. Danno, M. Koike, K. Doi, and M. Inoue, *Prog. Poly. Phys. Japan*, **5**:270 (1962).
39. C. Sperati and H. L. Starkweather, *Fortschritte der Hochpolymeren-Forschung*, **2**:465 (1961).
40. P. J. Flory, "Principles of Polymer Chemistry" Cornell University Press, Ithaca, N.Y., 1953, p. 568.
41. F. Kniess and O. Kratky, *Monatshefte*, **90**:506 (1959).
42. N. S. Tikhomirova, Yu. M. Malinskiĭ, and V. L. Karpov, *Vysokomol Soed.*, **2**:1335, 1349 (1960).
43. N. S. Tikhomirova, Yu. M. Malinskiĭ, and V. L. Karpov, *Doklady Akad. Nauk SSSR*, **130**:1081 (1960).
44. N. S. Tikhomirova, Yu. M. Malinskiĭ, and V. L. Karpov, Proc. of Tashkent Conf. on Peaceful Uses of Atomic Energy, 1961, translation AEC-tr-6398, TID-4500, U.S. Atomic Energy Commission, Vol. 1, p. 444.
45. P. E. Thomas, J. F. Lontz, C. A. Sperati, and J. L. McPherson, *J. Soc. Plastics Eng.*, **12**(6):89 (June 1956).
46. H. Kusumoto, *J. Phys. Soc. Japan*, **12**:826 (1957).
47. J. Burget and J. Sacha, *Czechosl. J. Phys.*, **9**:749 (1959).

48. M. Tsuchiya and K. Yamamoto, *Oyo Butsuri*, **34**(6):424 (1965); *Chem. Abstr.*, **64**: 11339c (1966).
49. K. L. Hsu, D. E. Kline, and J. N. Tomlinson, *J. Appl. Polymer Sci.*, **9**:3567 (1965).
50. K. G. McLaren and D. Tabor, *Wear*, **8**:3 (1965).
51. R. A. Weeks and D. Binder, *Power App. and Systems*, **Apr. 1959**:88.
52. J. C. Bresee, J. R. Flanary, J. H. Goode, C. D. Watson, and J. S. Watson, *Nucleonics*, **14**(9):75 (1956).
53. L. A. Wall, *J. Soc. Plastics Engr.*, **12**(3):17 (1956); ONR Symp. Report ACR-2 (Dec 14–15, 1954).
54. J. W. Ryan, *Modern Plastics*, **31**(2):152 (1953).
55. J. W. Ryan, *J. Soc. Plastics Engr.*, **10**(4):11 (1954).
56. R. E. Florin and L. A. Wall, *J. Research NBS*, **65A**:375 (1961).
57. R. E. Florin, M. S. Parker, and L. A. Wall, *J. Research NBS*, **70A**:115 (1966).
58. J. H. Golden, *J. Polymer Sci.*, **45**:534 (1960).
59. T. Watanabe, *J. Phys. Soc. Japan*, **13**:1063 (1958).
60. R. L. Abraham and D. H. Whiffen, *Trans. Faraday Soc.*, **54**:1297 (1958).
61. V. L. Karpov, Conf. of the Acad. of Sciences of the USSR on the Peaceful Uses of Atomic Energy, July 5, 1955, Div. of Chem. Sci., Translation, Consultants Bureau for U.S. Atomic Energy Commission, Washington, D.C. 1956, p. 13.
62. M. I. Bro and C. A. Sperati, *J. Polymer Sci.*, **38**:289 (1959).
63. N. A. Slovokhotova, "Deistvie Ioniziruĭushchikh Izlucheniĭ na Neorganicheskie i Organicheskie Sistemy," Akad. Nauk SSSR, Moscow, 1958, pp. 295–306.
64. D. G. Weiblen, "Fluorine Chemistry," Vol. 2, Ed. J. H. Simons, Academic Press, New York, 1954, p. 449.
65. M. Hauptschein, M. Brand, and A. H. Fainberg, *J. Am. Chem. Soc.*, **80**:851 (158).
66. A. Chapiro, Ref 3 p. 532.
67. B. L. Tsetlin, N. G. Zaitseva, V. M. Korbut, and V. A. Kargin, "Deistvie Ioniziruiush- chikh Izluchenii na Neorganicheskie i Organicheskie Sistemy," Akad. Nauk SSSR, Moscow, 1958, p. 362.
68. J. H. Simons and E. H. Taylor, *J. Phys. Chem.*, **63**:636 (1959).
69. R. F. Heine, *J. Phys. Chem.*, **66**:2116 (1962).
70. R. E. Florin, L. A. Wall, and D. W. Brown, *J. Research NBS*, **64A**:269 (1960).
71. J. R. Dacey and J. W. Hodgkins, *Can. J. Research*, **288**:173 (1950).
72. J. Fajer, D. R. MacKenzie, and F. W. Bloch, *J. Phys. Chem.*, **70**:935 (1966).
73. N. Colebourne and R. Wolfgang, *J. Chem. Phys.*, **38**:2782 (1963). See also discussion under "Free Radicals."
74. T. M. Reed, J. C. Mailen, and J. C. Askew, "Experimental Effects of Pile Radiation on Pure Fluorocarbons," AEC Report TID-22421, Sept. 14, 1965, p. 32.
75. T. M. Reed and J. C. Mailen, "Experimental Effects of Pile Irradiation on Pure Fluorocarbon," AEC Report TID-21576 (June 1, 1963).
76. D. R. Mackenzie, F. W. Bloch, and R. H. Wiswall, Jr., *J. Phys. Chem.*, **69**:2526 (1965).
77. F. W. Bloch, D. R. Mackenzie, and V. H. Wilson, Conference on Radiation, Argonne Natl. Lab. 1968, Adv. Chem. Ser. No. 82, Am. Chem. Soc., Washington, D.C., 1968, p. 546.
78. M. B. Fallgatter and R. J. Hanrahan, *J. Phys. Chem.*, **69**:2059 (1965).
79. A. Sokolowska and L. Kevan, *J. Phys. Chem.*, **71**:2220 (1967).
80. W. C. Askew, "Effects of Radiation on Mixtures Containing Fluorocarbons and the Identification of the Perfluoroheptane Isomers," Thesis, University of Florida 1966; *Diss. Abstr.*, **27**:1459-B.

81. J. C. Mailen, "Gamma and Neutron Irradiation of Pure Fluorocarbons," Thesis, University of Florida, 1964; *Diss. Abstr.*, **25**: 5812 (1965).
82. F. W. Block and D. R. MacKenzie, *J. Phys. Chem.*, **73**: 552 (1969).
83. S. Gordon, A. R. van Dyken, and T. F. Doumani, *J. Phys. Chem.*, **62**: 20 (1958).
84. R. O. Bolt and J. G. Carroll, Ed., "Radiation Effects on Organic Materials," Academic Press, New York, 1963, p. 85.
85. J. Fajer and D. R. MacKenzie, *J. Phys. Chem.*, **71**: 784 (1967).
86. J. Fajer and D. R. MacKenzie, Int. Conf. Radiation Chem., Argonne National Lab. 1968, Advances in Chemistry Series 82, Am. Chem. Soc. Washington, D.C., 1968, p. 469.
87. I. Haller, *J. Am. Chem. Soc.*, **88**: 2070 (1966).
88. G. Camaggi, F. Gozzo, and G. Cevidalli, *Chem. Commun.*, **1966**: 313.
89. D. Bryce-Smith, A. Gilbert, and B. H. Orger, *Chem. Commun.* **1966**: 512.
90. V. A. Khramchenkov, 2nd Tihany Symp. on Radiation Chem., Akademiai Kiado, Budapest, 1967, p. 443.
91. J. R. Majer, D. Phillips and J. C. Robb, *Trans. Faraday Soc.*, **61**: 110 (1965).
92. A. K. Basak, G. P. Semeluk, and I. Unger, *J. Phys. Chem.*, **70**: 1337 (1966).
93. J. R. Majer, *Trans. Faraday Soc.*, **61**: 118 (1965).
94. D. Phillips, *J. Chem. Phys.*, **46**: 4679 (1967).
95. D. L. Williams, USAEC unspecified T.I.D. Report, 1960, quoted in A. K. Barbour and P. Thomas, *Ind. Eng. Chem.*, **58**(1): 48 (Jan. 1966).
96. R. J. Woods and J. W. T. Spinks, *Can. J. Chem.*, **38**: 77 (1960).
97. J. C. Russell and K. J. McCallum, *Can. J. Chem.*, **44**: 243 (1966).
98. P. Y. Feng and L. Mamula, *J. Chem. Phys.*, **28**: 507 (1958).
99. P. Y. Feng, Proc. of 2nd United Nations Conf. on Peaceful Uses of Atomic Energy, Geneva, 1958, Vol. 29, p. 166.
100. A. V. Zimin, A. D. Verina, V. A. Khramchenkov, and S. V. Churmanteev, Proc. of 2nd All Union Conf. on Radiation Chemistry, Moscow, 1962, Trans AEC-tr 6228, 1964, p. 447.
101. L. A. Rajbenbach and U. Kaldor, *J. Chem. Phys.*, **47**: 242 (1967).
102. L. A. Rajbenbach, Int'l. Conf. on Radiation Chemistry Argonne National Lab., 1968, Advances in Chemistry Series No. 82, Am. Chem. Soc., Washington, D.C., 1968, p. 542.
103. E. Heckel and R. J. Hanrahan, Int'l. Conf. on Radiation Chemistry, Argonne National Lab., 1968, Advances in Chemistry Series No. 82, Am. Chem. Soc., Washington, D.C., 1968, p. 120.
104. M. B. Fallgatter and R. J. Hanrahan, quoted in Heckel, Ref. 99.
105. N. H. Sagert, *Can. J. Chem.*, **46**: 95 (1968).
106. N. H. Sagert, J. A. Reid and R. W. Robinson, *Can. J. Chem.*, **47**: 2655 (1969).
107. V. Caglioti, M. Lenzi, and A. Mele, *Nature*, **201**: 610 (1964).
108. V. Caglioti, A. Delle Site, M. Lenzi, and A. Mele, *J. Chem. Soc.*, **1964**: 5430.
109. D. Cordischi, M. Lenzi, and A. Mele, *Trans. Faraday Soc.*, **60**: 2047 (1964).
110. A. J. Pajaczkowski and J. W. Spoors, *Chem. and Ind.*, **Apr. 18, 1964**: 659.
111. J. L. Warnell (to E.I. duPont de Nemours and Co.), U.S. Patent 3,125,599 (March 17, 1964).
112. F. Gozzo and G. Camaggi, *Tetrahedron*, **22**: 2181 (1965).
113. P. Barnaba, D. Cordischi, M. Lenzi, and A. Mele, *Chim. Ind. (Milan)*, **47**: 1060 (1965).
114. A. Donato, M. Lenzi, and A. Mele, *J. Macromol Sci. (Chem.)*, **A1**: 429 (1967).
115. D. Sianesi, A. Pasetti, and C. Corti, *Makromol. Chem.*, **86**: 308 (1965).
116. D. Sianesi and R. Fontanelli, *Makromol. Chem.*, **102**: 115 (1967).

117. N. Cohen and J. Heicklen, *J. Phys. Chem.*, **70**:3082 (1966).
118. W. Stuckey and J. Heicklen, *Can. J. Chem.*, **46**:1361 (1968).
119. A. P. Modica and J. E. LaGraff, *J. Chem. Phys.*, **43**:3383 (1965).
120. T. Johnson and J. Heicklen, *J. Chem. Phys.*, **47**:475 (1967).
121. D. Saunders and J. Heicklen, *J. Am. Chem. Soc.*, **87**:4062 (1965).
122. J. Heicklen, V. Knight, and S. A. Greene, *J. Chem. Phys.*, **42**:221 (1965).
123. J. Heicklen and V. Knight, *J. Phys. Chem.*, **70**:3901 (1966); **70**:3893 (1966).
124. J. Heicklen and V. Knight, *J. Chem. Phys.*, **47**:4203 (1967).
125. W. C. Francis and R. N. Haszeldine, *J. Chem. Soc.*, **1955**:2151.
126. D. Marsh and J. Heicklen, *J. Phys. Chem.*, **70**:3008 (1966).
127. J. Heicklen, *J. Phys. Chem.*, **70**:112 (1966).
128. D. Marsh and J. Heicklen, *J. Phys. Chem.*, **69**:4410 (1965).
129. C. F. Bersch, R. R. Stromberg, and B. G. Achhammer, *Modern Packaging*, **32**:117 (Aug. 1959).
130. J. Byrne, T. W. Costikyan, C. Hanford, D. L. Johnson, and W. L. Mann, *Ind. Eng. Chem.*, **45**:2549 (1953).
131. C. D. Bopp, W. W. Parkinson, and O. Sisman, "Radiation Effects on Organic Materials," R. O. Bolt and J. G. Carroll, ed., Academic Press, New York, 1963, p. 722.
132. J. Goodman and J. H. Coleman, *J. Polymer Sci.*, **25**:253, 502 (1957).
133. L. A. Wall and S. Straus, *J. Research NBS*, **65A**:227 (1961).
134. G. H. Bowers and E. R. Lovejoy, *Ind. Eng. Chem. Product Res. and Dev.*, **1**(2):89 (June 1962).
135. S. Straus and L. A. Wall, *Soc. Plastics Engrs. Trans.*, **4**:61 (1964).
136. L. A. Wall, S. Straus, and R. E. Florin, *J. Polymer Sci. A1*, **4**:349 (1966).
137. T. Yoshida, R. E. Florin, and L. A. Wall, *J. Polymer Sci. A*, **3**:1685 (1965).
138. G. D. Sands and G. F. Pezdirtz, *Am. Chem. Soc. Polymer Preprints*, **6**(2):987 (Sept. 1965).
139. L. Mullins and A. G. Thomas, *J. Polymer Sci.*, **43**:13 (1960).
140. L. C. Case, *J. Polymer Sci.*, **45**:397 (1960).
141. A. R. Shultz, *J. Am. Chem. Soc.*, **80**:1854 (1958).
142. A. Charlesby, Ref. 2, pp. 167 and 146.
143. O. Saito, H. Y. Kang, and M. Dole, *J. Chem. Phys.*, **46**:3607 (1967).
144. L. Epstein, 1960 Ann. Rept. of Conf. on Electrical Insulation, National Academy of Science—National Research Council Publ. No. 842 (1960), p. 141.
145. R. Harrington, *Rubber Age*, **85**:963 (1959).
146. V. L. Lanza and E. C. Stivers (to Raychem Corp), U.S. Patent 3,269,862 (Aug. 30, 1966).
147. F. E. LaFetra, *Wire and Wire Products*, **41**:898 (1966).
148. J. S. Rugg and A. C. Stevenson, *Rubber Age*, **82**:102 (Oct. 1957).
149. J. S. Rugg, "Viton A, A Synthetic Rubber," DuPont Elastomer Chemicals Dept. Development Products Report No. 3, April 15, 1957.
150. S. Dixon, D. R. Rexford, and J. S. Rugg, *Ind. Eng. Chem.*, **49**:1687 (1957).
151. W. W. Jackson and D. Hale, *Rubber Age*, **77**:865 (1955).
152. F. A. Galil-Ogly, T. S. Nikitina, T. N. Dyumaeva, A. S. Novikov, and A. S. Kuz' minskii, *Atomnaia Energiia*, **6**:540 (1959).
153. J. F. Smith, Proc. Int'l. Rubber Conf., Washington, D.C., November 1959, pp. 575–581.
154. J. F. Smith and G. T. Perkins, *Rubber and Plastics Age*, **Jan. 1961**:59; *J. Appl. Polymer Sci.*, **5**:460 (1961).

155. K. L. Paciorek, L. C. Mitchell, and C. T. Lenk, *J. Polymer Sci.*, **45**:405 (1960).

156. D. K. Thomas, *J. Appl. Polymer Sci.*, **8**:1415 (1964).

157. F. S. Tolstukhina, B. Z. Berman, and A. S. Novikov, *Kauchuk i Rezina*, **25**:13 (1966); tr. *Soviet Rubber Tech.*, **25**:14 (1966).

158. A. S. Novikov, V. L. Karpov, F. A. Galil-Ogly, N. A. Slovokhotova, and T. N. Dyumaeva, *Vysokomol. Soedin.*, **2**:1761 (1960).

159. A. S. Novikov, V. L. Karpov, F. A. Galil-Ogly, N. A. Slovokhotova, and T. N. Dyumaeva, *Vysokomol. Soedin*, **2**:485 (1960); tr. *Polymer Sci. USSR*. **2**:329 (1961).

160. R. Harrington, *Rubber Age*, **83**:492 (1958).

161. Minnesota Mining and Mfg. Co. brochure, "Technical Data, Fluorel High Temperature Elastomers," about 1965.

162. A. R. Shultz, N. Knoll, and G. A. Morneau, *J. Polymer Sci.*, **62**:211 (1962).

163. C. V. Stephenson, B. C. Moses, and W. S. Wilcox, *J. Polymer Sci.*, **55**:451 (1961).

164. C. V. Stephenson and W. S. Wilcox, *J. Polymer Sci. A*, **1**:2741 (1963).

165. L. Ehrenberg and K. G. Zimmer, *Acta Chem. Scand.*, **10**:874 (1956).

166. P. Barnaba, D. Cordischi, A. Delle Site, and A. Mele, *J. Chem. Phys.*, **44**:3672 (1966).

167. A. Charlesby and R. H. Partridge, *Proc. Roy. Soc.*, **A283**:312 (1965).

168. V. G. Nikol'skii and N. Ya. Buben, *Doklady Akad. Nauk SSSR*, **134**:134 (1960).

169. J. F. Fowler, *Proc. Roy. Soc.*, **A236**:464 (1956).

170. S. Mayburg and W. L. Lawrence, *J. Applied Phys.*, **23**:1006 (1952).

171. R. A. Meyer, F. L. Bouquet, and R. S. Alger, *J. Applied Phys.*, **27**:1012 (1956).

172. S. E. Harrison, F. N. Coppage, and A. W. Snyder, *Inst. of Electrical and Electronics Engineers Transactions on Nuclear Science*, **10**:118 (1963).

173. S. E. Harrison, "A Study of Gamma Ray Photoconductivity in Organic Dielectric Materials," Thesis, University of New Mexico, 1962; AEC TID-15482 (July 24, 1962).

174. W. E. Loy, Jr., "Effects of Gamma Radiation on Some Electrical Properties of TFE-Fluorocarbon Plastics," Am. Soc. Testing Matls. Spec. Tech. Pub. No. 276, "Materials in Nuclear Applications," 1959, p. 68.

175. K. Yahagi, K. Shinohara, and K. Mori, *J. Inst. El. Engr. Japan* (*Denki Gakkai Zasshi*), **83**(3):242 (1963); tr. Inst. El. and Electronic Engrs. Inc., New York, p. 52.

176. K. Yahagi and A. Danno, *J. Appl. Phys.*, **34**:804 (1963).

177. V. M. Nesterov, E. S. Nesmelova, N. I. Ol'shanskaya, T. G. Mikhailova, and G. I. Potakhova, *Fiz. Tverdogo Tela*, **4**:3010 (1962); tr. *Soviet Physics-Solid State*, **4**(11):2206 (May 1963).

178. L. Heyne and O. Hauser, *Kolloid Z.*, **205**:39 (1965).

179. L. Heyne and O. Hauser, *Kolloid Z.*, **206**:20 (1965).

180. S. Ide, T. Urai, and H. Saito, *Chem. of High Polymers* (*Kobunshi Kagaku*), **20**:583 (1963).

181. V. Adamec, *Nature*, **200**:1196 (1963).

182. A. J. Warner, F. A. Muller, and H. G. Nordlin, *J. Applied Phys.*, **25**:131 (1954).

183. M. Kryszewski, A. Szymanski, and A. Włochowicz, *J. Polymer Sci. C*, **16**:3921 (1968).

184. E. Mathias and Glenn H. Miller, *J. Phys. Chem.*, **71**:2671 (1967).

185. P. B. Weisz, *J. Phys. Chem.*, **59**:464 (1955).

186. N. V. Thornton, A. B. Burg, and H. I. Schlesinger, *J. Am. Chem. Soc.*, **55**:317 (1933).

187. A. I. Akishin, L. N. Isaev, and Yu. I. Tyutrin, *Zh. Fiz. Khim.*, **39**:3067 (1967 5).

188. S. J. Tao and J. H. Green, *Proc. Phys. Soc.* (*London*), **85**(545):463 (1965).

189. F. A. Brisbout, C. F. Gauld, C. B. A. McCusker, *Nuovo Cimento*, **18**:400 (1960).
190. W. Friedlander, K. A. Neelakantan, S. Tokunaga, G. R. Stevenson, and C. J. Waddington, *Phil. Mag.*, **8**:1691 (1963).
191. I. M. Barkalov, V. I. Gol'danskii, B. G. Dzantiev, and E. V. Egorov, Vsesoiuznoe soveshchanie po radiatsionnoi khimii, 2nd, Moscow, 1962, tr. Israel Program Sci. Trans., Jerusalem, 1964, AEC tr-6228, p. 658.
192. W. J. Burlant and A. S. Hoffman, "Block and Graft Copolymers," Reinhold, New York, 1960.
193. A. Chapiro, *J. Polymer Sci.*, **34**:481 (1939).
194. D. Ballantine, P. Colombo, A. Glines, B. Manowitz, and D. Metz, Brookhaven National Lab. Report BNL-414, T-81 (1956).
195. J. Harwood, H. Hausner, J. Morse, and W. Rauch, "Effects of Radiation on Materials," Reinhold, New York, 1958, p. 287.
196. A. Restaino and W. J. Reed, *J. Polymer Sci.*, **36**:499 (1959).
197. N. Gaylord, U.S. Patent 2,907,675 (1959).
198. C. Rossi, S. Munari, and G. F. Tealdo, *Chim. Ind. (Milan)*, **45**:1494 (1963).
199. J. Dobo and P. Hedvig, *Makromol. Chem.*, **71**:289 (1964).
200. V. V. Korshak, K. K. Mozgova, and T. M. Babchinitser, *Doklady Akad Nauk SSSR*, **151**:1332 (1963).
201. Anon., *Chem. and Eng. News*, **37**(5):44 (Feb. 2, 1959).

12. THERMAL DECOMPOSITION OF FLUOROPOLYMERS

LEO A. WALL, *Polymer Chemistry Section, National Bureau of Standards, Washington, D.C.*

Contents

I. INTRODUCTION

The most complete previous review of investigations into the thermal decomposition of fluoropolymers is contained in a recent book on the thermal degradation of organic polymers by Madorsky (1). Important aspects of the field are briefly mentioned in various other reviews (2,3) having a broader scope. Since the previous review (1) a considerable number of new studies have been made on old and new fluoropolymers. In this chapter we discuss in detail studies on the decomposition of polytetrafluoroethylene, which is perhaps the most interesting of the fluoropolymers because its mechanism of decomposition is now well delineated and because of its outstanding thermal stability. It is one of the most stable of all linear polymeric structures, organic or inorganic.

II. POLYTETRAFLUOROETHYLENE

Polytetrafluoroethylene decomposes under vacuum above 500°C to give nearly pure monomer (4). Above 100 mm pressure and 500–700°C, thermal dimerization of the monomer occurs. Depending on residence time of the monomer vapor in the pyrolysis furnace and the temperature, the monomer can be decomposed to give high yields of perfluoropropylene and perfluoro-isobutylene. Thus the decomposition of large amounts of polytetrafluoro-ethylene can be hazardous. The fluorinated olefinic by-products, particularly the perfluoroisobutylene, are toxic when inhaled. In addition, the monomer at low temperatures can form with oxygen a polyperoxide, $+CF_2CF_2-OO+_x$, which is very unstable and can detonate. At ambient temperatures mixtures of oxygen and monomer can also explode. On several occasions when pyrolyzing large quantities of polymer in order to produce monomer for synthetic purposes, start up of a previously incompleted pyrolysis has produced minor explosions.

Another phenomenon usually observed when large quantities of polymer are pyrolyzed is the deposition of solid on cool portions of the vacuum train. Since the deposit will occur downstream of traps and filters, it is very likely the result of repolymerization of monomer. Thermal decomposition of the polymer in air produces mainly carbonyl fluoride, which is also highly toxic when inhaled. However, the amount of polytetrafluoroethylene coating on a cooking utensil is very small and comprises relatively little danger to anyone if accidentally decomposed.

Decomposition of polytetrafluoroethylene and other fluoropolymers in glass or quartz equipment can yield products such as silicon tetrafluoride, carbon monoxide, carbon dioxide, and water by reaction of the polymer or decomposition products with the walls of the vessels. Products other than monomer are very likely formed from the monomer and are primarily a result of furnace geometry and secondarily of reaction conditions. A conceivable primary product is the difluoromethylene radical, but the high yield of monomer rules this out as an important possibility.

A. Rate of Volatilization

The rate of thermal depolymerization of polytetrafluoroethylene has been studied by several investigators (5–9).

With finely divided powder or thin-film samples the rate of decomposition

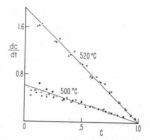

Fig. 1. Rate of thermal decomposition of polytetrafluoroethylene as a function of conversion.

is very closely proportional to the weight of the polymer. This is shown in Fig. 1. In this chapter we are primarily concerned with thermal degradation studies in which the rate of volatilization is the measure of polymer decomposition. The reaction conversion C is then defined by

$$C = 1 - \frac{W(t)}{W(0)} \tag{1}$$

where $W(0)$ is the initial weight of sample and $W(t)$ the weight at time t. The rate may be expressed as the fraction or the percentage volatilized per second or minute. In Fig. 1 then the rate has the behavior

$$\frac{dC}{dt} = K(1 - C) \tag{2}$$

For convenience we shall define for polytetrafluoroethylene the thermal rate r_0 as

$$r_0 = \frac{1}{(1 - C)} \frac{dC}{dt} = K = 4.7 \times 10^{18} e^{-80,500/RT} \mathrm{sec}^{-1} \tag{3}$$

The value for the fractional rate is based on work (5) covering a temperature range of from 400–500°C. Other work showed that at lower temperatures (7) and for thicker samples (8) slow diffusion of monomer out of the sample may lead to some differences in the character of the rate behavior with conversion. At low temperature the curves can at low conversion have a zero-order dependence on conversion and with thick samples a pseudo retarded character is seen in the rate or conversion curves.

An important criterion (2) for the mechanism of polymer decomposition is the dependence of the initial rate on the initial molecular weight of the polymer. If the polymer decomposes by the thermolysis of chain bonds only at the ends of the polymer molecules to give radicals that decompose to monomer with a certain probability, then the rate will be constant at low molecular weights when the large radicals completely decompose to monomer. At very high molecular weights the radical would not completely decompose before terminating and the rate would decrease with molecular weight. With random initiation the rate would be proportional to the molecular weight as long as the length of the polymer molecules were short compared to the zip length. If the polymer molecules were very long and the radicals only depropagated down a fraction of the chain, then the rate would become independent of molecular weight. Difficulties in determining molecular weights of polytetrafluoroethylene have limited definitive experiments measuring the rate as a function of molecular weight to only one (8) (Table 1). The rate was found to be independent of molecular weight.

TABLE 1

Rate of Depolymerization at 480°C of Polytetrafluoro-
ethylenes of Different Molecular Weights (8)

M.W.[a] $\times 10^{-6}$	$r_0 \times 10^5$ (sec^{-1})
10 to 40	2.48
1 to 10	2.49
0.3 to 1	2.48
1 to 10	2.52
8[b]	2.48
9	2.61

[a] By standard specific gravity.
[b] By counting radioactive sulfonic end groups.

Previous studies of the decomposition of numerous polytetrafluoroethylene samples prepared by various methods (6) and gamma irradiated in varying degrees showed no variation in rate of decomposition. On the basis of these simple alternative mechanisms, neither of which seems applicable to the polymer of interest, there seemed to be a dilemma (10,11). and also several inconsistencies between various articles in the literature. From observed decreases in molecular weight at 480°C and assuming termination by disproportionation, one can calculate a kinetic chain or zip length (2) of 720. With this relatively large zip length the rate should be at least detectably influenced by the variations in molecular-weight studies, provided the mechanism was one of the types discussed. Other suggestions advanced regarding various facets of the mechanism of depolymerization for polytetrafluoroethylene overlooked certain known characteristics of the process, such as the independence of the rate on molecular weight (12), or utilized assumptions valid only for gas-phase kinetic studies (13).

Resolution of these apparent inconsistencies required a new experimental measurement. Photochemical (14) or gamma-ray-initiated depolymerization could in theory supply an independent measurement of the kinetic chain length provided the rate of production of radicals could be determined. In the case of gamma-ray initiation this requires a determination of the yield of primary radicals, which are usually listed as $G(R)$ values, the number of radical produced per 100 eV of energy deposited in the material by the radiation. Ultraviolet light produces at elevated temperatures only slightly preceptible accelerations in the rate of volatilization of polytetrafluoroethylene, whereas gamma rays were found to produce measurable effects (9). With gamma rays the $G(R)$ for primary radical formation was determined by the amount of bound chlorine that was found in polymer irradiated in the presence of chlorine (15). It was assumed that $G(R)$ equalled the G-value for chlorine atoms fixed to the polymer.

Measurements at 327, 340, and 380°C gave 3.0 ± 0.1 atoms of chlorine per 100 eV absorbed (9). The relative independence of temperature observed suggests this value is applicable at considerably higher temperatures.

B. Depolymerization Mechanism

The following chain reaction mechanism can be applied to all polymers that thermally decompose to monomer, dimer, trimer, and other species requiring rupturing of the main chain (2):

Initiation
$$Q_n \rightarrow R_i + R_{n-i} \qquad\qquad 2k_1 n Q_n$$

Propagation
$$R_n \rightarrow R_{n-1} + M \qquad\qquad k_2 R_n$$

Transfer (intermolecular)
$$R_n + Q_m \rightarrow Q_n + Q_{m-e} + R_e \qquad k_3 R_n m Q_m$$

Termination

Disproportionation $\qquad R_i + R_j \rightarrow Q_i + Q_j \qquad\qquad 2k_{4_a} R^2$

Combination $\qquad\qquad R_i + R_j \rightarrow Q_{i+j} \qquad\qquad 2k_{4_c} R^2$

The symbols are defined as: Q_n moles of polymer molecules with n units in the chain, R_i moles of radicals in sample having i units, M moles of monomer. Other modes of initiation and termination are possible and may be utilized for other theoretical developments. We have limited the possibilities to those that we believe are involved in the depolymerization of polytetrafluoroethylene. It is anticipated that the transfer process is not operative, it is included because of the sensitivity of the residual molecular weight to this process. Thus end initiation, monomolecular and cage terminations are not depicted in the preceding mechanism.

C. Kinetics

By assuming random initiation and short or moderate zip length (2) and permitting only monomer to evaporate, on the basis of the preceding reaction scheme, the rate relation for the process of thermal decomposition is

$$\frac{dM}{dt} = k_2 R \tag{4}$$

where dM/dt is the moles of monomer evaporated per second from a sample containing R moles of polymer radicals each capable of releasing monomer. Since this is an open system, the sample volume is proportional to the weight $W(t)$ and decreasing. The steady-state condition is

$$2k_1 \sum nQ_n(t) = \frac{2k_4 R^2(t)}{V(t)} \tag{5}$$

where Q_n, R, and the volume V are all time dependent. The units of k_4 will follow from the choice of units used for the volume. Now to a very good approximation

$$k_1 \sum nQ_n(t) = k_1 \frac{W(t)}{m} = k_1 \frac{\rho V(t)}{m} \tag{6}$$

where m is the molecular weight of the monomer and ρ is the density of the polymer. The moles of radicals in the sample is then

$$R(t) = \left(\frac{k_1}{k_4 \rho m}\right)^{1/2} W(t) \tag{7}$$

and is proportional to the weight or volume of the sample. The rate of conversion dC/dt is now found to be given by

$$\frac{dC}{dt} = \frac{m \, dM}{W(0) \, dt} = k_2 \left(\frac{k_1 m}{k_4 \rho}\right)^{1/2} \frac{W(t)}{W(0)} \tag{8}$$

or

$$r_0 = \frac{1}{1 - C} \frac{dC}{dt} = k_2 \left(\frac{k_1 m}{k_4 \rho}\right)^{1/2} \text{sec}^{-1} \tag{9}$$

If conditions are such that initiation does not occur thermally but is induced by gamma rays, then the initiation reaction would be symbolically depicted as

$$Q_n \overset{\gamma}{\rightarrow} R_i + R_{n-i} \quad 2\phi In Q_n$$

If the character of the intermediate species is unchanged, then the steady-state relation becomes

$$\phi I \sum n Q_n(t) = \phi I \frac{\rho V(t)}{m} = \frac{k_4 R(t)^2}{V(t)} \tag{10}$$

This relation assumes that the radiation produces radicals also by random rupture of polymer chains.

The intensity of the radiation is denoted by I and ϕ is an efficiency factor for radical production related to the $G(R)$ value for radicals. Both these quantities are, or are taken to be, independent of temperature. The rate for the pure gamma-ray initiated process is then

$$r_\gamma = k_2 \left(\frac{\phi m}{k_4 \rho} \right)^{1/2} I^{1/2} \; \text{sec}^{-1} \tag{11}$$

Comparison of r_0 with r_γ by division of Eq. 9 by Eq. 11 gives us the relation

$$\frac{r_0}{r_\gamma} = \left(\frac{k_1}{\phi I} \right)^{1/2} \tag{12}$$

and

$$k_1 = \phi I \left(\frac{r_0}{r} \right)^2 \tag{13}$$

The numerical value for the thermal rate of initiation can then be easily calculated from known quantities provided ϕ is available. The factor ϕ is composed of a conversion factor for units and $G(R)$ the radiation yield of radicals for which values from 0.2 to 10 have been estimated (15–17).

D. Radiolytic Initiation of Depolymerization

Gamma rays or other forms of radiation can in theory greatly accelerate chain reactions. The situations that may operate when polymers are irradiated at different temperatures are illustrated in Fig. 2. Here we show the three different effects that exposure to radiation can have on the rate of depolymerization. At some low temperature T_1 the depolymerization would not be measurably accelerated because the temperature would be too low for even propagation to occur. At some very high temperature T_3 at which the thermal rate r_0 is high, radiation would again not measurably accelerate the process since the rate of initiation due to radiation would not be appreciable compared to the rate of thermal initiation. However, for every polymer there should exist an intermediate temperature where radiation would produce a maximum effect, $(r - r_0) \to$ maximum. If under certain experimental conditions the thermal rate r_0 is less than $0.01 r_\gamma$, then Eqs. 12 and 13 may be

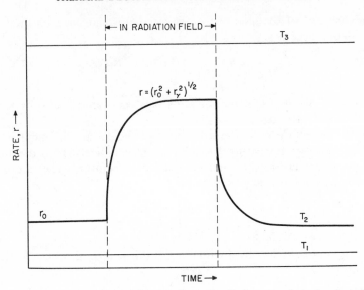

Fig. 2. Schematic plot demonstrating the effects of radiation on a chain depolymerization process at various temperatures, $T_3 > T_2 > T_1$.

utilized. However, if, as with polytetrafluoroethylene (9), r_0 cannot be neglected when compared to r_y, then we must derive and utilize the relations for combined thermal and radiolytic initiation.

The quantity most easily measured and plotted (9) is the increase $(r - r_0)$ in the rate when radiation was applied to the polymer, where r is the total rate under radiation and r_0 the thermal rate given by Eq. 9. The composite initiation rate is represented by

$$k_1 \sum nQ_n + \phi I \sum nQ_n = (k_1 + \phi I) \frac{W(t)}{m} \tag{14}$$

The total decomposition rate is

$$r = \frac{dC}{(1 - C) \, dt} = k_2 \left[\frac{(k_1 + \phi I) m}{k_4 \rho} \right]^{1/2} \tag{15}$$

and

$$r - r_0 = r_0 \left(1 + \frac{\phi}{k_1} I \right)^{1/2} - r_0 \tag{16}$$

This relation between the photoincrement $r - r_0$ and $I^{1/2}$ is a hyperbola which has an asymptote

$$r - r_0 = r_0 \left(\frac{\phi}{k_1} \right)^{1/2} I^{1/2} - r_0 \tag{17}$$

The asymptote has the slope $r_0(\phi/k_1)^{1/2}$ and intercepts at $I^{1/2} = (k_1/\phi)^{1/2}$ and $r - r_0 = -r_0$. Thus both the thermal rate and the pure radiation parameter, $(r_\gamma/I^{1/2}) = r_0(\phi/k_1)^{1/2}$, can be isolated (see Eq. 11).

Plots (9) of $r - r_0$ as a function of I or $I^{1/2}$ did not produce straight lines, suggesting that with the polytetrafluoroethylene depolymerization the kinetic chain length was relatively short and that the rates studied had appreciable contributions from both radiolytic and thermal initiation and some different method of analyzing the data is required. Since the rates are proportional to the square roots of the total rate of initiation, it can be shown that

$$r^2 = r_\gamma{}^2 + r_0{}^2 \tag{18}$$

or

$$r^2 - r_0{}^2 = r_\gamma{}^2 \tag{19}$$

Factoring the left side of the preceding relation gives

$$r + r_0 = \frac{r_\gamma{}^2}{r - r_0} \tag{20}$$

Subtracting $2r_0$ from both sides of Eq. 20 produces

$$r - r_0 = -2r_0 + \frac{r_\gamma{}^2}{r - r_0} = -2r_0 + \frac{r_\gamma{}^2}{I} \frac{I}{r - r_0} \tag{21}$$

or

$$r - r_0 = -2r_0 + \frac{k_2{}^2 m}{k_4 \rho} \frac{\phi I}{r - r_0} \tag{22}$$

According to Eq. 21 a plot of $r - r_0$ as a function of $I/(r - r_0)$ should be a straight line with slope r^2/I and intercept $-2r_0$. Thus the slope gives us the value of r_γ at any given intensity I of radiation.

Since it appears that $r_\gamma = sI^{1/2}$ where s, the slope (from Fig. 3) is a constant, it follows that termination is bimolecular, i.e., either combination, disproportionation, or some mixture of the two, but no evidence has yet been presented which distinguishes between the two modes of initiation mentioned. From slopes of plots such as that shown in Fig. 3 we may obtain (9) the reduced radiation rate as a function of temperature:

$$r_\gamma I^{-1/2} = 6.24 \, e^{-21,200/RT} \qquad \sec^{-1} Mr^{-1/2} hr^{1/2} \tag{23}$$

E. Thermal Initiation Rate of Polytetrafluoroethylene Depolymerization

Values of $r_\gamma I^{-1/2}$ can be used with Eq. 13 to deduce the thermal rate of initiation. Also needed are values for the factor ϕ, which is estimated from ρ

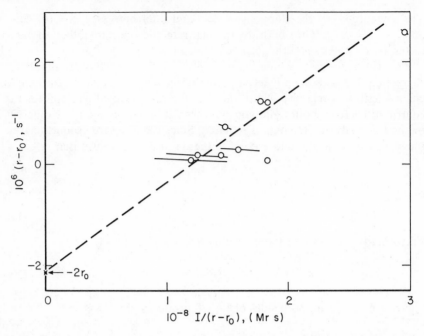

Fig. 3.   Variation in the photoincrement at 429.5°C with the intensity function for the gamma-ray induced depolymerization of polytetrafluoroethylene (9).

and m and the $G(R)$. The value of ρ/m as a function of temperature has been reported (8) as

$$\frac{\rho}{m} = 5.25 \times 10^{-3} e^{1393/RT} \text{ mol/cc} \qquad (650 < T°K < 750) \qquad (24)$$

at 400°C.

$$\frac{\rho}{m} = 14.38 \times 10^{-3} \text{ mol/cc} \qquad (25)$$

Using the experimental value $G(Cl) = 3$ for the value of $G(R)$ and taking the energy absorbed in the polymer per 1 Mr as 0.526×10^{20}eV/g (18), a value for ϕ is obtained:

$$\phi = 3.64 \times 10^{-8} \text{sec}^{-1}\text{Mr}^{-1}\text{hr} \qquad (26)$$

The rate constant for thermal initiation is then found to be given by

$$k_1 = 2.1 \times 10^{28} e^{-118,600/RT} \text{ sec}^{-1} \qquad (27)$$

F. Activation Energy for Propagation

The activation energy for propagation can be obtained from the relation

$$E_0 = E_2 + \tfrac{1}{2}(E_1 - E_4) \tag{28}$$

or, as has been pointed out (10, 11) from

$$E_2 = E_0 - \tfrac{1}{2}(E_1 - E_4) = E_0 - \tfrac{1}{2} D_{C-C} \tag{29}$$

The C—C bond dissociation energy has been estimated for fluorocarbons as 81.5 kcal/mol (19), and we find $E_2 = 39.75$ kcal/mol. The estimate of E_2 can also be made from the relation

$$E_\gamma = E_2 - \frac{E_4}{2} \tag{30}$$

since

$$E_2 = E_\gamma + \frac{E_4}{2} = E_\gamma - \frac{D_{C-C}}{2} + \frac{E_1}{2} = 39.75 \tag{31}$$

G. Termination

A value for E_4 of 37.10 is obtained from Eqs. 30 and 31; this is in agreement with the view that termination is diffusion controlled and limited by a very high melt viscosity. Melt-viscosity studies have reported activation energies for viscous flow of from 25 to 52 kcal/mol (20, 21), and it is believed that this variation is partly a consequence of degradation in the experiments near the region of 370°C (22,23).

The disappearance of radicals has been studied by electron spin resonance spectroscopy (24) at temperatures up to 220°C, and the results appear to follow second-order rate behavior with high activation energies ranging from 30 kcal/mol for the amorphous phase to 65 kcal for the crystalline phase. For our high-temperature depolymerization it seems reasonable to take the rate constant for radical disappearance (24) in the amorphous phase as the termination rate for depolymerization, thus

$$k_4 = 3 \times 10^{13} e^{-30,000/RT} \text{ l/mol sec} \tag{32}$$

In Eq. 32 $E_4 = 30$ kcal rather than the 37.1 kcal estimated in previously discussed work (9). Equation 11 permits (9) an estimate of the propagation constant to be obtained:

$$k_2 = 7 \times 10^{11} e^{-36,200/RT} \text{ sec}^{-1} \tag{33}$$

It is seen that k_2 and k_4 have relatively normal pre-exponential factors, whereas k_1 has a very abnormal one. Moreover, the E_1 and E_4 are very high. This behavior, which has been discussed in the distant past, is more readily explained (10,11) on the basis of the cage effect; i.e., both k_1 and k_4 are diffusion-controlled processes. The large pre-exponential factor for k_1 indicates a highly disordered transition state, which is very easily visualized as a cage containing more than one pair of radicals formed perhaps from different rather than the same polymer chains. The activation energy of thermal initiation is compatible with the concept that the radical escapes the cage by depropagating. For this process E_1 would be $\sim D_{CC} + E_2$ or ~ 120 kcal, which is close to the measured E_1. Both modes visualized for thermal initiation could of course be expected to compete depending on their relative energetics.

H. Zip Length

The kinetic chain length l for the thermal- and gamma-ray-initiated depolymerization is given by

$$l = \frac{r}{2(k_1 + \phi I)} \tag{34}$$

Numerical equations have been given for the kinetic chain length as a function of temperature and dose rate:

$$\log l_0 = -9.958 + \frac{8320}{T} \tag{35}$$

$$\log l_\gamma = 7.932 - \frac{4640}{T} - \frac{1}{2} \log I \tag{36}$$

The kinetic chain length or zip length for thermal depolymerization is from these relations much shorter than anticipated. At 480°C a value of 13 is produced, although on the basis of a different model the value of 720 at 480°C has been estimated (8). At higher temperatures the zip length gets smaller.

Although the hypothetical zip length for the radiolytic depolymerization increases as the temperature is lower, long zip lengths only occur at low radiation intensities where no measurable rate of depolymerization occurs. The short zip length coupled with the independence of rate on molecular weight is very strong evidence that thermal initiation is random.

I. Effect of Thickness

The thermal rate of depolymerization is always proportional to the weight of polymer. However, with radiolytic initiation thin specimens produced faster rates. Figure 4 shows the effect of thickness and that of radiation

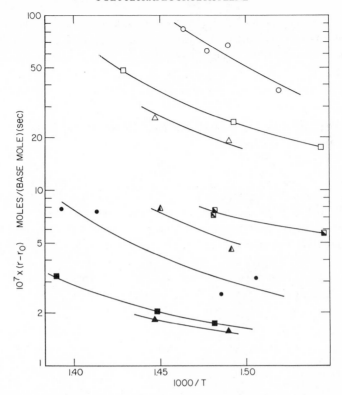

Fig. 4.   The increment $(r\text{-}r_0)$ in the rate of thermal decomposition of polytetrafluoroethylene produced by gamma rays as a function of film thickness ($\triangle$ 4 mil, $\square$ 1 mil, $\bigcirc$ $\frac{1}{4}$ mil),
intensity ($\square$ 7.66 Mr/hr, $\blacksquare$ 1.39 Mr/hr, $\blacksquare$ 0.22 Mr/hr), and temperature (9).

intensity. The lines show a suggestion of curvature. The overall activation
energy is quite small. The idea that this effect was the result of slower migration of monomer out of the sample, thus permitting a reversal of propagation,
was tested even though it is not apparent in thermal studies. The reverse
propagation reaction is written as

$$R_n + M \to R_{n+1} \qquad k_{-2} R_n M$$

Although the problem of diffusion-limited monomer evolution has been discussed before (14). the situation where reverse depolymerization occurs to
an appreciable extent as a result of a buildup of the monomer concentration
in the bulk polymer has been treated only recently (9). The differential equation
governing the monomer concentration in a slab of bulk polymer is

$$\frac{d(M)}{dt} = \frac{D\partial^2(M)}{\partial x^2} + k_2(R) - k_{-2}(R)(M) \qquad (37)$$

where D is the diffusion constant in $cm^2 \ sec^{-1}$. The solution of this equation is available in the literature (25). The particular solution we require is the time-independent steady-state result, which is

$$\frac{r}{k_2 R} = \frac{1}{100 \ \rho} \frac{\tanh Z}{Z} \tag{38}$$

The theoretical line in Fig. 5 presents a logarithmic plot (9) of the ratio $r/k_2 R$ as a function of Z according to Eq. 38 where

$$Z = l[k_{-2}(R)D^{-1}]^{1/2} \tag{39}$$

The experimental rate increments $(r - r_0)$ are also shown as a function of thickness in Fig. 5. Since the two curves have opposite curvatures, it is not possible to superimpose the curves, and it is concluded that reverse polymerization does not account for the effect observed.

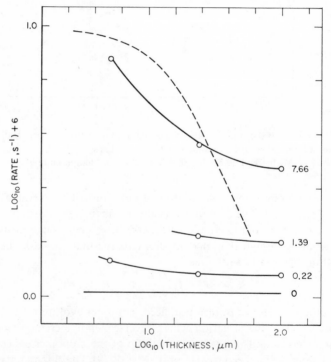

Fig. 5. Rate of depolymerization at 429.5°C as a function of sample thickness (27). Full lines: experimental for indicated dose rate, Mr/hr. Dashed line: Eq. 38 theory for rate limited by reverse polymerization. (Reprinted by permission of the copyright owner, The Am. Chem. Soc.)

J. Diffusion Influence on Radiolytic Initiation

The irradiation of methane with gamma rays has been shown to produce unequal concentrations of active species (26). With extent of radiation the difference between the concentrations of the methyl radicals and hydrogen atoms increased, and the observations were interpreted as an effect produced by the great difference in mobility between the two active species. This and the experimental result shown in Fig. 5 support the concept that with radiation initiation small radicals are formed in addition to the large polymer radicals which release monomer.

Deep in the bulk polymer these small radicals serve to terminate a sizable portion of the large polymer radicals. In the vicinity of the surface the small radicals can diffuse out and hence the large polymer radicals are relatively more numerous at the surface. This model (27) for the effect of thickness seems in qualitative agreement with experiment since at larger dose rates the effect would be more pronounced (see again Fig. 5). For polytetrafluoroethylene, theoretical treatment of this model is complicated owing to the fact that thermal initiation cannot be neglected, and numerical integration was necessary to obtain the theoretical curves (27) shown in Fig. 6. The experimental effect can apparently be reproduced by the system but the numerous parameters are very difficult to evaluate.

K. Molecular Weight as a Function of Conversion

Thermolytic changes in the molecular weight of the residual polymer have been followed at 480°C (8). These data are helpful for an evaluation of the importance of transfer and disproportionation in the degradation process. The square-root dependence of the radiation rate implies only that combination or disproportion may occur. Referring to the mechanism presented much earlier, we see that the differential equation for $Q(t)$, the sum of all the polymer molecules, as a function of time is

$$\frac{dQ}{dt} = \frac{k_{4_d}R(t)^2}{V(t)} + \frac{k_3 R(t)W(t)}{V(t)m} \tag{40}$$

since the system only gains a polymer molecule with each act of disproportionation and transfer. Dividing out $k_4 R^2/V(t)$ where $k_4 = k_{4_d} + k_{4_c}$, we find

$$\frac{dQ}{dt} = \frac{k_4 R^2(t)}{V(t)} \left\{ \frac{k_{4_d}}{k_4} + k_3 \left(\frac{m}{k_1 k_4} \right)^{1/2} \right\}$$

$$= \frac{k_1 W(t)}{m} \left\{ \frac{k_{4_d}}{k_4} + \sigma \right\} \tag{41}$$

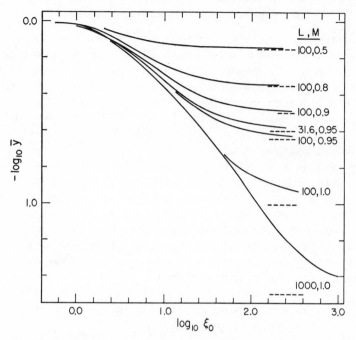

Fig. 6. Theoretical curves for rate of depolymerization as a function of thickness showing influence of inhomogeneous radical population (27). Variation of the log of average radical concentration $\bar{y}$ with log of thickness parameter ξ_0. Numbers on curves: Values of parameter for the termination and generation of small and large radicals. Broken line: interior limit. (Reprinted by permission of the copyright owner, The Am. Chem. Soc.).

The constant σ is the intermolecular transfer constant. Next, dividing by Eq. 8 and substituting $2l_0$ for $k_2 R(t)m/k_1 W(t)$, l being the kinetic chain length, we obtain

$$\frac{dQ}{dC} = \frac{W(0)}{2l_0 m}\left\{\frac{k_{4_d}}{k_4} \times \sigma\right\} \tag{42}$$

which is easily integrated to

$$Q(t) - Q(0) = \frac{W(0)}{2l_0 m}\left\{\frac{k_{4_d}}{k_4} + \sigma\right\}C \tag{43}$$

The number of moles of polymer molecules is the weight divided by the number average molecular weight, and we now write

$$\frac{W(0)(1 - C)}{m\bar{P}_n(t)} - \frac{W(0)}{m\bar{P}_n(0)} = \frac{W(0)C}{2l_0 m}\left\{\frac{k_{4_d}}{k_4} + \sigma\right\}$$

or

$$\frac{1-C}{\bar{P}_n(t)} - \frac{1}{\bar{P}_n(0)} = \left\{\frac{k_{4_d}}{k_4} + \sigma\right\}\frac{C}{2l_0} \tag{44}$$

where $\bar{P}_n(t)$ is the number average degree of polymerization at time t and $\bar{P}_n(0)$ the value at $t = 0$. Plotting the left-hand side of Eq. 44 as a function of C provided l is known, a value for the constants in parentheses can be obtained.

At 480°C some molecular-weight values of degraded polytetrafluoroethylene have been reported (8). On the assumptions $k_{4_d} = k_4$ and $\sigma = 0$, a value of 720 has been reported for l using a relationship slightly different from Eq. 44. One plot of the data according to Eq. 44 is shown in Fig (7. The thermal kinetic chain length is 13 at 480°C (see Eq. 35). This is a very small value compared to the previously mentioned value. As the temperature is increased the value of l_0 becomes even smaller. The value of the constants in parentheses obtained from the plot in Fig. 7 is 0.02. It is highly probable that $\sigma = 0$ and therefore

$$\frac{k_{4_d}}{k_4} \leq 0.02$$

It is thus interesting to note that although the observed molecular-weight decrease is appreciable, it can be the result of only a small amount of termination by disproportionation, termination being at least 98% combination. It is more than likely that termination is entirely by combination since only a small amount of hydrocarbon or other organic contamination is sufficient to produce the observed molecular-weight decrease.

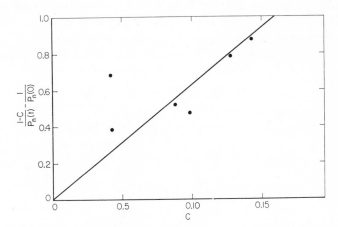

Fig. 7. Variation of degradation function with conversion for the thermal decomposition of polytetrafluoroethylene according to Eq. 43.

The absence of disproportionation is also consistent with studies on perfluoroalkyl radicals (28) and on the thermal decomposition of polytetrafluoroethylene in the presence of various gases (7,29). Equation 44 can be rewritten in the form

$$\frac{\bar{P}_n(t)}{\bar{P}_n(0)} = \frac{(1 - C)}{1 + [(k_{4_d}/k_4) + \sigma][\bar{P}_n(0)C/2l]} \tag{45}$$

Figure 8 presents theoretical curves calculated using Eq. 45 and varying the initial degree of polymerization $\bar{P}_n(0)$. The experimental points fall close to the appropriate curve and show the utility of the theory when plotted in this fashion. The plot in Fig. 7 is a more sensitive one in which the data show a relatively large degree of scatter. In Fig. 8 it is seen that the $\bar{P}_n(0)$ value influences greatly the appearance of the curves. From this figure it is seen how the mechanism can be tested to establish more definitely whether the lowering of the molecular weight below the linear diagonal is the result of disproportionation or transfer or whether the effect is the result of contamination. If impurities were responsible, then the result should be independent of $\bar{P}_n(0)$.

L. Degradation of Polytetrafluoroethylene in Various Gases

The influence of various gases and organic vapors on the rate of decomposition of polytetrafluoroethylene has been investigated (7,29). Gases were swept over the sample at atmospheric pressure. For substances boiling above room temperature the vapor was usually swept through the furnace in a nitrogen stream. Under pure nitrogen the rate of weight lost was the same as in vacuum experiments. Table 2 list the gases investigated according to the effect observed. It should be emphasized that the direction of the effect was determined over relatively small extents of conversion.

Figure 9 shows some results found with chlorine- and fluorine-containing

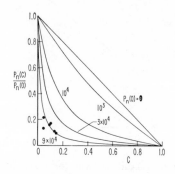

Fig. 8.   Variation of the relative degree of polymerization with conversion. Line: theoretical Eq. 44. Points: experimental at 480°C.

TABLE 2

Influence of Various Gases on the Thermal Decomposition
of Polytetrafluoroethylene (29)

Inhibitory	Catalytic	No Effect
IF_5	H_2S	N_2
ClF_3	H_2O	$C_6H_5CF_3$
H_2	O_2	Br_2
Cl_2	SO_2	CCl_3H
Cl_2/NO, 10/1	NO	$C_6H_5NH_2$
CCl_4	H_2/NO, 3/1	CCl_2F_2
$CClF_2H$	NH_3	CF_3I
$C_6H_5CH_3$	$\begin{array}{c} HC{=}CH \\ HC{\diagup}{\diagdown}N \\ HC{-}CH \end{array}$	CF_3H
$C_6H_5NO_2$		Cl_2/NO, 1/1
C_6H_5CHO		

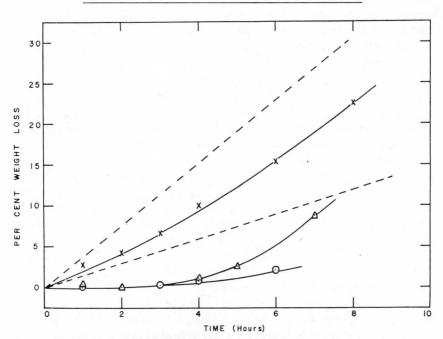

Fig. 9.   Pyrolysis of polytetrafluoroethylene under various halogen gases (29).

Upper dashed line:	*Lower dashed line:*
N_2 at 470°C	N_2 at 460°C
● Cl_2 at 470°C	○ IF_5 at 460°C
△ ClF_3 at 470°C	

gases. The weight lost under the gases is initially less but tends toward auto-catalytic behavior. Since inhibitors operate by terminating the depropagating radicals, the kinetic behavior of the decomposition under gases is different from that in a vacuum or under nitrogen. Under nitrogen in the absence of a catalytic on inhibitory gas the rate weight loss is first order, $dC/dt = K(1 - C)$. However, under the best inhibitors the kinetic behavior is approximately that of a simple random decomposition (30).

The rate of decomposition in the presence of hydrogen gas is seen in Fig. 10. Random decomposition has a rather complicated dependence on conversion, the rate going through a maximum with conversion. The rate at the maximum is to a good approximation related simply to the rate constant for random scission by the expression

$$\left(\frac{dC}{dt}\right)_{max} = \frac{k_1 L}{e} \tag{46}$$

where e is the base of the natural logarithms and L is the critical size for vaporization. In the theoretical model all molecules smaller than L vaporize without further degradation. In this situation the rate constant is the same as that for thermal initiation. By applying Eq. 46 to the data in Fig. 10, the following expression for k_1 is obtained:

$$k_1 = 10^{29} e^{-118,000/RT} \text{ sec}^{-1} \tag{47}$$

The value is in very good agreement with that previously deduced from the effect of gamma radiation on the rate. Under chlorine and toluene the polymer appeared to decompose also by random scission with the following apparent rate constants:

$$k(\text{chlorine}) = 10^{23} e^{-1000,000/RT} \text{ sec}^{-1}$$
$$k(\text{toluene}) = 10^{13} e^{-64,000/RT} \text{ sec}^{-1} \tag{48}$$

Thermolysis under chlorine produced a very clear fluid material upon removal of the polymer from the furnace. After 80% conversion the remaining polymer had an 0.6% chlorine content. Thermolysis under hydrogen and toluene produced a black fluid and the glass of the apparatus was etched, indicating hydrogen fluoride formation.

The incorporation of 10% by weight of carbon black in polytetrafluoroethylene also appears to shift the depolymerization away from pure monomer production and toward a random scission mechanism (31). Comparison of the full line curves in Fig. 11 with the dashed lines demonstrates the effect of carbon black on the rate of volatilization. Based on the maximum rates exhibited by the full lines in Fig. 11, a rate constant for random scission in the carbon black-filled polymer can be calculated:

$$k = 10^{14} e^{-70,000/RT} \text{ sec}^{-1} \tag{49}$$

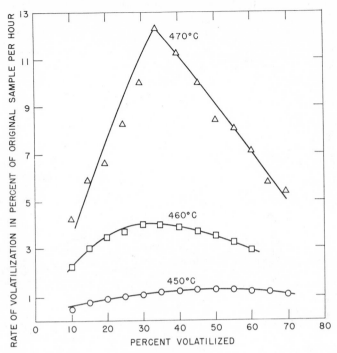

Fig. 10. Rate of pyrolysis of polytetrafluoroethylene under hydrogen gas as a function of conversion (7).

M. High-Pressure Decomposition

It was pointed out some time ago (10,11) that polytetrafluoroethylene, one of the most thermally stable polymers, is thermodynamically unstable. The experimentally observed decomposition to monomer is endothermic by 40 kcal. On the other hand, it is estimated that decomposition to graphite and perfluoromethane is exothermic to the extent of about 27 kcals. Under high-pressure, decomposition of polytetrafluoroethylene to carbon and presumably gaseous fluorocarbons occurs (32). The decomposition temperature is 700°C at 2 kbar and rises to 800°C at 30 kbar. It probably occurs rapidly once the critical conditions are attained. Decomposition of perfluorocyclobutene (33) and other fluorocarbons sometimes occurs explosively at high pressures to produce black carbonaceous products.

III. POLYTETRAFLUOROETHYLENE OXIDE

Polymers of this monomer having sizable molecular weights have recently been prepared (34). An initial study (35) of their thermal decomposition demonstrated thermal stability comparable or somewhat better than polytetrafluoroethylene. In Fig. 12 the two polymers are compared. The rate of

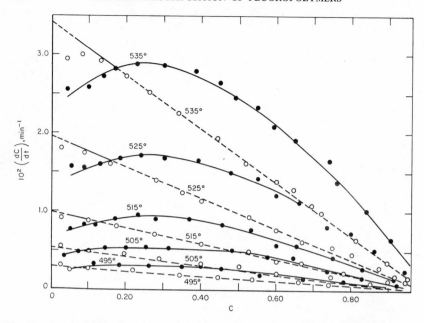

Fig. 11.   Influence of carbon black on the rate of volatilization of polytetrafluoroethylene (31). (●) with 10% Royal Spectra Black; (○) without additives.

weight loss peaks at a temperature that is approximately 60°C higher for the polytetrafluoroethylene oxide than for the polytetrafluoroethylene. However, the isothermal rates suggest only about a 30°C superiority for the oxide (compare Figs. 11 and 13). In Fig. 13 the isothermal rates of volatilization for polytetrafluoroethylene oxide are shown. The kinetic character of the curves shifts with temperature and conversion. From the low-temperature, zero-order plateaus, a 50 kcal activation energy is estimated. From the high-temperature, first-order curves, the rate as a function of temperature can be expressed by the relation

$$r_0 = 10^{17.27} e^{-83,000/RT} \text{ sec}^{-1} \tag{50}$$

This relation is similar to that for polytetrafluoroethylene, Eq. 3.

As with polytetrafluoroethylene, the rate is apparently independent of molecular weight (see Table 3). Presumably the mechanism of thermolysis is similar, i.e., random initiation, short zip length, and termination by combination. The molecular weight apparently changes little with conversion. The polymer melts near 50°C and is presumably an oily fluid at the temperature of pyrolysis. No data are available on the changes in molecular-weight distribution during degradation for either polytetrafluoroethylene or polytetrafluoroethylene oxide. It is anticipated, however, that provided the initial

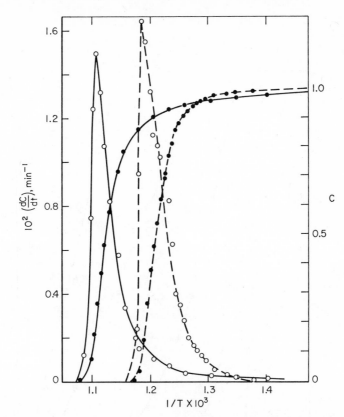

Fig. 12. Comparison of the thermal decomposition of polytetrafluoroethylene oxide (full lines) and polytetrafluoroethylene (dashed lines). Weight loss and rate of weight loss as a function of reciprocal programmed temperature, 0.5°K/min (35). (Reprinted by permission of the copyright owner, Marcel Dekker, Inc.)

TABLE 3

Rates of Decomposition for Polytetrafluoroethylene Oxides of Different Initial
Intrinsic Viscosities at 585°C

dl/g	$r_0 \times 10^3 (\text{sec}^{-1})$
0.585	1.58
0.475	2.22
0.365	1.63
0.195	1.65

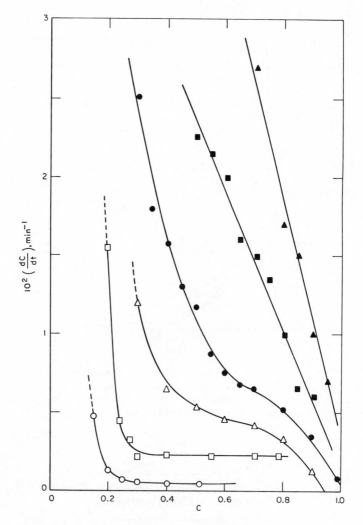

Fig. 13. Rate of thermal decomposition of polytetrafluoroethylene oxide as a function of conversion (35).

○	500°C	●	560°C
□	515°C	■	565°C
△	545°C	▲	575°C

(Reprinted by permission of the copyright owner, Marcel Dekker, Inc.)

polymer has a broad distribution with the ratio of weight to number average degree of polymerization exceeding 1.5, then the foregoing mechanism should lead to a progressive narrowing during the early stages of volatilization. Volatilization of small molecules will have a still further narrowing effect and in the limit of 100% conversion the distribution is monodispersed and the number average degree of polymerization is L, the smallest degree of polymerization requiring decomposition in order to vaporize.

The chief volatile products of polytetrafluoroethylene oxide decomposition are trifluoroacetyl fluoride, CF_3COF, carbonyl fluoride, CF_2O, and tetra-fluoroethylene, CF_2CF_2. The amount of the acetyl fluoride relative to the carbonyl fluoride decreases with temperature. Experiments have demonstrated that the carbonyl fluoride is also a primary product and not a secondary product of the decomposition of the acetyl fluoride. Apparently all three products result from the following depropagation of the polymer radicals:

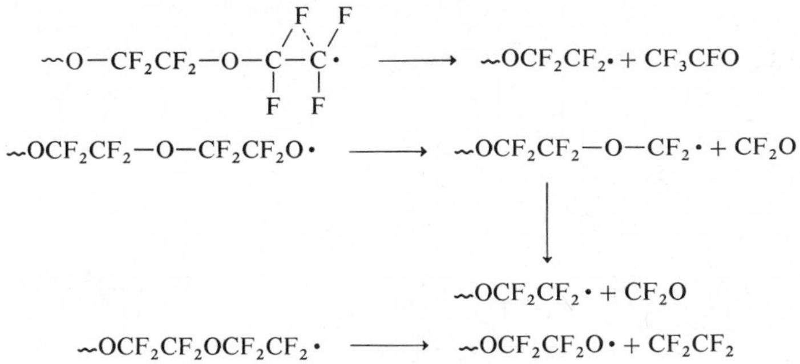

$$\sim O-CF_2CF_2-O-\underset{\underset{F}{|}}{\overset{\overset{F}{/}}{C}}-\underset{\underset{F}{|}}{\overset{\overset{F}{/}}{C}}\cdot \longrightarrow \sim OCF_2CF_2\cdot + CF_3CFO$$

$$\sim OCF_2CF_2-O-CF_2CF_2O\cdot \longrightarrow \sim OCF_2CF_2-O-CF_2\cdot + CF_2O$$

$$\downarrow$$

$$\sim OCF_2CF_2\cdot + CF_2O$$

$$\sim OCF_2CF_2OCF_2CF_2\cdot \longrightarrow \sim OCF_2CF_2O\cdot + CF_2CF_2$$

IV. POLYFLUOROHYDROETHYLENES

Polymers containing both hydrogen and fluorine in the repeating unit all decompose to produce hydrogen fluoride as an important product. The thermal decompositions of most of the various fluoroethylenes (5,36,37) have been investigated. These include trifluoroethylene, 1,1-difluoroethylene, 1,2-difluoroethylene, and vinyl fluoride, all of which have significant differences in decomposition kinetics. Polyfluorohydroethylenes that have received gamma irradiation exhibit enhanced rates for the decomposition process and enhanced production of hydrogen fluoride. The effect of irradiation is characteristic of the particular polymeric structure (see Figs. 14–16). Radiation produces cross-links and scissions in all the polyfluorohydroethylenes to about the same extent, cross-linking predominating with all the polymers of the type under discussion.

The thermal decomposition mechanisms of these polymers are so extremely complex that at best we only have a meager comprehension of the processes.

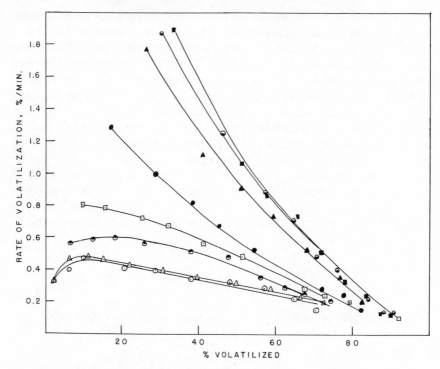

Fig. 14. Effect of prior gamma radiation on the rate of volatilization at 380°C of poly-trifluoroethylene (36). Radiation dose 10^{-20} eV/g.

○	original	●	19.38
△	0.51	▲	43.86
◖	5.0	◒	63.24
□	9.44	■	103.53

(Reprinted by permission of the copyright owner, The Soc. of Plastic Engineers, Inc.)

They can, however, be discussed in terms of two competing net processes: (1) decomposition by rupture of the chain carbon bonds; and (2) decomposition by the splitting off of adjacent pendant hydrogen and fluorine as hydrogen fluoride. Schematically,

$$\sim CF_2CH_2(CF_2CH_2)_iCF_2CH_2\sim$$

$$k_c \swarrow \qquad \searrow k_f$$

$$jCF_2CH_2 + yCF_2X(CH_2CF_2)_kCH_2X \qquad iHF + \sim CF_2CH_2(CF=CH)_iCF_2CH_2$$

where X = end groups or atoms.

If the polymer decomposes entirely by process k_c, then no residue would be formed and the fractional conversion or fraction weight loss would be

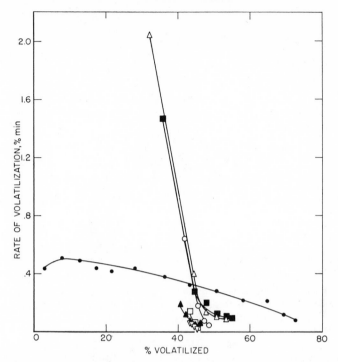

Fig. 15. Effect of prior gamma irradiation on the rates of volatilization at 365°C of polyvinyl fluoride (36). Radiation dose 10^{-20} eV/g.

●	0.00	□	43.9
△	0.51	▲	63.2
■	5.0	▽	104.0
○	9.44		

written as was done earlier. The process k_c includes all variations of mechanisms of the type treated in the first part of this chapter.

On the other hand, if the polymer decomposes by process k_f, there would be a finite residue and at complete conversion for the k_f reaction, as written above, the fractional weight loss would be 0.3125 for both vinylidene and vinylene difluoride. For polyvinyl fluoride, the fractional weight loss after the removal of one hydrogen fluoride unit per monomer would be 0.435 and for polytrifluoroethylene 0.244. The abscissas in the figures are often labeled percent volatilized instead of fractional weight loss.

The fluorohydropolymers studied decompose under most conditions by both mechanisms. Polyvinylidene fluoride produces more hydrogen fluoride than polyvinylfluoride, and a large amount of residue. At 400°C the reaction stops at very near to 62.5% volatilized, which is stoichiometric for the production of two molecules of hydrogen fluoride for each monomer unit in the

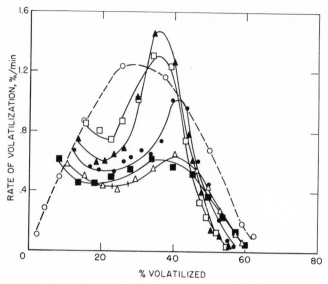

Fig. 16. Effect of prior gamma irradiation on the rate of volatilization at 410°C of polyvinylidene fluoride (36). Radiation dose 10^{-20} eV/g. (○) Experimental polymer prepared by gamma irradiation of monomer 0.00.
Commercial polymer:

■	0.00	▲	63.2
△	19.4	□	104.0
●	43.9		

polymer. Thus the decomposition of polyvinylidene fluoride is to a large extent by the k_f process, although the production (38) of some volatile hydrofluorocarbon and some hydrofluorocarbon wax, equal to about two-thirds by weight of the hydrogen fluoride produced, indicates that the k_c process is not negligible.

With polyvinylene difluoride (37) the residue is somewhat less at 400°C, i.e., the fraction weight loss appears to be closer to 70% At 900°C the weight loss is close to 90%. The two difluoroethylene polymers are thus comparable in their behavior, both producing mainly hydrogen fluoride with a small but significant tendency for carbon bond rupture.

The thermal volatilization of unirradiated polyvinylfluoride is found to be nearly 100%, even though hydrogen fluoride production is also nearly stoichiometric (5). Like the pyrolysis of polyvinyl chloride where benzene and some other aromatic compounds are found (1), the pyrolysis of polyvinyl fluoride produces some benzene.

As had been anticipated, radiation cross-linked polyvinyl fluoride forms a residue in about the expected stoichiometric amounts, a result not found previously with polytrifluoroethylene (36) (Fig. 14). There is also a greatly enhanced rate of the initial volatilization in which hydrogen fluoride is produced.

With unirradiated polyvinylidene fluoride, nearly the maximum weight loss for the volatilization of two hydrogen fluoride molecules is produced; however, an increased residue formation does occur with the irradiated polymer (Fig. 16). The rate of hydrogen fluoride evolution or, equivalently, the rate of char or residue formation is again greatly increased.

Extraction of the irradiated polymers demonstrated that the observed effects were not due to the rapid evaporation of low-molecular-weight fragments produced by scission during the radiolysis of the polymer though it completely changed the result with polyvinyl fluoride but not with polyvinylidene fluoride and polytrifluoroethylene (36). The k_f process is probably autocatalytic, i.e., the production of a few unsaturated sites in the polymer tends to accelerate the further release of hydrogen fluoride (39). For such an autocatalytic process, the type of plot shown in Figs. 14–16 would go through a maximum. On the other hand, if process k_c were a random one, a maximum in the rate would also occur (30).

The effect of extraction with hexamethyl phosphoramide on the pyrolysis of irradiated polyvinyl fluoride suggests that this treatment removes the unsaturated sites (36). This is not unlikely in view of the chemical nature of the solvent and the polymer. The pyrolysis of polytrifluoroethylene is altered by prior irradiation in a very different fashion from that found for the other two systems. The changes produced in volatilization of the polytrifluoroethylene do not suggest an enhancement of the rate of hydrogen fluoride evolution but rather a change from a linear chain to a branched chain structure, since the rate curves for volatilization before irradiation are similar to those for linear polyethylene, whereas upon irradiation the curves become similar to those found for branched polyethylene (40,41).

V. POLYCHLOROTRIFLUOROETHYLENE

In the thermal decomposition of polychlorotrifluoroethylene the yield of the monomer, CF_2CFCl, is 26% by weight (42). With the monomer there is also 1–2% each of mono and dichloropropylenes. The remaining product of decomposition is a halocarbon waxy mixture of number average molecular weight 904. The polymer is qualitatively known to rapidly decrease in molecular weight thereby readily forming low-molecular-weight greases. Although the monomer yield is sizable and indicative of some depropagation or radical unzipping, the majority of the data available, including the rate of volatilization curves shown in Fig. 17, emphasizes an appreciable component of random degradation in the mechanism. This random component in the mechanism is probably the result either of an intermolecular transfer process

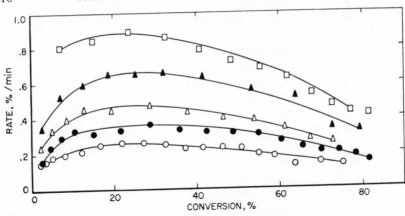

Fig. 17.   Pyrolysis of polychlorotrifluoroethylene, rate of volatilization as a function of conversion (42).

□	371°C	●	356°C
▲	366°C	○	351°C
△	361°C		

in which a primary polymer radical abstracts a chlorine atom from an adjacent chain and hence a chain scission is produced via a secondary radical or of a random initiation process in which the same secondary radicals result from the unimolecular dissociation of the carbon chlorine bonds along the chain. The rate curves in Fig. 17 go through a maximum in the region of 26–30% conversion and are generally similar to those produced by random scission (2). From the number average molecular weight of the waxy products and random theory, molecular vaporization of oligomers can be estimated to occur up to a molecular weight of ~ 1800 without appreciable decomposition. The activation energy of the thermal decomposition, which is probably initiated by the rupture of the carbon chlorine bond, is relatively low, ~ 50 kcal (43).

Unlike completely fluorinated polymers, thermal decomposition of polychlorotrifluoroethylene is greatly accelerated by ultraviolet light (43). The photoinitiated thermal decomposition has an activation energy of 13 kcal. From the postirradiation rate measurements it is concluded that termination is bimolecular. At the highest temperatures studied, the post-irradiation decay has an activation energy of -3 kcal (43). Unlike the radiolytic-induced thermal decomposition of polytetrafluoroethylene discussed in the beginning of this Chapter, no workable method has been found to estimate the quantum yield for the initiation and hence values of the individual rate constants cannot be obtained from the photostudies. However, from the three activation energies quoted earlier, the following values for activation energies of each step of the thermal decomposition mechanism are calculated:

Thermal initiation $\quad E_1 = 74$
Depropagation $\quad E_2 = 23$
Termination
(diffusion controlled) $\quad E_4 = 20$

For polychlorotrifluoroethylene considerable work remains to be done in order to more completely and quantitatively elucidate the detail mechanism of degradation. However, the basic kinetic features of the process seem clear.

VI. FLUORINATED POLYSTYRENES

The rates of thermal decomposition of poly-α,β,β-trifluorostyrene (42), poly-2,3,4,5,6-pentafluorostyrene (44), and polyperfluorostyrene have been investigated. A comparison of the results is shown in Table 4 along with similar data for polystyrene.

The presence of the fluorine substituents alters the decomposition of polystyrene considerably but not its thermal stability. Fluorination of the chain enhances the monomer yield. The lowest monomer yield, that of polypentafluorostyrene is probably a little lower than that given by polystyrene. The ring-fluorinated styrene is the most thermally stable (see decomposition temperatures listed in the last column of Table 4). In this table and others the decomposition temperature is taken as either that temperature where the isothermal rate of volatilization is 1% min^{-1} or where in a slow (1.5–4°C min^{-1}) temperature programmed experiment the 50% weight loss occurs. The two methods of determining decomposition temperature usually give nearly the same result. The 50% weight loss also corresponds

TABLE 4

Pyrolysis of Fluorinated Polystyrenes

Structure	Monomer yield (wt. %)	A	E_a	T_{dec} (°C)
$-CH_2CH-$ $\quad$ $\vert$ $\quad$ C_6H_5	42	10^{15}	55	355
$-CF_2CF-$ $\quad$ $\vert$ $\quad$ C_6H	70	10^{19}	64	330
$-CH_2CH-$ $\quad$ $\vert$ $\quad$ C_6F_5	<63	10^{17}	65	395
$-CF_2CF-$ $\quad$ $\vert$ $\quad$ C_6F_5	100		74	350

at least approximately to the temperature of the peak in a differential thermo-gravimetric plot. The rates go through a similar maximum with conversion for polystyrene and polypentafluorostyrene. The other two polymers tend to give rates which decrease nearly linearly with conversion.

All the fluorinated polymers have much better oxidation resistance than polystyrene. Since only a few samples of the fluorinated polymers have been investigated and the influence of molecular weight has not been ascertained for these materials, the results must be taken with some caution. However, the data appears to be reasonable and representative.

VII. EFFECT OF STRUCTURE ON THERMAL STABILITY

The thermal decomposition of a variety of fluoropolymers has been curso-rily investigated (43,45-47). Table 5 and Fig. 18 present comparisons of the thermal stability of a few of these. It is seen that branched, cyclic, and un-saturated perfluorostructures have considerably lower stabilities than those of the linear alkane or alkylene ether structures.

The degradation of fluoropolymers is always accelerated by oxygen. The effect is usually small in comparison to that observed when the polymer is nonfluorinated. In oxygen the relative stabilities of the fluoropolymers remain similar to that found *in vacuo* (48). Compared to their hydrocarbon analogs, they are almost always very resistant to oxidative attack.

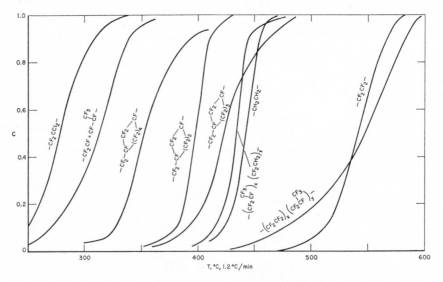

Fig. 18. Pyrolysis of various fluoropolymers, fraction volatilized as a function of tem-perature when temperature is programmed at 1.2°C/min.

TABLE 5

Pyrolysis of Fluoropolymers

Structure	T_{dec} (°C)	Ref.
$-CF_2CF_2-$	520	5

$-CF_2CF-$
    $\|$
    CF_3
	300	45

$-CF_2CF-$
    $\|$
    $CF_2(CF_2)_4F$
	250	45

$-CF_2-CF=CF-CF-$
              $\|$
              CF_3
	300	45

$-CF_2C\overset{\displaystyle CF_2}{\underset{\displaystyle CF_2}{\diagup\diagdown}}CF-$	400	45
$-CF_2C\overset{\displaystyle CF_2}{\underset{\displaystyle (CF_2)_2}{\diagup\diagdown}}CF-$	400	45
$-CF_2C\overset{\displaystyle CF_2}{\underset{\displaystyle (CF_2)_3}{\diagup\diagdown}}CF-$	420	45
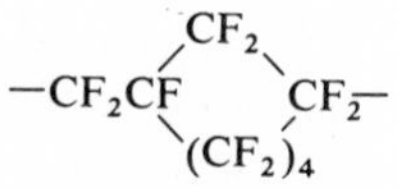 $-CF_2C\overset{\displaystyle CF_2}{\underset{\displaystyle (CF_2)_4}{\diagup\diagdown}}CF_2-$	354	45
$-CF_2CFCl-$	380	43
triazine ring: $-C(=N)-C-(CF_2)_3$ with $N-C(-CF_2CF_2CF_3)=N$	500	43
$-NO-CF_2CF_2-$ $\|$ CF_3	260	47

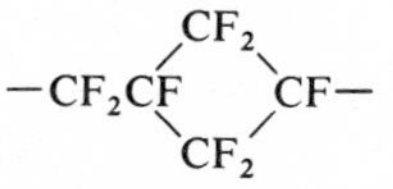

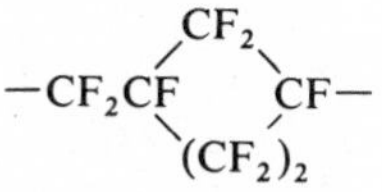

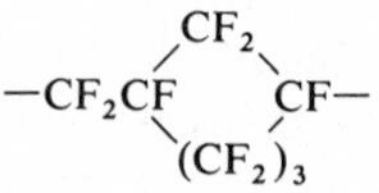

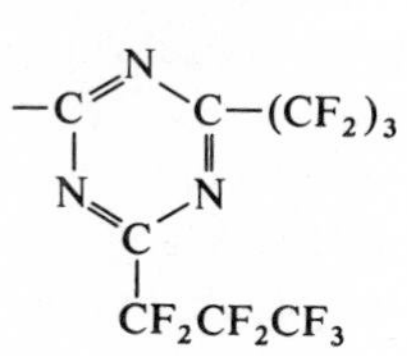

413

The decomposition of copolymers of tetrafluoroethylene and perfluoro-alkenes has recently been studied (49). In Fig. 18 a curve is shown for a commercial copolymer of tetrafluoroethylene and hexafluoropropylene. As more hexafluoropropylene is introduced into the tetrafluoroethylene polymer, the thermal stability decreases. However, as seen in Fig. 18, a small amount ($\sim$10–15%) is not especially detrimental. Introduction of 10–15% of per-fluorobutene-1 into the polymer further lowers the stability markedly. Co-polymerization with larger perfluoro-1-alkenes has little additional effect (49).

Many of the results in Table 5, Fig. 18, and elsewhere in this chapter are derived from research samples and apply only to such samples. It is always possible that the method of preparation, molecular weight, storage history, and impurities may determine the results of some degradation studies reported. Since the details of the degradation mechanisms for all but a few of the now known fluoropolymers have not yet been elucidated, very little information of a quantitative nature is available for correlation.

VIII. FLUORINATED ELASTOMERS

Certain copolymers of vinylidene fluoride with chlorotrifluoroethylene and hexafluoropropylene have elastomeric properties and are available commercially. The absence of the chlorine atoms in the hexafluoropropylene vinylidene fluoride copolymer leads to a considerably more thermally stable material (43,50–52). However, they have comparable stabilities in the presence of oxygen, which is the usual condition of use. The thermal stability of these elastomeric copolymers is of the same order as found for the fluorohydro-ethylene homopolymers (43). The amount of hydrogen fluoride production on thermal decomposition is relatively small (52). Organic compounding ingredients lead both to lower stability and larger amounts of hydrogen fluoride (51). Transfer reactions lead to the formation of products with larger molecular weights.

In the pyrolysis of the vinylidene fluoride-hexafluoropropylene copolymer, fluoroform is formed to some extent, usually mixed with the vinylidene fluoride fraction (52,53). The relative amount of fluoroform produced is increased when the copolymer is pyrolyzed as part of a mixture with poly-paraxylylene (53). The results appear to indicate that the carbon-carbon bond to the trifluoromethyl group is weak compared to the chain carbon-carbon bonds and also that trifluoromethyl free radicals are present during the thermal decomposition.

Copolymers of tetrafluoroethylene with 3,3,3-trifluoropropylene (54) and 3,3,4,4,5,5,5-heptafluoropentene-1 (55) have been prepared and their pyro-lyses have been investigated.

$$-(CF_2CF_2)_i(CH_2CH)_j- \quad -(CF_2CF_2)_i(CH_2CH)_j-$$

$$\begin{array}{ll} CF_3 & CF_2 \\ & CF_2 \\ & CF_3 \end{array}$$

The presence of only a small amount of the hydrogen-containing monomer greatly lowers the thermal stability of the polytetrafluoroethylene (see Figs. 19 and 20). Prior gamma irradiation of the copolymer also lowers the thermal stability greatly. Thermal stability greater than Viton is achieved when the copolymers contain greater than 50 mol % tetrafluoroethylene.

IX. PERFLUOROAROMATIC POLYMERS

Fully fluorinated high-molecular-weight aromatic polymers with useful physical properties have not yet been reported. The first material of polymeric size in this area was polyphenylene, having a number average degree of polymerization of ~ 11 (56). The synthesis of other polymers of this type are surveyed in Chapter 3. Polymers of this type are not yet as stable as might be expected from preliminary studies on model compounds (57–59). The following monomeric size compounds are listed in decreasing order of their thermal stability (43):

$$(C_6F_5)_2 > (C_6H_5)_2 > C_6F_5{-}C_6H_5 > (C_6H_5)_2O > Si(C_6H_5)_4 > (C_6F_5)_2O > Si(C_6F_5)_4$$
$$> C_6F_5O{-}C_6H_5 > P(C_6F_5)_3 > P(C_6H_5)_3 > P(C_6F_5)_3PO$$

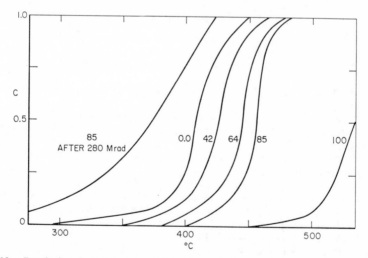

Fig. 19. Pyrolysis of copolymers of tetrafluoroethylene and 3,3,3-trifluoropropylene. Numbers on curves give percentage of tetrafluoroethylene in copolymers (54).

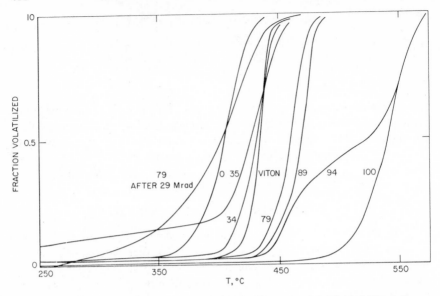

Fig. 20. Pyrolysis of copolymers of tetrafluoroethylene and 3,3,4,4,5,5,5-heptafluoro-pentene-1. Numbers on curves give percentage of tetrafluoroethylene in copolymers (55).

Discordant results reported in the literature (59) may be the consequence of different criteria or techniques. It should be noted that only very rarely can adequate comparisons be made with one criterion and one technique.

References

1. S. L. Madorsky, "Thermal Degradation of Organic Polymers," Interscience, New York, 1964, Chapter 5.
2. L. A. Wall, "Analytical Chemistry of Polymers," G. M. Kline, Ed., Interscience, New York, 1962, Chapter 5.
3. C. R. Patrick, "Advances in Fluorine Chemistry," Vol. 2, Butterworth, London, 1961, p. 1.
4. R. F. Lewis and A. Naylor, *J. Am. Chem. Soc.*, 69:1968 (1947).
5. S. L. Madorsky, V. E. Hart, S. Straus, and V. A. Sedlak, *J. Research NBS* 51:327 (1953).
6. R. E. Florin, L. A. Wall, D. W. Brown, L. A. Hymo, and J. D. Michaelson, *J. Research NBS*, 53:121 (1954).
7. L. A. Wall and J. D. Michaelson, *J. Research NBS*, 56:27 (1956).
8. J. C. Siegle, L. T. Muus, Tung-Po Lin, and H. A. Larsen, *J. Polymer Sci.* A2:391 (1964).
9. R. E. Florin, M. S. Parker, and L. A. Wall, *J. Research NBS*, 70A:115 (1966).
10. L. A. Wall, *SPE Trans.*, 16(8):1 (1960).
11. L. A. Wall, *SPE Trans.*, 16(9): (1960).
12. E. C. Penski and I. J. Goldfarb, *Polymer Letters*, 2:55 (1954).
13. H. L. Friedman, paper presented at the ACS Natl Meeting, Atlantic City, N.J., Sept. 1959; *Chem. Abstr.*, 55:26511 (1961).

14. P. R. E. J. Cowley and H. W. Melville, *Proc. Roy. Soc.*, **A210**:461 (1951); **A211**:320 (1952).

15. R. E. Florin and L. A. Wall, *SPE Trans.*, **3**:293 (1963).

16. R. E. Florin and L. A. Wall, *J. Research NBS*, **65A**:384 (1961).

17. A. Nishioka, K. Matsumae, M. Watanabe, M. Tajima, and M. Owaki, *J. Appl. Polymer Sci.*, **2**:117 (1959).

18. Report of the International Commission on Radiological Units and Measurements (ICRU) 1956; *NBS Handbook* 62, 1956, pp. 10–17.

19. W. M. D. Bryant, *J. Polymer Sci.*, **56**:277 (1962).

20. A. Nishioka and M. Watanabe, *J. Polymer Sci.*, **24**:298 (1957).

21. L. C. Case, *J. Appl. Polymer Sci.*, **3**:254 (1960).

22. A. V. Tobolsky, D. Katz, and A. Eisenberg, *J. Polymer Sci. A*, **1**:483 (1963).

23. M. Tekahashi and A. V. Tobolsky, *Polymer Letters*, **2**:129 (1964).

24. Yu. D. Tsvetkov, Ya. C. Lebedev, and V. V. Voevodsky, *Vysokomol. Soed.*, **3**:887 (1961).

25. H. S. Carslaw and J. C. Jaeger, "Conduction of Heat in Solids," 2nd ed., Clarendon Press, Oxford, 1959, p. 404.

26. D. W. Brown, R. E. Florin, and L. A. Wall, *J. Phys. Chem.*, **66**:2602 (1962).

27. R. E. Florin and L. A. Wall, *Macromolecules,* **3**:560 (1970).

28. G. O. Pritchard, G. H. Miller and J. R. Dacey, *Can. J. Chem.*, **39**:1968 (1961).

29. J. D. Michaelson and L. A. Wall, *J. Research NBS*, **58**:327 (1957).

30. R. Simha and L. A. Wall, *J. Phys. Chem.*, **56**:707 (1952).

31. J. Fock, *Polymer Letters*, **6**:127 (1968).

32. M. Tamayama, T. N. Andersen, and H. Eyring, *Proc. Natl. Acad. Sci.*, **57**:554 (1967).

33. D. W. Brown and L. A. Wall, *Am. Chem. Soc. Polymer Preprints*, **5**(2):907 (1964).

34. P. Barnaba, D. Cordischi, M. Lenzi, and A. Mele, *Chim. Ind. (Milan)*, **47**:1060 (1965).

35. A. Donato, M. Lenzi, and A. Mele, *J. Makromol. Sci. (Chem.)*, *A***1**(3):429; 1967).

36. L. A. Wall, S. Straus, and R. E. Florin, *J. Polymer Sci.,* **A1 4**:349 (1966);. S. Straus and L. A. Wall, *Soc. Plastics Engrs.*, **4**:61 (1964).

37. W. S. Durrell, G. Westmoreland, and M. G. Moshonas, *J. Polymer Sci.*, **3A**:2975 (1965).

38. S. L. Madorsky and S. Straus, *J. Research NBS*, **63A**:261 (1959).

39. R. F. Boyer, *J. Phy. Colloid Chem.*, **51**:80 (1947).

40. L. A. Wall, S. L. Madorsky, D. W. Brown, S. Straus, and R. Simha, *J. Am. Chem. Soc.*, **76**:7430 (1954).

41. L. A. Wall and S. Straus, *J. Polymer Sci.*, **44**:313 (1960).

42. S. L. Madorsky and S. Straus, *J. Research NBS*, **55**:223 (1955).

43. L. A. Wall and S. Straus, *J. Research NBS*, **65A**:227 (1961).

44. L. A. Wall, J. M. Antonucci, S. Straus, and M. Tryon, *Soc. Chem. Ind.*, Monogr. No. 13, 295 (1961).

45. S. Straus, D. W. Brown, and L. A. Wall, paper presented at 156th Natl. ACS Meeting, Div. of Fluorine Chemistry, Atlantic City, N.J., Sept. 1968.

46. W. W. Wright, *Soc. Chem. Ind.*, Monogr. No. 13, 248 (1961).

47. J. M. Cox, B. A. Wright, and W. W. Wright, *J. Appl. Polymer Sci.*, **8**:2935 (1964)

48. J. M. Cox, B. A. Wright, and W. W. Wright, *J. Appl. Polymer Sci.*, **8**:2951 (1964).

49. G. D. Dixon, W. J. Feast, G. J. Knight, R. H. Mobbs, W. K. R. Musgrave, and W. W. Wright, *European Polymer J.*, **5**:295 (1969).

50. T. G. Degteva, I. M. Sedova, and A. S. Kuz'minshii, *Vysokomol. Soyed.*, **5**(10):1485 (1963).

51. T. G. Degteva, A. S. Kuz'minskii, Kh. H. Khamidov, *Soviet Rubber Tech.*, **Sept. 1964**: 11.
52. T. G. Degteva, I. M. Sedova, Kh. A. Khamidov, A. S. Kuz'minskii, *Vysokomol. Soyed.* **7**(7):1198 (1965).
53. L. A. Oksent'evich and A. N. Pravednikov, *Doklady*, **182**:1 (1968).
54. D. W. Brown and L. A. Wall, *J. Polymer Sci.*, **6**:1367 (1968).
55. D. W. Brown, R. E. Lowry, and L. A. Wall, *J. Polymer Sci.*, A18: 2441 (1970).
56. M. Hellman, A. J. Bilbo, W. J. Pummer, *J. Am. Chem. Soc.*, **77**:3650 (1955).
57. L. A. Wall, R. F. Donadio, and W. J. Pummer, *J. Am. Chem. Soc.*, **82**:4846 (1960).
58. W. J. Pummer and L. A. Wall, *J. Chem. Eng. Data*, **6**:76 (1961).
59. G. A. Richardson and E. S. Balke *I & EC Prod. Res. and Devel.*, **7**:22 (1968).

13. SURFACE PROPERTIES OF FLUORO-CARBON POLYMERS

ALLEN G. PITTMAN, *Agricultural Research Service, U.S. Department of Agriculture, Albany, California.*

Contents

I. INTRODUCTION

The low intermolecular forces present in highly fluorinated organic compounds are widely recognized and account for the relatively low surface tension of fluorinated organic liquid compounds and low melting points of fluorinated organic solids. In fluorine-containing polymeric substances, low intermolecular forces at the air/solid interface can also give rise to surfaces with extremely low free energy. As a result, fluorine-containing polymers generally are difficult to wet with organic and aqueous liquids and, in addition, these surfaces tend to have nonadhesive character and low coefficients of friction.

Zisman's discovery of the unique nonwetting properties associated with a fluorochemical surface has led to a number of commercially interesting products which utilize this property. For example, fluorinated polymers are

419

available for imparting oil and water resistance to textiles, leather, paper, and various other substrates. These polymers evidently stem from derivatives of acrylic acid which contain perfluoroalkyl groups terminated by $-CF_3$ (1,2). Similar polymer compositions are finding use as barrier materials to prevent the spreading of liquids from points of contact on ball bearings, watches, and other devices (3).

It would be virtually impossible to consider thoroughly all aspects of the surface properties of fluorinated polymers. We have chosen to emphasize the surface-wetting phenomenon, since, although uncertainties still exist, the wetting phenomenon seems to be on a sounder theoretical footing than, for example, the phenomenon of adhesion or friction on polymer surfaces.

Studies on the wettability of solid surfaces have yielded a great deal of information, particularly regarding the influence of the surface chemical constitution on wetting properties. Zisman and his co-workers pioneered in this work and have done a great deal to clarify the relation between wetting, adhesion, and surface friction (4).

The terminology adopted in this chapter regarding low-surface-energy solid surfaces is that suggested by Zisman (4). That is, hard solid surfaces such as metals, metal oxides, and silica glass with surface free energies ranging from 5000 to 500 ergs cm^{-2} are considered high-energy surfaces. The relatively soft organic solids such as organic polymers and waxes, which generally have specific surface free energies under 100 ergs cm^{-2}, are termed "low-energy surfaces."

II. WETTABILITY OF FLUORINATED POLYMERS

A. General Consideration of the Contact Angle θ

Wetting properties of solids have been studied primarily using contact angle measurements of sessile drops of pure liquids on smooth solid surfaces. The contact angle θ, as illustrated in Fig. 1, can be obtained in a number of ways, through visual observations on liquid drops or by several indirect methods. Of the direct visual methods, perhaps the most commonly used employs the so-called contact angle goniometer originally described by Zisman and Fox (5). With this apparatus, which is now commercially available, the drop profile is observed through a low-power microscope equipped with a protractor (goniometer) eyepiece, and contact angles are obtained directly. An alternate procedure involves projection of an enlarged image of the drop on a screen and measuring the angle with a protractor (6). Fort and Patterson (7) have described an instrument for measuring contact angles up to 90° which employs light reflected from the surface of sessile drops and a special goniometer. Several techniques have been described which do not actually involve a sessile drop resting on a solid surface. These methods, e.g., the

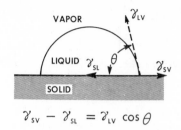

$$\gamma_{sv} - \gamma_{sl} = \gamma_{lv} \cos \theta$$

Fig. 1.   Contact angle (θ) for a pure liquid on a solid surface.

tilting plate method (8) or the wetting and dewetting balance of Guastalla (9), are somewhat more involved but can have certain advantages such as a more accurate measure of advancing and receding angles.

It is not uncommon to find that θ is larger when advanced over a surface (advancing angle) than when the liquid is retracted (receding angle) over the previously wet surface. This contact angle hysteresis can result from a variety of causes, but the primary cause appears to reflect surface inhomogenieties or surface roughness. Thus when a liquid is retracted from a surface, some of the liquid may remain in pores or crevices, leaving wet areas which will lower θ. Generally, however, with a sufficiently smooth solid surface little or no hysteresis will be observed (4).

The equation given in Fig. 1 (Eq. 1) was proposed by Thomas Young in 1801 (10):

$$\gamma_{LV} \cos \theta = \gamma_{SV} - \gamma_{SL} \qquad (1)$$

This equation defines the equilibrium contact angle in terms of the interfacial tension (surface tension) of the liquid-vapor interface, γ_{LV}, the solid-liquid interface, γ_{SL}, and the solid-vapor interface, γ_{SV}. Since the solid-vapor and solid-liquid interfacial tensions are not amenable to direct measurement, there has been considerable discussion concerning the validity of the Young equation, particularly at zero contact angle (11–13).

In order to relate the surface tension terms thermodynamically, Dupre introduced the reversible work of adhesion (W_A) for separating unit areas of solid and liquid phases in terms of free energies

$$W_A = \gamma_{SV} + \gamma_{LV} - \gamma_{SL} \qquad (2)$$

per unit surface area of solid-vapor, liquid vapor and solid-liquid interfaces.
Equation 1 is more accurately written in the form

$$\gamma_{LV}^{0} \cos \theta = \gamma_{SV}^{0} - \gamma_{SL} \qquad (3)$$

where γ_{LV}^{0} is the surface tension of the liquid which is, in principle, saturated with the solid species, and $\gamma_{SV}^{0} - \gamma_{SL}$ represents the free energy of exchanging unit geometric area of solid-vapor interface for solid-liquid interface.

Equation 2 is then written

$$W_A = \gamma_S^0 + \gamma_{LV}^0 - \gamma_{SL} \tag{4}$$

where γ_S^0 is the free energy at the solid-vacuum interface, and thus

$$W_A = (\gamma_S^0 - \gamma_{SV}^0) + \gamma_{LV}^0 (1 + \cos \theta) \tag{5}$$

The term $(\gamma_S^0 - \gamma_{SV}^0)$, the free energy decrease on immersion of the solid into the vapor phase, is often neglected to obtain the more useful expression

$$W_A = \gamma_{LV}^0 (1 + \cos \theta) \tag{6}$$

Adam (14) has pointed out that Eq. 6, which gives the work of adhesion to the film-covered solid surface, is of greater practical importance than Eq. 5 and, further, is the only work that can be directly measured. Adamson and Ling have recently reviewed the theoretical implications of the contact angle and the application of thermodynamics to the forces present in a sessile liquid drop on a solid surface (15).

B. Wettability as Determined by the Critical Surface Tension of Wetting γ_c

In examining the wetting properties of the two different solid surfaces of polyethylene and polytetrafluoroethylene, contact angles for water on these surfaces of 94 and 108°, respectively, are observed. The lower θ for polyethylene indicates that water has spread over this surface to a greater extent and thus reflects a more wettable surface than poly(tetrafluoroethylene) and greater adhesional forces between water and polyethylene. It is thus clear that as θ decreases, wettability increases. Conversely, Zisman has pointed out that the $\cos \theta$ is a direct reflection of wettability, since as $\cos \theta$ increases, wettability increases and as $\cos \theta$ decreases, wettability decreases (4). When $\cos \theta$ is 1, or the contact angle is 0 between a liquid and solid, the surface is presumed to be completely wetted by the liquid. It can also be said that when $\cos \theta$ is 1, the liquid-solid forces of adhesion exceed the cohesive forces of the liquid.

In reporting the unique low-surface-energy properties displayed by a polymeric fluorocarbon surface [poly(tetrafluoroethylene)], Fox and Zisman noted an interesting linear relationship between the cosine θ for a homologous series of n-alkanes and the surface tension of these liquids (16). They found that a plot of the cosine of the contact angle versus the surface tension for the liquid n-alkanes gave a straight line, which could be extrapolated to $\cos \theta = 1$ (or 0 contact angle) giving a surface tension value of about 18 dynes/cm (Fig. 2). As previously mentioned, at 0 contact angle complete wetting occurs. Fox and Zisman also found that similar plots were obtained with a number of organic liquids which were not necessarily homologous. Figure 3 illustrates

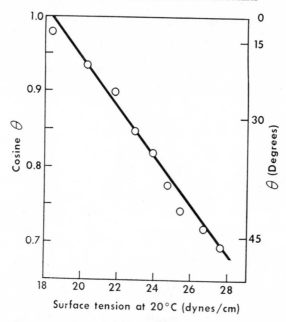

Fig. 2. Cosine θ versus surface tension of homologous n-alkanes on poly(tetrafluoro-ethylene) (16).

that a combined plot of the contact angle for various alkanes, esters, ethers, etc., versus their surface tensions gives an extrapolated surface tension value at $\cos \theta = 1$, the same that one obtains for the pure homologous n-alkanes (4,16). This extrapolated value of $\cos \theta = 1$ or 0 contact angle has been defined by Zisman as the critical surface tension of wetting or spreading γ_c. Obviously, liquids with surface tensions below the γ_c of a solid will wet and spread freely on the solid, whereas liquids with surface tensions above γ_c will form discrete drops with contact angles greater than zero.

Deviation from linearity does occur in Fig. 3 in liquids with surface tensions much above 35 dynes/cm. In the linear region of the plot, particularly with homologous n-alkanes where γ_{LV} ranges from about 18 to 27 dynes/cm (hexane to hexadecane), the forces of interaction between solid polytetrafluoroethylene and these liquids are presumed to arise primarily from London dispersion forces. In the higher surface tension region, the more polar liquids can interact with the solid surface through additional forces such as hydrogen bonding.

There have been a number of attempts to understand the critical surface tension of wetting on a nonempirical basis. Gardon (17) correlated γ_c with solubility parameters of polymers. Lee (18) recently proposed a relationship between the polymer glass transition temperature and γ_c. Schonhorn and

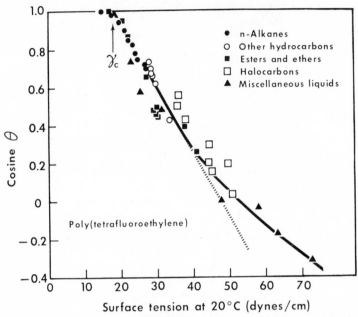

Fig. 3.   Cosine θ versus surface tension of a variety of liquids on poly(tetrafluoroethylene) (4,16).

Ryan (19) developed equations relating polymer wetting to surface density. Several attempts have been made to relate γ_c to the solid surface tension (20–22). Schonhorn and Sharpe (21,23) found that the extrapolated values for the surface tension of amorphous samples of polyethylene and polypropylene at 20°C were 35 and 29 dynes/cm, respectively. These extrapolated surface tensions are in close agreement with the critical surface tensions for these polymers. Based partly on these findings, Schonhorn (21) postulated that the γ_c of an amorphous polymer is equal to the surface tension of the amorphous polymer. A recent study was reported in which Johnson and Dettre (13) are in disagreement with Schonhorn's postulate on the relation of γ_c to polymer surface tension. Johnson and Dettre examined the wettability of low-energy *liquid* fluorocarbon surfaces by a series of alkanes. Data obtained on surface tensions, interfacial tensions, and spreading pressures were then compared with solid surfaces for which these measurements cannot be directly obtained. This approach appears attractive since, as Johnson and Dettre point out, a test can be made on any general theory of wettability that does not depend explicitly on the nondeformability of the substrate; moreover, the complicating factors of surface roughness and surface inhomogeneity are essentially eliminated. They found that the critical surface tensions for the (amorphous) liquids which were employed as substrates were not equal to the surface tensions of these liquids. From this they concluded that the critical surface tension for a polymer could not be the same as the polymer surface tension.

One of the most interesting approaches to an understanding of γ_c has come from Fowkes (24–26). He separated the forces present at a liquid/air interface into contributions arising from London dispersion forces and polar interactions, such as hydrogen bonding, i.e.,

$$\gamma_{total} = \gamma^d + \gamma^h$$

where γ^d is the contribution to surface tension arising from dispersion forces and γ^h the force arising from polar interactions. Interfacial tensions were examined between hydrocarbons, whose surface tension γ_1 arises solely from dispersion forces γ_1^d, and other liquids which have a surface tension γ_2 arising from both γ_2^d and γ_2^h. Using the relation

$$\gamma_{12} = \gamma_1 + \gamma_2 - 2(\gamma_1^d \gamma_2^d)^{1/2} \tag{7}$$

where γ_{12} is the interfacial tension, γ_1 is the hydrocarbon surface tension, and γ_2 is the surface tension of the second liquid. He determined, for example, γ_2^d for mercury and for water with a variety of alkanes.

For solid-liquid systems interacting by dispersion forces, Fowkes used Eq. 7 to predict γ_{SL} in the Young equation. The Young equation (Eq. 1) can be written in the form

$$\gamma_L \cos \theta = \gamma_S - \gamma_{SL} - \pi_e \tag{8}$$

where π_e is the equilibrium film pressure of adsorbed vapor on the solid surface. Fowkes thus obtained

$$\gamma_L \cos \theta = -\gamma_L + 2(\gamma_L^d \gamma_S^d)^{1/2} - \pi_e \tag{9}$$

For liquid-solid interactions where the solid is a low-energy polymer and a finite contact angle exists, π_e is commonly neglected and the equation simplifies to

$$\cos \theta = -1 + 2 (\gamma_S^d)^{1/2} \frac{(\gamma_L^d)^{1/2}}{\gamma_L} \tag{10}$$

The dispersion-force component of the surface energy of the solid (γ_S^d) is then obtained by plotting $\cos \theta$ versus $(\gamma_L^d)^{1/2}/\gamma_L$.

This component (γ_S^d) has been shown to be equal to the critical surface tension of wetting of a solid surface if liquids having only dispersion force interactions have been utilized in the determinations of γ_c and long extrapolations have not been made.

The use of saturated hydrocarbon liquids (e.g., n-alkanes) as reference liquids has been particularly recommended since they are subject almost solely to London dispersion force interactions. Unfortunately, with solid fluorocarbon surfaces, the use of n-alkanes as reference liquids leads to γ_s^d values which are lower than when fluorinated liquids are employed (25, 26). The assumption has been made that with all solid surfaces other than a fluorocarbon surface, γ_s^d obtained using liquid hydrocarbons is a more accurate measure of the dispersion force component of the surface energy. With solid fluorocarbon surfaces, γ_s^d obtained using fluorinated materials as test liquids is presumed to more accurately reflect the dispersion force contribution to its surface energy. Fowkes has pointed out that weaker than predicted interactions also occur between fluorocarbon liquids and hydrocarbon liquids and that generally the fluorocarbon-hydrocarbon interactions are not well understood (25,26).

Even though the use of saturated liquid hydrocarbons as test liquids on fluorocarbon polymer surfaces may lead to a γ_c value which is lower than the true γ_s^d, they are nevertheless quite useful in studies of wetting behavior. As a practical matter, it is difficult to obtain pure fluorinated liquids that do not have zero contact angle on most fluorinated polymeric surfaces. In addition, a majority of problems concerning the wettability of a polymer surface involve nonfluorinated or hydrocarbon-type liquids. Since most liquid hydrocarbons display finite contact angles on fluorocarbon surfaces, it is possible to use the homologous n-alkanes as test liquids and work in the linear region of a γ_c plot. In general this circumvents long extrapolations to $\cos \theta = 1$; furthermore, slight changes in γ_c resulting from constitutional and/or surface morphological changes can be detected.

Thus even though γ_c is empirical in nature, it has been extremely useful in examining the wetting properties of many types of polymers, both fluorinated and nonfluorinated (4,27,28). Table 1 lists some representative polymers and γ_c values ranging from a very low surface energy fluoroalkyl acrylate (γ_c, 10.4 dynes/cm) to a relatively higher surface energy polymer of urea-formaldehyde (61 dynes/cm).

TABLE 1

Critical Surface Tensions (γ_c) for Selected Polymers

Polymer	γ_c (dynes/cm)	Ref.
Poly (1,1-dihydroperfluorooctyl acrylate)	10.4	37
Poly(tetrafluoroethylene)	18.5	16
Polyethylene	31	49
Polystyrene	33–35	75
Poly(hexamethylene adipamide)	46	75
Urea-formaldehyde resin	61	76

III. EFFECT OF COMPOSITION ON POLYMER WETTABILITY

A. General Considerations

Shafrin and Zisman (29) emphasized the influence of surface constitution on wetting behavior in their "constitutive law of wettability." This law states that "in general, the wettability of organic surfaces is determined by the nature and packing of the surface atoms or exposed groups of atoms of the solid and is otherwise independent of the nature and arrangements of the underlying atoms and molecules." The constitutive law is based partly on Langmuir's "principle of independent surface action" (30,31), which pointed out the extreme localization of surface forces. In arriving at this concept, Zisman and co-workers examined the wetting properties of a number of solid surfaces (4). It was found that a variety of high-energy solid surfaces (e.g., platinum, glass) could be coated with condensed packed monolayers of organic compounds and the wettability (as defined by θ and γ_c) was always a reflection of the outermost atoms of the monolayer without regard to the composition of the substrate. For example, a monolayer of perfluorolauric acid on platinum gave a γ_c value of 6 dynes/cm (32) (Table 2). This value represents the least wettable surface known and is a result of a highly allineated $-CF_2-$ chain terminated by $-CF_3$ at the air/solid interface. The importance of the terminal group (i.e., the group concentrated at the air/solid interface) can be appreciated by considering that a similar fluorinated acid terminated by $-CF_2H$ gave a γ_c value of 15 dynes/cm, which is more than

TABLE 2

γ_c for Various Surfaces

Surface constitution	γ_c	Organic Material	Ref.
Fluorinated			
$-CF_3$	6	Monolayer of perfluorolauric acid	32
$-CF_2H$	15	Monolayer of ω-monohydro-perfluoroundecanoic acid	33
$+CF_2-CF_2+$	18.5	Poly(tetrafluoroethylene)	16
$+CF_2-CFH+$	22	Poly(trifluoroethylene)	35
$+CF_2-CFCl+$	31	Poly(chlorotrifluoroethylene)	77
Hydrocarbon			
$-CH_3$	24	Fatty amine monolayer	78
$+CH_2-CH_2+$	31	Polyethylene	49
Chlorocarbon			
$+CClH-CH_2+$	39	Poly(vinyl chloride)	35
$+CCl_2-CH_2+$	40	Poly(vinylidene chloride)	35

twice the value for the $-CF_3$-terminated acid (33). Wettability studies of this type were also conducted on a diverse group of polymeric materials, and Table 2 gives γ_c values for a number of surfaces, both polymeric substances and monolayers of acids on high-energy surfaces. It is evident upon examination of Table 2 that substitution of elements such as hydrogen or chlorine for fluorine in a polymer increases the wettability of the polymer. Recent publications indicate that the surface energy decrease which occurs in the order $-CH_2->-CH_3>-CF_2->-CF_3$, is primarily due to increasing group size. The larger volume occupied by $-CF_3$, as compared to $-CF_2-$, results in fewer interactions per unit area and a lower surface energy (34).

In examining the wetting properties of vinyl polymers in which there is an increasing amount of hydrogen substituted for fluorine ($-CF_2CF_2-$, $-CF_2CFH-$, $-CF_2CH_2-$, $-CFHCH_2-$, $-CH_2CH_2-$) (16,35), Shafrin and Zisman noted an increase of approximately 3 dynes/cm for each substitution of hydrogen for fluorine (29). Replacement of fluorine with chlorine caused a much greater increase in γ_c, 18.5 dynes/cm for polytetrafluoroethylene vs. 31 dynes for polychlorotrifluoroethylene ($-CF_2CFCl-$). The substitution of a second chlorine seems to cause only a slight additional increase in γ_c (compare $-CHClCH_2-$ to $-CCl_2CH_2-$, $\Delta\gamma_c = 1$ dyne/cm).

In determining the contribution of certain groups to the overall polymer wettability it was necessary to assume that the surface composition of the polymeric solids was statistically equivalent to that of the monomer repeating unit stretched out in a horizontal configuration at the surface (28). Although this appears to be a valid assumption at least for such vinyl polymers as polytetrafluoroethylene, polyvinylidine fluoride, and the like, a number of complicating factors appear in attempting to make this assumption with polymers containing pendant groups, such as polyfluoroalkyl acrylates and polyfluoroalkyl ethers. It may be anticipated that the wetting properties of polymers containing these side chains would be influenced to some extent by the main chain composition and surface exposure of the main chain. However, the primary contribution to γ_c in polymers of this type should be dependent on side-chain composition and the ability of the pendant groups to align in some fashion. In addition to side-chain composition, the crystallinity, tacticity, and side-chain branching would also be expected to influence side-chain alignment and hence wettability.

In order to examine additional aspects of the effects of polymer structure on wettability, Pittman et al. devised a synthetic route which leads to a wide variety of low-surface-energy polymers with variations in the structure and composition of the polymer main chain and pendant chains (36–42).

The synthetic route to these polymers is illustrated in the preparation of certain fluoroalkyl acrylates (36,37):

$$(CF_2X)_2C = O + MZ \longrightarrow (CF_2X)\, CZO^- M^+ \tag{11}$$

$$(CF_2X)_2CZO^- M^+ + CH_2 = CHCOCl \longrightarrow CH_2 = CHCOOCZ(CF_2CX)_2 \tag{12}$$

where X is halogen, hydrogen, or alkyl and Z is F^-, CN^-, OR^-, etc., and M is an alkali metal.

Other variations in structure were achieved in preparing fluoroalkyl glycidyl ethers (Eq. 13), (38,39), fluoroalkyl allyl ethers (Eq. 14) (40), fluoro-alkyl vinyl ethers (Eq. 15) (40), fluoroalkyl vinyl esters (Eq. 16) (41), and fluoroalkyl silanes (Eq. 17) (42):

Glycidyl ethers

$$(CF_2X)_2CFO^- M^+ + H_2C\overset{O}{\overset{\triangle}{-}}CHCH_2Br \longrightarrow H_2C\overset{O}{\overset{\triangle}{-}}CHCH_2OCF(CF_2X)_2 + MBr \tag{13}$$

Allyl ethers

$$(CF_2X)_2CFO^- M^+ + CH_2 = CHCH_2Br \longrightarrow CH_2 = CHCH_2OCF(CF_2X)_2 + MBr \tag{14}$$

Vinyl ethers

$$(CF_2X)_2CFO^- M^+ + X'CH_2CH_2X' \longrightarrow X'CH_2CH_2OCF(CF_2X)_2 + MX'$$

$$X'CH_2CH_2OCF(CF_2X)_2 \xrightarrow{\text{KOH}} CH_2 = CHOCF(CF_2X)_2 + KX' \tag{15}$$

Vinyl esters

$$(CF_2X)_2CFO^- M^+ + CH_2 = CHOOCCH_2Br \longrightarrow CH_2CHOOCCH_2OCF(CF_2X)_2 \tag{16}$$

Silanes

$$CH_2 = CHCH_2OCF(CF_2X)_2 + HSiCl_3 \longrightarrow Cl_3SiCH_2CH_2CH_2OCF(CF_2X)_2 \tag{17}$$

With these model polymer systems some of the following problems were examined:

1. What is the effect of side-chain composition on polymer wettability?
2. What is the effect of polymer main-chain composition on wettability?
3. What is the effect of the length or size of the side chain on wettability?

B. Pendant-Chain Composition

Figure 4 gives a critical surface tension plot for polyfluoroalkyl acrylates in which hydrogen or chlorine have been substituted for fluorine on the fluoroalkyl side chain. In general, the change of γ_c with fluorine substitution agrees with earlier work on adsorbed fluorocarbon acid monolayers and on fluorinated vinylic polymers (4). That is, substitution of hydrogen for fluorine increases γ_c slightly, whereas chlorine promotes a larger increase in γ_c.

In Table 3, the magnitude of change in γ_c with changes in composition for vinylic and acrylate polymers is compared. In examining the overall increase in γ_c (Table 3, column 2), a larger increase in γ_c is observed for both hydrogen and chlorine substitution in the vinyl polymer systems than for the acrylate polymers.* However, if the increase in γ_c is calculated in each case per 1%-decrease in fluorine content for the system (Table 3, column 3), it can be seen that the increase in γ_c does not change greatly whether the substituting atom is hydrogen or chlorine. This calculation is made, for example, in the first case, which considers going from poly(tetrafluoroethylene) to poly(trifluoro-ethylene), by substracting the percentage fluorine content of poly(trifluoro-ethylene) from the percentage fluorine content of poly(tetrafluoroethylene) and dividing into the overall increase in γ_c (in this case, 3.5 dynes/cm). It should be emphasized that γ_c in a particular fluorinated polymer is not necessarily dependent on the total fluorine content but rather on the arrangement of the fluorine atoms. For example, poly(tetrafluoroethylene) with a overall 76% fluorine is more wettable (higher γ_c) than the polyacrylate $A-CF(CF_3)_2$, which has a 55% fluorine content. However, calulations of the sort indicated here could be important in estimating changes in γ_c in a given polymer system with changes in constitution (37).

It is possible to compare the γ_c values for the acrylates, $A-CF(CF_3)_2$ and $A-CF(CF_3)CF_2Cl$, with those values obtained for acid monolayers with similar fluorocarbon segments in order to gain some appreciation of the surface composition of these acrylate polymers. For monolayers of the terminally branched acids $(CF_3)_2CF(CF_2)_nCOOH$, where $n = 1-11$, Bernett and Zisman (43) obtained γ_c values ranging from 13.3 (where $n = 11$) to 15.2 dynes/cm (where $n = 1$). The acrylate $A-CF(CF_3)_2$ has an intermediate γ_c of 14.1 dynes/cm and the surface constitution of this polymer must

---

* In order to simplify nomenclature and polymer identification, A refers to the repeat-ing acrylate unit

$$+CH_2CH+$$
$$|$$
$$C=O$$
$$|$$
$$O$$
$$|$$

and M refers to the methacrylate repeating unit

$$+CH_2C(CH_3)+.$$
$$|$$
$$C=O$$
$$|$$
$$O$$
$$|$$

For example, poly(heptafluoroisopropyl acrylate) would be written $A-CF(CF_3)_2$.

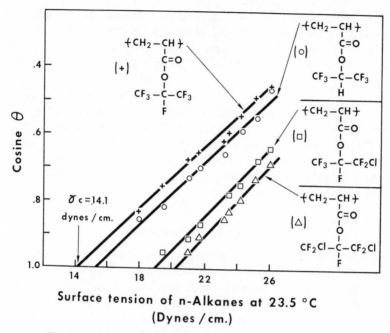

Fig. 4. Cosine θ versus surface tension of n-alkanes at 23.5°C.

TABLE 3

Comparison of Vinylic and Acrylate Polymers

	Polymers[a]		$\Delta \gamma_c$ Increase in γ_c (dynes/cm)	Increase in γ_c per 1%[b] decrease in fluorine content (dynes/cm)
(A)		(B)		
$+CF_2CF_2+_n$	$\longrightarrow$	$+CF_2CFH+_n$	3.5	0.55
$A-CF(CF_3)_2$	$\longrightarrow$	$A-CH(CF_3)_2$	1.3	0.33
$+CF_2CF_2+$	$\longrightarrow$	$+CF_2CFCl+_n$	12.5	0.44
$A-CF(CF_3)_2$	$\longrightarrow$	$A-CF(CF_3)(CF_2Cl)$	4.8	0.47

[a] A refers to acrylate repeating unit

$$+CH_2-CH+$$
$$\qquad | $$
$$\qquad C=O$$
$$\qquad | $$
$$\qquad O$$
$$\qquad | $$

[b] $\dfrac{\Delta \gamma_c}{\% \text{ F in A} - \% \text{ F in B}}$

431

therefore be very similar to the branched acid monolayers. A similar result is observed for the acid monolayers of $CF_2Cl(CF_3)CF(CF_2)_nCOOH$, where $n = 1–9$. Here the acid monolayers ranged from 17.2 (where $n = 9$) to 20 dynes/cm (where $n = 1$), and the acrylate, $A—CF(CF_3)CF_2Cl$, was found to have a γ_c of 18.9 dynes/cm.

It is also of interest to note that the carbon-oxygen linkage in the acrylates [e.g., $(CF_3)_2CF—O—$] did not materially affect surface wetting. In fact, in this structural arrangement, it appears that the oxygen is virtually equivalent in its effect on surface wetting to a difluoromethylene $(—CF_2—)$ group. This might not have been predicted on the basis of the dramatic increase in γ_c when a difluoromethylene-dihydromethylene junction $(—CF_2—CH_2—)$ occurs near the surface region (44).

C. Main-Chain Composition

In addition to compositional changes in the polymer side chain and the resulting changes in wetting properties, it is of interest to consider what effect if any, can be expected from compositional changes in the polymer main chain. In studies of the wetting properties of adsorbed unbranched fluoro-carbon acid monolayers, the substrate on which the acid was adsorbed had no effect on the final wetting properties (4,28). If it were assumed, for example, in a fluoroalkyl acrylate, that the polymer backbone was analogous to a substrate on which fluoroalkyl groups were adsorbed and extended outward, no significant changes in wetting properties would be expected to arise from alteration in the make up of the polymer backbone. Obviously this is not a valid assumption since polymer chains in, for example, the random coil configuration should result in exposure, at least to a limited extent, of the polymer main chain at the air/solid interface. This problem was examined using polymers with similar side chains, fluorine content, and fluorine atom arrangement, but differing in main-chain composition (45). The three polymer-types shown in Table 4 include, in one case, a polyacrylate containing only carbon and hydrogen in the main chain (polymer I); a polyacylaziridine which contains nitrogen in addition to carbon and hydrogen in the polymer main chain (polymer II); and a polyglycidyl ether which contains carbon, hydrogen, and ether-oxygen in the main chain (polymer III).

The contact angle for hexadecane increases slightly as the percentage fluorine increases from polymers I to III; however, all of these polymer surfaces would be considered to be highly "oil-repellent." These three polymer surfaces behave quite differently when exposed to water. A high water contact angle (100°) was observed on the polyacrylate surface (polymer I). This is a typical contact angle for water on a "water-repellent" fluoro-carbon surface. Polymer II, however, displays much greater adhesional

TABLE 4

Influence of Backbone Composition on Wetting

Polymer	Percentage Fluorine	Contact angle (θ)	
		Hexadecane	Water
I. 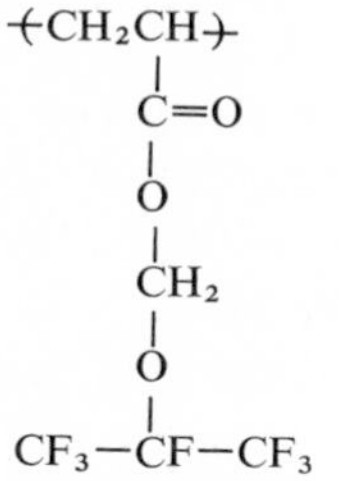 $+CH_2CH+$ $\quad\quad\mid$ $\quad\quad C=O$ $\quad\quad\mid$ $\quad\quad O$ $\quad\quad\mid$ $\quad\quad CH_2$ $\quad\quad\mid$ $\quad\quad O$ $\quad\quad\mid$ $CF_3-CF-CF_3$	46.82	54°	100°
II. 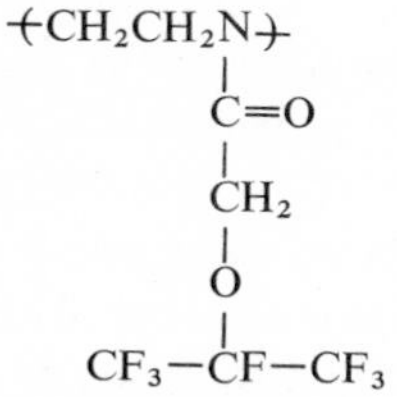 $+CH_2CH_2N+$ $\quad\quad\quad\mid$ $\quad\quad\quad C=O$ $\quad\quad\quad\mid$ $\quad\quad\quad CH_2$ $\quad\quad\quad\mid$ $\quad\quad\quad O$ $\quad\quad\quad\mid$ $CF_3-CF-CF_3$	49.40	56°	56°
III. 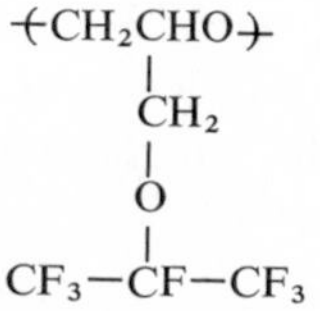 $+CH_2CHO+$ $\quad\quad\quad\mid$ $\quad\quad\quad CH_2$ $\quad\quad\quad\mid$ $\quad\quad\quad O$ $\quad\quad\quad\mid$ $CF_3-CF-CF_3$	54.96	63°	105°———→ <55°[a]

[a] Initial angle, 105°; angle decreased steadily to <55° during a 30-min period.

forces for water ($\theta = 56°$) and indeed a polymer of this type could not be utilized in applications where water repellency is desired. Polymer III, the polyglycidyl ether, displays an initial hydrophobic surface, which gradually becomes somewhat hydrophilic. The marked difference in wetting behavior of these polymers to the two different wetting liquids is a good illustration of the distinct types of liquid-solid interactions that can occur at a polymer-liquid interface.

Hexadecane is presumed to interact with the solid fluorocarbon surfaces primarily through dispersion forces (γ^d); hence, if the slight increase in fluorine content is taken into consideration, there is no significant difference in values of the contact angle for this liquid on the three polymer surfaces. However, since water can interact with the solid surface through dispersion forces and polar forces such as hydrogen bonding (i.e., $\gamma_{H_2O} = \gamma_{H_2O}^d + \gamma_{H_2O}^h$) (24), a significant difference in θ can be observed for these polymer systems. Both polymers II and III contain groups in the main chain capable of hydrogen bonding with water (amide units and ethylene oxide units), thus both display rather low water contact angles, which do not seem typical of a fluorocarbon surface. The high initial contact angle of water on polymer III followed by a slow decrease to a low angle suggests some kind of surface hydration or perhaps molecular reorientation at the polymer surface. This phenomenon has not yet been explained satisfactorily.

"Oleophobic-hydrophilic" polymers of types II and III are of great interest in textile applications where it is desirable to have an oil-(stain)-resistant surface but a surface which, during laundering, will be sufficiently wetted by water so that ground-in, oily stains will be displaced.

Several oleophobic-hydrophilic polymers have been described recently which incorporate oleophobic-hydrophobic fluorinated monomers and hydrophilic nonfluorinated monomers (46–48). For example fluorinated alkyl acrylates copolymerized with acrylic acid and block polymers containing polyethylene oxide segments and fluoroalkyl acrylate segments exhibit oleophobic-hydrophilic properties. Sherman and Smith have explained the oil-repellent, hydrophilic properties of the block polymers on the basis of what was termed a "flip-flop" surface reorientation mechanism (46). In effect this mechanism described the surface in terms of its environment. Thus in an air environment the surface molecules are oriented so as to give an oil repellent fluorochemical surface. In a water environment, however, the surface molecules presumably reorient to give the predominantly hydrophilic ethylene oxide surface.

D. Pendant-Chain Length

Hare, Shafrin, and Zisman (32) have shown that with monolayers of perfluorinated acids on high-energy surface, γ_c is affected by the length of the perfluorocarbon chain. Thus in progressing from perfluorobutyric acid to perfluorolauric acid the γ_c value decreases from about 9 to 6 dynes/cm. This effect of decreasing γ_c with increasing perfluoro chain length has been attributed to an increasing efficiency in the allineation of the perfluoroalkyl chain, so that the longer-chained fluorocarbon acids present a higher concentration of $-CF_3$ groups at the air/solid interface.

The question arises then as to whether or not the chain has to be completely fluorinated or, whether a γ_c as low as 6-9 dynes/cm could be achieved by utilizing a long-chain hydrocarbon acid terminated by a trifluoromethyl group. A study was conducted by Hare, Shafrin, and Zisman which revealed that simply employing a long-chain fatty acid with a terminal trifluoromethyl group [e.g., $CF_3(CH_2)_{16}COOH$] did not result in low γ_c values (see Fig. 5)

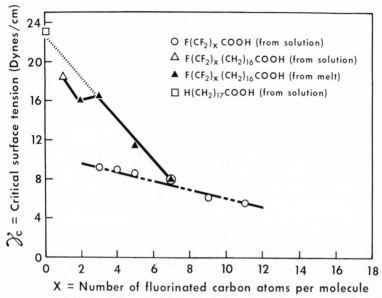

Fig. 5.   Relation of γ_c to chain length for acid monolayers (44). (Reprinted by permission of the copyright owner, The American Chemical Society.)

(44). It was found in fact that in order to achieve the low γ_c values associated with completely fluorinated acids, a fairly long fluorinated chain terminated by $-CF_3$ was required [e.g., $CF_3(CF_2)_6(CH_2)_{16}COOH$]. This behavior has been explained in terms of the uncompensated dipole that exists at the junction $-CF_2-CH_2-$. Presumably, with the shorter fluoroalkyl units, the uncompensated dipole arising at the perfluoromethylene-dihydromethylene junction can increase the dipole at the terminal $-CF_3$ and can also adversely affect chain allineation.

Pittman et al. (37,45) examined the relationship between wetting properties of fluorinated acid monolayers and fluoroalkyl acrylate polymer films. Table 5 gives a comparison of the γ_c values for fluorinated acid monolayers and fluoroalkyl acrylate polymers. In the first two cases it can be seen that the acid monolayers give lower γ_c values than the acrylates with a comparable fluorocarbon side chain. This is not too surprising since chain allineation would be expected to occur with greater ease in a monolayer than with the fluorocarbon chain covalently attached to alternate carbon atoms along the polymer backbone. It is interesting to note, however, that one of the polyacrylates reported in which the fluoroalkyl group is terminated by a $-CF_2H$, gave a slightly lower value than that reported for an ω-hydroperfluoroundecenoic acid monolayer. The low γ_c values for the $-CF_2H$-terminated acrylates is believed to be due to the crystallinity associated with these polymers. This

TABLE 5

Wetting Properties of Acid Monolayers versus Comparable
Acrylate Films

No. C atoms	Terminal group	γ_c (dynes/cm)	
		Monolayer acid	Acrylate film
4	$-CF_3$	9.2	15.5
8	$-CF_3$	7.9	10.3
9	$-CF_2H$	—	13
11	$-CF_2H$	15	14.5

point will be discussed in a subsequent section; however it should be pointed
out that all of the other polymeric materials containing bulky fluoroalkyl
side chains mentioned previously were noncrystalline, amorphous polymers.

Another aspect of chain length in amorphous polyacrylates has been
examined and is illustrated in Fig. 6. In this study the fluoroalkyl portion was
kept constant while the length of the alkylene chain was increased. A plot of

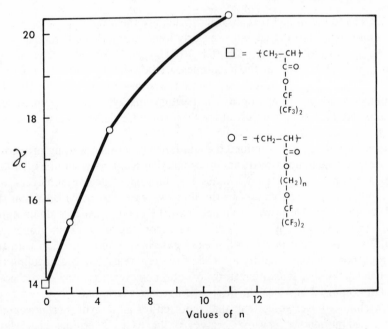

Fig. 6.   Effect of chain length on γ_c for polyacrylates containing a heptafluoroisopropyl
group.

γ_c versus alkylene chain length gave a fairly regular increase in γ_c in proceeding from the parent polymer, poly(heptafluoroisopropyl acrylate) (14.1 dynes/cm) to a polymer containing eleven $-CH_2-$ units interspersed between the perfluoroisopropoxy group and the main chain. The critical surface tension increased by about 0.7 dyne/cm per methylene group from the parent polymer up to 5 $-CH_2-$ units and then by about 0.4 dyne per methylene unit up to 11 methylene groups. Lengthening the nonfluorocarbon portion of the side chain did not result in improved alignment of the fluorocarbon segment since there was a continual increase in γ_c as the alkylene chain length was increased. Increased wettability with increasing alkylene chain length in this case appears to be due to an overall increase in hydrocarbon content (or decrease in fluorine content) per monomer unit (37).

IV. EFFECT OF POLYMER CRYSTALLINITY ON WETTABILITY

Since the nature and packing of groups or atoms present at the air/solid interface of a polymeric solid determines the wetting properties of the polymer, we might logically expect a highly crystalline polymer to display different wetting properties from its amorphous counterpart.

The influence of crystallinity (or chain packing) on γ_c was first demonstrated by Fox and Zisman in a comparison of the γ_c values for a monolayer of octadecylamine and a crystal of hexatriacontane. The lower γ_c associated with the crystal (20–22 dynes/cm compared to 22–24 dynes/cm) was attributed to a tighter packing of the alkyl chains with a concomitant greater concentration of $-CH_3$ groups at the air/solid interface (49).

Until recently, studies on the wetting properties of polymers had not been concerned with the crystalline or amorphous nature of the polymers. Schonhorn and Ryan (19) examined the wettability of a compressed aggregate of polyethylene crystals and found a γ_c value of 53.6 dynes/cm, which is considerably higher than that for ordinary melt-crystallized polyethylene (31 dynes/cm). Schonhorn has also reported γ_c values for nylon 6–6, polychlorotrifluoroethylene, and isotactic polypropylene which are considerably higher than values previously reported (50). These higher values were obtained after a slow melt-crystallization of the polymers against a thin gold film. Subsequent dissolution of the film reportedly left a surface region of high crystallinity.

This work has raised a question as to the effect of increasing or decreasing the surface density of particularly groups on surface wettability. The decrease in γ_c previously found in comparing single crystals of n-hexatriacontane to an adsorbed monolayer of hexadecylamine suggests that increased packing or density of $-CH_3$ groups at the air/solid interface causes a lowering of surface wettability (49). However, the increase in γ_c found when comparing a polyethylene single crystal mat to ordinary bulk polyethylene (19) suggests the opposite effect for the methylene $(-CH_2-)$ group.

Additional studies on the effect of polymer crystallinity on polymer wettability have been reported (45) in which fluoroalkyl acrylates were used containing the alkyl group $HCF_2(CF_2)_nCH_2-$ (where $n = 1, 3, 5, 7, 9$). In order to simplify identification of these acrylate polymers, they will be referred to in terms of the number of carbon atoms present in the alkyl side chain. For example, the acrylate polymer derived from the monomer $HCF_2(CF_2)_9CH_2$ $OOCCH{=}CH_2$ will be referred to as the C-11 acrylate (11 carbon atoms in the alkyl side chain).

The C-9 and C-11 acrylates form polymers which are partially crystalline at room temperature; however, the shorter-chained polyalkyl acrylates did not display crystallinity at room temperature on examination by X-ray diffraction analysis. Since these polymers were prepared under conditions in which stereoregularity would not be expected to develop, it is assumed that the crystallinity in the C-9 and C-11 polyacrylates arises from alignment and packing of the fluoroalkyl side chains. Stereoregular polymers containing long unbranched side chains (e.g., polyoctadecene-1) have been reported in which crystallization can involve both the side chain and the main chain (51,52).

A critical surface tension plot for these fluoroalkyl acrylates is presented in Fig. 7. The curvature in the γ_c plots for the noncrystalline C-3, C-5, and C-7 polymers is unusual for hydrocarbon test liquids. Film solubility in the

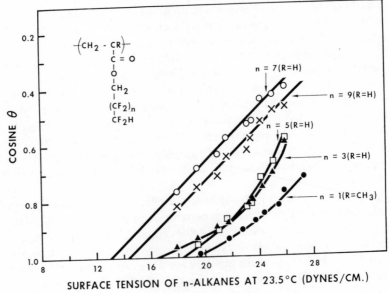

Fig. 7.   CST plot of ω-hydrofluoroalkyl acrylates.

hydrocarbon test liquids may be suspected when curvature appears; however, this does not seem to be the case here since preconditioning of the test liquids with polymer did not affect the results. The cause of this curvature is still unclear.

Mention was previously made (Table 5) that the crystalline C-9 and C-11 acrylates containing the terminal $-CF_2H$ group gave γ_c values as low or lower than the value reported for a long-chain, $-CF_2H-$ terminated fluorocarbon acid. A comparative γ_c plot is given in Fig. 8 for the C-9 and C-11 polyacrylates and an adsorbed monolayer of ω-hydroperfluoroundecenoic acid. The low γ_c values displayed by the C-9 and C-11 acrylates is believed to be due to polymer crystallinity, which gives improved packing of the fluoroalkyl side chains. In the C-9 polyacrylate, there appears to be more efficient packing than in the acid monolayer or the C-11 polyacrylate, and a higher concentration of $-CF_2H$ occurs at the air/solid interface. This phenomenon is analogous to Zisman's observation that a lower γ_c can be obtained with a crystal of hexatriacontane than a monolayer of octadecylamine. These results imply that it is possible to obtain crystalline fluorocarbon polymers with critical surface tensions less than the 6 dynes/cm reported for a long-chain CF_3-terminated acid monolayer.

The influence of crystallinity on wettability was further demonstrated by an examination of the contact angles for hexadecane and heptane on polymer films which had been heated above the crystalline melting point and then rapidly quenched. The polymer films were then subsequently re-examined after annealing below the melting point to induce crystallization. Quenched films for the C-9 acrylate polymer gave contact angles of 50 and 17° for

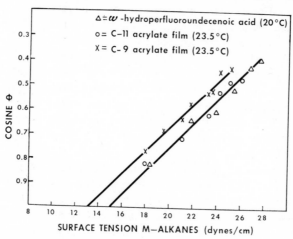

Fig. 8. Comparative γ_c plots for C-11 acid monolayer, C-11 acrylate polymer, and C-9 acrylate polymer.

hexadecane and heptane, respectively. It is interesting to note that these angles are quite close to the angles given for a $-CF_2-$ surface of polytetra-fluoroethylene (hexadecane 46°, heptane 21°). Annealing the quenched polymer film resulted in increased contact angles for both hexadecane and heptane (61 and 45°, respectively). Thermal analysis of the polymer indicated that quenching could, in some cases, completely eliminate the crystalline melting endotherm of the C-9 polymer, whereas the annealing process caused an increase in overall crystallinity and a sharpening of the melting endotherm.

The findings of Zisman (49) and the results on the $-CF_2H-$-terminated acrylates, as opposed to the findings of Schonhorn and Ryan (19), at first seem contradictory. That is, γ_c has been found to *decrease* with substances presenting either CH_3- or HCF_2- at the air/solid interface as a closer packed more crystalline arrangement is achieved; however, Schonhorn and Ryan found that γ_c *increases* as the surface crystallinity of polyethylene is increased.

With a methyl (CH_3-) surface, an increase in the number of methyl groups at the air/solid interface should result in a lower γ_c as a result of the decrease in $-CH_2-$ exposure. A surface comprised of methylene groups is known to have a higher γ_c than one made up of methyl groups. The same argument can apply to perfluoroalkyl chains terminated by $-CF_2H$. That is, as the concentration of $-CF_2H$ increases at the air/solid interface as a result of side-chain packing during crystallization, an attendant decrease in the number of $-CF_2-$ groups will occur, which should cause an overall decrease in γ_c. In both instances the substitution of $-CH_3$ for $-CH_2-$ or of $-CF_2H$ for $-CF_2-$ at the air/solid interface should result in fewer interacting groups per unit area which could account for the reduction in γ_c (34).

In the case of polymers comprised only of secondary carbon atoms, such as polyethylene, different considerations are involved. We might expect an increase in the surface energy (hence also γ_c) with increasing surface packing in polyethylene where there is only the possibility of methylene groups occurring at the air/polymer interface. Although a higher γ_c was found for a crystalline mat of polyethylene than for bulk-crystallized polyethylene, it is not clear whether the increase in γ_c is in this case the result of increased surface packing of the methylene groups. It seems that the alternate possibility exists; that of decreased methylene packing at the surface of a crystalline polyethylene mat.

It is well established that the lamellae surface of polymer single crystals consists of some type of folded chain in which the polymer chains exit from the crystal lattice, fold, and re-enter the crystal lattice (53). The exact nature of the fold surface is controversial and numerous models have been proposed (54–57).

However, regardless of the model used to describe the exit and re-entry of the polymer chains into the crystal lattice, there is considerable evidence indicating that there is an amorphous component associated with the surface region which amounts to *ca.* 10–20% of the total weight (58,59). The increased internal energy associated with sharp folds may be sufficient to account for the increased wettability (higher γ_c) found for the polyethylene single-crystal aggregate. In any case, a liquid in contact with a polyethylene single-crystal aggregate would have minimum exposure to a close-packed array of minimum energy planar zig-zag methylene groups. Instead, the wetting liquid would presumably be exposed to a disordered, loosely packed methylene surface and/or a sharp folded methylene surface in a higher energy state than found in the planar zig-zag conformation. In order to assess the wetting behavior of close-packed methylene groups in a state of low internal energy, it will be necessary to examine a surface composed of methylene groups in the extended-chain planar zig-zag conformation. This may be possible since, in certain cases, extended-chain crystals reportedly occur in polyethylene after special melt crystallization techniques (60,61) and special techniques involving simultaneous polymerization and crystallization (62).

In spite of the uncertainties discussed here, it has been clearly demonstrated that the wetting properties of crystalline polymeric surfaces can vary considerably depending on a number of factors which determine the nature of the polymer/air interface.

The effect of possible changes in wetting properties arising from alterations in stereoregular side-chain ordering evidently has not yet been studied.

V. EFFECT OF COPOLYMERIZATION ON THE CRITICAL SURFACE TENSION

The fluorinated polymers available for application to textiles and other substrates generally consist of copolymers of a fluorine-containing monomer and one or more nonfluorinated monomer(s). Nonfluorinated monomer(s) may be introduced partly for the alteration of polymer bulk properties but are added primarily to reduce the overall cost of the polymer (63). In this regard, it would be helpful to be able to predict γ_c values for polymers with a wide variety of compositions.

Bernett and Zisman found that this could be done with reasonable accuracy with copolymers of hexafluoropropylene and tetrafluoroethylene (64). They found that a graph of contact angle versus mole % hexafluoropropylene in copolymers of hexafluoropropylene and tetrafluoroethylene gave a straight line up to 23 mole % hexafluoropropylene. Extrapolation to 100% hexafluoropropylene gave a predicted γ_c value of 15.5 dynes for pure poly(hexafluoropropylene), which was in reasonably good agreement with the value later found for the homopolymer (65).

Lee (66) proposed an empirical equation for the calculation of the critical surface tension for a multicomponent polymer system as follows:

$$\gamma_c \begin{pmatrix} \text{copolymers or} \\ \text{a mixture} \end{pmatrix} \begin{array}{l} = N_1\,\gamma_{c_1} + N_2\,\gamma_{c_1} + N_3\,\gamma_{c_3} + \cdots \\ \\ = \sum_1^\infty N_i\,\gamma_{ci} \end{array}$$

where N_i is the mole fraction of the individual monomer in the copolymer, γ_{c_i} is the critical surface tension of each homopolymer. A number of copolymers are given in Table 6 with the values of γ_c calculated using this equation compared with the experimentally determined values of γ_c. Although this equation has not been tested sufficiently to determine its general validity, it can be seen that in the random cases given in Table 6 there is reasonably good agreement between the calculated and experimentally determined values. A test of this equation with a three or more component polymer system would be of particular interest.

VI. MODIFICATION OF THE SURFACE PROPERTIES OF NONFLUORINATED POLYMERS WITH FLUORINE-CONTAINING ADDITIVES

Jarvis, Fox and Zisman (67) found that it is possible to produce fluorocarbonlike low-surface-energy polymers by the addition of small amounts of selected fluorocarbon additives to such nonfluorinated polymers as

TABLE 6

Calculation of γ_c From Lieng-Huange Lee Empirical Equation (30)

Copolymer	γ_c		Ref.
	Calcd.	Found	
Poly(tetrafluoroethylene-co-chlorotrifluoroethylene)			
80:20	21	20	77
60:40	23.5	24	77
Poly(tetrafluoroethylene-co-ethylene)			
50:50	24.8	26–27	77
Poly(tetrafluoroethylene-co-hexafluoropropylene)			
84:16	18.1–18.3[a]	18	64
77:23	18.0–18.2[a]	17.8	64
Poly(styrene-co-2,2,3,3-tetrafluoropropyl methacrylate)			
20:80	21.8	20	79
50:50	26	25	79
80:20	30.2	28	79

[a] Calculated using 16.2 and 17.1 dynes/cm as the γ_c for poly(hexafluoropropylene) (65).

poly(vinylidene chloride), poly(methyl methacrylate), and polyacrylamide. Generally the amount of additive varied from approximately 0.2–1.0% and the method of addition to polymer involved either addition to monomer before polymerization or addition to a polymer solution with subsequent solvent evaporation. The structure of the additives was found to be of marked importance and a proper "organophobic-organophilic balance" was required to effect a lowering of the polymer γ_c. Dramatic changes in the wetting properties and frictional properties were demonstrated. For example, addition of 0.6% of the inner salt of N,N,N-dimethyl-3-(n-perfluoroheptan-carboxamido)propyl-3-aminopropionic acid to polyacrylamide decreased the γ_c from 35–40 to about 10.4 dynes/cm. They also showed that a poly(tetra-fluoroethylene)like surface could be obtained by the addition of 0.5% of tris(1H,1H,-pentadecafluorooctyl) tricarballylate to poly(methylmethacry-late); i.e., the polymer γ_c was lowered from 40 to 19 dynes/cm. Bowers, Jarvis, and Zisman (68) reported the reduction of the coefficient of friction of poly(methyl methacrylate) and poly(vinylidene chloride) to 0.1 or less by the incorporation of *ca*. 1% of a fluorocarbon additive. In addition to the decreased wettability and the reduction in the coefficient of friction, the surface effects produced by the use of a fluorocarbon additive are said to be self-healing. That is, on scraping or abrading away the surface layer of a treated polymer, the wetting properties became the same as those of the pure, untreated polymer. However, with time or the application of heat to speed the diffusion process, the additive again concentrates at the air/solid interface giving a fluorocarbonlike surface.

This approach to lowering the surface energy of polymers by the addition of selected fluorocarbon additives appears economically attractive and will undoubtedly be useful in many applications where decreased wettability, adhesive, and frictional properties are of importance.

VII. ADHESIVE, ABHESIVE, AND FRICTIONAL PROPERTIES OF FLUORINE-CONTAINING POLYMERS

The subject of adhesion has been given extensive treatment in a number of publications (69–71) and no attempt will be made here to review the theories advanced to account for the adhesion phenomenon. A review of the various adhesion theories has recently been given by Huntsberger (72).

It is generally agreed that London dispersion forces are sufficient for obtaining strong adhesion provided intimate intermolecular contact is achieved between adhesive and adherend. In order for intermolecular contact to be achieved and for dispersion forces to be operative, wetting of the substrate by the adhesive must occur. Because of their low surface energy and low wettability, fluorine-containing polymers are of special interest in regard to problems dealing with adhesives.

The role of fluorine-containing polymers in adhesion problems should be considered both with respect to their ability to serve as an adhesive and in regard to the difficulty of obtaining good adhesion to fluorinated polymers, i.e., the abhesive character of fluorinated polymers. Sharpe and Schonhorn (73) suggested that highly fluorinated polymers could make useful adhesives because of the low γ_c of these polymers and the ability of these polymers to spread over other polymers with a higher γ_c. However, Jarvis and Zisman pointed out a possible limitation in the utility of a fluoropolymer as an adhesive on an adherend other than another fluoropolymer (74). The cohesive strength (W_c) of a liquid adhesive is also decreased by lowering the surface tension since $W_c = 2\gamma_{LV}$. Because of this, the application of a fluorinated adhesive to a nonfluorinated substrate would probably produce a low-strength bond, subject to cohesive failure within the adhesive.

In contrast to what appears to be limited utility for fluoropolymers as adhesives, they are used extensively as abhesives or antistick coatings. Applications of this sort involve a fluoropolymer film or coating over another solid surface to prevent or reduce adhesion when another material is brought into contact with the solid. The action of the fluoropolymer film is mainly that of reducing the surface energy at the air/solid interface. Thus when liquids are placed on the fluoropolymer-coated surface less wetting occurs, hence the adhesional forces will be smaller. Zisman has pointed out that the easy parting action derived from abhesive films cannot be solely accounted for on the basis of the reduced γ_c (71). In such applications as polymer injection molding, the surface roughness of the mold, viscosity, and rate of solidification of the material to be molded will also greatly affect the ultimate parting action. In any event, it is clear that the lower the γ_c of the abhesive film, the better it will be as a release agent. Copolymers of tetrafluoroethylene and hexafluoropropylene and homopolymers of tetrafluoroethylene are presently used as mold release agents and presumably other fluorine-containing polymers with even lower γ_c values will be employed as they become available.

The friction observed when one solid object is slid over a solid polymer surface is also interrelated to the adhesion and wetting properties of the polymer. Friction is commonly expressed in terms of the coefficient of friction (μ) and results from adhesion between molecules of the two sliding surfaces. The coefficient of friction is a dimensionless term defined by Amontons' law as $F = \mu W$, where F is the force of friction and W is the normal load. Such polymer bulk properties as the elastic modulus and shear strength influence the coefficient of friction; however, it has also been demonstrated that lowering the polymer γ_c generally also lowers μ. Table 7 gives the coefficients of kinetic friction (μ_k, the coefficient of friction measured after sliding commences) for a number of common plastics where there is a gradual increase in μ_k as γ_c increases. These values of μ_k were taken from a tabulation by

TABLE 7

Increase of Kinetic Coefficient of Friction with Increasing γ_c

Polymer	γ_c	μ_k
Poly(tetrafluoroethylene)	18.5	0.05
Poly(vinylidene fluoride)	25	0.25
Polyethylene (low density)	31	0.26
Poly(chlorotrifluoroethylene)	31	0.33
Poly(vinyl chloride)	39	0.40
Poly(vinylidene chloride)	40	0.45

[a] Steel on polymer (28).

Bowers and Zisman in which the measurements were conducted under identical conditions.

As mentioned earlier, Bowers, Jarvis and Zisman (73) demonstrated that it is possible to lower polymer wettability and boundary friction by the addition of certain fluorinated additives to a nonfluorinated polymer. The changes in surface properties that were reported were brought about without significant changes in polymer hardness.

VIII. SUMMARY

From the preceding discussion we can see that γ_c is an extremely useful measure of polymer wettability and is closely related to polymer adhesion and polymer friction properties.

In considering what is known about the effects of structure and composition on the γ_c of polymers, many unanswered questions appear. For example, it would seem desirable to investigate changes in γ_c with changes in side-chain ordering brought about by stereoregular polymerization. It would also be of interest to examine the surface behavior of polymers containing crystalline fluoroalkyl groups terminated by $-CF_3$. Although the overall effects on γ_c with changes in constitution have been established, there remain questions in regard to the effects of side-chain branching and the placement of atoms (other than fluorine) on a fluorinated side chain in positions other than at the terminal chain end.

Further work appears desirable on the nature of the surface forces present in "oleophobic-hydrophilic" polymers. These polymers, which are prepared by proper introduction of fluorocarbon and hydrogen-bonding groups, are quite interesting from both a practical and a theoretical standpoint. We can imagine a number of unique uses for materials of this type, e.g., adhesives with selective bonding properties.

Table 8 presents a compilation of γ_c values for fluorine-containing polymers. The availability of fluoropolymers having similarities in structural features but small changes in γ_c should be useful in future studies related to wetting, spreading, and adhesion.

TABLE 8

Critical Surface Tension of Fluorinated Polymers in Increasing Order of γ_c
(dynes/cm)

Polymer[a]	γ_c	Ref.
$CF_3(CF_2)_6CH_2-A$	10.4	37
$CF_3(CF_2)_6CH_2-M$	10.6	80
$CF_3(CF_2)_7SO_2N(CH_2CH_2CH_3)CH_2CH_2-A$	11.1	80
$HCF_2(CF_2)_7CH_2-A$	13.0	45
$(CF_3)_2CF-A$	14.1	37
Poly(heptafluoroisopropyl vinyl ether)	14.2	40
$HCF_2(CF_2)_9CH_2-A$	14.5–15.0	45
$(CF_3)_2CH-M$	14.8–15.4	41
Poly(3-heptafluoroisopropoxypropyl trichlorosilane)	14.9–15.9	41
$(CF_3)_2CH-A$	15.0–15.4	41
$CF_3(CF_2)_2CH_2-A$	15.2	37
$(CF_3)_2CFO(CH_2)_2-A$	15.5	37
Poly(vinyl heptafluoroisopropoxy acetate)	16.1	41
Poly(hexafluoropropylene	16.2–17.1	65
$HCF_2(CF_2)_3CH_2-A$	17.0	45
$(CF_3)_2CFO(CH_2)_5-A$	17.7	37
Poly(tetrafluoroethylene)	18.5	16
$HCF_2(CF_2)_5CH_2-A$	18.5	45
$(CF_3)(CF_2Cl)CF-A$	18.9	37
$HCF_2CF_2CH_2-A$	19.0	79
$(CF_3)(CF_2Cl)CF-M$	19.1	37
$(CF_2Cl)_2CF-A$	20.3	37
$(CF_3)_2CFO(CH_2)_{11}-A$	20.3	37
Poly(trifluoroethylene)	22	35
Poly(vinylidene fluoride)	25	35
Poly(vinyl fluoride)	28	35
Poly(chlorotrifluoroethylene)	31	77

[a] The symbols A and M refer to the acrylate and methacrylate repeat units respectively.

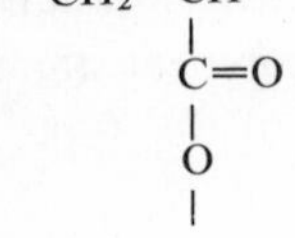

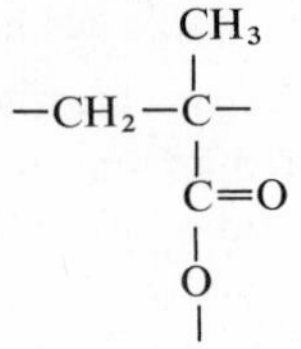

$$
\begin{array}{cc}
-CH_2-CH- & \quad\quad CH_3 \\
\qquad | & \qquad\quad | \\
\qquad C=O & \quad -CH_2-C- \\
\qquad | & \qquad\quad | \\
\qquad O & \qquad\quad C=O \\
\qquad | & \qquad\quad | \\
& \qquad\quad O \\
& \qquad\quad |
\end{array}
$$

Reference to a company or product name does not imply approval or recommendation of the product by the U.S. Department of Agriculture to the exclusion of others that may be suitable.

References

1. H. A. Brown, U.S. Patent 2,995,542 (1961).
2. E. J. Grajeck and W. H. Petersen, *Textile Res. J.*, **32**: 320 (1962).
3. M. K. Bernett and W. A. Zisman, "Prevention of Liquid Spreading or Creeping," in *Advances in Chemistry*, No. 43, Am. Chem. Soc., Washington, D.C., 1964, p. 332.
4. W. A. Zisman, "Relation of Equilibrium Contact Angle to Liquid and Solid Constitution," in *Advances in Chemistry*, No. 43, Am. Chem. Soc., Washington, D.C., 1964, p. 1.
5. H. W. Fox and W. A. Zisman, *J. Colloid Sci.*, **1**: 513 (1946).
6. E. Kneen and W. W. Benton, *J. Phys. Chem.*, **41**: 1195 (1937).
7. T. Fort, Jr., and H. T. Patterson, *J. Colloid Sci.*, **18**: 217 (1963).
8. F. M. Fowkes and W. D. Harkins, *J. Am. Chem. Soc.*, **62**: 3377 (1940).
9. J. Guastalla, *J. Colloid Sci.*, **11**: 623 (1956).
10. T. Young, *Phil. Trans. Roy. Soc. (London)*, **95**: 65 (1805).
11. J. C. Melrose, "Evidence for Solid-Fluid Interfacial Tensions from Contact Angles," in *Advances in Chemistry*, No. 43, Am. Chem. Soc., Washington, D.C., 1964, p. 158.
12. J. J. Bikerman, *Kolloid-Z. und Z. Polymere*, **218**: 52 (1967).
13. R. E. Johnson, Jr., and R. H. Dettre, *J. Colloid Sci.*, **21**: 610 (1966).
14. N. K. Adam, "The Chemical Structure of Solid Surfaces as Deduced from Contact Angles," in *Advances in Chemistry*, No. 43, Am. Chem. Soc., Washington, D.C., 1964, p. 52.
15. A. W. Adamson and I. Ling, "The Status of Contact Angle as a Thermodynamic Property," in *Advances in Chemistry*, No. 43, Am. Chem. Soc., Washington, D.C., 1964, p. 57.
16. H. W. Fox and W. A. Zisman, *J. Colloid Sci.*, **5**: 514 (1950).
17. J. L. Gardon, *J. Phys. Chem.*, **67**: 1935 (1963).
18. L. H. Lee, *J. Appl. Polymer Sci.*, **12**: 719 (1968).
19. H. Schonhorn and F. W. Ryan, *J. Phys. Chem.*, **70**: 3811 (1966).
20. V. R. Gray, *New Scientist*, **19**: 143 (1963).
21. H. Schonhorn, *J. Phys. Chem.*, **69**: 1084 (1965).
22. E. Wolfram, *Kolloid Z.*, **75**: 182 (1962).
23. H. Schonhorn and L. H. Sharpe, *J. Polymer Sci. B*, **3**: 235 (1965).
24. F. M. Fowkes, *J. Phys. Chem.*, **67**: 2538 (1963).
25. F. M. Fowkes, *Ind. Eng. Chem.*, **56**(12): 40 (1964).
26. F. M. Fowkes, "Surface Chemistry," in *Treatise on Adhesion and Adhesives*, Vol. I, Marcel Dekker, New York, 1967, p. 325.
27. E. G. Shafrin, "Critical Surface Tension of Polymers," in *Polymer Handbook*, J. Brandrup and E. H. Immergut, Eds., Interscience, New York, 1966, Chapter III, p. 113.
28. W. A. Zisman, *Record of Chemical Progress*, **26**(1): 13 (1965).
29. E. G. Shafrin and W. A. Zisman, *J. Phys. Chem.*, **64**: 519 (1960).
30. I. Langmuir, *J. Am. Chem. Soc.*, **38**: 2286 (1916).
31. I. Langmuir, "Third Colloid Symposium Monograph," Chem. Catalog Co., New York, 1925.
32. E. F. Hare, E. G. Shafrin, and W. A. Zisman, *J. Colloid Sci.*, **58**: 236 (1954).
33. A. H. Ellison, H. W. Fox and W. A. Zisman, *J. Phys. Chem.* **57**: 622 (1953).

34. F. M. Fowkes, *J. Colloid Sci.*, **28**:493 (1968); R. E. Johnson, Jr. and R. H. Dettre, "Wettability and Contact Angles," in Surface and Colloid Science, Vol. 2, E. Matijevic, Ed., Wiley-Interscience, New York, 1969, p. 85.
35. A. H. Ellison and W. A. Zisman, *J. Phys. Chem.*, **58**:260 (1954).
36. A. G. Pittman, D. L. Sharp, and R. E. Lundin, *J. Polymer Sci. A*1, **4**:2637 (1966).
37. A. G. Pittman, D. L. Sharp, and B. A. Ludwig, *J. Polymer Sci.*, **6**:1729 (1968).
38. A. G. Pittman and D. L. Sharp, *J. Polymer Sci. B*, **3**:379 (1965).
39. A. G. Pittman and D. L. Sharp, *J. Polymer Sci. B*, **4**:159 (1966).
40. A. G. Pittman, D. L. Sharp, and B. A. Ludwig, *J. Polymer Sci. A*1, **6**:1741 (1968).
41. A. G. Pittman and W. L. Wasley, unpublished.
42. A. G. Pittman and W. L. Wasley, *Am. Dyestuff Rep.*, **56**(12): 808 (1967).
43. M. K. Bernett and W. A. Zisman, *J. Phys. Chem.*, **71**:2075 (1967).
44. E. G. Shafrin and W. A. Zisman, *J. Phys. Chem.*, **66**:740 (1962).
45. A. G. Pittman, "Interrelationships Between Polymer Structure and Surface Wettability," presented at the Gordon Conference on Textiles, July 1968, New London, New Hampshire.
46. P. O. Sherman, S. Smith and B. Johannessen, *Textile Res. J.*, **39**:449 (1969).
47. A. G. Pittman, J. N. Roitman and D. Sharp, *Textile Chemist and Colorist*, **3**:175 (1971).
48. W. L. Wasley and A. G. Pittman, Factors Affecting the Relative Surface Energy of Polymer Coated Fibers," American Chemical Society, April 1969, Minneapolis, Minn.
49. H. W. Fox and W. A. Zisman, *J. Colloid Sci.*, **7**:428 (1952).
50. H. J. Schonhorn, *J. Polymer Sci. B*, **5**:919 (1967).
51. A. T. Jones, *Makromol Chem.*, **71**:1 (1964).
52. D. W. Aubrey and A. Barnatt, *J. Polymer Sci. A*2, **6**:241 (1968).
53. P. H. Geil, "Polymer Single Crystals," Interscience, New York, 1963; D. H. Reneker and R. H. Geil, *J. Appl. Phys.* **31**(11):1916 (1960).
54. A. Keller, *Polymer*, **3**:393 (1962).
55. P. J. Flory, *J. Am. Chem. Soc.*, **84**:2857 (1962).
56. E. W. Fisher and R. Lorenz, *Kolloid Z.*, **189**:97 (1963).
57. D. J. Blundell, A. Keller, and T. Connor, *J. Polymer Sci. A*2, **5**:991 (1967).
58. P. Ingram and A. Peterlin, "Morphology," "Encyclopedia of Polymer Science and Technology," Vol. 9, Interscience, New York, 1968, p. 204.
59. A. Keller, *Kolloid Z.*, forthcoming paper presented at Conference on Crystallization of Polymers, Garmisch-Partenkirchen, Germany, September 27–29, 1967.
60. P. H. Geil, F. R. Anderson, B. Wunderlich, and T. Arakawa, *J. Polymer Sci. A*, **2**:3707 (1964).
61. F. R. Anderson, *J. Appl. Phys.*, **35**:64 (1964).
62. H. Chanzy, A. Day, and R. H. Marchessault, *Polymer*, **8**:567 (1967).
63. R. E. Johnson, Jr., and S. Raynolds, U.S. Patent 3,256,230 (1966).
64. M. K. Bernett and W. A. Zisman, *J. Phys. Chem.*, **64**:1292 (1960).
65. M. K. Bernett and W. A. Zisman, *J. Phys. Chem.*, **65**:2266 (1961).
66. L. H. Lee, *J. Polymer Sci. A*2, **5**:1103 (1967).
67. N. L. Jarvis, R. B. Fox, and W. A. Zisman, "Surface Activity at Organic Liquid Interfaces," in *Advances in Chemistry*, No. 43, Am. Chem. Soc., Washington, D.C., 1964, p. 317.
68. R. C. Bowers, N. L. Jarvis, and W. A. Zisman, *Ind. Eng. Chem. Prod. Res. Develop.*, **4**:86 (1965).

69. R. L. Patrick, Ed., "Treatise on Adhesion and Adhesives," Vol. I, Marcel Dekker, New York, 1967.
70. P. Weiss, Ed., "Adhesion and Cohesion," Elsevier, New York, 1962.
71. W. A. Zisman, *Ind. Eng. Chem.*, **55**(10):19 (1963).
72. J. R. Huntsberger, "The Mechanisms of Adhesion," in "Treatise on Adhesion and Adhesives," Vol. I, Marcel Dekker, New York, 1967, p. 119.
73. L. H. Sharpe and H. Schonhorn, "Surface Energetics, Adhesion and Adhesive Joints," in *Advances in Chemistry*, No. 43, Am. Chem. Soc., Washington, D.C., 1964, p. 189.
74. N. L. Jarvis and W. A. Zisman, "Surface Chemistry of Fluorochemicals," in "Encyclopedia of Chemical Technology," Vol. 9, Interscience, New York, 1966, p. 707.
75. A. H. Ellison and W. A. Zisman, *J. Phys. Chem.*, **58**:503 (1954).
76. H. D. Feldtman and J. R. McPhee, *Tex. Res. J.*, **34**:634 (1964).
77. H. W. Fox and W. A. Zisman, *J. Colloid Sci.*,**7**:109 (1952).
78. E. G. Shafrin and W. A. Zisman, *J. Colloid Sci.*, **7**:166 (1952).
79. K. Tamaribuchi, *Am. Chem. Soc. Polymer Preprints*, **8**(1):631 (1967).
80. M. K. Bernett and W. A. Zisman, *J. Phys. Chem.*, **66**:1207 (1962).

14. CONFIGURATIONAL CHARACTERISTICS OF PERFLUOROALKANES AND POLYTETRAFLUOROETHYLENE

Terence W. Bates*, *Department of Chemistry, Dartmouth College, Hanover, New Hampshire.*

Contents

I. INTRODUCTION

Many of the important physical properties of polymers depend on the configurational characteristics or spatial form of the individual polymer chains. For example, chain configuration strongly influences the performance of a polymer as a rubber, fiber, film, or plastic. The useful physical properties of polytetrafluoroethylene, PTFE, enhance the importance of a detailed study of the configurational characteristics of this particular linear chain molecule.

Crystalline polymers are characterized by chains in which, over long stretches, the backbone or skeletal atoms are held in a rigid configuration. On dissolving or melting the polymer, however, the restraints imposed on the configuration by the crystal lattice are removed; the chain then adopts an irregular shape and becomes flexible.

* Present Address: Shell Research Limited, Carrington Plastics Laboratory, Manchester, England.

The flexibility of the chain is a consequence of the ability of each main-chain bond i to rotate through a variable rotation angle ϕ_i (Fig. 1). Bond angles θ and lengths l are fixed for a given chain and do not contribute to chain flexibility. In general, rotation about single bonds is not free but is hindered by a bond rotation potential. Such potentials, examples of which are shown in Fig. 5, consist of a number of minima of unequal depths. The bonds tend to adopt conformations close to the rotation angles locating these minima. For example, the bond rotation potential for polyethylene (1–8), shown in Fig. 5b, contains three minima corresponding to the *trans* (t) conformation (for which ϕ is arbitrarily taken as zero) and the two equivalent *gauche* conformations, g^+ ($\phi \cong 120°$) and g^- ($\phi \cong -120°$). The t and $g^\pm$ conformations are shown in Fig. 2.

A given configuration of the chain is determined when the rotation angles ϕ_i about each skeletal bond are specified. Each configuration occurs with a specific statistical weight and spatial form and may be characterized by the distance r between the ends of the chain. Light-scattering and viscosity measurements on dilute polymer solutions at the so-called theta temperature allow the experimental evaluation of the end-to-end distance (2). Such measurements yield the squared magnitude $\langle r^2 \rangle_0$ of the unperturbed end-to-end distance averaged over all possible configurations of the chain. Mathematical methods have been developed for computing $\langle r^2 \rangle_0$ in terms of the rotation angles ϕ_i and energies $E_i(\phi)$ of the individual bond conformations (3, 4, 9–14). Comparison of the calculated and experimental values of $\langle r^2 \rangle_0$ allows the conformational energies to be evaluated. The configurational characteristics of the chain are then known.

Experimental methods for determining $\langle r^2 \rangle_0$ are applicable only to very long chains in dilute solution (2). The insolubility of PTFE means that its

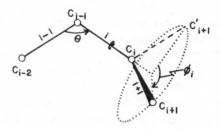

Fig. 1.   The rotation angle ϕ_i for a carbon skeleton with fixed valence angles θ. Rotation about bond i allows carbon atom C_{i+1} to occupy any position on the dotted circle. The angle ϕ_i defines the position of C_{i+1} and is the angle between the planes defined by successive bond pairs $i-1$, i and i, $i+1$. The *trans* conformation, for which $\phi_i = 0$, places atom C_{i+1} in the position denoted by C'_{i+1}; i.e., in the plane defined by bonds $i-1$ and i.

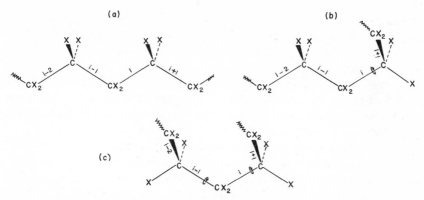

Fig. 2. Schematic representation of conformations in a $+CX_2+_n$ chain where $X = H$ or F. Solid, broken, and filled lines denote bonds in, below, and above plane of paper respectively. (a) Planar *trans* form $t_{i-1}t_i$. (b) *Gauche* conformation of bond i, $t_{i-1}g_i^+$. (c) Successive *gauche* conformations of opposite sign, $g_i^- {}_{-1}g_i^+$, which result in severe steric repulsions between CX_2 groups separated by four intervening C—C bonds.

unperturbed dimensions cannot be evaluated directly. However, the short-chain α,ω-dihydroperfluoroalkanes, $H(CF_2)_nH$, are soluble in nonpolar solvents such as benzene. The experimental dipole moments of these compounds are determined by the configurational characteristics of the CF_2-CF_2 bonds separating the terminal CF_2-H bonds. As Stockmayer first pointed out, a direct determination of the energies of the various conformations adopted by the CF_2-CF_2 bonds may be obtained by comparing experimental and theoretical calculated dipole moments of $H(CF_2)_nH$. Such studies have now been carried out (14,15).

Configurational information about the PTFE chain is also available from experimental quantities such as: (1) X-ray diffraction measurements on crystalline PTFE (16); (2) the temperature coefficient, $dln\langle r^2\rangle_0/dT$, of the unperturbed dimensions (17,18); and (3) the entropy of melting of PTFE (17). Semiempirical calculations (19–22) of conformational energies have provided additional information about the shape of the bond rotation potentials in perfluoroalkanes.

The PTFE chain $+CF_2+_x$ is structurally one of the simplest of chain molecules and has a repeat unit analogous to that of polyethylene $+CH_2+_x$. The van der Waals radius of the fluorine atom (1.4 Å) is, however, considerably larger than that of the hydrogen atom (1.1–1.2 Å). As a result, van der Waals interactions between nonbonded atoms are more repulsive in PTFE than in polyethylene.

These differences in intramolecular interactions are reflected by significant differences in the configurational characteristics of the two molecules. Thus the PTFE molecule adopts a helical configuration in the crystalline state (16), whereas the polyethylene chain assumes the planar all-*trans* form (23), as shown in Fig. 3. The greater steric repulsions between nonbonded fluorine atoms also cause the potential barrier (4.0 kcal mole^{-1}) to rotation in hexafluoroethane (24), $CF_3—CF_3$, to be considerably larger than that for ethane (25) (3.0 kcal mole^{-1}). It is to be expected, therefore, that the PTFE chain will be somewhat less flexible than the polyethylene chain. Qualitative indications of this reduced flexibility in PTFE chains are the very high melting point (26) (327°C compared with 140°C for polyethylene) and the high melt viscosity of the polymer. As we shall see, however, the perfluoroalkane chain is by no means completely rigid and rotational isomerism imparts flexibility to the chain when in the dissolved or molten state.

Unfortunately, PTFE appears to be the only fluorine-containing polymer for which detailed configurational information is available. Attention will be concentrated, therefore, on straight-chain perfluoroalkanes.

II. THE CONFIGURATION OF *n*-PERFLUOROALKANE CHAINS IN CRYSTALS

PTFE possesses a high degree of crystallinity. Bunn and Howells (16) and later Pierce, Clark, Whitney, and Bryant (27) studied its crystal structure by X-ray diffraction techniques. A summary of the results is given in the review article by Sperati and Starkweather (28). The X-ray diffraction pattern shows that the crystal structure of PTFE undergoes two reversible transitions at about 19 and 30°C. Below 19°C the repeat unit consists of 13 CF_2 groups with a repeat distance of 16.8 Å; the chain is given a 180° twist in this repeat distance (16). The C—C—C valence angle is 116°. Above the 19°C transition, a slight unwinding of the spiral occurs. Thus at 25°C the repeat distance is 19.5 Å and there is a 180° twist per 15 CF_2 groups (27). The molecules pack like cyclindrical rods in an arrangement that is almost hexagonal.

These configurations may be generated by twisting each $CF_2—CF_2$ bond, in the same sense, through about 17° from the planar *trans* position (16,19), i.e., $\phi_2 = \phi_3 = \cdots \phi_n \cong 17°$. This chain configuration is shown in Fig. 3*b*.

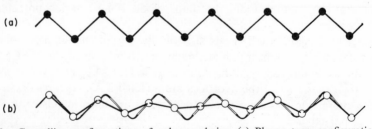

Fig. 3. Crystalline configurations of polymer chains. (a) Planar *trans* configuration of polyethylene. (Filled circles denote CH_2 groups.) (b) Helical configuration of PTFE obtained by twisting each C—C bond, in the same sense, through about 17° from the planar *trans* position. (Circles denote CF_2 groups.)

III. THE CONFIGURATIONAL CHARACTERISTICS OF ISOLATED *n*-PERFLUOROALKANE CHAINS

As mentioned earlier, the configurational characteristics of an isolated chain molecule depend on the positions and relative energies of the minima in the bond rotation potentials. In order to describe the spatial behavior of isolated perfluoroalkane chains, information about the rotation potential of the $-CF_2-CF_2-$ bonds is required. The crystalline configuration of PTFE already described and semiempirical calculations of conformational energies have provided such information.

The most important factors (29) which influence the configuration of the PTFE chain in the crystal are: (1) intramolecular interactions, which determine the most stable configuration of an isolated molecule; and (2) intermolecular interactions, such as packing effects, which may distort the configuration favored by the isolated molecule. Bunn and Holmes (29) argued that intramolecular rather than intermolecular interactions play the dominant role in determining the crystalline configuration of PTFE. These authors pointed out that the PTFE helical structure packs inefficiently. Thus intermolecular interactions are unlikely to be important in determining the crystal structure of PTFE since they would favor a configuration that results in a lower packing energy. The crystalline configurations of a large number of other polymers are also stongly influenced by intramolecular interactions (20,29).

It is therefore reasonable to assume that the helical configuration of the PTFE chain in the crystal is also the most stable configuration of an isolated molecule. Consequently, the bond rotation potential for $-CF_2-CF_2-$ bonds in perfluoroalkanes will contain minima with rotation angles ϕ close to $+17$ and $-17°$. The conformations corresponding to these most stable minima may be regarded as distorted planar *trans* forms and will be denoted by t^+ and t^-.

A. Semiempirical Calculations of Conformation Energies

Further insight into the nature of the intramolecular interactions and the bond rotation potentials of isolated PTFE chains has been obtained from semiempirical calculations of conformation energies carried out by Iwasaki (19), De Santis, Giglio, Liquori, and Ripamonti (20), McMahon and McCullough (21), and Bates (22). These semiempirical calculations are based on the idea that the intramolecular energy of a molecule can be divided into additive contributions from various sources. The most important contributions to the energy come from steric interactions between nonbonded atoms, electrostatic interactions between the polar $C-F$ bonds, and an "intrinsic" torsion potential associated with each $C-C$ bond undergoing rotation.

The energy, $E_{nb}(\phi)$, resulting from attractive and repulsive steric inter-actions between nonbonded atoms depends on the distance d between the atoms and hence on the set $\{\phi\}$ of rotation angles about the bonds separating the two interacting atoms. Two expressions for E_{nb} have been used for the n-perfluoroalkanes. McMahon and McCullough (21) represented E_{nb} by a Lennard-Jones " 6-12 " potential of the form

$$E_{nb}(\phi) = \varepsilon\left[\left(\frac{A}{d}\right)^{12} - \left(\frac{A}{d}\right)^{6}\right] \tag{1}$$

Other authors (19,20,22) have used a Slater-Kirkwood " 6-exp " function

$$E_{nb}(\phi) = a \exp(-bd) - cd^{-6} \tag{2}$$

The parameters a, b, c, ε, and A characterize the nonbonded interactions. They were evaluated for each atom pair either by semiempirical adjustment (21,22) (using experimental conformation energies such as the barrier to rotation in C_2F_6) or from inert gas data (19,20,30). Summing E_{nb} over all nonbonded atom pairs yields the total steric interaction energy.

The magnitude of the C—F dipole moment is at least 1.2 D (19,22), a sufficiently large value to warrant inclusion of a term $E_{dip}(\phi)$ to account for electrostatic repulsions and attractions between the C—F dipoles. Some (19,22) but not all (20,21) authors have included this term in their energy equations. The electrostatic energy between a given pair of C—F dipoles depends not only upon their distance apart but also upon their relative orien-tation; e.g., parallel alignment of the two dipole vectors results in repulsion, whereas antiparallel alignment results in attraction. Expressions for calculat-ing E_{dip} as a function of ϕ may be found elsewhere (19,22). The total electro-static energy is of course the sum over every pair of C—F dipoles not attached to the same carbon atom.

An energy equation that includes a contribution from an intrinsic torsion potential $E_t(\phi)$ has also been considered (22). This torsion potential is in no way connected with nonbonded interactions but is associated with each bond undergoing rotation. Semiempirical energy calculations for ethane and its substituted derivatives show that nonbonded and dipolar interactions alone are unable to account completely for the energy barrier to rotation about the C—C bond (7,31–34). It is necessary to account for the remainder of the barrier in ethane by ascribing an intrinsic torsion potential to the C—C bond itself (32). This potential is a periodic function of the rotation angle ϕ and might be expected to play an important role in energy calculations for the perfluoroalkanes. It has been treated as a threefold potential for these com-pounds (22), i.e., it shows three maxima and three minima for ϕ in the range $-\pi$ to π:

$$E_t(\phi) = \frac{E_0}{2} (1 - \cos 3\phi) \tag{3}$$

The parameter E_0 may be regarded as the barrier height, in the absence of nonbonded interactions, for rotation about a single C—C bond (31–33). The origin of the intrinsic torsion potential is poorly understood; consequently, it is not possible to make an *a priori* prediction of the magnitude of E_0 for C—C bonds in perfluoroalkanes. It is therefore preferable to treat E_0 as an empirically adjustable parameter. This has been done for the perfluoroalkanes (22), E_0 being varied between zero and 2.8 kcal mole^{-1}.

Calculations of the distances $d(\phi)$ between various nonbonded atoms allows the total intramolecular energy of a molecule to be computed as a function of the rotation angle ϕ.

The most stable configuration of PTFE has been studied by calculating the intramolecular energy as a function of the bond rotation angle ϕ of a PTFE helix. The helix is formed by making equivalent rotations about each C—C bond. The results obtained by the various authors are summarized in Table 1. The first column of this table shows the type of contributions, E_{nb}, E_{dip}, and E_t, which the different authors considered in their energy equations.

TABLE 1

Predicted Most Stable Configuration of PTFE Helices Obtained by Rotating Each C—C
Bond Through an Equivalent Rotation Angle ϕ
[Energies are relative to a zero energy for the planar *trans* form ($\phi = 0$)]

Contributions to intramolecular energy	ϕ for most stable minima	Energy per CF_2—CF_2 unit (kcal mole^{-1})	Reference
$E_{nb}{}^{a,c}$ (Eq. 1)	$\pm 17°$	-1.8	21
$E_{nb}{}^{b,d}$ (Eq. 2)	$\pm 15°$	-0.3	20
$E_{nb}{}^{a,d}$ (Eq. 2)	$\pm 18°$	-0.4	19
$E_{nb}{}^{a,d} + E_{dip}$	$\pm 19°$	-1.0	
$E_{nb}{}^{a,c}$ (Eq. 2)	$\pm 15°$	-0.3	
$E_{nb}{}^{a,c} + E_{dip}$	$\pm 17°$	-0.53	22
$E_{nb} + E_{dip} + E_t$ ($E_0 = 1.5$)	$\pm 11°$	-0.05	
$E_{nb} + E_{dip} + E_t$ ($E_0 = 2.8$)	$0°$	zero	

[a] Only F · · · F nonbonded interactions were included.

[b] Contributions from F · · · F, F · · · C, and C · · · C nonbonded interactions were included.

[c] Nonbonded interaction parameters were empirically adjusted.

[d] Nonbonded interaction parameters were deduced from inert gas data (30) and were not empirically adjusted.

The intramolecular energy is most sensitive to the magnitude of the parameter E_0 in the intrinsic torsion potential E_t. In the absence of any contribution from the torsion potential both Eqs. 1 and 2 predict that the configuration having the lowest energy is a helix with $\phi = 17 \pm 2°$; i.e., the crystalline configuration. This result is relatively insensitive to the choice of parameters appearing in these equations (19–22). Table 1 shows that the helix is predicted to be between 0.3 and 1.8 kcal mole^{-1} per monomer unit, $-CF_2-CF_2-$, more stable than the planar *trans* form when $E_t = 0$.

Figure 4b shows that both the dipole-dipole, E_{dip}, and the nonbonded atom, E_{nb}, interaction energies have a minimum at $\phi > 0°$. The reason the helical form minimizes E_{nb} is clear from a study of models. Thus the planar zig-zag *trans* conformation places nonbonded F atoms separated by two C—C bonds at a distance (29) of 2.54 Å. This is considerably less than twice the van der Waals radius (1.4 Å) of the fluorine atom. A rotation of 17° about the C—C bonds reduces the van der Waals repulsions by increasing the F . . . F distance to 2.7 Å.

Table 1 and Fig. 4a show the effects of including a contribution to the intramolecular energy from the intrinsic torsion potential term. When the parameter E_0 is greater than 1.5 kcal mole^{-1}, the most stable configuration

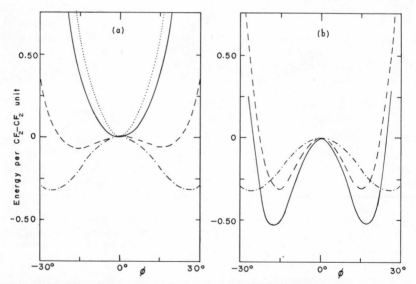

Fig. 4. The intramolecular energy (in kcal mole^{-1}) of PTFE helices plotted against the rotation angle ($\phi_2 = \phi_3 = \cdots \phi_n = \phi$) according to (21). Contributions from E_{nb}, Eq. 2, (– – –), E_{dip} (– · – ·), and E_t (· · · · ·) are shown. The solid line is the total energy $E_{nb} + E_{dip} + E_t$. The parameter E_0 (see Eq. 3) is 2.8 and zero kcal mole^{-1} in (a) and (b), respectively. The parameter a in Eq. 2 for F · · · F nonbonded pairs was empirically adjusted so that the energy equation reproduced the experimental barrier height in C_2F_6.

predicted is the planar *trans* form ($\phi = 0°$), and not a helix. This is due to the dominance of the E_t term when $E_0 > 1.5$ kcal mole^{-1}. Since the available evidence suggests that the helix is the most stable form of an isolated PTFE chain, E_0 must apparently be less than 1.5 kcal mole^{-1}. It thus appears that the intrinsic torsion potential plays only a minor or even negligible role in semiempirical energy equations for perfluoroalkanes. The *n*-alkanes (7,31–34), on the other hand, require a value of E_0 close to 3 kcal mole^{-1}.

Semiemprical calculations have also been used to predict the positions and relative energies of other minima in the bond rotation potential of perfluoroalkanes. Conformational energies have been calculated for *n*-perfluorobutane, F(CF$_2$)$_4$F, in the neighborhood of its *gauche*, $\phi \cong \pm 120°$, configuration, using a semiempirical expression with contributions from E_{nb}, E_{dip}, and E_t terms (22). The calculations show that there is a well-defined minimum in the rotation potential at $|\phi| \cong 120°$. This minimum corresponds to a $t^+g^+t^+$ (or $t^-g^-t^-$) conformation of *n*-C$_4$F$_{10}$. When $0 \leqslant E_0 \leqslant 1.5$ kcal mole^{-1}, the energy of this conformation is predicted to be 2.0 ± 0.3 kcal mole^{-1} above that of the $t^+t^+t^+$ conformation. This energy difference will be representative of the energy difference between equivalent conformational sequences in chains of greater length.

Calculations by De Santis and co-workers (20) for PTFE helices confirm the presence of a minimum in the rotation potential at $\phi = \pm 120°$. The semiempirical calculations of these authors and those of McMahon and McCullough (21) also predict two additional minima in the bond rotation potential for a PTFE helix. These minima are equivalent and occur at rotation angles close to $\pm 90°$. Similar minima, which may be regarded as distorted *gauche* states, have also been predicted for the *n*-alkanes (7, 34). McMahon and McCullough (21) have pointed out, however, that the semi-empirical method becomes unreliable in regions of high conformational energy such as in the neighborhood of the minima at $\pm 90°$. Consequently, the predicted minima at $\pm 90°$ for PTFE helices should not be taken too literally. In any event, configuration-dependent properties of the PTFE chain, such as the unperturbed dimensions, $\langle r^2 \rangle_0$, are probably not very sensitive to the presence of these minima. This conclusion is justified to some extent by calculations carried out by Abe, Jernigan, and Flory (7). These authors found that a bond rotation potential containing five minima at $\phi = 0°$, $\pm 80°$, and $\pm 115°$ resulted in values of $\langle r^2 \rangle_0$ for polyethylene which were essentially the same as those predicted on the basis of a potential containing only three minima at $\phi = 0°$ and $\pm 115°$.

In summary, the results of the semiempirical energy calculations are consistent with the conclusion of Bunn and Holmes (29) that the helix is the most stable configuration of an isolated PTFE chain. These calculations show that there are four well-defined minima in the bond rotation potential. These minima correspond to t^+ and t^- conformations at $\phi \cong \pm 17°$, and g^+ and g^- conformations at $\phi \cong \pm 120°$. Such a rotation potential is shown in Fig. 5a and will be referred to as the four-state model. A realistic treatment of the configurational characteristics of perfluoroalkane chains requires the use of such a four-state model. This situation is to be contrasted with that for the n-alkanes and polyethylene whose bond rotation potentials (see Fig. 5b) contain only three minima (1–8) at $\phi = 0°$ and $\phi \cong \pm 120°$. The use of this three-state model for studying the configurational behavior of the perfluoro-alkanes is likely to be unrealistic.

B. Quantitative Determination of Conformation Energies

A detailed description of the configurational characteristics of the per-fluoroalkane chain requires not only the rotation angles characterizing the minima in the bond rotation potential but also a quantitative determination of the relative energies of the minima. The semiempirical calculations dis-cussed in the previous section indicate that the g^+ minimum lies 2 kcal mole^{-1} above that of the t^+ minimum. *Gauche* conformations should there-fore be accessible to the C—C bonds in PTFE.

The occurrence of conformational sequences such as $t^+t^+t^-t^-$ will impart additional flexibility to the chain. Such sequences involve a reversal in the pitch of the helix such as occurs when a right-handed helix changes to a left-handed helix. The semiempirical calculations of Brown (35) predict that the energy difference between $t^+t^+t^-t^-$ and $t^+t^+t^+t^+$ sequences is between 1.4 and 2.5 kcal mole^{-1}, depending on the type of potential function used to calculate the F . . . F nonbonded interactions. Another semiempirical calcu-lation (22) yielded about 0.7 kcal mole^{-1} for this energy difference.

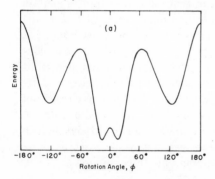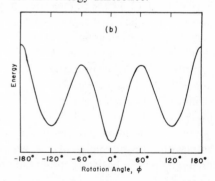

Fig. 5.   Schematic diagram of bond rotation potentials. Plot of potential energy against rotation angle ϕ. (a) Four-state model for CF_2—CF_2 bonds in isolated PTFE chains with minima at $\phi(t^\pm) \cong \pm 17°$ and $\phi(g^\pm) \cong \pm 120°$. Three-state model for CH_2—CH_2 bonds in polyethylene with minima at $\phi(t) = 0°$ and $\phi(g^\pm) \cong \pm 120°$.

The energy difference between $tg^{\pm}$ and tt sequences in the n-alkanes (1, 36) and polyethylene (5–8) is about 0.5 ± 0.2 kcal mole^{-1}. There is therefore little doubt that the PTFE chain should have a greater persistence of direction than the polyethylene chain. The magnitudes of the foregoing energy differences indicate, however, that the isolated PTFE chain cannot be treated as a rigid molecule. This conclusion is also supported by experimental measurements on perfluoroalkanes. Thus electron diffraction studies on n-C_5F_{12}, n-C_7F_{16}, and iso-C_5F_{12} by Bastiansen and Hadler (37) clearly show that restricted rotation occurs about the C—C bonds. Szasz (38) obtained values of 0.3 to 0.7 kcal mole^{-1} for the energy differences between unspecified rotational isomers of liquid n-C_5F_{12}, n-C_6F_{14}, and n-C_7F_{16}. These energies were obtained from the temperature dependence of the infrared absorption spectra. No attempt was made by Szasz (38) to associate these energy differences with particular conformations. The energies would seem to be too low to be associated with *gauche* conformations aad probably represent the energy difference between t^+t^+ and t^+t^- sequences.

A direct quantitative determination of the conformational energies of CF_2—CF_2 bonds has been carried out by analyzing the dipole moments of the α,ω-dihydroperfluoroalkanes, $H(CF_2)_nH$, in terms of chain configuration (14, 15). Experimental dipole moments of these compounds with $n = 4, 6, 7, 8, 10$ were determined from dielectric constant and refractive index measurements on dilute ($\sim 1\%$ by weight) benzene solutions at 25°C. Comparison of the experimental and theoretically calculated dipole moments provides the required conformational energies.

The theoretical calculations follow closely the methods developed by Flory (9) and Flory and Jernigan (13) and are based on two important concepts: the rotational-isomeric-state approximation to chain flexibility and the interdependence of bond rotational potentials.

The former concept was developed by Mizushima (1) and Volkenstein (3) and is based on the fact that the bonds tend to assume rotation angles ϕ close to the minima in the bond rotation potentials. The rotational-isomeric-state model approximates this situation by assuming that each bond adopts only a small number, s, of precisely specified rotation angles corresponding to the positions of the minima. According to this model therefore, chain flexibility arises solely from the ability of each skeletal bond to adopt one of several discrete rotation angles, bond lengths and angles being fixed. This model considerably simplifies the mathematical treatment of chain flexibility (3,4, 9–13).

Any realistic treatment of chain configurational statistics must take account of the interdependence of bond rotation potentials. This interdependence arises because the energy of a given state of a particular bond is frequently influenced by the rotational states of its neighbor bonds. For example, when two adjacent bonds $i-1$ and i of polyethylene adopt *gauche* states of opposite sign, severe repulsive interactions, referred to as "pentane interferences" by Taylor (39), occur between CH_2 groups separated by four $C-C$ bonds. Such $g_{i-1}^{+} g_i^{-}$ sequences are shown in Fig. 2c. They are almost completely suppressed in polyethylene (5–9). Only first-neighbor dependence of bond rotation potentials need be considered for the perfluoroalkane chain since second- and higher-neighbor dependence is weak.

The concepts of rotational isomerism and the interdependence of bond rotation potentials allow us to define a statistical weight $u_i(\phi_{i-1}^{\alpha}, \phi_i^{\beta})$ for each bond pair $i-1, i$ of the $H(CF_2)_nH$ chain. The superscripts $\alpha, \beta = 1, 2, \ldots, s$ indicate that the ϕ_i are restricted to one of s discrete values. For the four-state model discussed earlier, $s = 4$.

In formulating the molecular dipole moment μ of the $H(CF_2)_nH$ chain, it was assumed that the only contributions to μ come from the dipole moments μ_1 and μ_{n+1} of the terminal $H-CF_2$ and CF_2-H bonds 1 and $n+1$. (The enumeration of the bonds in the $H(CF_2)_nH$ chain is shown in Fig. 6.) A given configuration of the chain is defined by specifying the rotation angles ϕ_i^{α} for each $C-C$ bond i. Each configuration is characterized by a specific value of the molecular dipole moment $\mu = \mu_1 + \mu_{n+1}$ and occurs with a statistical weight given by the product $u_2(\phi_2^{\alpha})u_3(\phi_2^{\alpha}, \phi_3^{\beta})u_4(\phi_3^{\alpha}, \phi_4^{\beta}) \cdots u_n(\phi_{n-1}^{\alpha}, \phi_n^{\beta})$.

Treating the experimental dipole moment as a statistical mechanical average over all possible configurations requires the calculation of the mean-squared magnitude of μ:

$$\langle \mu^2 \rangle = \sum_{i=1}^{n+1} \sum_{j=1}^{n+1} \langle \mu_i \cdot \mu_j \rangle = \mu_1^2 + \mu_{n+1}^2 + 2\langle \mu_1 \cdot \mu_{n+1} \rangle$$

The exact mathematical methods of Flory (9) and Flory and Jernigan (13), applicable to chains of any length, were used to compute $\langle \mu^2 \rangle$. The calculation follows closely the method used by Leonard, Jernigan, and Flory (8) for computing $\langle \mu^2 \rangle$ for α,ω-dibromoalkanes. The computation was carried out in terms of the statistical weights u_i for the various bond pairs, the rotational angles ϕ_i^{α}, valence angles θ, and the scalar magnitudes μ_1 and μ_{n+1} of the vectors μ_1 and μ_{n+1}. The experimental dipole moment (1.60 D) of 1-hydroperfluoroheptane, $H(CF_2)_7F$ (15), in benzene at $25°C$ was used for $\mu_1 = \mu_{n+1}$, and θ was taken as $116°$ (16). For the four-state model, $\phi_i^{\alpha} \cong \pm 15°$ and $\pm 120°$.

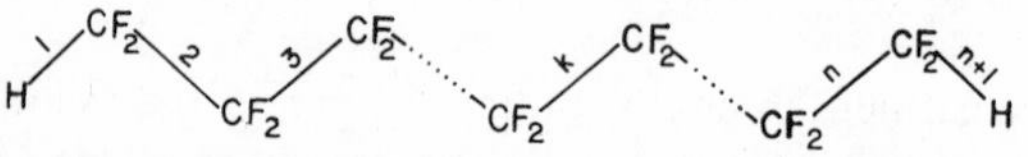

Fig. 6. Method of numbering the bonds in the α,ω-dihydroperfluoroalkane chain $H(CF_2)_nH$. The "internal" bonds are denoted by k where $4 \leqslant k \leqslant n - 1$.

Only the formulation of the statistical weight parameters $u_k(\phi^\alpha_{k-1}, \phi^\beta_k)$ applicable to the internal bonds k of the $H(CF_2)_nH$ chain need concern us here [$4 \leqslant k \leqslant (n-2)$]. Two such statistical weights, denoted by σ and ω, are required for a four-state model. They are defined relative to an arbitrary statistical weight of unity for the helical configuration, $t^+_{k-1}t^+_k$ or $t^-_{k-1}t^-_k$. The parameter σ is equal to $u_k(t^+_{k-1}g^+_k) = u_k(t^-_{k-1}g^-_k)$. It represents the statistical weight for a g^+ or g^- rotational state of bond k when bond $k-1$ is in a t^+ or t^- state, respectively. The statistical weight $\omega = u_k(t^+_{k-1}t^-_k)$ $= u_k(t^-_{k-1}t^+_k)$ takes into account the unfavorable steric interactions precipitated by such sequences. The parameters σ and ω are related to the corresponding energies E_σ and E_ω by

$$\sigma = \exp\left(\frac{-E_\sigma}{RT}\right) \quad \text{and} \quad \omega = \exp\left(\frac{-E_\omega}{RT}\right) \tag{5}$$

These equations assume that entropy differences between the bond rotational states are negligible.

Neighbor dependence of the bond rotation potentials are taken into account by excluding the sequences $g^+_{k-1}g^-_k$, $g^-_{k-1}g^+_k$, $t^-_{k-1}g^+_k$, $t^+_{k-1}g^-_k$, in view of the large steric repulsions resulting from such conformations. These considerations are conveniently summarized by defining a statistical weight matrix U_k in which the rotational states of bond $k-1$ are indexed on the rows and the states of bond k are indexed on the columns:

$$U_k = \begin{array}{c} \\ (t^+) \\ (g^+) \\ (g^-) \\ (t^-) \end{array} \overset{\begin{array}{cccc} (t^+) & (g^+) & (g^-) & (t^-) \end{array}}{\begin{bmatrix} 1 & \sigma & 0 & \omega \\ 1 & \sigma & 0 & 0 \\ 0 & 0 & \sigma & 1 \\ \omega & 0 & \sigma & 1 \end{bmatrix}} \tag{6}$$

for the four-state model.

Good agreement (within 0.05 D) between observed and calculated dipole moments at 25°C was obtained for $\sigma = 0.2$–0.06 and $\omega = 0.5$–0.05. The corresponding energies calculated by Eq. 5 with $T = 298°K$ are $E_\sigma = 1.35 \pm 0.35$ and $E_\omega = 1.1 \pm 0.7$ kcal mole^{-1}. The results are summarized in Table 2. Small changes in μ_1, $\phi(t^\pm)$, $\phi(g^\pm)$, and θ do not alter the energies significantly.

A similar calculation was carried out for the more simple but less realistic three-state-model. Defining a statistical weight $\sigma' = u_k(t_{k-1}g^\pm_k)$ relative to a statistical weight of unity for the planar *trans* conformation, and putting $u_k(g^+_{k-1}g^-_k) = u_k(g^-_{k-1}g^+_k) = 0$ yields the statistical weight matrix,

$$
U_k = \begin{array}{c} \\ (t) \\ (g^+) \\ (g^-) \end{array}
\begin{array}{ccc} (t) & (g^+) & (g^-) \\ \left[\begin{array}{ccc} 1 & \sigma' & \sigma' \\ 1 & \sigma' & 0 \\ 1 & 0 & \sigma' \end{array}\right] \end{array}
\tag{7}
$$

for the three-state model. The calculated dipole moments were fairly insensitive to the choice of model, the three-state model yielding $E_{\sigma'} = 1.2 \pm 0.2$ kcal mole^{-1} for the *gauche* energy relative to the planar *trans* energy.

As Table 2 shows, the magnitudes of the bond conformational energies obtained from the dipole moment calculations are consistent with the values predicted by the semiempirical calculations. Starkweather and Boyd (17) obtained the much larger value of 4.3 kcal mole^{-1} for $E_{\sigma'}$ by analyzing, in terms of a three-state model, the experimental temperature coefficient, $dln\langle r^2\rangle_0/dT$, of the unperturbed dimensions of PTFE. An energy difference of this magnitude is incompatible with the experimental dipole moments of $H(CF_2)_nH$. However, the figure of 4.3 kcal mole^{-1} should be treated with caution because experimental difficulties cause the measured $dln\langle r^2\rangle_0/dT$ for PTFE to be unreliable. This question is discussed more fully in Section IVA.

TABLE 2

Energies for Various Conformational Sequences about the CF_2—CF_2 Bonds in
n-Perfluoroalkanes
(Unless otherwise stated, energies refer to a four-state model with $\phi(t^\pm) \simeq \pm 15°$ and $\phi(g^\pm) \simeq \pm 120°$ and are relative to a zero energy for a t^+t^+ sequence. Numbers in parentheses are reference numbers.)

		Energy (kcal mole^{-1})		
Conformation	Energy parameter	From μ of $H(CF_2)_nH$ (ref. 15)	From semi-empirical calculations	From $dln\langle r^2\rangle_0/dT$ (ref. 17)
t^+g^+, t^-g^-	$E_\sigma = E_{t^+g^+} = E_{t^-g^-}$	$1.35 \pm 0.35,$ 1.2 ± 0.2^a	2.0 ± 0.3 (22)	$\sim 4.3^a$
t^+t^-, t^-t^+	$E_\omega = E_{t^+t^-} = E_{t^-t^+}$	1.1 ± 0.7	~ 0.7 (22), 1.4 to 2.5 (35)	
g^+g^+, g^-g^-	$2E_\sigma = E_{g^+g^-} = E_{g^-g^+}$	2.7 ± 0.7		
g^+g^-, g^-g^+		∞ (assumed)		
t^+g^-, t^-g^+		∞ (assumed)		

[a] These energies are equal to $E_{tg}\pm - E_{tt}$ and were obtained by assuming a three-state model with $\phi(t) = 0°$.

IV. PREDICTION OF CONFIGURATION-DEPENDENT
PROPERTIES OF POLYTETRAFLUOROETHYLENE

Some of the physical properties of PTFE that depend upon the configurational characteristics of the perfluoroalkane chain will now be discussed. Theoretical calculations have been made (40) for the unperturbed dimensions, the temperature coefficient of this quantity, and the entropy of melting of PTFE. The calculations were based on the four-state model previously described with a statistical weight matrix for each $C-C$ bond given by Eq. 6. Conformational energies obtained from the dipole moment analysis described in Section IIIB were used; i.e., $E_{t^+g^+} = E_{t^-g^-} = E_\sigma = 1.35 \pm 0.35$ kcal mole^{-1} and $E_{t^+t^-} = E_{t^-t^+} = E_\omega = 1.1 \pm 0.7$ kcal mole^{-1}. These energies, although they apply essentially to short-chain perfluoroalkanes at 25°C, can be taken as representative of the conformational energies in PTFE chains. This assumption is justified by analogy with polyethylene. Thus the energy of a $tg^\pm$ sequence in the lower n-alkanes (0.5 to 0.8 kcal mole^{-1}) (1, 36) is essentially the same as that in polyethylene (0.4 $\pm$ 0.1 kcal mole^{-1}) (7). Calculations of configurational properties of PTFE at temperatures above 25°C can be carried out by use of Eq. 5.

A. The Characteristic Ratio and its Temperature Coefficient for
Polytetrafluoroethylene

The characteristic ratio, $\langle r^2 \rangle_0/nl^2$, for a PTFE chain containing n bonds each of length l may be calculated by methods analogous to those used for computing the mean-square dipole moment of $H(CF_2)_nH$. An expression derived by Hoeve (10) for $n \to \infty$ has been used for this purpose (40). The calculation is based on the rotational-isomeric-state model with first-neighbor dependence of the bond rotation potentials. The results are shown in Table 3. The predicted characteristic ratio for PTFE is 30 $\pm$ 15 at the melting point, 600°K. This result is to be compared with the experimental value (41) of 6.7 $\pm$ 0.3 for the characteristic ratio of polyethylene at temperatures near its melting point, 413°K. The PTFE chain therefore has a greater persistence of direction than the polyethylene chain. There are three reasons for this reduced flexibility in PTFE:

1. The *gauche* conformational energy is greater than that for polyethylene. Thus calculations show that less than 23% of the $C-C$ bonds in PTFE at 600°K are in g^+ and g^- states compared with about 40% for polyethylene at 413°K.

2. The four-state model excludes not only g^+g^- and g^-g^+ sequences but also t^+g^- and t^-g^+ sequences, whereas only $g^\pm g^\mp$ pairs are excluded for polyethylene.

TABLE 3

Configuration-Dependent Properties of PTFE (Predictions were made for four-state model with conformational energies as shown in column 3 of Table 2)

Property	Predicted (600°K)	Experimental
$p_{g^+} + p_{g^-}$[a]	0.16 ± 0.07	—
$p_{t^+t^-} + p_{t^-t^+}$[b]	0.25 ± 0.10	—
$\langle r^2 \rangle_0 / nl^2$	30 ± 15	~20 to 64
$[\eta]_\theta$ for $M = 10^6$ (ml g^{-1})	160 to 750	—
$-10^3 dln\langle r^2 \rangle_0 / dT$ (deg^{-1})	0.9 ± 0.5	~5 (17)
ΔS_g (cal deg^{-1} mole^{-1})	0.8 ± 0.1	0.76 (17)

[a] $p_{g^+} + p_{g^-}$ is the fraction of C—C bonds in g^+ and g^- conformations.

[b] $p_{t^+t^-} + p_{t^-t^+}$ is the fraction of bond pairs in t^+t^- and t^-t^+ sequences.

3. The occurrence of t^+t^- and t^-t^+ sequences (about 25% of bond pairs are found in such sequences at 600°K) still leaves the chain with a "sense of direction" extending over several bonds.

The last conclusion is supported by the results of X-ray diffraction measurements on molten PTFE. Kilian and Jenckel (42) observed an intensity maximum in the X-ray diffractometer scan, which indicated that the chains are nearly straight over several C—C bonds.

Accurate intrinsic viscosities $[\eta]$ are not available for PTFE. The polymer is, however, sufficiently soluble in certain fully fluorinated hydrocarbons to allow approximate values of $[\eta]$ to be measured. Doban and co-workers and Sherratt (44) obtained $[\eta] = 800$–2000 ml g^{-1} with $\overline{M}_n \sim 10^6$ and for temperatures near 300°C. Experimental difficulties and uncertainties in $\overline{M}_n$ account for the uncertainty in $[\eta]$. This $[\eta]$ can be used to calculate the characteristic ratio of PTFE. As shown by Flory (2), the intrinsic viscosity $[\eta]_\theta$ in a theta solvent is given by

$$[\eta]_\theta = \Phi\left(\frac{\langle r^2 \rangle_0}{M}\right)^{3/2} M^{1/2}$$

$$= \Phi\left(\frac{\langle r^2 \rangle_0}{nl^2}\right)^{3/2} \left(\frac{l^2}{50}\right)^{3/2} M^{1/2} \tag{8}$$

The constant $\Phi = 2.5 \times 10^{23}$ (43). Because fluorinated solvents are likely to be thermodynamically good solvents for PTFE, we may write $[\eta] = [\eta]_\theta \alpha^3$, where α is an expansion factor arising from intermolecular interactions (2). For polyethylene in n-alkanes, Flory, Ciferri, and Chiang (41) obtained values of α^3 between 1.7 and 2.8. By analogy, PTFE in perfluoroalkanes might be expected to have α^3 values in a similar range, say 1.5–3.5. These values, in conjunction with the experimental $[\eta]$ and taking $l = 1.54$ Å, $M = 10^6$, yield $\langle r^2 \rangle_0 / nl^2 = 20$–64 at 300°C by use of Eq. 8. Such a calculation is of course very approximate. Nevertheless, the characteristic ratio so obtained is of the magnitude to be expected from the theoretical calculations for the four-state model.

The temperature coefficient, $dln\langle r^2 \rangle_0 / dT$, of the unperturbed dimensions of PTFE has also been calculated in terms of the parameters σ and ω (40). We may write

$$\frac{dln\langle r^2 \rangle_0}{dT} = -\frac{1}{T} \left[ln\sigma \frac{\partial ln\langle r^2 \rangle_0}{\partial ln\sigma} + ln\omega \frac{\partial ln\langle r^2 \rangle_0}{\partial ln\omega} \right] \tag{9}$$

The predicted value is $dln\langle r^2 \rangle_0 / dT = -(1.2 \pm 0.5) \times 10^{-3}$ at 600°K. The predicted temperature coefficient is significantly lower in magnitude than the value of about -5×10^{-3} deg^{-1} deduced by Starkweather and Boyd (17) from experimental (18) stress-temperature measurements on PTFE above its melting point. Interpretation of this result in terms of a three-state model yielded $E_{\sigma'} = 4.3$ kcal mole^{-1} (17), a figure which is inconsistent (14, 15) with the interpretation of the experimental dipole moments of $H(CF_2)_nH$. Unfortunately, an accurate experimental determination of $dln\langle r^2 \rangle_0 / dT$ for PTFE is by no means easy. Thus there is a considerable amount of scatter on the experimental stress-temperature measurements (18) from which $dln\langle r^2 \rangle_0 / dT$ was calculated. Moreover, it is possible that degradation of the polymer occurred at the temperatures at which the experimental measurements were made. It is reasonable therefore to suggest that uncertainties in the experimental $dln\langle r^2 \rangle_0 / dT$ are a major contribution to the discrepancy between the predicted and experimental values since the latter could well be uncertain by a factor as large as 5.

B. Crystalline Transitions in Polytetrafluoroethylene

The existence of two crystalline transitions in PTFE at about 19 and 30°C has been confirmed by calorimetry measurements of Rigby and Bunn (45), Furakawa, McCoskey and King (46), and Marx and Dole (47) and by dilatometry measurements of Quinn, Roberts and Work (48) and Leksina and Novikova (49). X-ray studies of Bunn and Howells (16), Pierce, Clark, Whitney, and Bryant (27), and Clark and Muus (50) show that these transitions produce a diffuseness in the X-ray fiber patterns. This diffuseness has been attributed to the onset of disorder in the structure. Rigby and Bunn (45) and Clark and Muus (50) have pointed out that the X-ray measurements are consistent with an essentially perfect three-dimensional order below the 19°C transition. The structure above 19°C is not, however, a new well-defined structure but rater a less ordered form of the structure below 19°C. The 19°C transition has been discussed in terms of order-disorder theories by Mark and Dole (47) and Miyake (51).

Clark and Muus (50) derived mathematical expressions for calculating the effect of various types of disorder on the X-ray diffraction pattern. These authors concluded that the disorder between 19 and 30°C is caused by small torsional oscillations of CF_2 groups about the C—C bonds. Nuclear magnetic resonance experiments by Hyndman and Origlio (52) confirm that the disorder is dynamic rather than static.

According to Clark and Muus (50), low-frequency torsional oscillations become active above the 19°C transition because there is a reduction in the intermolecular forces caused by an increase in the interchain distance at this temperature. The oscillations of a given chain are correlated with those of its neighbors so that the angular displacements of the CF_2 groups are symmetrically distributed about a preferred direction in a plane perpendicular to the chain axes. Thus the mean position of the CF_2 groups is that of an orderly array. At the 30°C transition a further reduction in intermolecular interactions is caused by an additional small increase in the interchain distance. The X-ray evidence indicates that there is a reduction in the correlation of motions between neighboring chains. The CF_2 groups carry out large-amplitude, low-frequency oscillations without any preferred direction in the plane perpendicular to the chain axis. The interchain distance increases linearly with temperature up to 220°C allowing the degree of angular disorder about the long axes to increase.

Semiempirical calculations of conformational energies (19–22) discussed earlier show that the minimum characterizing the PTFE helix is fairly shallow; consequently, torsional oscillations about the C—C bonds are entirely reasonable. As Clark and Muus (50) pointed out, the slope of the potential curve is lower for rotation angles *less* than the angle locating the minimum (see Fig. 4). Torsional oscillations will therefore tend to unwind the helix toward the planar *trans* form.

Brown (35) has also discussed the type of conformational changes that may be responsible for the disorder in the PTFE crystal. This author has shown that the temperature dependence of the infrared absorption of PTFE is consistent with the presence of a defect for which the energy of formation is about 1.2 kcal mole^{-1}. An energy increase of this magnitude accompanies the occurrence of t^+t^- conformations, as shown in Table 2. It is reasonable to associate disorder in the crystal structure with the presence of such t^+t^- sequences. Unlike t^+g^+ sequences, t^+t^- conformations do not involve a large change in the shape of the molecule and could be incorporated in the crystal lattice. The energy of formation of such a defect is sufficiently large so that only a small number of such reversals in the pitch of the helix will occur at room temperature; consequently, the preferred crystallographic direction will not be lost. Raising the temperature will increase the statistical weight for a t^+t^- sequence, thereby increasing the degree of disorder in the crystal.

A third crystalline transition has also been observed [by Bridgman (53), Weir (54), and Beecroft and Swenson (55)] for PTFE at about 4500 atm and 70°C. It has been suggested by Brown (35) and Stockmayer (40) that the high-pressure transition involves a change in configuration from the helical to the planar *trans* form. The infrared absorption spectrum of PTFE is consistent with such a configurational change (35).

The helical model of the PTFE chain therefore provides a reasonable explanation of the crystalline transitions of the polymer.

C. Entropy of Melting of Polytetrafluoroethylene

Starkweather and Boyd (17) have discussed the entropy change, $(\Delta S_m)_v$, accompanying the melting at constant volume of crystalline PTFE. These authors consider three contributions to $(\Delta S_m)_v$: (1) an increase in entropy, ΔS_g, due to rotational isomerism of an isolated chain; (2) a decrease in entropy equal to 0.86 due to packing of molecules on a crystal lattice; and (3) a contribution, ΔS_D, common to all liquids, due to long-range disorder. We therefore may write

$$(\Delta S_m)_v = \Delta S_g - 0.86R + \Delta S_D \qquad (10)$$

Starkweather and Boyd assume that ΔS_D is similar to the value calculated for metals (56); i.e., $\Delta S_D \cong 1.4 \pm 0.4 \cong 0.86R$. Thus the experimental value of $(\Delta S_m)_v$ may be identified with the entropy change ΔS_g due to rotational isomerism of an isolated molecule.

For a four-state PTFE chain which can adopt only a regular $t^+t^+t^+$ (or $t^-t^-t^-$) helix in the solid phase, the conformational entropy, S_g (liq), of melting per mole of bonds is given by (40):

$$S_g(\text{liq}) = R\{ln\lambda - [\sigma(\lambda - \omega)ln\sigma + \omega(\lambda - \sigma)ln\omega][\lambda(2\lambda - 1 - \sigma - \omega)]^{-1}\} \qquad (11)$$

where λ is the maximum eigenvalue of the matrix U_k in Eq. 6. The ranges of σ and ω allowed by the dipole moment analysis yield $S_g(\text{liq}) = 1.7 \pm 0.1$ cal deg^{-1} mole^{-1} at 600°K. We must not, however, identify $S_g(\text{liq})$ with ΔS_g. Such a procedure would imply that the chain is rigidly held in the helical configuration in the solid just below the melting point. As we have seen in the discussion of the crystalline transitions in PTFE, t^+t^- sequences probably occur at room temperature and almost certainly are present near the melting point in view of the degree of disorder present. Therefore the entropy change ΔS_g due to rotational isomerism will be equal to the difference between $S_g(\text{liq})$ (given by Eq. 11) and the conformational entropy $S_g(\text{solid})$ in the solid due to t^+t^- helical reversals. This latter entropy (40) is just

$$S_g(\text{solid}) = R\left[ln(1 + \omega) - \frac{\omega ln\omega}{(1 + \omega)}\right] \quad (12)$$

The parameter ω in Eq. 12 would be expected to be somewhat less than the ω occurring in Eq. 6 for an isolated molecule. This is because intermolecular as well as intramolecular interactions will determine the magnitude of ω in the solid. Using the value of 1.8 kcal mole^{-1} (deduced from a thermodynamic treatment (40) of the solid-solid phase transitions) for the defect energy $E_\omega = -RT ln\omega$ in the solid, $S_g(\text{solid})$ is found to be 0.9 cal deg^{-1} mole^{-1}. Hence

$$\Delta S_g = S_g(\text{liq}) - S_g(\text{solid}) = 1.7 - 0.9 = 0.8 \pm 0.1 \text{ cal mole}^{-1}$$

This result is in excellent agreement with the experimental (17) constant-volume entropy of melting of 0.76 cal deg^{-1} mole^{-1}.

The melting point of a polymer is inversely proportional to the change in entropy at the melting temperature (2). The high melting point of PTFE is a consequence of its low entropy of melting, as Sperati and Starkweather (28) pointed out. As shown previously, the low entropy of melting results not only from the relatively high energy difference between t^+ and g^+ conformations [i.e., from "chain stiffness," as Bunn (26) first suggested] but also from the presence of t^+t^- rotational states in the solid near the melting point.

V. CONFORMATIONS OF CRYSTALLINE POLYHEXAFLUOROPROPYLENE

Isotactic polyhexafluoropropylene (PHFP), in which the stereochemistries of adjacent asymmetric carbon atoms are the same, appears to be the only other perfluoroalkane chain for which information about conformational properties is available. Before discussing PHFP, a summary of the conformational properties of isotactic polypropylene [$CH(CH_3)-CH_2$] will be given. This chain adopts a helical conformation in the crystal, there being three monomer units per turn of the helix (57). The analysis by Borisova and Birshtein (58), Natta, Corradini, and Ganis (59), De Santis and co-workers (20), and Flory, Mark, and Abe (60), involving semiempirical calculations of conformational energies, indicates that this 3_1 helix is generated by a repetition of rotation angles $(\phi', \phi'')(\phi', \phi'') \ldots$ about the $CH(CH_3)-CH_2$ and $CH_2-CH(CH_3)$ bonds. Two equivalent-energy conformations occur, corresponding to right- and left-handed helices; the rotation angles characterizing these helices are $(\phi', \phi'') = (t, g^-)$ or $(\phi', \phi'') = (g^+, t)$. Interactions between nonbonded atoms may displace the *trans* and *gauche* minima from their symmetrical $0 \pm 120°$ positions. Thus $(t, g^-) = (\Delta\phi, -120° + \Delta\phi)$ and $(g^+, t) = (120° - \Delta\phi, -\Delta\phi)$ where $\Delta\phi$ may be as large as $15°$ (60).

The X-ray crystallographic study by Sianesi and Caporiccio (61) on fibers of isotactic PHFP prepared by a Ziegler catalyst show that the most stable conformation of this chain corresponds to four monomer units per turn of the helix. The change from a 3_1 to a 4_1 helix on passing from isotactic polypropylene to isotactic PHFP has been rationalized, in terms of differences in nonbonded interactions between hydrogen and fluorine, by semiempirical calculations (62) of conformational energies of PHFP as a function of its internal rotation angles. The energy equation employed was the same as that developed (22) for investigating the conformational energies of PTFE (see Section IIIA). These semiempirical calculations predict that the most stable conformation of isotactic PHFP is a helix generated by a repetition of rotation angles $(\phi', \phi'')(\phi', \phi'')\ldots$, with (ϕ', ϕ'') in the range $(40° + \Delta\phi, -100° - \Delta\phi)$, $\Delta\phi < 10°$; the CCC valence angle is predicted to be $113 \pm 2°$. Repetition of $(100° + \Delta\phi, -40° - \Delta\phi)$ results in the equivalent-energy helix with a reversal of pitch. The number of monomer units per turn which characterize this helix lie in the range 3.87 ± 0.04, in satisfactory agreement with the experimental value of 4.0.

VI. SUMMARY

A realistic treatment of the configurational characteristics of isolated PTFE chains requires the use of a four-state rotational-isomeric-state model with neighbor dependence of the bond rotation potentials. The model is based on a bond rotation potential with minima at $\phi \cong \pm 15°$ and $\pm 120°$. This potential is suggested by X-ray diffraction meaurements on PTFE and by semiempirical calculations of conformation energies.

Conformational energies, obtained from an analysis of the dipole moments of $H(CF_2)_nH$ (see Table 2), indicate that the PTFE chain has a greater persistence of direction than the polyethylene chain. Rotational isomerism occurs about the CF_2-CF_2 bonds, however, and the PTFE chain is by no means a rigid structure. This conclusion is supported by electron diffraction and infrared measurements on the lower n-perfluoroalkanes and by the results of the semiempirical energy calculations.

The four-state model with $E_\sigma = 1.35 \pm 0.35$ and $E_\omega = 1.1 \pm 0.7$ kcal mole^{-1} is consistent with a number of physical properties of PTFE such as the entropy of melting, the high melting point, and the crystalline order-disorder transitions. The only experimental quantity that cannot be reproduced by the model is the temperature coefficient, $dln\langle r^2\rangle_0/dT$, of the unperturbed dimensions. Experimental difficulties in determining $dln\langle r^2\rangle_0/dT$ are the most likely source of this discrepancy.

The configurational characteristics of the PTFE chain differ significantly from those of the structurally analogous polyethylene chain. The configurational behavior of the latter has been successfully studied in terms of a three-state model. The available evidence suggests that such a model should not be used to interpret the configurational behavior of PTFE. Theoretical calculations show that predicted configuration-dependent properties of PTFE are often quite sensitive to the choice of model (40). Thus, although a three-state model with $E_{tg\pm} = E_{\sigma'} = 1.2 \pm 0.2$ kcal mole^{-1} is consistent with the dipole moments of $H(CF_2)_nH$, the characteristic ratio, $\langle r^2 \rangle_0 / nl^2$, predicted by this model is only 11 ± 2 at $600°K$. Such a value would indicate that PTFE at its melting point is almost as flexible as polyethylene at its melting point. Moreover, the entropy of melting for PTFE predicted by the three-state model is much too high (1.6 cal deg^{-1} mole^{-1}) and is inconsistent with the high melting point of PTFE. The four-state model therefore presents a more consistent picture of the configurational characteristics of PTFE and the n-perfluoroalkanes.

Acknowledgment

The author is indebted to Professor W. H. Stockmayer, who initiated the dipole moment analysis and under whose guidance the work was carried out. It is also a pleasure to thank Dr. R. A. Orwoll for many helpful discussions.

References

1. S. Mizushima, "Structure of Molecules and Internal Rotation," Academic Press, New York, 1954.
2. P. J. Flory, "Principles of Polymer Chemistry," Cornell University Press, Ithaca, N.Y., 1953.
3. M. V. Volkenstein, "Configurational Statistics of Polymer Chains," Interscience, New York, 1963.
4. T. M. Birshtein and O. B. Ptitsyn, "Conformations of Macromolecules," Interscience, New York, 1966.
5. C. A. J. Hoeve, *J. Chem. Phys.*, **35**:1266 (1961).
6. K. Nagai and T. Ishikawa, *J. Chem. Phys.*, **37**:496 (1962).
7. A. Abe, R. L. Jernigan, and P. J. Flory, *J. Am. Chem. Soc.*, **88**:631 (1965).
8. W. J. Leonard, Jr., R. L. Jernigan, and P. J. Flory, *J. Chem. Phys.*, **43**:2256 (1965).
9. P. J. Flory, *Statistical Mechanics of Chain Molecules*, Interscience, New York, 1969.
10. C. A. J. Hoeve, *J. Chem. Phys.*, **32**:888 (1960).
11. K. Nagai, *J. Chem. Phys.*, **37**:490 (1962).
12. P. J. Flory, *Proc. Natl. Acad. Sci. U.S.*, **51**:1060 (1963).
13. P. J. Flory and R. L. Jernigan, *J. Chem. Phys.*, **42**:3509 (1965).
14. T. W. Bates and W. H. Stockmayer, *J. Chem. Phys.*, **45**:2231 (1966).
15. T. W. Bates and W. H. Stockmayer, *Macromolecules*, **1**:12 (1968).
16. C. W. Bunn and E. R. Howells, *Nature*, **174**:549 (1954).
17. H. W. Starkweather, Jr., and R. H. Boyd, *J. Phys. Chem.*, **64**:410 (1960).
18. A. Nishioka and M. Watanabe, *J. Polymer Sci.*, **24**:298 (1957).
19. M. Iwasaki, *J. Polymer Sci.* **A1**:1099 (1963).
20. P. De Santis, E. Giglio, A. M. Liquori, and A. Ripamonti, *J. Polymer Sci.* **A1**:1383 (1963).

21. P. E. McMahon and R. L. McCullough, *Trans. Faraday Soc.*, **61**:197, 201 (1965).
22. T. W. Bates, *Trans. Faraday Soc.*, **63**:1825 (1967).
23. C. W. Bunn, *Trans. Faraday Soc.*, **35**:482 (1939).
24. D. A. Swick and I. L. Karle, *J. Chem. Phys.*, **23**:1499 (1955).
25. K. S. Pitzer, *Disc. Faraday Soc.*, **10**: 66 (1951); D. R. Lide, Jr., *J. Chem. Phys.*, **29**:1426 (1958).
26. C. W. Bunn, *J. Polymer Sci.*, **16**:323 (1955).
27. R. H. H. Pierce, Jr., E. S. Clark, J. F. Whitney, and W. M. D. Bryant, Meeting of the American Chemical Society, Atlantic City, N.J., Sept. 1956.
28. C. A. Sperati and H. W. Starkweather, Jr., *Adv. Polymer Sci.*, **2**:465 (1961).
29. C. W. Bunn and D. R. Holmes, *Disc. Faraday Soc.*, **25**:95 (1958).
30. E. A. Mason and M. M. Kreevoy, *J. Am. Chem. Soc.*, **77**:5808 (1955); M. M. Kreevoy and E. A. Mason, *J. Am. Chem. Soc.*, **79**:4851 (1957).
31. L. Pauling, *Proc. Natl. Acad. Sci. U.S.*, **44**:211 (1958).
32. E. B. Wilson, Jr., *Adv. Chem. Phys.*, **2**:367 (1959).
33. R. A. Scott and H. A. Scheraga, *J. Chem. Phys.*, **42**:2209 (1965).
34. R. A. Scott and H. A. Scheraga, *J. Chem. Phys.*, **44**:3054 (1966).
35. R. G. Brown, *J. Chem. Phys.*, **40**:2900 (1964).
36. G. J. Szasz, N. Sheppard and D. H. Rank, *J. Chem. Phys.*, **16**:704 (1948); N. Sheppard and G. J. Szasz, *J. Chem. Phys.*, **17**:86 (1949); S. Mizushima and H. Okazaki, *J. Am. Chem. Soc.*, **71**:3411 (1949); W. B. Person and G. C. Pimentel, *J. Am. Chem. Soc.*, **75**:532 (1953); L. S. Bartell and D. A. Kohl, *J. Chem. Phys.*, **39**:3097 (1963).
37. O. Bastiansen and E. Hadler, *Acta Chem. Scand.*, **6**:214 (1952).
38. G. J. Szasz, *J. Chem. Phys.*, **18**:1417 (1950).
39. W. Taylor, *J. Chem. Phys.*, **16**:257 (1948).
40. T. W. Bates and W. H. Stockmayer, *Macromolecules*, **1**:17 (1968).
41. P. J. Flory, A. Ciferri, and R. Chiang, *J. Am. Chem. Soc.*, **83**:1023 (1961); R. Chiang, *J. Phys. Chem.*, **69**:1645 (1965); C. J. Stacy and R. L. Arnett, *J. Phys. Chem.*, **69**:3109 (1965); A. Nakajima, F. Hamada, and S. Hayashi, *J. Polymer Sci.* **C15**:285 (1966).
42. H. G. Kilian and E. Jenckel, *Z. Electrochem.*, **63**:308 (1959).
43. D. McIntyre, A. Wims, L. C. Williams, and L. Mandelkern, *J. Phys. Chem.*, **66**:1932 (1962).
44. R. C. Doban, A. C. Knight, J. H. Peterson, and C. A. Sperati, Meeting of the American Chemical Society, Atlantic City, N.J., 1956; S. Sherratt, Kirk-Othmer, "Encyclopedia of Chemical Technology," 2nd ed., Vol. 9, Interscience, New York, 1966.
45. H. A. Rigby and C. W. Bunn, *Nature (London)*, **164**:583 (1949).
46. G. T. Furukawa, R. E. McCoskey, and G. J. King, *J. Research NBS.*, **48**: 273 (1952).
47. P. Marx and M. Dole, *J. Am. Chem. Soc.*, **77**:4771 (1955).
48. F. A. Quinn, Jr., D. E. Roberts, and R. N. Work, *J. Appl. Phys.*, **22**:1985 (1951).
49. I. E. Leksina and S. I. Novikova, *Sov. Phys. Solid State*, **1**:453 (1959).
50. E. S. Clark and L. T. Muus, *Z. Krist.*, **117**:108, 119 (1962); L. T. Muus and E. S. Clark, *Am. Chem. Soc. Polymer Preprints*, **5**(1):17 (1964).
51. A. Miyake, *Chem. High Polymers (Tokyo)*, **15**:153 (1958).
52. D. Hyndman and G. F. Origlio, *J. Appl. Phys.*, **31**:1849 (1960).
53. P. W. Bridgman, *Proc. Am. Acad. Arts and Sci.*, **76**:71 (1948).
54. C. E. Weir, *J. Research NBS*, **50**:95 (1953).
55. R. I. Beecroft and C. A. Swenson, *J. Appl. Phys.*, **30**:1793 (1959).
56. R. A. Oriani, *J. Chem. Phys.*, **19**:93 (1951).

57. G. Natta and P. Corradini, *J. Polymer Sci.*, **20**:251 (1956).
58. N. P. Borisova and T. M. Birshtein, *Vysokomol. Soedin.*, **5**:279 (1963).
59. G. Natta, P. Corradini, and P. Ganis, *J. Polymer Sci.*, **58**:1191 (1962).
60. P. J. Flory, J. E. Mark, and A. Abe, *J. Am. Chem. Soc.*, **88**:639 (1966).
61. D. Sianesi and G. Caporiccio, *Makromol. Chem.*, **60**:213 (1963).
62. T. W. Bates, *Trans. Faraday Soc.*, **64**:3180 (1968).

15. DIELECTRIC PROPERTIES OF FLUORINE-CONTAINING POLYMERS*

ANTHONY J. BUR, *Polymer Dielectrics Section, National Bureau of Standards, Washington, D.C.*

Contents

I. INTRODUCTION

Dielectric studies generally are concerned with the electrical behavior of insulators.† An insulator is defined as matter that has a high electrical resistivity, in the range of 10^{12} to 10^{22} ohm-cm. This is compared to 10^{-6} ohm-cm for a good conductor. Polymers fall into the category of insulators because they consist of chemical structures containing H, C, Si, N, F, and O in which the electrons are not mobile.‡ In spite of the negativeness of the definition of a dielectric as matter that does not conduct electricity, it is possible to obtain

* "Contribution of the National Bureau of Standards, not subject to copyright."

† General descriptions of dielectric behavior may be found in references 1 to 5. In addition, "The Digest of Literature on Dielectrics," an annual survey of published literature (6), is a valuable guide to literature sources.

‡ Dielectric properties of polymers have been reviewed by Curtis (7) and McPherson (8).

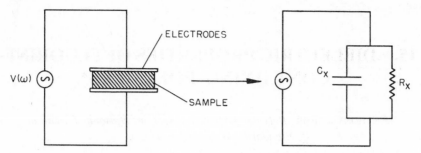

Fig. 1.   The equivalent circuit of a dielectric material is an ideal capacitor C_x in parallel with an ideal resistance R_x.

valuable information about the molecular character of a dielectric by observing the effects of applying a voltage across a dielectric specimen. Whereas the study of electrical conductors is concerned with the interaction of an applied field with monopolar charged electrons, this review of dielectric materials is primarily concerned with the interaction of an applied field with permanent dipoles which exist on the microscopic scale.

In the usual dielectric measurement, a sinusoidally alternating voltage is applied across a sample, as in Fig. 1. The experimental setup includes a sample which entirely fills the space between two parallel-plate electrodes, where the electric field is homogeneous in space. As the equivalent circuit in Fig. 1 indicates, the sample may be viewed as an ideal capacitor in parallel with an ideal resistance. The parallel capacitance and conductance are determined as a function of frequency, temperature, pressure, thermodynamic history of the sample, etc. The measurements are expressed as the complex dielectric constant, $\varepsilon^* = \varepsilon' - i\varepsilon''$, where

$$\varepsilon' = \frac{C_x}{C_0}$$

and

$$\varepsilon'' = \frac{1}{\omega\, C_0\, R_x} = \frac{G_x}{\omega\, C_0}.$$

C_0 is the capacitance of the parallel-plate electrodes without the sample; C_x, R_x, and G_x are the capacitance, resistance, and conductance of the sample; ω is the frequency of the applied field.* The real part of the dielectric constant, ε', is called the relative permittivity, ε'' is called the dielectric loss, and tan $\delta = \varepsilon''/\varepsilon'$ is the loss tangent.

Our prime consideration here is to establish what is happening on the

---

* The measurements of C_x and R_x are carried out by using conventional bridge circuits, resonant circuits, and microwave lines and cavities (9–11).

microscopic scale when an electric field is applied to a sample. Such a field interacts with electric charge in the sample, essentially displacing the positive charge in the direction of the field and the negative charge in the opposite direction. This polarization of charge takes place in three distinct ways:*

1. Orientation of permanent dipoles (orientation polarization).
2. Displacement of positive and negative atomic cores (atomic polarization).
3. Displacement of electrons with respect to their positively charged parent nuclei (electronic polarization).

The total polarization P_t in a uniform electric field is the algebraic sum of three polarizations:

$$P_t = P_o + P_a + P_e \qquad (1)$$

where P_t is the total dipole moment per unit volume and P_e, P_a, and P_o are the electronic, atomic, and orientation polarizations, respectively. That these three polarizations are physically distinct can be seen from the frequency dependence of the dielectric constant. The three polarization sources have different mechanical time constants and hence under the influence of an alternating electric field there are three distinct regions of the frequency spectrum in which absorption of electrical energy occurs. This absorption of energy is manifested by a maximum in the ε'' versus $\log f$ curve. Loss maxima at audio or microwave frequencies are due to orientation polarization; loss maxima in the infrared region (infrared absorption) arise from atomic polarization; and ultraviolet absorption is due to electronic polarization. Our main concern here is the audio and microwave region because the orientation polarization reflects movement of the entire polymer molecule or a segment of it.

An example of loss due to orientation polarization is schematically displayed in Fig. 2, where ε' and ε'' are plotted against log frequency. The following phenomenological equations describe the frequency dependence of ε' and ε'':

$$\varepsilon'(\omega) = \varepsilon_\infty + (\varepsilon_S - \varepsilon_\infty) \int_0^\infty \frac{\Phi(\tau)\,d\tau}{1 + \omega^2\tau^2} \qquad (2)$$

$$\varepsilon''(\omega) = (\varepsilon_S - \varepsilon_\infty) \int_0^\infty \frac{\Phi(\tau)\,\omega\tau d\tau}{1 + \omega^2\tau^2} \qquad (3)$$

where τ is a relaxation time, $\Phi(\tau)$ is a distribution of relaxation times, $\omega = 2\pi f$, ε_s is the static dielectric constant for this material, and ε_∞ is the

* We are neglecting the presence of ionic molecules which could cause a Maxwell-Wagner polarization at low frequencies.

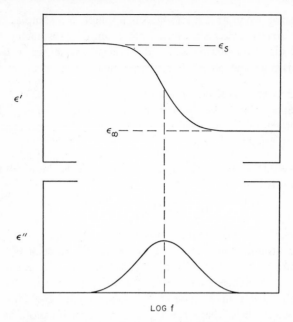

Fig. 2. A schematic diagram of ε' and ε'' versus frequency in the region of a relaxation dispersion is shown. ε'' is proportional to the energy absorbed by the dielectric material.

level to which the dielectric constant drops at the high frequencies. If the microscopic motion can be described by a single relaxation time, then $\Phi(\tau)$ is a delta function. In polymers and in low-molecular-weight materials as well, a single relaxation time is rarely observed. There may, in general, be several loss maxima present, reflecting different modes of motion on the microscopic scale. This is particularly true in polymers where many molecular configurations mean that there exist many possible modes of motion, which in turn implies a distribution of relaxation times describing these motions (12–14). Figure 7, which shows the data of Scott et al. (15) for polychlorotrifluorothylene, is an example of the structured loss spectrum which may be obtained with polymer samples.

The dielectric increment, $(\varepsilon_s - \varepsilon_\infty)$, is proportional to $\sum_i N_i \mu_i^2$ where μ is the permanent dipole moment of the ith species and N_i is the number of molecules of the ith species per unit volume. If $\mu_i = 0$, then $\varepsilon_s - \varepsilon_\infty = 0$ and there is no dielectric loss or dispersion region. Thus, from a dielectrics point of view, polymer molecules are either polar or nonpolar, i.e., they possess a permanent dipole moment or none. Relaxation phenomena may be observed dielectrically only if the material under observation possesses a microscopic dipole moment. A nonpolar dielectric is characterized by a low dielectric constant, practically no dielectric loss, and negligible frequency dependence of these quantities in the audio and microwave frequency regions. Whether or not a molecule possesses a permanent dipole moment is determined by the distribution of charge in the molecule. An asymmetric spatial distribution of positive and negative charge will yield a dipole moment; a symmetric distribution of charge has no permanent dipole moment. The bonding of dissimilar atoms spatially situated in an asymmetric manner will yield a permanent dipole moment. Examples of polar and nonpolar molecules are shown in Fig. 3.

Ideally, polyethylene and polytetrafluoroethylene are nonpolar polymers. However, any sample of these two polymers does display a spectrum of dielectric loss. The molecular origin of these loss regions is not known in all cases, but it can be recognized that it is nearly impossible to manufacture a polymer molecule with zero dipole moment. Dipoles arise from many

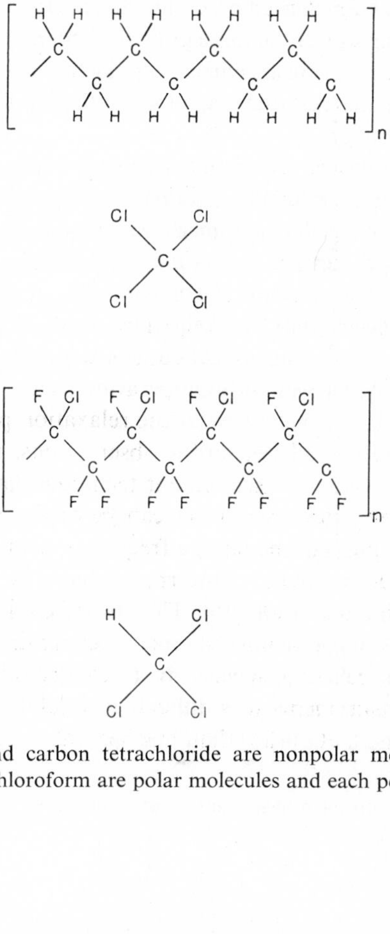

Fig. 3. (a) Polyethylene and carbon tetrachloride are nonpolar molecules. (b) Polychlorotrifluoroethylene and chloroform are polar molecules and each possess a permanent dipole moment.

asymmetries: oxidation, end groups, kinks in the chain, side groups, branching, etc. Moreover, the relaxation spectrum of some "nonpolar" molecules has been viewed dielectrically by slightly oxidizing the sample. The oxidized groups provide a polar handle by which the molecular modes of motion may be examined dielectrically.

To be complete, a dielectric investigation must include both frequency and temperature dependence. These parameters complement each other. A loss maximum which is observed at high frequencies at room temperature will be observed at low frequency and low temperature. If an observer has a limited frequency range, a variation in temperature may bring an unseen relaxation within this frequency capability. Dielectric measurements carried out over a broad frequency and temperature range may yield the following information: (1) the dielectric increment, $\varepsilon_s - \varepsilon_\infty$; (2) the spectrum of relaxation loss regions and the frequencies at which maximum dielectric loss occurs; (3) the shape of the relaxation dispersion regions and the distribution of relaxation times $\Phi(\tau)$; and (4) the temperature dependence of the preceding quantities. These quantities help characterize the relaxation and establish its relationship with other loss regions in the same polymer and in other polymers. The relative relaxation effects that occur when a small chemical change is introduced into a polymer molecule are often the most valuable information obtained from a relaxation experiment.

Some relaxations are associated with real thermodynamic transitions; i.e., in the limit of zero frequency a transition temperature will be manifested in specific volume or specific heat measurements. Consequently, some relaxations are identified with certain transition temperatures, e.g., the glass temperature T_g, and it has become the custom to view relaxations on the temperature scale rather than the frequency scale. This custom also arises from the fact that dynamic mechanical data usually have a very limited frequency range and must be viewed with temperature as the independent variable.

For an overall, complete view of the relaxation processes in a polymer it is necessary to refer to all relaxation observations, mechanical and NMR as well as dielectric. Each measurement technique has its advantages and disadvantages. Dielectric observation can be made only on a polar molecule. For mechanical measurements, the frequency range is limited and there is an inherent unsolved problem* of interpretation of the dynamic mechanical data on semicrystalline materials (16). This interpretation problem arises because there is no knowledge of how the macroscopic driving force is connected to the microscopic relaxing species in a semicrystalline material. Thus from mechanical measurements it is difficult to establish the molecular origins of these relaxations. Data other than mechanical, e.g., dielectric, NMR, specific

---

* Recently, this problem has been discussed by R. W. Gray and N. G. McCrum (80).

volume, specific heat, must be used in order to establish the molecular nature of a relaxation process in a semicrystalline polymer.

A. Polymer Morphology

Since the character of the relaxation processes in polymers depends on the morphological condition, we must distinguish between morphological states, between amorphous, semicrystalline, and crystalline polymers.

The relaxation spectra of amorphous polymers are fairly well understood (17,18). Generally, there is a relaxation region above the glass temperature T_g which is associated with the movement of the entire polymer molecule and whose energy of activation is in the range 40–100 kcal/mole. The temperature dependence of the relaxation times, according to the WLF equation (19), is

$$\log \frac{\tau}{\tau_0} = \frac{C_1(T - T_0)}{C_2 + T - T_0}$$

where τ and τ_0 are the relaxation times at temperatures T and T_0 and C_1 and C_2 are constants. Below T_g there are two or three other relaxation or loss regions (17,20) associated with segmental movement of the polymer chain; these are characterized by low energies of activation (ca. 5–10 kcal/mole) and their temperature dependence of relaxation times follow the Arrhenius equation, $\tau = A \exp \Delta H/kT$ where A is a constant, ΔH is the activation energy, k is the Boltzman constant, and T is absolute temperature.

In crystalline and semicrystalline polymers, there are many possible modes of motion. The relaxing species may be in the environment of a crystalline field, at the site of a crystalline defect, at a grain boundary site, or in an amorphous environment. The relaxing species may be a whole chain, a chain fold, a chain segment, a chain end; or some other mode of motion may exist, e.g., the distortion of a crystal habit which depends on the size and shape of the crystal. Most of the experimental studies that may isolate or eliminate the possible modes of relaxation in semicrystalline polymers are still forthcoming.

Recently, some success has marked the attempts to quantitatively describe relaxation phenomena in semicrystalline polymers, notably work by Takayanagi (21), Sinnot (22), and Hoffman, Williams, and Passaglia (23). Sinnott made dynamic mechanical measurements on solution-grown crystals of polyethylene as a function of the annealing temperature. He attributed the γ (low-temperature) relaxation peak in polyethylene to stress-induced reorientation effects within the lamellae of the chain-folded crystal and attributed the α (high-temperature) relaxation to reorientation of the folds at the surfaces of the lamellae. Hoffman et al. (23) expanded quantitatively the models of Sinnott and introduced several other theoretical models to explain other relaxation phenomena (e.g., chain twisting) in polychlorotrifluoroethylene and polyethylene. Some of their suggestions remain to be verified experimentally.

Concerning the α relaxation, Colson and Eby (22) have shown that Sinnott's assignment (orientation of the folds) is not unique. Other quantities, e.g., the area of the basal plane of the crystal and residual solvent content, also correlate with the intensity of the α relaxation.

B. Nomenclature

The loss regions of relaxation spectra (mechanical, dielectric and NMR), which have been published during the last 20 years, have been given various names. Among these names we have glass peak, glass transition, high-temperature process, low-temperature process, α, β, and γ peak (process or transition), and crystal peak. In many instances different names are used to describe the same loss region. On the one hand, there is a need for consistent terminology, but on the other hand there are still many unknown factors concerning relaxation in polymers and confusion exists.

For the sake of consistency we have chosen the same nomenclature used by Hoffman, Williams, and Passaglia (23). This system of nomenclature is shown in Fig. 4, where a schematic outline of the relaxation spectra observed in semicrystalline and amorphous polymers is presented. We start with the most general case, semicrystalline polymers. When the relaxation peaks of a semicrystalline polymer are viewed on the temperature scale at a low frequency (audio or below), the first peak below the melting point T_m will be α, the second peak β, then γ, δ, and so on, in the order of descending temperature.

As indicated in Fig. 4, the relaxation spectrum of a semicrystalline polymer has an α peak close to T_m, a β peak immediately above T_g, and a γ peak below T_g. A δ peak at lower temperatures has also been observed. From a molecular point of view the β relaxation is associated with the rubber-to-glass transition temperature T_g and has its origin in the amorphous phase of the polymer.

In supercooled amorphous polymers the well-known relaxation associated with T_g will also be called the β peak. Designating this T_g peak "β" in both amorphous and semicrystalline polymers aligns the terminology being used for the crystalline and non-crystalline polymers. Generally an amorphous polymer will not have the α peak that occurs in semicrystalline polymers, which is attributed to the crystalline phase of the polymer. Refinements to this terminology will be necessary. For example, γ_c is the crystalline component of the γ relaxation and γ_a is the amporhous component.

Beside illustrating the nomenclature we have adopted here, Fig. 4 also gives a general outline of the relaxation spectra that may be observed in polymers (17,23). Figure 4 remains an ideal schematic. In practice a clear

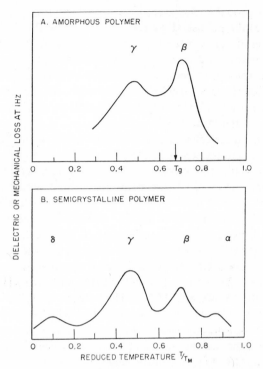

Fig. 4. Dielectric or mechanical loss at 1 Hz versus reduced temperature T/T_m, where T_m is the melting point temperature, for a linear polymer molecule [see (23)].

separation of the relaxation loss peaks on the temperature or frequency scale is not always the case. Nor are all relaxation losses seen with any one method of observation. Often there is considerable overlap of the loss regions, even to the point where one relaxation loss region completely absorbs the other. By variation of temperature, pressure, frequency, and degree of crystallinity of the sample, overlapping relaxation loss regions can sometimes be separated.

II. POLYTETRAFLUOROETHYLENE

The electrical properties of PTFE have been the subject of many published investigations (24–28). A good composite view of its electrical properties is difficult to obtain simply by gathering all of this information into one brief review. This is because of the lack of purification standards for polymers and because the electrical properties of PTFE are very sensitive to its thermodynamic history. Since PTFE is a nominally nonpolar polymer, ε'' is small (less than 10^{-3}) and any maxima observed in the loss spectrum (ε'' versus log f curve) are probably due to the presence of impurity polar groups or end groups. In all of the published dielectric literature on PTFE there is no mention of the level of purity or end-group concentration; and in many publications there is inadequate description of the thermodynamic history of the samples or the amount of crystallinity. We must proceed with caution when comparing dielectric data for PTFE which have been taken in different laboratories, because the level of purity and crystallinity of the different samples are probably not the same.

Whether or not a dielectric relaxation peak is observed in PTFE depends on three factors: (1) the sensitivity of the measurement; (2) the concentration of impurity dipoles and end groups; and (3) the thermodynamic history and degree of crystallinity; PTFE can be prepared in states of semicrystallinity which may vary from 40 to 98 % crystalline according to the recipe of quenching and annealing which is used. Lack of definition of these three factors in the published literature has led to confusion. Ehrlich (24) in 1953, with a sensitivity in ε'' of 10^{-4}, observed no loss maximum in the temperature range from room temperature to 314°C and frequency range from 10^2 to 10^5 Hz. Krum and Müller (25), with a sensitivity of 10^{-5} in ε'' at 1 kHz, observed loss maxima at -80, 15, 80, and 157°C. The ε'' peak at 157°C (1 kHz) had a maximum of 3×10^{-3} (within Ehrlich's capability). Reddish and co-workers (26), with the most sensitive measurements to date (sensitivity of 10^{-6} in ε''), could distinguish the peak at -80°C (1 kHZ) plus a peak of very low intensity at 150°C (1 kHz). In the temperature range from -40 to $+250$°C, ε'' did not rise above 9×10^{-5}, 100 times less than the maximum observed by Krum and Müller in the same temperature range. Other observers who have seen only the peak at -80°C (1 kHz) are Mikhailov (27) and Eby and Wilson (28).

Little information about the dispersion regions in PTFE is obtained from the real part of the dielectric constant ε' because the changes in ε' are too small to be observed. The change in ε' in any dispersion is proportional to the value of ε'' at its maximum. In PTFE, where ε_{max} is no larger than 10^{-3}, significant changes in ε' can be observed only if a sensitivity of 0.1 % in ε' is available. None of the authors mentioned claim such precision. Consequently, within the limits of experimental error and precision, no dispersion regions have been seen in the curves of ε' versus log f or ε' versus T for PTFE. At a given temperature, the value of ε' at any frequency is the same as the static dielectric constant ε_s. As a function of temperature ε_s for PTFE follows the classical Clausius-Mossotti equation, i.e., the polarization (the number of induced dipoles per unit volume) is a linear function of the density (24).

There is common agreement in the dielectric literature only about the γ peak at -80°C (1 kHz). It is characterized by an activation energy of approximately 15 kcal/mole and by a decrease in intensity with increasing crystallinity (26,27). Reddish and co-workers (26) have shown that a sample of 60 %

crystallinity has a γ loss maximum 13 times higher than a sample of 93%
crystallinity. At room temperature this relaxation occurs at 4×10^8 Hz.
Figure 5 shows the PTFE dielectric loss data in a plot of tan δ versus log f at
room temperature and tan δ versus T at 1 kHz.

If dynamic mechanical and NMR data are brought into the picture, a
better view of the relaxation spectrum of PTFE can be obtained. Mechanical
data by McCrum (31) and by Eby and Sinnott (30) along with recent NMR
data by McCall and co-workers (32) yield a complete but complicated
relaxation pattern in PTFE. The relaxations occur in both the crystalline and
amorphous phases. Accordingly, three main relaxation areas have been
found: $-80°$C (1 kHz); 71°C (1 kHz); and 157°C (1 kHz). These relaxations
are summarized in Fig. 6 in a plot of log f versus $1/T$. The 71°C (1 kHz)
relaxation is associated with the well-known first-order crystalline phase
transitions at 19 and 30°C (33) and will be called the room-temperature
crystal relaxation.

The relaxations at $-80°$C (1 kHz) and 157°C (1 kHz) were observed by
McCrum (31), who used a torsion pendulum at 0.8 Hz and consequently
observed the relaxations at lower temperatures, -97 and 127°C, respectively.
McCrum called these relaxation peaks Glass II $(-97°$C) and Glass I (127°C)

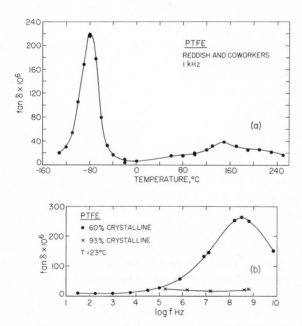

Fig. 5.   Dielectric data of Reddish and co-workers (26) on PTFE. (a) Dielectric loss tangent
at 1 kHz versus temperature. (b) Dielectric loss tangent versus frequency at room temperature for a 60% crystalline and a 93% crystalline sample.

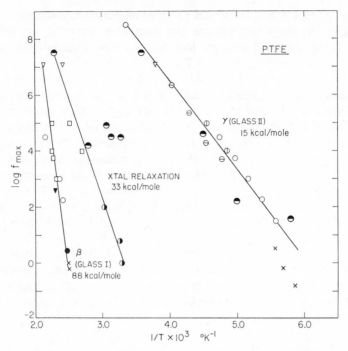

Fig. 6. Log f_{max} versus $1/T$ for PTFE. Mechanical, dielectric, and NMR data of various observers:

- ○   Reddsherd, co-workers (26)
- ⊖   Mikhailov, Kabin, and Smolianski (27)
- ⓘ   Eby and Wilson (28)
- ×   McCrum (31,37)
- ▽   Eby and Sinnott (30)
- ◓   McCall, Douglass, and Falcone (32)
- □   Krum and Müller (25)
- ◑   K. H. Illers and E. Jenckel, *Kolloid-Z*, **160**: 97 (1958)
- ▼   J. A. Sauer and D. E. Kline, *Int. Congress Appl. Mech. Brussels*, **5**: 368 (1956)
- ●   K. Schmieder and K. Wolf, *Kolloid-Z*, **134**: 149 (1953)

because the intensity of the mechanical loss tangent is proportional to the amorphous content in PTFE. But, as mentioned in the introduction, it is difficult to obtain information about the molecular origins of these relaxations from dynamic mechanical data (16). Dielectric (26,27) and NMR (32) data do, however, show that McCrum was correct in assigning the Glass I and Glass II relaxations to amorphous regions of the semicrystalline sample (80). According to the nomenclature of Hoffman et al (23) which we have adopted here, the Glass II and Glass I relaxations are, respectively, the γ and β relaxations in PTFE. The γ relaxation is the "low-temperature peak" and the β relaxation is the loss that occurs above the classical glass transition temperature; we discuss this in more detail later. Between the β (Glass I) and γ (Glass II) loss maxima on the temperature scale is the room-temperature crystalline relaxation. The relaxations observed by Eby and Sinnott (30), 140°C (12 mHz), by McCall (32), 84°C (12 kHz), and by McCrum (31), 22°C (0.4 Hz) are one and the same and appear to be associated with the "19 and 30°C" crystalline-phase transition (30,33). From X-ray work by Clark and Muus (33) this crystalline-phase transition has been identified with a change in the pitch of the helix of the polymer chain from 13 carbon atoms to 15 carbon atoms per 180° twist.

At room temperature and above the observation of dielectric relaxations in PTFE is rare. Only the data of Krum and Müller (25) and Reddish and co-workers (26) show indications of dielectric loss above room temperature. Even though the data of Krum and Müller do fit on a straight-line plot with mechanical data in Fig. 6, the Krum-Müller data must be held suspect because the dielectric loss arises from impurity dipoles in the sample. That impurity dipoles are present can be seen from the loss level of their samples at room temperature (1 kHz), which are an order of magnitude or more higher than the data of Koizumi (34), Reddish and co-workers (26), Eby and Wilson (28), Mikhailov (27), Johnson (35), and Scott and Kinard (75). In addition, one relaxation process that was observed by Krum and Müller has not been universally observed by others, that at $-13°C$ (1 kHz). Since the level of impurities containing dipoles is probably high in the Krum-Müller samples, the presence of this additional relaxation indicates that the inherent molecular character of PTFE has been changed. On the other hand, the extremely low level of loss for the PTFE samples of Reddish and co-workers establishes that the level of purity is high in these samples. In Fig. 5 the relaxation which occurs at 150°C (1 kHz) is the β (Glass I) relaxation. Unfortunately, the dependence of this peak on crystallinity has not been studied in detail.

A. The γ (Glass II) Relaxation

That the molecular origins of the γ relaxation, $-80°C$ (1 kHz), are situated in the amorphous regions of the polymer can be seen from dielectric and NMR data. Work by Reddish (26) (Fig. 5) and by Mikhailov (27) shows that the dielectric γ loss maximum decreases with increasing crystallinity or the dielectric loss is proportional to the amorphous fraction. The NMR data of McCall (32) are particularly useful in determining the molecular nature of the relaxations. From the width of the NMR peak it is possible to distinguish between relaxation processes which have their origins in the amorphous phase and those that occur in the crystalline phase. Thus in NMR experiments, where the crystalline and amorphous relaxation times are not the same, the source of the relaxation can readily be determined.

The dielectric measurements (26–28,35) of the γ relaxation show a loss maximum of $\varepsilon_{max} \approx 5 \times 10^{-4}$ for a sample which is approximately 60% crystalline. The agreement among observers of this loss maximum is good and suggests that the dipole moment and dipole concentration are the same in each case. Where does the dipole moment come from in PTFE? We rule out end groups because different molecular-weight samples would have different end-group concentration and various loss intensities would result in the literature. Eby (28,30) has shown that the volume of the relaxing species in the γ process involves 5–13 monomer units. The segment in motion is small. Such a small segment may have a small dipole moment because of the presence of the helix in the main chain of PTFE. Any segment shorter than the helical repeat distance, 13 atoms per 180° twist, will possess a dipole moment because symmetry will be absent. The exact motion of the segment and the exact number of atoms involved in the motion is not known. The crankshaft model of Schatzski (20), which has been used to explain the γ relaxation in other polymers, may also apply here.

Eby and Wilson (28) attempted to gain insight into the molecular nature of the γ relaxation in PTFE by examining dynamic mechanical properties of tetrafluoroethylene and hexafluoropropylene (HFP) copolymers.* At low HFP content the copolymer is essentially PTFE with an occasional CF_3 side group attached. It was their contention that the CF_3 group entered the PTFE crystalline lattice as a defect and hindered the motion of a PTFE crystalline defect; this PTFE crystal defect was the suggested source of the γ relaxation in PTFE. That CF_3 groups are included in the crystal lattice is acceptable (36). But the relationship between the CF_3 crystal defects and the γ relaxation is not clear. The CF_3 groups also alter the amorphous phase by their presence there, and the change in the γ relaxation (28,37) that occurs in going from PTFE to the TFE-HFP copolymer can have its origin in the amorphous phase. Dielectric (26,27) and NMR (32) data on the homo-polymer PTFE indicate that the γ relaxation has its source in the amorphous phase and not in the crystalline phase as proposed by Eby and Wilson (28).

B. The β (Glass I) Relaxation

Because of the nonpolar nature of PTFE, the β relaxation, 127°C (1 Hz), has been observed only in the most sensitive measurements of Reddish and co-workers (26). Since the β relaxation is out of the range of NMR experimental capabilities, mechanical and thermal expansion data account for most of our knowledge of the character of this relaxation. It has an activation energy of 88 kcal/mole, several times higher than the activation energy of the γ process (30).

* Also called fluorinated ethylene propylene (FEP) copolymer.

The β relaxation is associated with the glass transition in PTFE and although there has been some confusion and controversy concerning T_g of PTFE (38), recent thermal expansion data by Araki (39) show that $T_g = 123°C$ (396°K). Other published works (40,41,44) disagree with Araki's value of T_g. In one case, Durrel et al. (40) obtain $T_g = -50°C$ from the extrapolation of DTA measurements on copolymers of TFE but no measurements were made on the homopolymer itself. In another published work, Ohzawa and Wada (41) rely on the questionable existence of a relaxation peak at $-13°C$ (1 kHz) to obtain $T_g \approx -50°C$. The existence of this peak is questionable because Ohzawa and Wada can only corroborate their findings with the questionable dielectric data of Krum and Müller (25) and mechanical measurements of Becker (24). We also pointed out the difficulty in interpreting mechanical relaxation data in terms of molecular concepts for semicrystalline substances (16).

Two criteria that can be used to decide whether or not a given transition temperature is T_g are: (1) the temperature dependence of the relaxation times in the temperature range $T_g < T < T_g + 100°C$ is according to the WLF equation (19); and (2) the difference between the coefficients of thermal expansion above and below T_g, $(\alpha_1 - \alpha_2)$, obeys the relationship $T_g(\alpha_1 - \alpha_2)$ = either 0.08 (43) or 0.113 (44). The concept that underlies these semiempirical criteria is that the glass state is an isofree volume state. The WLF criterion has not been applied to the β relaxation in PTFE because of considerable overlap of the β and room-temperature crystal relaxation. In 1965 Araki (39) published data on the thermal expansion of nine samples of PTFE which varied in crystallinity from 52 to 76%. He observed a second-order transition in each sample in the vicinity of 396°K. A quantitative analysis of the data showed that $T_g(\alpha_1 - \alpha_2) = 0.116$ where $T_g = 396°K$. The 396°K transition also follows another empirical rule, $T_g \approx 2/3\ T_m$ (45) where T_m, the melting point temperature, is 600°K for PTFE (46). In addition, unpublished specific volume versus temperature data of Quinn and Roberts (47) show a second-order transition at 400°K.

Mechanical studies of the TFE-HFP (FEP) copolymer show that the β loss peak of PTFE shifts to lower temperatures with small additions of HFP. McCrum (37), who maintains that the β peak arises in the amorphous phase, attributes this shift in temperature to the plasticizing effect of the CF_3 side groups, which lowers T_g. Eby and Wilson (28), however, suggest that the molecular origin of the temperature shift is in the crystalline phase; the CF_3 side groups, which are incorporated in the PTFE crystal lattice as defects, decrease the intermolecular interaction and lower the potential barrier for relaxation. Since the β relaxation has been associated with T_g and the amorphous phase by thermal expansion measurements, McCrum's explanation of this shift in temperature is most plausible.

C. Summary of Polytetrafluoroethylene Relaxations

We have attempted to establish a similarity between the relaxation spectrum of PTFE and that spectrum which is seen in other semicrystalline polymers (15,16a,48–50). In order to do this it is necessary to establish which one of the relaxation processes is associated with T_g. For PTFE we have relied on Araki's data ($T_g = 396°K$) and rejected other data which show that $T_g \approx -50°C$. Araki's data alone probably will not settle the confusion regarding T_g. It has been suggested (41) that PTFE may possess two glass transitions, one associated with a liquid crystalline phase and the other associated with the amorphous liquid. Continuing NMR (22) and X-ray (51) studies of the molecular nature of PTFE are necessary.

As it is usually observed the activation energy of the γ (Glass II) process is several times less than the activation energy of the β (Glass I) process. The crystal relaxation is associated with the room-temperature crystal-crystal transition in PTFE. The presence or absence of an α peak at high temperatures has not been resolved, although indications of a high-temperature relaxation which overlaps with the β relaxation have been noted by McCrum (31), Eby and Sinnott (30), and McCall (32).

III. POLYCHLOROTRIFLUOROETHYLENE

Polychlorotrifluoroethylene, PCTFE, is a stable, crystallizable, polar molecule which is well suited for dielectric studies. In going from PTFE to PCTFE, the chlorine-for-fluorine substitution creates an unsymmetric polar molecule which does not prevent crystallization but considerably lowers the melting point from 327°C for PTFE to 225°C for PCTFE (52). Observations of the glass transition temperature are not in agreement (15, 23, 53, 54). From specific volume and relaxation experiments T_g has been reported to be in the temperature region between 45 and 90°C. Hoffman, Williams, and Passaglia (23) report that T_g varies with degree of crystallinity.

Since the polymer is stable and since the amount of crystallinity can be controlled over a wide range, 12–90%, PCTFE presents a good opportunity to study dielectric relaxation phenomena in semicrystalline polymers. Ultimately, a description of the mechanisms of relaxation in terms of the morphological state of the polymer is desired. Studies of PCTFE morphology include observations of the crystallization kinetics and electron microscope studies. Crystallization kinetics show that the crystal growth, as in polyethylene, is spherulitic (55) and consists of folded-chain crystallites. Electron microscope studies by Khoury and Barnes have shown that the solution grown PCTFE crystals consist of multisectional disk shaped lamellae with a substantial number of voidlike regions encapsuled between adjacent sectors (56).

Next to PTFE, the most extensive dielectric study of a fluoropolymer is that of PCTFE. This is mostly due to the efforts of Scott, Hoffman, and co-workers (15), who made an exhaustive study of the dielectric properties of bulk (unfractionated) PCTFE as a function of temperature, frequency, thermodynamic history, and degree of crystallinity. Other dielectric investigations of PCTFE have been carried out by Mikhailov and Sazhin (57), by Hartshorn, Parry, and Rushton (58), by Nakajima and Saito (59), and by Hara (60). Only the investigation by Scott et al. considered the effects of crystallinity in detail.

Relaxation experiments on PCTFE follow the pattern which has been observed in many other semicrystalline polymers (16, 48, 49): α, β, and γ relaxations are observed in accordance with the schematic diagram of Fig. 4. Figure 7 shows the data of Scott and co-workers (15) for the 80% crystalline

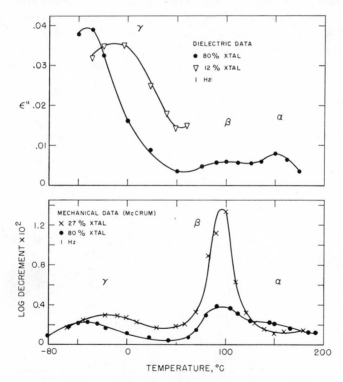

Fig. 7.   (a) Dielectric loss versus temperature for PCTFE at 1 Hz for an 80% and a 12% crystalline sample, data of Scott et al. (15,23). (b) Mechanical loss tangent (proportional to log decrement) versus temperature for PCTFE at 1 Hz, data of McCrum (50).

sample in which the three loss regions are clearly separated. These three relaxations, α, β, and γ, occur approximately at 150°C (1 Hz), 90°C (1 Hz), and -37°C (1 Hz) for the isothermally crystallized 80% crystalline sample. The temperatures of these 1 Hz loss maxima were observed to be slightly dependent on crystallinity (15,23). The mechanical loss data of McCrum (50), which are shown below the dielectric data in Fig. 7, also display the three loss regions in the 80% crystalline sample. In samples of lower crystallinity the α and β loss peaks either merge or one of them is absent. These mechanical data agree with those of Crissman and Passaglia (16). Figure 8 shows the frequency-temperature characteristics for the 80% crystalline sample. Activation energies calculated from the slopes of these curves are 14 kcal/mole, 40–80 kcal/mole, and 80 kcal/mole for the γ, β, and α process.

We concentrate our attention here on the work of Scott, Hoffman, and co-workers (15) because these workers were able to separate the α, β, and γ relaxations and because their studies covered a broad range of crystallinity and frequency. Four samples with different crystallinities were prepared: (1) quenched from the liquid phase, 12% crystalline; (2) quenched from the liquid phase, 44% crystalline; (3) quenched from the liquid phase and then annealed at 190°C, 73% crystalline; (4) isothermally crystallized at 200°C and 190°C, 80% crystalline. These samples retained a constant degree of crystallinity as long as they were stored below T_g.

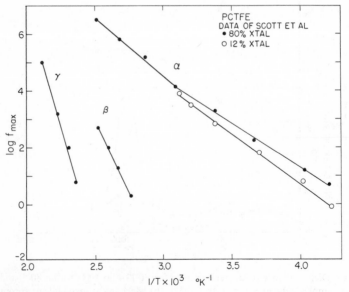

Fig. 8.   Log f_{max} versus $1/T$ from dielectric data of Scott et al. (15,23) on PCTFE.

A. The α Relaxation

Initially, Scott, Hoffman, and co-workers (15) called the α relaxation the "high-temperature process". Later, the nomenclature was changed to "α process" (23). The main characteristics of this relaxation are: (1) it can only be resolved in the isothermally crystallized sample, 80% crystalline; (2) the energy dissipated is small in comparison with the γ relaxation; (3) the activation energy is 80 kcal/mole; (4) its molecular origin is in the crystalline phase.

Hoffman, Williams, and Passaglia (23) have associated the α relaxation in PCTFE with the α relaxation in polyethylene and have assumed that the relaxation arises from the same source in both polymers. The high-temperature relaxations in both polymers are of the α type and the crystalline phase in both polymers consists of chain-folded lamellae. In a set of ingenious dynamic mechanical experiments, Sinnott (22) studied the α and γ relaxations in crystalline polyethylene. Since the polyethylene samples were solution-grown single crystals there was no question that the molecular origin of the observed relaxations were in the crystalline phase. The two relaxations Sinnott observed corresponded in temperature and frequency to the α and γ relaxations of bulk polyethylene. Thus the connection is made between single crystals of polyethylene and bulk PCTFE.

Sinnott (22) found that as he annealed the polyethylene single crystals at progressively higher temperatures the intensity of the α peak decreased and the loss maximum shifted to higher temperatures. The molecular interpretation of this effect (decrease in intensity) is ambiguous. Sinnott (22a) attributed the α relaxation to reorientation of the folds at the surface of the lamellae; but Colson and Eby (22b) recently showed that Sinnott's assignment was an arbitrary one. The source of the α relaxation in polyethylene has not been firmly established.

The polyethylene experiments also showed that, in extrapolation to the limit of zero number of folds, the intensity of the α process did not approach zero. Consequently, Hoffman, Williams, and Passaglia (23) viewed the α relaxation as two overlapping mechanisms, the reorientation of a fold and the rotation and twisting of a long polyethylene chain in the interior of the paraffinlike crystal lattice. An attempt was made to relate the α relaxation with those relaxations which are observed in the crystalline state of low-molecular-weight, long-chain paraffins, esters, and ketones (61).

B. The β Relaxation

The β relaxation occurs at approximately 90°C (1 Hz) (15). It has been attributed to the relaxation in the amorphous phase above the glass transition temperature T_g, because the temperature dependence of the relaxation times is according to the WLF equation (15,23). Specific volume-temperature measurement (53,54) have placed the glass transition between 45 and 60°C, considerably below the 1 Hz relaxation at 90°C. The difference between the 1 Hz relaxation temperature and T_g from specific volume measurements is usually not larger than 15°C. This suggests that T_g for the dielectric samples of Scott et al. is somewhat higher than 60°C (23).

The intensity of the dielectric loss of the β relaxation is small by comparison with the γ relaxation. It is less than what is expected from a polar polymer at the glass transition, even though the polymer is part crystalline. A chemically similar polymer, polyvinylchloride $\sim(100\%$ amorphous), has a loss maximum which is 100 times higher than the ε''_{max} of PCTFE at T_g (62). This large difference between the two β loss intensities cannot be explained by the difference in the crystallinities and/or dipole moment. An explanation may be found in the tacticity and flexibility of the two polymer molecules.

C. The γ Relaxation

The γ relaxation occurs at $-40°C$ (1 Hz) for the 80% crystalline sample and at $-10°C$ (1 Hz) for a 12% crystalline sample (15). From the data of Scott et al., the main characteristics of the relaxation are: (1) it decreases in intensity with increasing crystallinity but does not approach zero intensity at 100% crystallinity; (2) the distribution of relaxation times is very broad (half width of the loss curve = 4 decades), which indicates many microscopic modes of motion; (3) the temperature of the loss maximum increases with decreasing crystallinity; (4) the activation energy of the relaxation is approximately 15 kcal/mole and varies slightly with crystallinity (15). Moreover, on the low-temperature side of the peak ($T < -40°C$), the complex mechanical modulus at 1 Hz is independent of the crystallinity (16b).

On the basis of items (1) and (3), Hoffman et al. (23) proposed that the γ relaxation was composed of two parts, one arising in the amorphous phase γ_a and the other in the crystalline phase γ_c.

The presence of an amorphous relaxation in semicrystalline PCTFE below T_g is expected since this relaxation is commonly observed in totally amorphous polymers (18). The molecular nature of the γ_a peak was for some time associated with the motion of rotatable side groups of amorphous polymers. However, further observation has shown that this relaxation occurs in polymers which have no side groups, e.g., PCTFE, PTFE, polyethylene, polyethylene oxide. In light of this information the crankshaft model of Schatzski (17,20) provides the best view of the molecular origins of the γ_a relaxation. In this model a short segment of the main chain (four or five monomer units) forms a double kink and moves in a crankshaft motion about the main chain.

For the γ_c relaxation, Hoffman, Williams, and Passaglia (23) calculated the properties of a model which involved the motion of the molecule in the vicinity of a crystal defect. The defect they proposed was a chain end in the crystal interior. The vicinity of the chain end provides a region of low-energy relaxation which extends over various lengths of polymer molecules according to the position of the defect with respect to the lamellar surfaces. Such a distribution in lengths would bring about a distribution of relaxation times.

D. The Status of Molecular Models

The Hoffman, Williams, and Passaglia models of relaxation in polyethylene and PCTFE have provided a basis for establishing a connection between relaxation phenomena and polymer morphology. At this time, however, these models remain quasi-theoretical descriptions whose validity and uniqueness has not been determined. For example, electron microscope data of Khoury and Barnes (56) indicate a large difference in the PCTFE crystals compared with polyethylene crystals grown under similar conditions. It thus appears that modes of relaxation (perhaps at grain boundary sites) would be present in PCTFE which were not present in polyethylene. It is not clear that the relaxations in both polymers arise from exactly the same sources or that some other model will describe the data as well.

As a test of these models Hoffman (63) has proposed that mechanical and dielectric measurements be made on narrow fractions of PCTFE. In this way the end-group concentration can be controlled and its influence on the relaxations can be determined. The dependence of crystallinity on molecular weight, if any, could be isolated.

IV. POLYVINYLIDENE FLUORIDE

Polyvinylidene fluoride (PVF2) is a semicrystalline polymer with a melting point in the range 165–171°C (64). An extensive investigation of its dielectric behavior as a function of frequency, temperature, and degree of crystallization has been made by Sasabe et al (65). Dielectric measurements have also been made by Wentink (66), Ishida (67), and Peterlin and Holbrook-Elwell (68, 69). Other data, NMR, X-ray and IR spectra, have been published (64, 65, 70).

The dielectric, NMR, and dilatometric data of Sasabe et al. (65) give a full picture of the relaxation spectrum of PVF2. The spectrum follows the general character of the schematic in Fig. 4 for a semicrystalline polymer. The α, β, and γ relaxations data of Sasabe et al., are shown in Fig. 9. The magnitude of the dielectric loss for the α and β peak is large, $\varepsilon'' = 1.0$–2.0, ten times larger than the γ peak. The intensity of the loss peaks can be contrasted with those for PCTFE in Fig. 7; there is no simple explanation for this difference.

We note in Fig. 9 that the dielectric loss is a function of crystallinity; the intensity of the α peak increases as χ increases, but the reverse is true for the β peak. Thus the α relaxation has been assigned to the crystalline phase of the

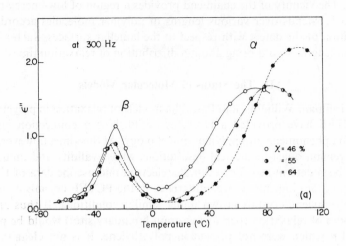

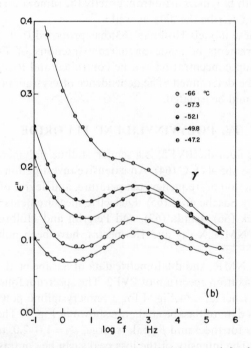

Fig. 9. (a) Dielectric loss ε'' versus temperature for the α and β relaxations in PVF2. χ is % crystallinity. (b) Dielectric loss ε'' versus log f for the γ relaxation in PVF2. Data of Sasabe et al. (65).

496

polymer and the β relaxation to the micro-Brownian motion of amorphous chain segments. The β peak occurs at the rubber-to-glass transition, immediately above T_g on the temperature scale, where T_g, as measured by Sasabe et al., is $-52°C$.

Other published dielectric data for PVF2 are in essential agreement and they are shown in the $\log f_{max}$ versus $1/T$ plot of Fig. 10. There has been some conflict regarding the assignment of the β peak, however. Peterlin and Holbrook-Elwell (68) incorrectly designated the high-temperature or α peak ($86°C$ at 1 kHz) as the β peak and these authors claimed that their unpublished data indicated that the glass transition occurred at $13°C$. The evidence that $T_g = -52°C$ is convincing: (1) the β peak of Sasabe et al. obeys the WLF equation; (2) NMR and dielectric data (65) show that the β peak is associated with the amorphous phase; (3) the dilatometric data of Sasabe et al. shows a distinct second order transition at $-52°C$. The volume-temperature data of Fig. 11 do show a second-order anomaly at $T_b = 13°C$, but this is not the primary glass transition (65). Furthermore, the α relaxation is of the Arrhenius type, a straight line in the plot of Fig. 10, whereas the data for the β relaxation traces out a curve of nonzero curvature in accordance with the WLF equation. Other dilatometric data of Mandlekern, Martin, and Quinn (71) give $T_g = -35°C$.

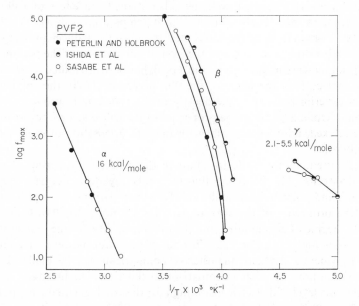

Fig. 10.   Log f_{max} versus $1/T$ for the α, β, and γ relaxations in PVF2; data of Peterlin and Holbrook (68), Ishida et al. (67), and Sasabe et al. (65).

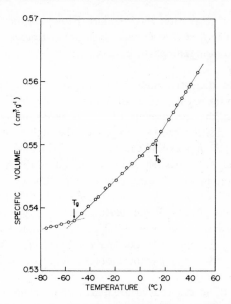

Fig. 11.   Volume-temperature data for PVF2; data of Sasabe et al. (65).

Peterlin and Holbrook-Elwell (69) also studied the effect of rolling the PVF2 samples into sheets of thin film at temperatures above T_g. Rolling introduces orientation of the molecules and crystals, reduces the magnitude of the α relaxation, and increases the overall polarization of the polymer. The significant increase in intensity of the β peak has been attributed to orientation of the molecules in the plane of the film perpendicular to the applied field. Within experimental error, the density of the rolled and unrolled sample is the same. Upon annealing, the properties of the rolled sample revert to those of the relaxed, unrolled material with random orientation of molecules and crystallites.

Often attempts are made to compare the dielectric properties of various vinyl halogens (72), polyvinyl fluoride, polyvinylidene fluoride, polyvinyl chloride, and polyvinylidene chloride. These comparisons are legitimate only if all the properties of the various samples are normalized with respect to density, degree of crystallinity, and the difference between the temperature of observation and T_g and/or T_m. When all of this information is available for each polymer, a detailed quantitative comparison can be made. Lack of information prevents such a comparison at this time. Qualitatively, it can be said that the level of dielectric loss is high for these halogenated vinyl polymers (tan $\delta_{max} \approx 0.1$) and that they possess large DC conductivity at elevated temperatures.

V. FLUORINATED ETHYLENE-PROPYLENE COPOLYMER

The copolymer of hexafluoropropylene and tetrafluoroethylene is called fluorinated ethylene propylene copolymer (FEP). The commercially available FEP consists of a low percentage of hexafluoropropylene (15–20%) so that the polymer can be considered to be PTFE with an occasional side group of CF_3 attached. There are several advantages of adding the CF_3 group to the PTFE molecule. First, the FEP retains the good qualities of PTFE (low dielectric loss, chemical stability, moisture resistance, and heat resistance) along with the added quality of being moldable. Second, the bothersome room-temperature volume transition in PTFE moves to lower temperatures with the addition of the CF_3 side groups (37) until the relaxation associated with the transition blends with the γ relaxation ($-85°C$, 1 Hz or $-60°C$, 1 kHz) at 14% hexafluoropropylene. This shift in the temperature of relaxation at 1 Hz can be seen in the mechanical data of McCrum (37) shown in Fig. 12. There is some sacrifice of high-temperature qualities since the melting point of FEP containing 15% hexafluoropropylene is $270 \pm 10°C$ compared with $327°C$ for PTFE (73).

A general outline of the dielectric properties of FEP copolymer can be obtained by combining the high-temperature measurements of Koizumi et al. (34) (16.8% hexafluoropropylene) with the low-temperature measurements of Eby and Wilson (28) (10.7% hexafluoropropylene). A quantitative comparison of these two sets of data is not possible because the degree of

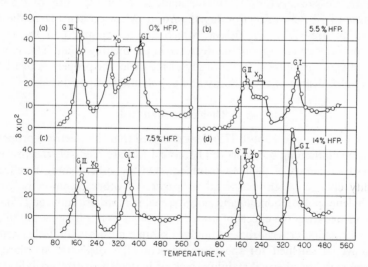

Fig. 12. Mechanical loss tangent (proportional to log decrement) versus temperature of PTFE (a) and three FEP copolymers (Reprinted from (37) N. G. McCrum, Die Makro molekulare Chemie **34**, 53 (1959)) by permission of the publisher, Hüthig & Wepf.)

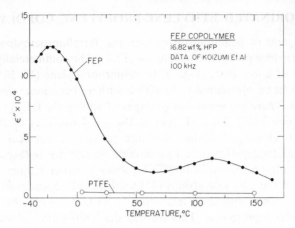

Fig. 13.   Dielectric loss ε'' versus temperature for FEP copolymer, 16.82% HFP, and for PTFE; data of Koizumi et al. (34).

crystallinity is not known in each case. From the dielectric data of Koizumi et al., shown in Fig. 13, along with the mechanical data of McCrum (14% HFP), it can be seen that no audio frequency dispersion region is present at room temperature. The relaxations seen in FEP are essentially the inherent β and γ relaxations of PTFE which have been shifted somewhat in temperature and intensity.

Eby and co-workers (36,73,74) have made an extensive study of the crystallization of FEP as a function hexafluoropropylene content. Here the CF_3 side groups are viewed as defects which are incorporated in the crystal lattice and cause a corresponding decrease in the melting point. This concept will have application for a future study of the dielectric properties of FEP as a function of degree of crystallinity and copolymer composition.

A. A Dielectric Standard

Work by Scott and Kinard (75) demonstrated that the electrical properties of fluorine-containing polymers (PTFE and FEP) are relatively insensitive to changes in humidity of their environment. On the other hand, the effect of humidity change on the electrical properties of other common polymers, polyethylene, polycarbonate, and polystyrene, is more pronounced (75). The independence from the effects of humidity changes over a long period of time is one of the first criteria that a standard sample must satisfy. Other criteria that must be satisfied for the establishment of a dielectric standard are chemical stability, thermal stability, lack of hysteresis or irreversibility after reasonable changes of environment, dimensional stability, and low dielectric loss. By stability we mean the following: the properties of the sample must be constant to the limit of the state of the measurement art. For ε' and ε'', the measurement capabilities at this time are such that for solid specimens ε' can be measured to $\pm 0.1\%$ and ε'' as small as 5×10^{-6} can be measured.

The intention of Scott and Kinard (75) was to make a preliminary investigation of several commercially available polymers in order to decide which could be used for the manufacture of standard samples. Although PTFE had the best electrical stability, it was ruled out as a standard because it lacked dimensional stability at room temperature. FEP was the next best electrically and satisfied all other criteria. Table 1 shows the maximum changes observed in six polymer specimens over a one-year period after they were subjected to a change in humidity.

The effect of humidity change was studied by storing a sample in a constant humidity environment for a long time (several hundred days) and then subjecting the sample to an environment that was either much lower or higher in humidity than the stored humidity. At this second humidity the dielectric constant and loss were measured for several hundred days. [Scott and Kinard (75) worked at two extremes of humidity, $<1.5\%$ and 52% relative humidity.] Figures 14 and 15 show the results of these long time studies of PTFE and FEP. Here the dielectric constant and dissipation factor at 1000 Hz are plotted against time from humidity change for the two polymers which had the best performance under this test. From the point of view of Fig. 14, PTFE is remarkably insensitive to the presence or absence of water. Unfortunately, it is not dimensionally stable.

The FEP copolymer Scott and Kinard used contained 16% hexafluoropropylene. It may be possible to diminish the change of humidity effect in FEP by lowering the amount of hexafluoropropylene to a value of 8%, at which level the copolymer is closer in composition to PTFE and dimensional stability is still present at room temperature (37).

TABLE 1

Maximum Change in the Real, ε', and Imaginary, ε'', Parts of the Complex Dielectric Constant Occurring Over a Period of One to Two Years After an Abrupt Change in Humidity (75)

Polymer	Change in ε'	Change in ε''
PTFE	0.003	10×10^{-6}
FEP	0.005	30×10^{-6}
Polystyrene	0.010	35×10^{-6}
Polycarbonate	0.080	240×10^{-6}
Polyethylene	<0.002	150×10^{-6}
Polycyclohexylenedimethyleneterephalate	0.10	400×10^{-6}

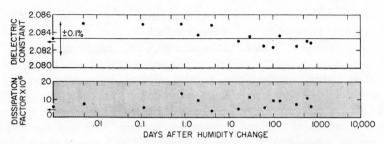

Fig. 14.   The real part of the dielectric constant ε' and tan δ versus days after humidity change for PTFE going from <1.5% relative humidity to 52% relative humidity; data of Scott and Kinard (75).

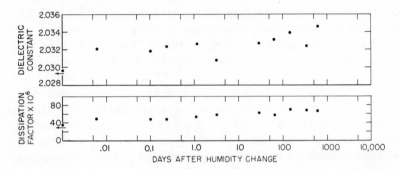

Fig. 15.   The real part of the dielectric constant ε' and tan δ versus days after humidity change for FEP going from <1.5% relative humidity to 52% relative humidity; data of Scott and Kinard (75).

From the dielectric and mechanical data in Figs. 12 and 13, we can see that at room temperature and audio frequencies the loss is small. A low loss level is essential in dielectric standards in order that the energy absorbed during the measurement of the sample does not cause significant temperature change. As a high-frequency standard, FEP will have its limitations since the loss at room temperature increases as frequency increases above the audio region (34).

VI. SUMMARY

The foregoing discussion nearly exhausts the available published information on dielectric properties of fluorine-containing polymers. There are a few other papers which contain a limited amount of dielectric data such as Ishida's paper on polyvinylfluoride (76) and published work by Sorkin and co-workers on some poly(fluoroalkyl vinyl ethers) (77). Thin films of polyvinylfluoride have been measured by Sacher (78); these measurements show the high DC conductivity above room temperature, which is common in vinyl halogen polymers. The electrical properties of irradiated PTFE also have been of interest (79).

Since there have been only half a dozen polymers which we could review in detail, there remain many fluorine-containing polymers whose dielectric properties are still unknown. There is a large volume of dielectric data forthcoming as these polymers become available.

References

1. P. Debye, "Polar Molecules," Dover, New York, 1947.
2. C. J. F. Bottcher, "Theory of Electric Polarization," Elsevier, New York, 1952.
3. C. P. Smyth, "Dielectric Behavior and Structure," McGraw-Hill, New York, 1955.
4. H. Fröhlich, "Theory of Dielectrics," 2nd ed., Oxford University Press, London, 1958.
5. R. H. Cole, *Ann. Rev. Phys. Chem.*, **11**:149 (1960); C. P. Smyth, *Ann. Rev. Phys. Chem.*, **17**:433 (1966).
6. *Digest of Literature on Dielectrics*, **1** (1936) to date, National Academy of Sciences, National Research Council, Washington, D.C.
7. A. J. Curtis, "Progress in Dielectrics," Vol. 2, edited by J. B. Birks and J. H. Schulman, Heywood, London, 1960, p. 29.
8. A. T. McPherson, *Rubber Chem. Technol.* **36**:1230 (1963).
9. L. Hartshorn, "Radio Frequency Measurements," John Wiley and Sons, New York, 1941.
10. A. R. Von Hippel, "Dielectric Materials and Applications," John Wiley and Sons, New York, 1954.
11. M. G. Broadhurst and A. J. Bur, *J. Research NBS*, **69C**:165 (1965).
12. J. G. Kirkwood and R. M. Fuoss, *J. Chem. Phys.*, **9**:329 (1941).
13. F. Bueche, *J. Chem. Phys.*, **22**:603 (1954).
14. B. Zimm, *J. Chem. Phys.*, **24**:269 (1956).
15. A. H. Scott, *et al.*, *J. Research NBS*, **66A**:269 (1962).
16. (a) E. Passaglia and G. M. Martin, *J. Research NBS*, **68A**:519 (1964); (b) J. M. Crissman and E. Passaglia, *J. Polymer Sci. C*, **14**:237 (1966).
17. R. F. Boyer, *J. Polymer Sci. C*, **14**:3 (1966).
18. J. D. Ferry, "Viscoelastic Properties of Polymers," John Wiley and Sons, New York, 1961.
19. M. L. Williams, R. F. Landel, and J. D. Ferry, *J. Am. Chem. Soc.*, **77**:3701 (1955).
20. T. F. Schatzki, *Am. Chem. Soc. Polymer Preprints*, **6**:646 (1965); also *J. Polymer Sci. C* **14**:139 (1966).
21. M. Takayanagi, *Mem. Faculty of Eng.*, *Kyushu Univ.*, **23**:41 (1963).
22. (a) K. M. Sinnott, *J. Polymer Sci. C*, **14**:141 (1966); (b) J. P. Colson and R. K. Eby, *Bull. Am. Phys. Soc.*, **15**:351 (1970).
23. J. D. Hoffman, G. Williams, and E. Passaglia, *J. Polymer Sci. C* **14**:173 (1966).
24. P. Ehrlich, *J. Research NBS*, **51**:185 (1953).
25. F. Krum and F. Müller, *Kolloid-Z.* **164**:81 (1959).
26. W. Reddish, C. B. Cresswell, and P. J. Hyde, Technical Service Note F12, Imperial Chemical Industries Ltd., Plastics Div., Welwyn Garden City, Hertfordshire, England, p. 20.
27. G. P. Mikhailov, S. P. Kabin, and A. L. Smolianskii, *Z. Tech. Phys. (USSR)* **25**:2179 (1955).

28. R. K. Eby and F. C. Wilson, *J. Appl. Phys.*, **33**:2951 (1962).
29. W. Reddish, J. G. Powles, and B. I. Hunt, *J. Polymer Sci. B* **3**:671 (1965).
30. R. K. Eby and K. M. Sinnott, *J. Appl. Phys.*, **32**:1765 (1961).
31. N. G. McCrum, *J. Polymer Sci.*, **34**:355 (1959).
32. D. W. McCall, D. C. Douglass, and D. R. Falcone, *J. Phys. Chem.*, **71**:998 (1967).
33. F. A. Quinn, D. E. Roberts, and R. N. Work, *J. Appl. Phys.*, **22**:1085 (1951); also E. S. Clark and L. T. Muus, *Z. Kristallographie*, **117**:119 (1962).
34. N. Koizumi, S. Yano, and F. Tsuji, *J. Polymer Sci. C*, **23**:499 (1968).
35. G. R. Johnson, 1966 Ann. Rpt. Conf. Elect. Insulation and Dielectric Phenomena, National Academy of Sciences—National Research Council Publication No. 1484, p. 78.
36. L. H. Bolz and R. K. Eby, *J. Research NBS*, **69A**:481 (1965).
37. N. G. McGrum, *Makromol. Chemie*, **34**:50 (1959).
38. R. F. Boyer, *Rubber Chem. Technol.*, **36**:1303 (1963).
39. Y. Araki, *J. Appl. Polymer Sci.*, **9**:421 (1965).
40. W. S. Durrell, E. C. Stump, and P. D. Schuman, *J. Polymer Sci. B*, **3**:831 (1965).
41. Y. Ohzawa and Y. Wada, *Jap. J. Appl. Phys.*, **3**:436 (1964).
42. G. W. Becker, *Kolloid-Z*, **167**:44 (1959).
43. N. Hirai and H. Eyring, *J. Polymer Sci.*, **37**:51 (1959).
44. R. Simha and F. R. Boyer, *J. Chem. Phys.*, **37**:1003 (1962).
45. R. G. Beaman, *J. Polymer Sci.*, **9**:472 (1952).
46. W. E. Hanford and R. M. Joyce, *J. Am. Chem. Soc.*, **68**:2082 (1946).
47. D. E. Roberts, private communication.
48. Y. Wada, *Phys. Soc. Japan*, **16**:1226 (1961).
49. N. G. McCrum, *J. Polymer Sci.*, **54**:561 (1961).
50. N. G. McCrum, *J. Polymer Sci.*, **60**:S3 (1962).
51. E. S. Clark, *J. Macromol. Sci.-Phys.*, **B1**:795 (1967).
52. J. D. Hoffman and J. J. Weeks, *J. Research NBS*, **66A**:13 (1961).
53. L. Mandelkern, G. M. Martin, and F. A. Quinn, *J. Research NBS*, **58**:137 (1957).
54. J. D. Hoffman and J. J. Weeks, *J. Research NBS*, **60**:465 (1958).
55. J. D. Hoffman and J. J. Weeks, *J. Chem. Phys.*, **37**:1723 (1962).
56. F. Khoury and J. D. Barnes, National Bureau of Standards, private communication.
57. G. P. Mikhailov and B. I. Sazhin, *Soviet Phys. Tech. Phys.*, **1**:1670 (1956). (Translated from *Z. Teck. Fiz.*)
58. L. Hartshorn, J. V. L. Parry, and E. Rushton, *J. I. E. E.* **100**, Pt. 11A, No. 3 (1953).
59. T. Nakajima and S. Saito, *J. Polymer Sci.*, **31**:423 (1958).
60. T. Hara, *Jap. J. Appl. Phys.*, **6**:135 (1967).
61. R. J. Meakins, "Progress in Dielectrics," Vol. 3, edited by J. B. Birks and J. Hart, Heywood, London, 1961, p. 151.
62. R. M. Fuoss, *J. Am. Chem. Soc.*, **63**:378 (1941).
63. J. D. Hoffman, private communication.
64. T. Wentink, L. J. Willwerth, and J. P. Phaneuf, *J. Polymer Sci.*, **55**:551 (1961).
65. H. Sasabe et al. *J. Polymer Sci.*, **7**:1405 (1969).
66. T. Wentink, *J. Appl. Phys.*, **32**:1063 (1961).
67. Y. Ishida, M. Watanabe, and K. Yamafuji, *Kolloid-Z*, **200**:48 (1964).
68. A. Peterlin and J. D. Holbrook, *Kolloid-Z*, **203**:68 (1965).
69. A. Peterlin and J. Holbrook-Elwell, *J. Materials Sci.*, **2**:1 (1967).
70. (a) J. B. Lando, H. G. Olf, and A. Peterlin, *J. Polymer Sci. A1*, **4**:941 (1966); (b) S. Enomoto, Y. Kawai, and M. Sugita, *J. Polymer Sci. A2* **6**:861 (1968).
71. L. Mandelkern, G. M. Martin, and F. A. Quinn, *J. Research NBS*, **58**:137 (1957).

72. Y. Ishida, M. Yamamoto, M. Takayanagi, *Kolloid-Z*, **168**:124 (1960).
73. J. P. Colson and R. K. Eby, *J. Appl. Phys.*, **37**: 3511 (1966).
74. R. K. Eby, *J. Appl. Phys.* **34**: 2442 (1963).
75. A. H. Scott and J. R. Kinard, *J. Research NBS*, **71C**:119 (1967).
76. Y. Ishida and K. Yamafuji, *Kolloid-Z*, **200**: 50 (1964).
77. H. Sorkin, et al. 1964, Ann. Rpt. Conf. Elect. Insulation, National Academy of Sciences—National Research Council Publ. 1238, p. 34.
78. E. Sacher, *J. Polymer Sci. A*2, **6**:1813 (1968).
79. K. Yahagi and A. Danno, *J. Appl. Phys.*, **34**: 804 (1963).
80. R. W. Gray and N. G. McCrum, *J. Polymer Sci. A*2, **7**:1329 (1969).

16. STRUCTURE AND MECHANICAL PROPERTIES OF FLUOROPOLYMERS

GEORGE P. KOO,* *Allied Chemical Corporation, Plastics Division, Morristown, New Jersey*

Contents

I. INTRODUCTION

Thorough investigations of structure and mechanical properties of a polymer, unlike the synthesis and characterization aspects, are not begun until the polymer is well established and in all probability is commercially available. The fluorine-containing polymers of commercial importance are polyvinyl fluoride (PVF), polyvinylidene fluoride (PVF$_2$), polychlorotrifluoroethylene (PCTFE), polytetrafluoroethylene (PTFE), and copolymers of tetrafluoroethylene and hexafluoropropylene, commercially known as FEP. The extent to which the structure and properties of these polymers have been investigated ranges from quite extensive for PTFE [see one recent literature compilation (1)] to few scattered literature reports on PVF and PVF$_2$ (2). Consequently, this chapter reflects the uneven abundance of information available.

* Present address: Pharmetrics Inc., Palo Alto, Calif.

The mechanical properties of a polymer represent one aspect of macroscopic behavior of the polymer. Although the relationships between mechanical properties and the molecular structure of a polymer are not always clear cut, it is always the ultimate desire and goal to explain the mechanical behavior in terms of particular structural and morphological features of the polymer. Therefore it is appropriate that we first briefly review the structure and morphology of the fluoropolymers in the solid state.

II. STRUCTURE AND MORPHOLOGY

Due to the larger van der Waals radius of the fluorine atom compared to the hydrogen atom, the substitution of fluorine for hydrogen on the carbon-carbon chain leads to the gradual twisting from the planar, fully extended, zig-zag conformation of polyethylene to the helical coil of polytetrafluoroethylene. The substitution of one fluorine atom (as in PVF) can be accommodated by the planar zig-zag conformation (3,4). Polyvinylidene fluoride with two fluorine atoms on the same carbon atom shows the planar zig-zag conformation when stretched between room temperature (4) and 50°C (5), which is commonly known as the β form. However, the polymer also has another less well-defined, apparently twisted or helical conformation under other conditions (4, 5), which is known as the α form. At room temperature, the α form appears to be thermodynamically the most stable form (6). The two crystalline forms of PVF_2 can also be isolated by solution-grown crystals using different solvents for each form (7); the two forms can be clearly identified by their infrared spectra (8).

When all the hydrogen atoms are substituted, as in the case of PCTFE and PTFE, the polymer chains exhibit a helical conformation. In particular, Bunn and Howell (9) pointed out that the helical twist of the C—C chain of PTFE results in a nearly perfect cylinder with an outer sheath of fluorine atoms. This is most easily seen by considering the end view of the molecule (Fig. 1). The outstanding chemical resistance and the low cohesive energy density of this

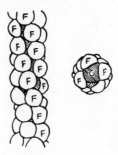

Fig. 1.   Schematic representation of the polytetrafluoroethylene helix (9).

polymer are attributed to the presence of this inert, completely "fluorinated" outer layer. Weak intermolecular attraction due to the fluorinated outer sheath coupled with an effectively rigid, rodlike configuration, due to the helical shape, facilitates the slip of the chains past each other: this is the reason given to account for the low friction of the PTFE surface, the tendency to cold flow, and the high ductility of this material at low temperatures. The same argument could be applied to the similar properties of PCTFE and FEP, except the presence of the chlorine atoms and of the side groups disturbs the perfect symmetry and results in lower chemical resistance, thermal stability, and melting point compared to PTFE.

The helical conformation of polytetrafluoroethylene has been well characterized (9–11). Below 19°C, the helix has a period of 13 carbon atoms per 180° turn, and the helices are arranged in a triclinic lattice. The helix untwists slightly to 15 carbon atoms per 180° turn at 19°C and is accompanied by a lattice expansion. This is a first-order transition and its effect on mechanical properties will be discussed subsequently. The chains rearrange into a hexagonal packing above 19°C up to 30°C; above 30°C, the chain helix becomes irregularly twisted. The degree of disorder about the chain axis increases with temperature. However, up to the melting point, the lateral hexagonal packing of the chains in the crystalline region persists, perhaps as a reflection of remarkable chain stiffness.

There is no completely satisfactory explanation for the high chain stiffness of PTFE; it is usually attributed to the high bond energy of the C—F bond and the mutual repulsion of fluorine atoms, as both factors tend to resist the bending of chain backbone.

In addition to X-ray characterization and the inferences from mechanical behavior, other indications of the chain stiffness come from the polymer behavior in the molten state. Bunn and co-workers (12) pointed out that the PTFE melt has extremely low melt strength and does not exhibit the rubberlike behavior common to high-molecular-weight polymer melts. This is a clear indication that in the molten state, the chains are not flexible and coiled enough for chain entanglement to occur; randomly coiled chains with chain entanglement are generally agreed to be the cause of the rubberlike properties of high-polymer melts.

Moreover, in order for entanglement to occur, the chain conformation would have to change drastically from the helical shape upon melting. This clearly does not occur, since the unusually high melting point ($T_m = 327°C$) combined with a low enthalpy of fusion (comparable only to polyethylene) can be thermodynamically consistent only with a low entropy of fusion:

$$T_m = \frac{\Delta H_f}{\Delta S_f}$$

The low entropy of fusion implies that the molecular chains in the melt change very little from the helical shape and high degree of order and do not form random coils. An interesting sidelight is that PTFE does melt fracture (13) (at 10^{-5} sec^{-1}, 380°C), indicating that chain entanglement is not a requisite for melt instability.

The structure of PCTFE has not been as thoroughly examined as that of PTFE. The available data indicate that the fiber axis of the PCTFE helix, unlike PTFE, appears to vary with different experimental conditions (14). The repeat distance has been reported to be 35 Å (15), equivalent to 13 monomer units (similar to PTFE), and also to be 43 Å (11), equivalent to 16 monomer units.

The commercial FEP copolymer is assumed to have the same basic helical conformation as PTFE, but it is probably irregularly twisted due to the interference of the hexafluoropropylene (HFP) units on the chain. The characteristic chain rigidity of PTFE is lost with increasing content of HFP; an indication of considerable chain flexibility is the millionfold decrease in melt viscosity of commercial FEP compared to PTFE (16).

In spite of the common helical conformation between PCTFE and PTFE, the morphology of the crystalline regions is distinctly different. Spherulites are observed when PCTFE is cooled from the melt (17). On the other hand, the spherulitic structure typical of virtually all crystalline polymers is not observed in PTFE. Examination by electron microscopy of PTFE fracture surfaces reveals that the crystalline regions contain long bands whose width ranges from 0.2 to 1.0 μ (10^4 Å), and the bands contain parallel striations which are perpendicular to the long axis of the band (12). Figure 2 is a typical example. The width of the bands is a direct function of the cooling procedure applied to the melt; slow cooling yields wide bands, whereas rapid quenching produces narrow bands (18). At sufficiently low degree of crystallinity, the band formation disappears altogether (19).

Optical studies of drawn PTFE fibers showed that they were positively birefringent (12), the refractive index along the fiber axis being 0.12 higher than the lateral refractive index ($\Delta n = 0.12$). Therefore for PTFE positive birefringence is associated with the axis of orientation of the chain backbones. The crystalline bands, on the other hand, are negatively birefringent, indicating that the chain orientation must be perpendicular to the band axis. Bunn and co-workers (12) concluded that PTFE chains lie parallel and straight in the direction of the striations for several thousand chain atoms. Since the length of the polymer chain is at least ten times the width of the band, the chain is considered to be folded along the edges of the band. This conclusion, although tentative, preceded the discovery of chain folding in lamellae and subsequently was frequently cited as evidence linking chain folding in single crystals to similar expectation of crystals in bulk.

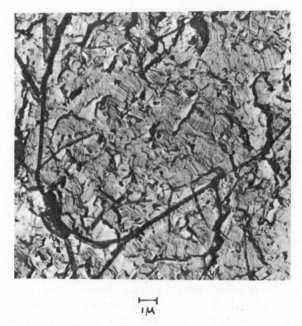

Fig. 2. Electron micrograph of fractured surface showing the crystalline morphology of polytetrafluoroethylene slowly cooled from 380°C.

III. MOLECULAR WEIGHT AND CRYSTALLINITY

Both molecular weight and crystallinity have profound effects on the mechanical properties of a given polymer. The relationship between properties and these parameters are known to varying degrees for commercial polymers, depending on how readily and accurately average molecular weight distribution and degree of crystallization can be measured.

In the case of fluorine-containing polymers, molecular-weight measurements are particularly difficult because of the difficulties in dissolving the polymers. Although solvents exist for PVF and PVF_2, data on molecular-weight determinations have not come to our attention. Solution viscosities on PCTFE have been measured at 130°C in 2,5-dichlorobenzotrifluoride and correlated with molecular weight (20). Polytetrafluoroethylene is known to dissolve under rather exotic conditions, e.g., in perfluorokerosene at 350°C (21), which is not suitable for molecular-weight measurements. The number average molecular weight of PTFE has been measured by end-group analysis using radioactive sulfur (22), and a correlation between the maximum relaxation time of the melt and weight average molecular weight has been proposed (23). All the aforementioned techniques are not convenient to perform routinely and are subject to large errors. Under most circumstances, the molecular weight of the commercial fluoropolymer is assumed to be sufficiently high that the properties are relatively unaffected by any variation in molecular weight.

Indirect methods of measuring the molecular weight of PCTFE and PTFE are used in the industry for routine, quality-control purposes. A so-called ZST (zero strength test) is used to determine the molecular weight of PCTFE (20). Samples are compression-molded and quenched in water to minimize crystallinity. Notched strips with a static load are placed in an oven at 250°C and the time-to-fail is measured. The log of time-to-fail is linear with the solution viscosity measurements (20).

The molecular weight of PTFE can be routinely determined from the specific gravity of samples molded by a standard procedure (22). Under a standard fabricating cycle, the rate of crystallization should vary inversely with molecular weight. Therefore, since specific gravity depends on degree of crystallinity, the specific gravity of PTFE samples from such a cycle should be related to the number average molecular weight, $\overline{M}_n$. Actual data show that the specific gravity is linearly related (with a negative slope) to log $\overline{M}_n$.

The degree of crystallinity can be measured by X-ray diffraction (9,14) infrared spectroscopy (24,25), or NMR (26). However, for multiple and repeated measurements, the techniques are too time-consuming to be practical; furthermore, they are not sensitive or reliable enough to distinguish small differences in crystalline content. In addition, the IR and X-ray techniques require very thin films (~ 1 mil); therefore, to measure the average crystallinity of thick sections, a series of samples taken at different depths is necessary in order to take account of the gradient in rate of cooling through the bulk of the material.

Since the simple assumption of a two-phase, crystalline-amorphous structure appears to be valid, it is possible to calculate the degree of crystallinity from density measurements. Because the volumes of the two phases are additive, the specific volume $\overline{V}$ of a sample is given by

$$\overline{V} = (1 - \chi)\overline{V}_a + \chi \overline{V}_c$$

where χ is the degree of crystallinity in volume fraction and the subscripts a and c refer to the amorphous and crystalline phases, respectively. In terms of densities, the equation for χ is

$$\chi = \frac{\rho - \rho_a}{\rho_c - \rho_a}\left(\frac{\rho_c}{\rho}\right)$$

Table 1 provides the reported values of ρ_a and ρ_c for PCTFE and PTFE.

Density measurements are, of course, simple and easy to carry out. A number of techniques are available for precise determination of density, the most popular being by displacement (ASTM Method D792) or by use of a

TABLE 1

Polymer	ρ_a	ρ_c	Reference
PCTFE	2.077	2.187	27
PTFE	2.00	2.30	24, 25, 28

density gradient tube (ASTM Method D1505). Careful sectioning of thick samples automatically averages any gradient of crystallinity from the surface to the core resulting from nonuniform cooling. However, because polytetrafluoroethylene parts are fabricated by sintering, the presence of voids randomly distributed in the part can be minimized but usually cannot be eliminated completely. Hence crystallinity determination by measurement of density tends to give a low value for this polymer. The magnitude of the error depends on the void content of the sample. Normal sound processing procedure easily keeps the void content below 1 %.

A technique of measuring crystallinity of PTFE which is independent of void content is to relate the degree of crystallinity to the shear modulus G' as measured by the torsion pendulum (29). The pendulum requires specimens in the form of flat strips of uniform thickness (cylindrical rods are less practical). The method is most useful when the void content is appreciable, rendering the density correlation unreliable. However, the relationship between shear modulus and degree of crystallinity has not been firmly established. Owing to the presence of the first-order transition, the shear modulus at room temperature changes rapidly with the slightest variation in temperature. To minimize this sensitivity to temperature, the shear modulus measured at both 15 and 23°C and correlated with degree of crystallinity (30) is shown in Fig. 3. The shear modulus data at 15°C are slightly more sensitive to crystallinity differences and show less scatter.

IV. TRANSITIONS, RELAXATIONS, AND DYNAMIC MECHANICAL SPECTRUM

It is now taken for granted that any discussion of properties of a polymer must include the effect of time and temperature. The effect of time will be discussed in a later section; the focus of this section is the effect of temperature. The most significant changes in mechanical properties occur at the temperature where a transition or relaxation takes place. Therefore the first- and second-order transitions of a polymer need to be located. However, common agreement among investigators on the location of these transitions does not always exist, mainly because of the different experimental techniques they use. The "static" measurements such as by dilatometry and calorimetry usually give a lower temperature for the same transition than dynamic methods

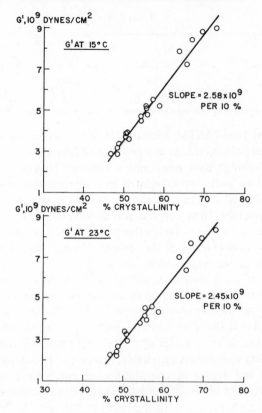

Fig. 3. The correlation of dynamic shear modulus at 15 and at 23°C with crystallinity of polytetrafluoroethylene.

(mechanical, dielectric, and NMR). When determined by dynamic measurements, the transition will shift to higher temperature with higher frequency. To confuse the issue further, static methods occasionally can locate a transition that cannot be detected by dynamic measurements and vice versa.

Dynamic measurements are preferred for studies of mechanical properties because the data more closely correspond with actual mechanical response of a polymer. Also, glassy transitions and relaxations in crystalline polymers are frequently obscured by the presence of crystalline regions when measured by static means.

The reported transitions of the fluorine-containing polymers are tabulated in Table 2. For dynamic measurements, the test frequency is given in parenthesis when available. The transitions are given in roughly increasing order of temperature from left to right, with the clearly major transitions in the T_g and T_m columns, where T_g and T_m are the glass transition and the melting point, respectively. The T_c column denotes a transition below T_m known to be due to the crystalline region.

The first two polymers of Table 2 seem " well-behaved." In any case, there are not enough (conflicting) data on the transitions of PVF and PVF$_2$ to

TABLE 2

Reported Transitions of Fluoropolymers

Polymer	Experimental methods	Other transitions (°C)	T_c (°C)	T_g (°C)	T_m (°C)	Ref.
PVF	n.a.				230	4
	dyn. mech. (1.7–5.4 cps)	−20		38	>192	31
	n.a.				198	32
	DTA				182–192	33
	n.a.	−75, 125		45		34
PVF$_2$	n.a.				185	4
	X-ray				171	5
	DTA				160–167	33
	dilatometric			−35		35
	dielectric (10 cps)	−80	80	−20		36
PCTFE	dilatometric			52	224	27
	dyn. mech. (3.3–12 cps)	0		104	>202	31
	dilatometric			45		35
	dyn. mech.	−10	−40	90		37
	dielectric (1 cps)	−13.5 to −41				38
	dyn. mech.	−32		90		39
PTFE	X-ray		20, 30		330	9
	dyn. mech. (2–12 cps)	−70	31	130		31
	DTA				324–344	33
	calorimetric	−113	20, 30			40
	calorimetric		20, 30			41
	dilatometric		20, 30			42
	dyn. mech. (400–730 cps)	−75	47	140		43
	dyn. mech. (∼1 cps)	−97	27	127	327	44
	dilatometric			130		45
	dyn. mech.	−87	30	115	310	46
FEP	n.a.				270	16
	dyn. mech. (∼1 cps)	−96	<30	<127		47

create any controversy. Only recent dielectric measurements (36) on PVF_2, which located three dielectric relaxations (at 10 cps) near 80, -20, and $-80°C$, need to be discussed. The dielectric loss at $80°C$ was observed to increase with increasing crystallinity, an indication that the transition is due to molecular motion in the crystalline region. This is apparently the first report of a crystalline transition below the melting point of PVF_2. The dielectric data about the transition at $-20°C$ conform to the WLF relationship, and the transition most likely corresponds to the T_g from static measurements. The transition located at $-80°C$ is very faint and very little is known.

On the other hand, the data on PCTFE, PTFE, and FEP are more complex and will be discussed further.

A. Transitions of Polychlorotrifluoroethylene

Dilatometric measurements of quenched PCTFE samples indicate that the T_g is around $50°C$ (27,35). This temperature corresponds roughly to a minimum between two loss peaks when compared with dynamic-mechanical data (Fig. 4) (31,37,39). Since loss peaks or maxima (though somewhat arbitrary) are generally accepted as the transition temperatures by dynamic measurement and correlate fairly well with static measurements [see Lewis' empirical correlation (48), for example], PCTFE appears to be a deviation from such a correlation. Actually, the correspondence to a loss minimum is coincidental, due to a loss peak below $50°C$ which is *not* detected by static experimental methods. Most evidence suggests that the relaxation represented by the dynamic loss peak in the region of $90°C$ is associated with the static T_g at $50°C$.

Dynamic measurements reveal a broad dispersion below $0°C$ that is not found by dilatometry. The region of dispersion seems to shift to lower temperatures with increasing degree of crystallinity (37,38). Closer examination indicates that the broad dispersion is the result of a closely spaced doublet, the higher temperature peak disappearing at high degrees of crystallinity (39). McCrum (37) proposed that the doublet is the result of a crystalline peak at $-40°C$ and an amorphous peak at $-10°C$.

Dynamic mechanical measurements by Crissman and Passaglia (39) also indicate the possible existence of a crystalline transition at $133°C$. The transition is observed only for the highest crystalline sample.

B. Transitions of Polytetrafluoroethylene

There are probably more transitions reported for PTFE than for any other polymer, and they provide ample fuel for controversy. A comprehensive bibliography on the reported transitions can be found in Araki's paper (49). Most of the investigations generally support the existence of two amorphous relaxations and one crystalline transition below the melting point. McCrum

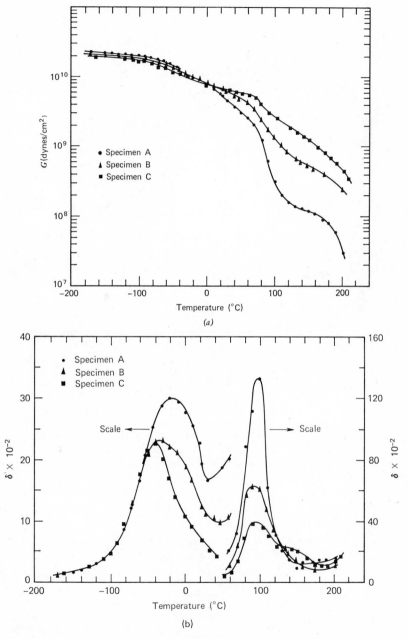

Fig. 4. The dynamic mechanical spectrum of polychlorotrifluoroethylene of various crystallinities. (a) Shear (torsion) modulus versus temperature. (b) Internal damping versus temperature. Specimen A is 27% crystalline, specimen B is 42%, and specimen C is greater than 57% (37).

517

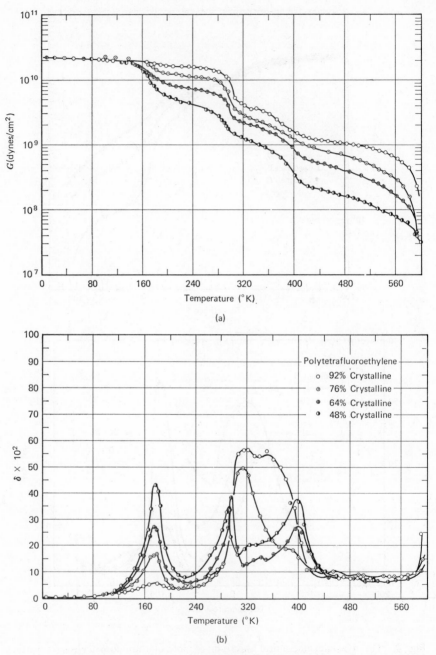

Fig. 5. The dynamic mechanical spectrum of polytetrafluoroethylene of various crystallinities. (a) Shear modulus. (b) Internal damping (44).

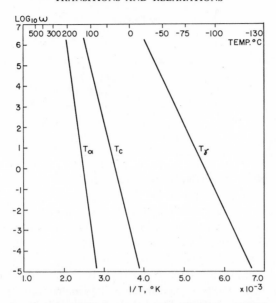

Fig. 6. The frequency dependence of the transitions of polytetrafluoroethylene plotted from data in reference 50.

(44) was among the first to report all three transitions from dynamic mechanical measurements (Fig. 5). After altering the original nomenclature, they are $T_\alpha = 400°K$, $T_c = 300°K$, and $T_\gamma = 176°K$. Eby and Sinnott (50) pointed out that other reported transitions from dynamic measurements are actually the same as these but shifted in temperature due to the frequency dependence of these transitions. The relationship of frequency and temperature for these transitions is plotted in Fig. 6; the substantial dependence of the transition temperature (defined as the temperature of the loss peak) on frequency is clearly evident. Table 3 shows the activation energy values found (50) for the transitions.

Whereas the two room-temperature crystalline-phase changes observed from X-ray crystallographic data can also be detected by dilatometric (42) and calorimetric techniques (40,41), only one peak is observed by dynamic measurements. Eby and Sinnott (50) reasoned that dynamic mechanical response will detect only the crystalline relaxation that takes place at 30°C but not the lattice change at 19°C.

Which of the two glassy transitions can be considered as the "true" glass transition has been a subject of considerable controversy in the past. However, it is difficult to justify assigning the glass transition with the T_γ from dynamic measurements for several reasons. The usual molecular interpretation of T_g is that it is the temperature at which significant segmental or chain

TABLE 3

	T_γ	T_c	T_α	Reference
1 cps	−97°C	27°C	127°C	44
12 MC	−10°C	140°C	197°C	50
$\Delta H^{\neq}$ (cal/mole)	18.0	34.0	88.0	50
$\Delta S^{\neq}$ (cal/°C/mole)	46.0	59.0	162.0	50

motion takes place. Such suggestion of molecular mobility below room temperature is contrary to the X-ray data showing a rigid chain and a rigid crystalline lattice. Moreover, considering that the barrier to internal rotation of a completely fluorinated C—C bond is significantly higher than that of a partially fluorinated bond, which in turn is higher than that of a paraffin hydrocarbon linkage, Tobolsky has argued that the T_g's of the corresponding polymers should be in the same order (51). Brown and Wall (52) recently measured the T_g of various experimental fluorine-containing polymers by differential scanning calorimetry and their data confirm the preceding argument. As shown by Table 4 increasing substitution of fluorine in the basic propylene unit increased the T_g. The intrinsic viscosities of the fluorinated polymers in hexafluorobenzene were comparable, indicating approximately equivalent molecular weight. The same authors also reported that increasing the n-perfluoroalkyl side group from C_1 to C_3 also increases the T_g, which is contrary to the behavior reported for hydrocarbon polymers. The argument that the glass transition of PTFE should be in the proximity of that of polyethylene is not consistent with the foregoing considerations.

TABLE 4

Effect of Fluorine Substitution on the Glass
Transition (52)

Repeating unit	$T_g(°C)$

—CH$_2$CH— atactic
    |
   CH$_3$
	−20

—CH$_2$CH— isotactic
    |
   CF$_3$
	−10
	27

—CH$_2$CF—
    |
   CF$_3$
	42

—CF$_2$CF—
    |
   CF$_3$
	152

Another argument for assigning T_g to the higher amorphous transition temperature T_α is that it would be consistent with what we know about polymers. Since chain stiffness is the usual explanation for the high melting point of polytetrafluoroethylene, it is reasonable to assume that chain stiffness would also retard segmental motion and would result in a high T_g. This is consistent with the suggestion of Andrews (53) that the glass transition of a polymer is in many respects similar to the crystalline melting point, and molecular factors influencing one transition tend to have the corresponding effect on the other. Long ago, Beaman (54) pointed out that the T_g of most polymers is about 2/3 of T_m. For polytetrafluoroethylene with $T_m = 327°C$ (600°K), the calculated T_g would be 127°C (400°K), in agreement with the higher of the two experimentally observed amorphous transitions T_α.

Interestingly enough, the lower transition, T_γ in Table 3, also seems to follow the 2/3 relationship with respect to the *room-temperature* crystalline-phase change [as pointed out originally by Araki (55)]. Following Andrews' rationale for amorphous-crystalline correspondence, this result would suggest that T_γ is the amorphous analog of T_c. That is, at T_c, the helical chain in the crystalline lattice untwists slightly, whereas at T_γ, a similar chain rotation occurs in the amorphous region. Such a chain rotation mechanism for T_γ has been postulated by others (44,50,56).

A third crystalline phase for PTFE exists at very high pressures (57,58). The triple point between the three phases is located at 343°K and 4500 atm (58). Of the polymers tested, including polyethylene, polychlorotrifluoro-ethylene, polyvinyl fluoride, polyvinylidene fluoride, and polyvinyl alcohol, only PTFE exhibits polymorphism from 1 to 10,000 atm and 20 to 80°C (59).

There is also some evidence of a transition in the PTFE melt above the melting point from melt relaxation measurements (60,61).

C. Transitions of Tetrafluoroethylene-Hexafluoropropylene Copolymer

The composition of commercial FEP resin is not generally known. Based on various publications (47,56,62), it appears that FEP contains no more than 15 mole % of hexafluoropropylene; Wilson (63) by NMR analysis found 9 mole % in his sample. Introduction of HFP into the PTFE backbone lowers the melting point by nearly 60°C and reduces the melt viscosity to the extent that the polymer becomes melt processable (16).

The presence of HFP also has significant effect on the other transitions normally found in PTFE. Dynamic measurements with a torsion pendulum were made by McCrum (47) on samples containing 0–14 mole % of HFP. He observed the following:

1. Increasing HFP content lowers the temperature of T_c, the crystalline transition.

2. Increasing HFP content also lowers the T_α transition from 127°C to as low as 75°C, and at the same time increases the intensity of the loss peak.

3. Increasing HFP content does not appear to shift the T_γ transition remaining at −96°C but lowers the intensity of the loss peak.

These results indicate that the hexafluoropropylene units on the backbone reduce the stiffness of the chain and increase the internal mobility so that both T_c and T_α occur at lower temperatures. Since T_α is an amorphous transition, the increase in the T_α loss peak is an indication that HFP tends to lower the degree of crystallinity under comparable cooling conditions and molecular weight. On the other hand, these HFP units also tend to hinder rotation about the chain axis and therefore reduce the loss peak of T_γ transition. This interference with rotation is more clearly shown by ultrasonic measurements at 12 Mc (56), where, at this higher frequency, the T_γ transition shifts to higher temperature with increasing HFP content.

D. Dynamic Mechanical Spectra

Dynamic mechanical data not only help us locate transitions and hypothesize molecular mechanisms [see, e.g., a review by Woodward and Sauer (64) and Nielsen's book (65)] but also provide an overall view of how the mechanical properties of a polymer depend on temperature. Therefore the spectra of the fluoropolymers are included to introduce and serve as a reference point for the following section on the mechanical properties as well as to complete our discussion here.

The dynamic modulus, measured from the in-phase dynamic response, depends on the degree of crystallinity, generally increasing with increasing crystallinity above the lowest transition temperature, as demonstrated by PTFE in Fig. 5. The out-of-phase component of the dynamic measurements, called dynamic loss dispersion or internal damping or internal friction among other nomenclature, increases in magnitude with decreasing crystallinity when the transition or relaxation is due to molecular response in the amorphous region. Conversely, the internal damping increases with crystallinity in the temperature region of a crystalline transition. From the modulus spectrum, we see how the rigidity of a polymer changes with temperature. The internal damping curve along with the modulus curve tells us whether the polymer is amorphous or crystalline and enables us to hypothesize the possible molecular mechanisms governing the various transitions. Examples of the usefulness of dynamic mechanical data were demonstrated in previous sections.

The effect of crystallinity on the dynamic mechanical spectrum of PCTFE and of PTFE has been investigated by McCrum (37,44); his results are presented in Fig. 4 and 5. The data indicate that the effect of crystallinity on the torsion modulus of PTFE becomes significant above approximately $-130°C$ (143°K), which is also the onset of the T_y transition judging from the internal friction curve. The same effect of higher modulus with higher degree of crystallinity is observed above 0°C for PCTFE. Below 0°C, the modulus of PCTFE appears to undergo a small but real inversion, being slightly lower for the higher crystalline sample. This curious effect has been confirmed by some stress-strain testing in the cryogenic environment to be discussed.

The damping curves of PCTFE and PTFE (Figs. 4b and 5b, respectively) illustrate very nicely the amorphous nature of various transitions. The damping peaks of PTEE at $-97°C$ (T_y) and at 127°C (T_α) increase in magnitude with decreasing crystallinity, as do the two transitions of PCTFE.

The crystalline transition peak of PTFE near room temperature not only becomes higher with higher crystallinity but also shifts to higher temperature and broadens with increasing crystallinity. The damping peak of the 92% crystalline sample is particularly broad to the extent that the T_α transition is completely obscured. This behavior may not be due completely to high crystallinity but may also involve some intrinsic contribution due to the considerably lower molecular weight of the sample necessary to achieve such a high degree of crystallinity.

The lower damping peak of PCTFE shifts to lower temperature with increasing crystallinity and suggests the possibility of a crystalline transition at $-40°C$ (37).

The dynamic mechanical properties of copolymers of TFE and HFP were also reported by McCrum (47). The effect of HFP is difficult to isolate from the contributing effects of different degrees of crystallinity; the subject is too involved to go into in great detail. Figure 7, from McCrum's data, shows the conflicting effects of HFP content and crystallinity on the torsion modulus. Increasing HFP content shifts the transition and the drop in modulus to a lower temperature; it is most clearly seen as the curves approach melting above 227°C (500°K). At lower temperatures, the effect of HFP is not as clear because the crystallinity of the samples is also varying. Figure 8 shows the effect of the HFP content on the T_α transition discussed in the preceding section.

The effect of crystallinity on the dynamic mechanical properties of PVF and PVF_2 has not been reported. Figure 9 contains the dynamic mechanical spectra of the two polymers of presumably typical crystallinity. Further discussions of transitions, dynamic mechanical measurements, and (hardly discussed here) complementary dielectric measurements are now available in an excellent book by McCrum, Read, and Williams (66).

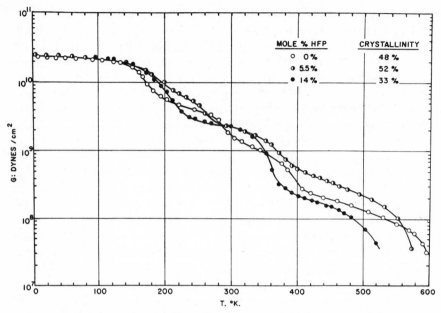

Fig. 7. The variation with temperature of the shear modulus of polytetrafluoroethylene and two tetrafluoroethylene-hexafluoropropylene copolymers. (Reprinted from (47) N. G. McCrum, Die Makromolekulare Chemie **34**, 56 (1959) by permission of the publisher, Hüthig & Wepf.)

V. MECHANICAL PROPERTIES

The mechanical properties of the fluorine-containing polymers have some general characteristics in common. They maintain some strength at high temperatures where most polymers are no longer solid, and they are still somewhat ductile at low temperatures where most polymers become brittle. Within this group, PTFE extends farthest at both ends of the temperature range.

Polytetrafluoroethylene is also the only polymer of this group whose mechanical behavior has been investigated to some degree. Properties listed in the data sheets (such as Table 5), although readily available on all the polymers from such sources as manufacturers' bulletins and Modern Plastics Encyclopedia, are no more than an initial reference point for guiding investigations in depth. Unfortunately, except for some dynamic mechanical studies described in previous sections, the mechanical behavior of the fluoropolymers other than PTFE has by and large not been reported. Some fragmental data on the effect of crystallinity, time, and temperature on strength and creep of PVF_2 and PCTFE have been summarized (67). In this section, we review primarily the properties of PTFE.

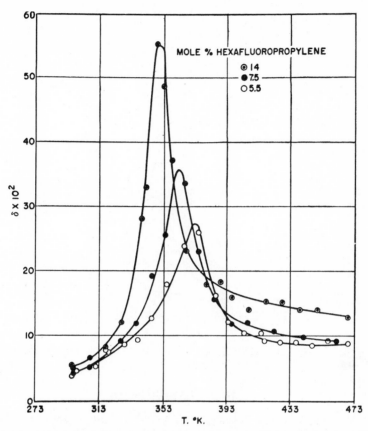

Fig. 8. The effect of hexafluoropropylene content on the T_α loss peak of tetrafluoro-ethylene-hexafluoropropylene copolymers. (Reprinted from (47) N. G. McCrum, Die Makromolekulare Chemie **34**, 54 (1959) by permission of the publisher, Hüthing & Wepf.)

TABLE 5

Property at 25°C (typical values)	PVF[a]	PVF$_2$	PCTFE	PTFE	FEP
Density	1.38–1.39	1.74–1.78	2.10–2.20	2.14–2.24	2.12–2.17
Tensile strength (10^3 psi)	8.0–18.0	7.0	4.5–6.5	3.0–4.5	2.7–3.1
Elongation (%)	110–250	100–300	80–250	200–400	250–330
Tensile modulus (10^5 psi)	2.0–2.8	1.2	1.5–3.0	0.6	0.5

[a] PVF properties are measured from film samples and the values are dependent on testing direction with respect to molding direction and extent of biaxial orientation.

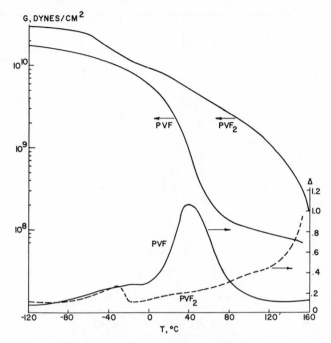

Fig. 9.   The dynamic mechanical spectrum of polyvinyl fluoride (31) and polyvinylidene
fluoride (30).

A. Stress-Strain Properties

Unlike most crystalline polymers, including PVF_2 and PCTFE, PTFE does
not neck and draw in a tensile stress-strain test at room temperature (18,68).
Instead, the polymer will draw uniformly (i.e., with homogenous elongation)
up to several hundred percent as shown by the typical family of stress-strain
curves in Fig. 10. Although the stress-strain curve is also smooth and without
the characteristic yield at elevated temperatures, localized necking has been
visually observed to occur at 100°C (18). This necking phenomenon appears
to depend on the crystalline band size; its occurrence is favored by larger
band widths (resulting from annealing) and by slower testing speeds. A
specimen that has necked and drawn will not completely return to its original
geometry even after heating above the melting point, whereas one that draws
uniformly will resume its original unstressed shape when heated above T_m,
even though the total extensions from drawing were comparable in the two
cases. On the other hand, the same investigators (18) did not observe any
necking when the polymer was strained at -196 and $-70°C$. Yet the true
stress versus true strain curves at these subzero temperatures show a pronounced yield peak or stress drop, a common characteristic associated with

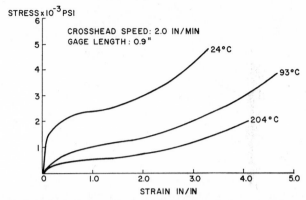

STRESS x 10⁻³ PSI

CROSSHEAD SPEED: 2.0 IN/MIN
GAGE LENGTH: 0.9"

STRAIN IN/IN

Fig. 10. Typical stress-strain curves of polytetrafluoroethylene at various temperatures (68).

yielding and necking (see Fig. 11). However, comparable data at −75, −120, and −196°C, plotted as engineering stress and strain (based on original cross section), did not show any yield peak (69).

Unusual stress-strain behavior of PTFE has also been reported in the temperature vicinity just below the room-temperature transition (68). A stress drop was observed around 100% elongation, the magnitude of which depended on the cross-head speed and the temperature. Faster speed of extension results in a steeper drop, and a similar effect is observed with lower test temperature (Figs. 12, 13). It is unlikely that a true yield could occur at such a large elongation, and no necking was observed. The investigators (68) attributed the phenomenon to localized heating caused by the externally applied stress about microscopic defects and inhomogeneity within the specimen.

The effect of degree of crystallinity becomes quite pronounced at low

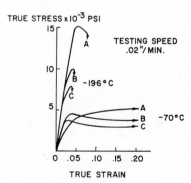

TRUE STRESS x 10⁻³ PSI

TESTING SPEED
.02"/MIN.

TRUE STRAIN

Fig. 11. The true stress versus true strain curves of polytetrafluoroethylene at very low temperatures (18).

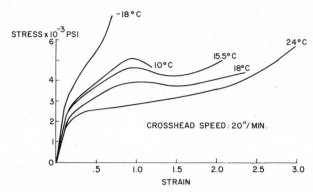

Fig. 12. The unusual stress-strain behavior of polytetrafluoroethylene in the vicinity of the room-temperature first-order transition (68).

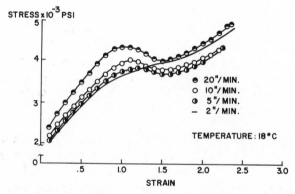

Fig. 13. The effect of crosshead speed on the stress-strain behavior of polytetrafluoro-ethylene at 18°C (68). Note the stress drop disappears at the lowest speed.

temperatures. The data at −196°C for PTFE in Fig. 11 indicate that at very low temperatures, the tensile strength and elongation increases with decreasing crystallinity. The same effect is also apparent for PCTFE below −130°C (70). Although this phenomenon has not been generally investigated, the effect could be general for most polymers. Part of the explanation is that the more amorphous samples are more ductile and can extend to higher elongation and hence to correspondingly higher stresses. However, ductility cannot account for the inversion of the dynamic modulus of PCTFE below 0° as shown by Fig. 4; it is clear that a more complete explanation is not yet available.

Speerschneider and Li (18) related the two general types of deformation of the crystalline bands observed to the stress-strain data of PTFE. As schematically shown in Fig. 14, at low stresses after only a small extension, the first type

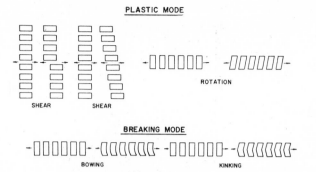

Fig. 14. Observed deformation modes of the crystalline bands of polytetrafluoroethylene: bond distortion denoted as plastic mode and striation distortion denoted as the breaking mode. Arrows indicate the direction of applied stress (18).

of deformation occurs by the lateral sliding of striations away from the band axis; some striations become slanted by rotating in the direction of the applied stress. At higher stress levels, a second type of deformation is observed where the striations appear to bow and kink into chevrons pointing in the direction of the applied stress. At −196°C, only the last mode of deformation was observed. The authors hypothesized that the first mode (band distortion) is due to noncrystalline deformation and hence occurs at lower stresses; whereas the second mode (striation distortion) is the actual straining of the crystals and would take place only at high stresses. At very low temperature, such as −196°C, the noncrystalline regions are frozen and are as rigid as the crystalline regions and therefore do not deform by the first mode. They attribute the higher strength and elongation of the quenched samples to the smaller crystalline band size.

The combined effect of temperature and strain rate on tensile properties of PTFE was reported by Lohr (71) in the range from −100 to 150°C and 0.1 to 100 in./min strain rates. Apparently because of the lack of a clearly defined yield point in the stress-strain curve, he correlated the stress at 2% offset, which he considers as the yield stress, with the cross-head speed and temperature. The choice of offset stress values is unfortunate because the offset stress depends on the Young's modulus and on the selection of the reference strain (see ASTM Method D638 for exact definition), and this parameter does not have a precise, quantitative correlation with the actual yield process. In any case, this stress value was found to be linearly related to the log of the product (cross-head speed × temperature shift factor). Figure 15, taken from Lohr's publication, shows two linear relationships, one above and the other below 50°C. This is somewhat surprising since such discontinuity has not been observed by other measurements.

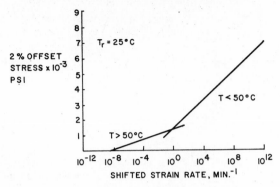

Fig. 15. The master curve of 2% offset stress over the testing speed range of 0.1 to 100 in./min and temperature range of −100 to 150°C (71).

An earlier publication (72) indicated that the log of tensile strength and yield strength is linear with the reciprocal of absolute temperature for PTFE. There have been no subsequent investigations that confirmed such a relationship.

B. Creep and Stress Relaxation

The time-dependent rigidity (i.e., creep and stress relaxation properties) of these fluoropolymers is significantly dependent on the level of crystallinity. At room temperature, the creep deformation of a crystalline sample could be as much as four times *less* than that of an amorphous sample at the same stress. Various manufacturers' data indicate that up to at least 70°C, PVF_2 and PCTFE are considerably more rigid than PTFE; however, further increase in temperature results in more rapid softening of the partially fluorinated polymers relative to PTFE. The creep resistance of PTFE also appears to vary with crystallinity in an interesting way. The review by Sperati and Starkweather (22) contains data indicating that optimum rigidity of PTFE corresponds to 75–80% crystallinity, above which creep resistance decreases with further increase in crystallinity. Since very low-molecular-weight samples are usually used to achieve crystallinity greater than 80%, the data strongly suggest that the increase in creep is due to the presence of shorter polymer chains. However, the anomalous effect of crystallinity on creep has been reported even at constant molecular weight; this is discussed in the following section on cold drawing.

It should be pointed out that PTFE's low rigidity, as indicated by creep and stress relaxation measurements, is consistent with the highly ductile behavior shown in stress-strain measurements. Unlike more rigid polymers, the stress and/or strain response of PTFE deviates from linear viscoelastic behavior nearly from the outset. Stress relaxation data (71) from −50 to 50°C were nonlinear at strains as small as 0.65%.

It is now recognized that time-temperature superposition of creep and stress relaxation data of crystalline polymers require a vertical shift (73) to adjust for thermal softening of the crystallites in addition to the usual horizontal translation. One of the earliest suggestions for the vertical correction came from some stress relaxation measurements on PCTFE (74). However, accurate stress relaxation data on PTFE (75) in the vicinity of room temperature showed that a master curve can be constructed by the simple horizontal superposition without involving any vertical correction. The implication is either that in the case of PTFE, crystalline softening within this temperature range is negligible, or that the superposition is fortuitous and another perhaps more accurate master curve could have been constructed including the vertical correction. The reported master curve (Fig. 16) revealed a discontinuity at 19°C corresponding to the first-order crystalline phase change.

Engineering data on creep in tension (68) and in compression (76,77) of PTFE have also been generated. Since the creep data were generally linear on log-log plots, empirical correlations involving parameters of time, temperature, and stress were presented as potentially useful relationships.

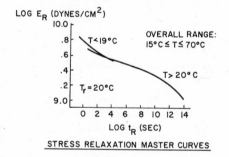

STRESS RELAXATION MASTER CURVES

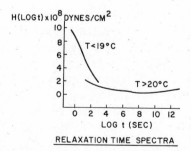

RELAXATION TIME SPECTRA

Fig. 16. The stress relaxation master curves and relaxation time spectra of polytetra-fluoroethylene from 15 to 70°C (75).

C. Cold-Drawing Behavior of Polytetrafluoroethylene

Cold drawing generally refers to the plastic yielding of polymers in the solid state. Most typically, the onset of cold drawing is marked by the neck formation (except PTFE) of a given specimen under a tensile load, and the plastic yielding generally continues until the specimen fails. Since this mechanical response extends from around 5% strain to several hundred percent strain and occurs whenever the polymer deforms in a ductile manner, cold drawing is a major and fundamental aspect of mechanical behavior of high polymers. Recently, the cold drawing behavior of PTFE has been reported (78–80) and is now briefly summarized.

The investigations concentrated on the transitions around −97, 19, and 127°C, relying mostly on standard stress-strain and tensile creep experimental methods. By very slow cooling and rapid cooling of identical molecular weight samples from the melt, two groups of test specimens were attained which differ in crystalline level (60 versus 49%) and significantly in morphology. The characteristic band width from electron micrographs of the slow cooled, more crystalline specimens were approximately fourfold larger than the less crystalline specimens.

The most unusual result was the discovery that the more crystalline sample was not necessarily more resistant to deformation than the less crystalline sample around all three transitions. Figure 17 shows the stress-strain behavior around the γ transition. As we would expect, the less crystalline PTFE is more ductile, whereas the more crystalline sample appears initially more rigid at small deformation. However, it is surprising and not at all obvious why the stress-strain curves intersect and the less crystalline samples exhibit significantly higher stress at higher extension. This behavior also appeared in some previous data (18). Apparently, failure in this temperature region begins in the crystalline domains initiated by the presence of defects. The less crystalline material containing smaller domains is therefore more resistant to onset of failure.

The data in Fig. 17 also indicate that the stress-strain curves of the less crystalline sample drop faster with increasing temperature. The greater sensitivity to temperature of the sample containing greater fraction of amorphous domains is consistent with the conclusion that the γ transition involves only the amorphous region (81). Furthermore, the steady-state drawing (after the knee) of the stress-strain curves appear curiously parallel until fracture. Indeed, plotting the stress at any arbitrary extension greater than 10% revealed that from −140 to −90°C the stress drops linearly at a rate of 135 psi/°C and 100 psi°/C for the less and more crystalline samples, respectively. Vincent (82) has reported that the yield stress (which marks the onset of cold drawing) of PTFE from −250 to −140°C drops linearly at a typical rate of

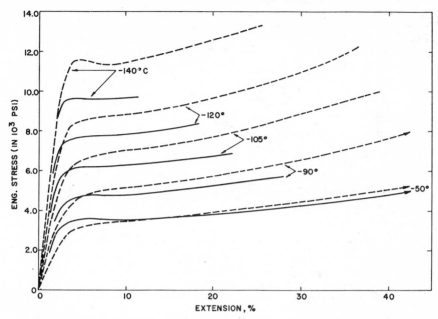

Fig. 17. Stress-strain curves of polytetrafluoroethylene around the γ transition at 2.22%/ min (0.02-in./min); (———) more crystalline, (– – –) less crystalline (78).

120 psi/°C, which seems consistent with the preceding results if we assume that the linear range extends from −250 to −90°C.

Perhaps the results (79) about the room-temperature transition are of even greater interest. The cold-drawing behavior has been investigated around the second-order transition of various polymers; PTFE presents the opportunity to examine plastic deformation around a first-order transition. The stress-strain data in the room-temperature region (Fig. 18) also show a curious mechanical "inversion" similar to the behavior at the low-temperature transition region. The important difference is that in this case, the inversion is approached from the opposite direction, i.e., the intersection of stress-strain curves correspond to more rapid softening of the *more* crystalline specimen with increasing temperature. Figure 19 shows the creep behavior of the two crystalline samples around room temperature. Effects analogous to the stress-strain data can be observed. Comparison of the master curves in Fig. 20 obtained by superposition of data in Fig. 19 discussed elsewhere (78,79) leads to the following generalization: At a common stress level, the more crystalline sample is initially more resistant to deformation than the less crystalline sample. However, the creep deformation of the more crystalline sample increases more rapidly with time and eventually reaches significantly higher level of deformation intersecting the lower crystalline curve enroute. Data

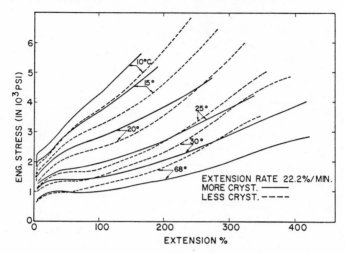

Fig. 18. Stress-strain curves of polytetrafluoroethylene around the room temperature transition at 22.2%/min (0.2 in./min); (——) more crystalline, (– – –) less crystalline (79).

around the α transition reveal no further unusual characteristics but only a continuation and exaggeration of the behavior observed at room temperature (78).

Based on deformational modes of the crystalline bands described by Speerschneider and Li (18) and the results condensed here and described in detail elsewhere (78), it was concluded that the less crystalline PTFE sample is more rigid in cold drawing than the more crystalline sample primarily because the crystalline domains of PTFE as shown by Speerschneider and Li are plastic rather than rigid and yield readily under stress. The chains in the amorphous domains, on the other hand, orient and align in the direction of stress. Conceivably, very small crystallites may also orient so as to resist plastic deformation. Therefore the mechanical crossover behavior observed for PTFE at and above room temperature is because the less crystalline material strain hardens more readily by orientation, whereas the more crystalline material draws more readily due to plastic deformation within the larger crystalline domains. An elastic-plastic model to describe the behavior has been proposed (80).

The anomalous mechanical crossover on cold drawing need not be unique behavior of PTFE. The phenomenon is readily observed for this polymer because the crystalline domains are plastic over a wide temperature range. Although crystalline domains of semicrystalline polymers are usually regarded as rigid (at least relative to the amorphous regions), it is reasonable to assume that a region of temperature and stress exists for every polymer where

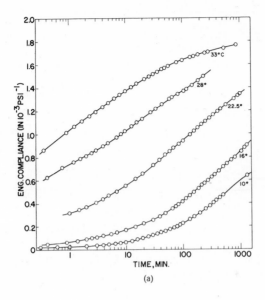

(a)

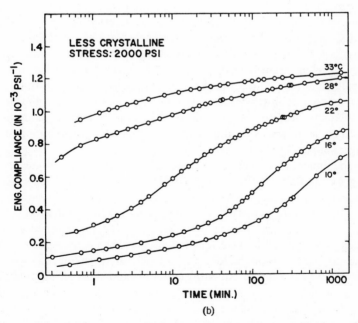

(b)

Fig. 19.   Effect of temperature on creep behavior of polytetrafluoroethylene at engineering stress of 2000 psi. (a) More crystalline. (b) Less crystalline (79).

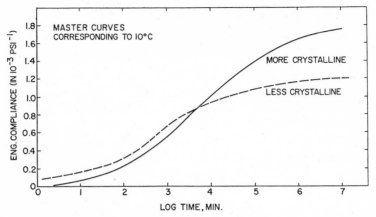

Fig. 20.   Comparison of the master curves from superposition of creep data (79).

the crystalline domains can be considered as plastic. Under proper conditions analogous behavior may be observed.

D. Fatigue Behavior

There are many facets to fatigue, even to mechanical fatigue. The nature of response of the polymer is a function of such external factors as the cyclic frequency, temperature and effectiveness of heat transfer, amplitude and type of stress, specimen geometry and presence of stress concentrations, and whether stress or strain is held constant. On one extreme, at low frequency (<1 cps) and at temperatures far below the T_g, a polymer tends to fail in the classical manner by brittle fracture. On the other hand, under high frequency most polymers heat up due to the relatively high internal damping; the specimen sometimes fails in fatigue from extreme thermal softening and can no longer support the cyclic load.

The phenomenon of hysteretic heating in polymers under fatigue has been investigated on PVF_2, PCTFE, and PTFE (83) and will be briefly discussed in this section.

The samples were subjected to constant amplitude, alternating flexural stresses at 30 cps in the manner detailed in ASTM D671, method B. The specimen is considered to have failed in fatigue when it fractures or when it becomes too soft to bear the load without actually breaking. In either case, the test machine reaches a resonating condition and shuts off terminating the test. The fatigue-life curves (commonly referred to as S—N curves) of the fluoropolymers, such as in Fig. 21, show an elbow and an endurance limit (which is the asymptotic stress), characteristic of many S—N curves. Parenthetically, we should point out that the data are taken on a changing

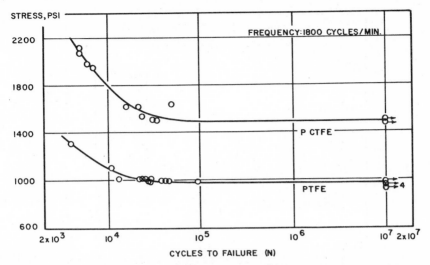

Fig. 21. The fatigue-life (SN) curves of polychlorotrifluoroethylene and polytetrafluoroethylene at 30 cps and 23°C ambient (83).

temperature base so that the endurance limit as such is a convenient but somewhat arbitrary reference.

At each stress level, the temperature rise of the specimen undergoing fatigue is also measured using an infrared thermometer. As long as the ambient conditions and heat transfer around the specimen are kept constant, the temperature rise is directly related to the energy dissipation from damping. The energy dissipated per cycle, ΔW, is proportional to the loss compliance of the polymer,

$$\Delta W = \pi \sigma_0^2 J''$$ (1)

where σ_0 is the peak stress and J'' is the loss compliance (84).

The actual temperature rise in the specimen resulting from hysteretic heating corresponds to the manner in which the loss compliance varies with temperature as indicated by Eq. 1. The relationship is necessarily qualitative since J'' is determined from separate torsion pendulum measurements, rather than from actual fatigue specimens under test. As an example of the data reported, Fig. 22 shows the temperature rise during fatigue and the variation in loss compliance over the same temperature range for PTFE.

On the basis of the published findings (83), the following generalizations were established for the polymers:

1. When stressed above the endurance limit, a sharp and rapid temperature increase occurs in the specimen just before fatigue failure. The temperature at failure is in the temperature region where J'' is large and increasing rapidly

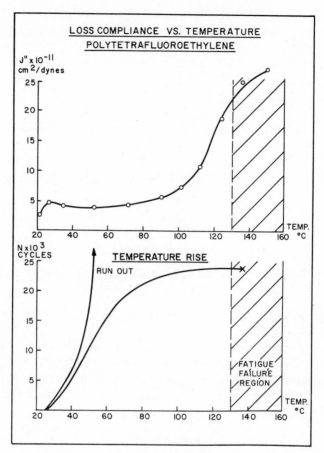

Fig. 22. The relationship of temperature rise during fatigue to the loss compliance of polytetrafluoroethylene (83).

with temperature and roughly corresponds with the reported secondary transitions from dynamic measurements.

2. When stressed at or below the endurance limit, the specimen will not fail (by definition) and the temperature levels off after an initial rise. As expected, this stabilized or steady-state temperature is also related to J''. The steady-state temperature usually corresponds to the region where J'' is at a minimum, as is the case for PCTFE and PTFE. Where J'' increases with temperature without a minimum, the steady-state temperature is dependent on the balance between the heat generated by the cyclic stress and the heat transferred to the surrounding.

3. For most polymers the internal damping at ambient conditions is

sufficient to result in hysteretic heating, which in turn increases the damping of the polymer, until the specimen becomes too soft to withstand the applied load and fails. This was the behavior observed for PVF_2, PCTFE, PTFE, nylon 6, polyethylene, and polymethyl methacrylate. Unplasticized polyvinyl chloride was the only polymer (with such low damping) that did not heat up under the conditions imposed.

4. Permanent fatigue damage to the specimen does not occur until just approaching failure where the temperature is far above the steady-state temperature and is in the "runaway" region. Therefore the usual laws of cumulative damage cannot apply without modification.

In addition, PTFE appears to have a unique feature in that the polymer will not fracture under the test conditions described. A PTFE specimen can be repeatedly failed in fatigue and still retain physical integrity and appear to maintain a "residual" fatigue strength. This peculiar property is probably related to the structure and morphology of the polymer and the extremely high molecular weight, but a clear explanation is not available.

A limited examination of the morphological changes in PTFE due to cyclic stresses was done by electron microscopy (85). The replicate of a fractured surface of a sample before fatigue is similar to Fig. 2. After fatigue failure, the fractured surface shows a disappearance of nearly all the crystalline bands and the apparent presence of some voids as shown by Fig. 23. Although a more definitive investigation is required, it would appear that the combination of mechanical stress with the accompanying hysteretic heating destroys the crystalline bands, and that, if voids result from fatigue, they do not propagate onward to macroscopic rupture.

VI. SUMMARY

It is clear that much of the structure-properties relationship of fluorine-containing polymers is still unclear and has not been firmly established. Some of the hazy areas for clarification depend on future advances in polymer science in general, whereas others are rather specific to this class of polymers.

For example, it is difficult to discuss the impact resistance of fluoropolymers to any degree because the very criteria of impact resistance are not clearly defined for polymeric materials. From the conventional impact tests, the fluoropolymers invariably turn out to be among the most impact-resistant polymers, as expected of semicrystalline polymers, particularly those with noted ductility at low temperatures. On the other hand, the existing impact tests do not provide insight on how impact resistance might be related to structure and morphology, nor do they enable us to predict the variability over a range of temperatures and/or rate of loading. A study of impact under projectile speeds of 136,800 in./min (86) illustrates this point. The toughness

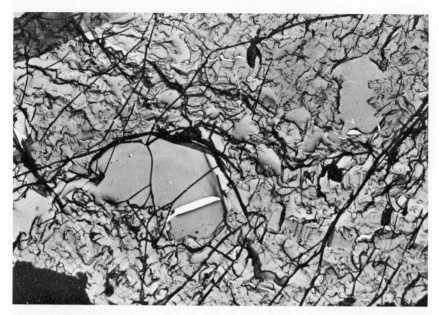

Fig. 23. Morphology of a polytetrafluoroethylene fracture surface after mechanical fatigue. Electron micrograph is courtesy of Dr. L. G. Roldan, Central Research Laboratory, Allied Chemical Corp.

and ductility of PTFE are second to no other polymer. Yet under this particular projectile impact, the sample crumbles into many fragments, whereas polyethylene under same loading draws in a ductile manner. Recent studies (87,88) of the tensile response under hydrostatic pressure of PTFE and polyethylene also indicated that PTFE fails in an abrupt, brittle manner, whereas polyethylene fails by gradual, ductile drawing. There seems to exist some intriguing relationship between the impact of projectiles and the response under hydrostatic pressure for the two polymers. Is the difference in behavior due to morphological differences or chain structure or chain stiffness or chain entanglements or an interaction of these parameters? Polyethylene crystallized under pressure (89,90) yields extended-chain crystals showing bands and striations characteristic of those of PTFE; an examination of the mechanical behavior of such polyethylene crystals should provide worthwhile clarifications.

Not enough is known of some of the fluoropolymers even to allow us to formulate specific questions. However, there is one question of great interest: Is the peculiar behavior of PTFE merely different in degree or is it unique? Let us enumerate the chief differences of PTFE. The polymer has a sharp, clearly defined first-order transition, which, unlike others such as nylon or polybutene, is reversible and not " monotropic " in the thermodynamic sense, i.e., the transition is independent of time and history. The melting point is higher than most crystalline polymers, yet the cohesive energy density is the lowest. The morphological feature of crystalline bands found in PTFE is also unlike the morphology of other crystalline polymers. The polymer appears ductile over a wide range of conditions, yet the polymer does not seem to neck during drawing, which is typical of crystalline polymers. The lack of rigidity, in spite of unusually high density, is also unlike crystalline polymers.

Which aspects of the behavior of PTFE can be ascribed to its chain stiffness, helical conformation, rodlike shape, extremely high molecular weight, and presence of fluorine atoms on the chain? What are the intertwined relationships of these factors? Clarification of these questions would lead to a better understanding of fluorine-containing polymers and a more fundamental understanding of the mechanical behavior of polymers in general.

Acknowledgment

The author is indebted to Allied Chemical Corporation for the use of its library facilities and the permission to use previously unpublished data. The author is appreciative of many stimulating discussions with numerous colleagues at the Plastics Division of Allied Chemical during the preparation of this manuscript. The author also thanks Professor Rodney D. Andrews of the Stevens Institute of Technology for his helpful criticisms and comments. Finally, gratitude is due to the Plastics Institute of America for the fellowship grant at the Stevens Institute of Technology; part of this chapter was prepared during the time of this grant.

References

1. J. T. Milek, H. E. Wilcox and M. Bloomfield, "A Bibliography on Polytetrafluoro-ethylene Plastics," U.S. Government OTS Report, AD 633 579, November 1965.
2. C. A. Barson and C. R. Patrick, *Brit. Plastics*, **36**: 70 (Feb. 1963).
3. R. C. Golike, *J. Polymer Sci.*, **42**: 583 (1960).
4. G. Natta, G. Allegra, I. W. Bassi, D. Sianesi, G. Caporiccio, and E. Torti, *J. Polymer Sci. A*, **3**: 4263 (1965).
5. J. B. Lando, H. G. Olf, and A. Peterlin, *J. Polymer Sci. A1*, **4**: 941 (1966).
6. R. P. Teulings, J. H. Dumbleton, and R. L. Miller, *Polymer Letters*, **6**: 441 (1968).
7. K. Okuda, T. Yoshida, M. Sugita, and M. Asahina, *Polymer Letters*, **5**: 465 (1967).
8. S. Enomoto, Y. Kawai, and M. Sugita, *J. Polymer Sci.*, *A2*, **6**: 861 (1968).
9. C. W. Bunn and H. R. Howells, *Nature*, **174**: 549 (1954).
10. E. S. Clark and L. T. Muus, *Z. Krist*, **117**: 119 (1962).
11. C. Y. Liang and S. Krimm, *J. Chem. Phys.*, **25**: 563 (1956).
12. C. W. Bunn, A. J. Cobbold, and R. P. Palmer, *J. Polymer Sci.*, **28**: 365 (1958).
13. J. P. Tordella, *Trans. Soc. Rheol.*, **7**: 231 (1963).
14. L. G. Roldan and H. S. Kaufman, *Norelco Reporter*, **10**(1): 11 (1963).
15. H. S. Kaufman, *J. Am. Chem. Soc.*, **75**: 1477 (1953).
16. A. P. Cox, *Plastics (London)*, **30**: 75 (Oct. 1965).
17. F. P. Price, *J. Am. Chem. Soc.*, **74**: 311 (1952).

18. C. J. Speerschneider and C. H. Li, *J. Appl. Phys.*, **33**:1871 (1962); *J. Appl. Phys.*, **34**:3004 (1963).
19. H. W. Starkweather, *SPE Trans.*, **3**:57 (1963).
20. H. S. Kaufman, C. O. Kroncke, and C. R. Giannotta, *Modern Plastics*, **32**(2):146 (Oct. 1954).
21. N. K. J. Symonds, *J. Polymer Sci.*, **51**:S21 (1961).
22. C. A. Sperati and H. W. Starkweather, *Fortschr. Hochpolym.-Forsch.*, **2**:465 (1961).
23. A. V. Tobolsky, D. Katz, and A. Eisenberg, *J. Appl. Polymer Sci.*, **7**:468 (1963).
24. R. G. J. Miller and H. A. Willis, *J. Polymer Sci.*, **19**:485 (1956).
25. R. E. Moynihan, *J. Am. Chem. Soc.*, **81**:1045 (1959).
26. C. W. Wilson and G. E. Pake, *J. Polymer Sci.*, **10**:503 (1953); *J. Chem. Phys.*, **27**:115 (1957).
27. J. D. Hoffman and J. J. Weeks, *J. Research NBS*, **60**:465 (1958).
28. P. E. Thomas, J. F. Lontz, C. A. Sperati, and J. L. McPherson, *SPE J.*, **12**(6):89 (June 1956).
29. N. G. McCrum, *ASTM Bull.*, *No. 242*, 80 (1959).
30. G. P. Koo, unpublished data.
31. K. Schmieder and K. Wolf, *Kolloid Z.*, **134**:149 (1953).
32. G. H. Kalb, D. D. Coffman, T. A. Ford, and F. L. Johnson, *J. Appl. Polymer Sci.*, **4**:55 (1960).
33. K. L. Paciorek, W. G. Lajiness, R. G. Spain, and C. T. Lenk, *J. Polymer Sci.*, **61**:S41 (1962).
34. R. F. Boyer, *J. Polymer Sci.*, **14**:3 (1966).
35. L. Mandelkern, G. M. Martin, and F. A. Quinn, *J. Research NBS*, **58**:137 (1957).
36. N. Koizumi, S. Yano, K. Tsunashima, *Polymer Letters*, **7**:59 (1969).
37. N. G. McCrum, *J. Polymer Sci.*, **60**:S3 (1962).
38. J. D. Hoffman, G. Williams, and E. Passaglia, *J. Polymer Sci., C*, **14**:173 (1966).
39. J. M. Crissman and E. Passaglia, *J. Polymer Sci. C*, **14**:237 (1966).
40. G. T. Furukawa, R. E. McCoskey, and G. J. King, *J. Research NBS*, **49**:273 (1952).
41. P. Marx and M. Dole, *J. Am. Chem. Soc.*, **77**:4771 (1955).
42. R. K. Kirby, *J. Research NBS*, **57**:91 (1956).
43. J. A. Sauer and D. E. Kline, *J. Polymer Sci.*, **18**:491 (1955).
44. N. G. McCrum, *J. Polymer Sci.*, **27**:555 (1958); **34**:355 (1959).
45. T. Satogawa and H. Koizumi, *J. Ind. Chem. (Japan)*, **65**(8):71 (1962).
46. S. G. Turley and H. Keshkula, *J. Polymer Sci. C*, **14**:69 (1966).
47. N. G. McCrum, *Makromol. Chem.*, **34**:50 (1959).
48. A. F. Lewis, *J. Polymer Sci. Bl*, 649 (1963); A. F. Lewis and O. G. Lewis, *Proc. 4th Int'l. Conq. on Rheol.*, *Pt. 2*, 505 (1963).
49. Y. Araki, *J. Appl. Polymer Sci.*, **9**:3585 (1965).
50. R. K. Eby and K. M. Sinnott, *J. Appl Phys.*, **32**:1765 (1961).
51. A. V. Tobolsky, *J. Polymer Sci.*, **35**:555 (1959).
52. D. W. Brown and L. A. Wall, *J. Polymer Sci. A2*, **7**:601 (1969).
53. R. D. Andrews, *J. Polymer Sci. C*, **14**:261 (1966).
54. R. G. Beaman, *J. Polymer Sci.*, **9**:470 (1952).
55. Y. Araki, *J. Appl. Polymer Sci.*, **9**:421 (1965).
56. R. K. Eby and F. C. Wilson, *J. Appl. Phys.*, **33**:2951 (1962).
57. C. E. Weir, *J. Research NBS*, **46**:207 (1951).
58. R. I. Beecroft and C. A. Swenson, *J. Appl. Phys.*, **30**:1793 (1959).
59. C. E. Weir, *J. Research NBS*, **53**:245 (1954).
60. L. C. Case, *J. Appl. Polymer Sci.*, **3**:254 (1960).

61. A. V. Tobolsky, D. Katz, and M. Takahashi, *J. Polymer Sci. A*, **1**:483 (1963).
62. K. M. Sinnott, *J. Appl. Phys.*, **29**:1433 (1958).
63. C. W. Wilson, III, *J. Polymer Sci.*, **56**:S12 (1962).
64. A. E. Woodward and J. A. Sauer, *Fortschr. Hochpolym.-Forsch.*, **1**:114 (1958).
65. L. E. Nielsen, "Mechanical Properties of Polymers," Reinhold, New York, 1962.
66. N. G. McCrum, B. E. Read, and G. Williams, "Anelastic and Dielectric Effects in Polymeric Solids," Wiley, New York, 1967.
67. Staff Report, *Mat'ls. in Design Eng.*, p. 93 (Feb. 1964).
68. G. P. Koo, E. D. Jones, M. N. Riddell, and J. L. O'Toole, *SPE J.*, **21**:1100 (1965).
69. J. Dyment and H. Ziebland, *J. Appl. Chem. (London)*, **8**:203 (1958).
70. R. E. Mowers, "Program of Testing Nonmetallic Materials at Cryogenic Temperature," U.S. Government OTS Report, AD 294-772 (Dec. 1962).
71. J. J. Lohr, *Appl. Polymer Sym.*, **1**:55 (1965).
72. R. C. Doban, C. A. Sperati, and B. W. Sandt, *SPE J.*, **11**(9):17 (1955).
73. T. Takemura, *J. Polymer Sci.*, **38**:471 (1959).
74. A. V. Tobolsky and J. McLoughlin, *J. Phys. Chem.*, **59**:989 (1955).
75. K. Nagamatsu, T. Yoshitomi, and T. Takemoto, *J. Colloid Sci.*, **13**:257 (1958).
76. E. D. Jones, G. P. Koo, and J. L. O'Toole, *Modern Plastics*, **45**(3):137 (Nov. 1967).
77. D. A. Thomas, "SPE ANTEC," Vol. XV, Chicago, Ill., May 1959.
78. G. P. Koo, "Cold Drawing Behavior of Polytetrafluoroethylene," D.Sc. Thesis, Stevens Institute of Technology, May, 1969.
79. G. P. Koo and R. D. Andrews, *Polymer Eng. & Sci.*, **9**:268 (1969).
80. R. D. Andrews and G. P. Koo, forthcoming.
81. R. W. Gray and N. G. McCrum, *Polymer Letters*, **6**:691 (1968).
82. P. I. Vincent, "Fracture, Short Term Phenomena," in *Encyclopedia of Polymer Science and Technology*, Vol. 7, Wiley, New York, 1967, p. 292.
83. G. P. Koo, M. N. Riddell, J. L. O'Toole, *Polymer Eng. & Sci.*, **7**:182 (1967).
84. G. P. Koo, *S.P.E., P.A.G. Technical Conference*. Chicago, Ill., Sept. 1968.
85. G. P. Koo and L. G. Roldan, forthcoming.
86. E. A. Ripperger, "Stress-Strain Characteristics of Materials at High Strain Rates," Univ. of Texas, Struc. Mech. Res. Lab. Rept., Austin, Texas, Aug. 1958.
87. K. D. Pae and D. R. Mears, *Polymer Letters*, **6**:269 (1968).
88. W. I. Vroom and R. F. Westover, *SPE J.*, **25**(8):58 (1969).
89. B. Wunderlich and T. Arakawa, *J. Polymer Sci. A*, **2**:3697 (1964).
90. P. H. Geil, F. R. Anderson, B. Wunderlich, and T. Arakawa, *J. Polymer Sci. A*, **2**:3707 (1964).

INDEX